REFLECTION
SEISMOLOGY

REFLECTION SEISMOLOGY

A Tool for
Energy Resource Exploration

Third Edition

KENNETH H. WATERS

Consultant—Applied Seismology
Former Research Fellow—CONOCO, Inc.
Former Phoebe Apperson Hearst Lecturer
University of California—Berkeley

A WILEY-INTERSCIENCE PUBLICATION
JOHN WILEY AND SONS
New York · Chichester · Brisbane · Toronto · Singapore

Library of Congress Cataloging in Publication Data:

Waters, Kenneth Harold, 1913–
    Reflection seismology.

    "A Wiley-Interscience publication."
    Includes bibliographies and index.
    1. Seismic reflection methods. I. Title.
TN269.W37 1986   622'.159   86-15809
ISBN 0-471-01144-4

Printed in the United States of America

10 9 8 7 6 5 4 3 2 1

Preface

The French, with their rather hackneyed bon mot *Plus ça change, plus c'est la même chose*, might have been describing reflection seismology. Certainly our scientific—and economic—aims have not changed but the methods and instruments seem to be in a constant state of flux. If these changes were mere details, it would be possible, even prudent, to ignore them as immaterial to the overall description of the seismic reflection method. But in five years since the inauguration of the second edition most of the changes have been important enough to modify the procedures that we have relied on and penetrating enough to search out more reliable and more valuable exploration information.

A textbook must mirror in its pages the latest substantial advances in the discipline it describes. Otherwise it may be consigned to everlasting perdition. In an era of rapid changes it has not been a sinecure to keep oneself up to date.

This substantial revision entails, in addition to the replacement of obsolescent material by equivalent modern facts, the internal rearrangement of some subject matter. The former appendix on Vibroseis® methods has been spliced into the mainstream of Chapter 3. Elastic modeling, once a rather rare operation, is now much more commonplace and has been moved to join other modeling methods in Chapter 4.

Much of the material has been updated, and this is particularly true for sonic logging methods, which have, over the last 10 years, undergone revolutionary development. It is also true for vertical seismic profiling—a valuable tool that originated in the 1950s but was brought to full fruition in the last five years.

Shear-wave prospecting has now grown up to occupy a substantial place in reflection seismology and for this reason has now been treated in much more detail, with more examples. Moreover, it is beginning to exhibit offshoots that may be as important as the parent branch a few years hence. But all of these daughters of the original reflection seismology family are still maturing, and it was thought best to leave them for careful evaluation in the final chapter—as interesting and fruitful offspring should be.

Much has also happened to data processing, particularly in filtering operations in the *p*-tau domain, available from conventional seismic displays through the *slant stack* operation. These operations are described in an extension of Chapter 6.

In the period of nearly five years since the second edition was presented I have been fortunate enough to have worked in a number of different locations (Cape Town, Houston, Johannesburg). In each of these cities there have been new problems to address. In South Africa the use of seismic methods to provide structural information for the gold mining industry emphasized the need for a detailed description of field procedures, problems, and solutions. This has now been done and I hope that this edition will be more generally valuable by its inclusion.

I now place on record my gratitude to W. E. (Bill) Laing for welcoming me back to CONOCO, Inc., for a short period from September 1981 to May 1983. He, and others at CONOCO—Barry Coon, John Hopkins, Hans Sheline, and Fenton Morrison—have provided information, examples of some new techniques, and drafting help. In South Africa Jan Fatti and George Smith of SOEKOR, Emil Weder of RAND MINES, and Geoff Campbell of J.C.I. were all most accommodating in describing their special and unique problems. Although somewhat isolated from the mainstream of seismic technology, all have tackled their work diligently (and with evident success). Tight competition in the gold mining industry has prevented the release of much information on this most interesting application.

During my present stay in Houston (1985 to 1986) I have also profited a great deal from discussions with Jim Frasher and his colleagues at Teledyne Exploration Company—George Parker, Kevin Barry, Art Pollett, and John Lucas. They have helped materially with detailed and down-to-earth discussions of their data collection and data-processing techniques. They, too, arranged some drafting help for me.

In the later years of life when we begrudge any time taken away from happy pursuits with those close to us, I can only record again, thankfully, the typing efforts and moral support given to me by my dear wife, Frankie. Her skill, selflessness, and forebearance in dealing with my peccadillos have been a source of continual wonder. As it should be, this edition is also dedicated to her.

KENNETH. H. WATERS

Houston, Texas,
and Henwood, Cornwall, England
November 1986

Prologue to the First Edition

Although the title gives the dominant theme of this book, a philosophical subtitle, suggesting the manner of treatment, might well be "The more ye know, the more ye need to know." With that as a benchmark my task will, for some readers, remain unfinished when the book has been read.

My goal has been to present an up-to-date framework of modern reflection seismology. Exploration seismology is a multidisciplinary science. A detailed description of all instruments, physical principles, geological methods, and the mathematical background is too extensive to be covered in one volume and too diverse to be adequately described by one author. At the level I have chosen I hope that it will be understandable by senior or first-year graduate students in geophysics and geology. Practicing exploration personnel who have worked for some time in their field may also find the book a useful supplement to their practical knowledge by making them aware of some of the strengths and limitations, or the doubts and hopes for the future, or reflection seismology. More advanced students may be disappointed, because the mathematical level has been truncated short of the highly useful methods and a full discussion of z-transform usage. I can only hope that this thirst for additional knowledge will be assuaged by the references given for each chapter or by a few books listed in the bibliography of Chapter 11.

There is an enormous literature. Two scientific journals (*Geophysics* and *Geophysical Prospecting*) devote themselves entirely to the science. Applicable papers abound in the *Journal of Geophysical Research*, the *Bulletin of the Seismological Society of America*, the *Geophysical Journal of the Royal Astronomical Society*, and many others. New material continues to pour out, ranging from highly sophisticated mathematical treatises on elastic wave propagation and data processing principles to case histories of oil field discoveries. Competition in the industry makes innovation a necessity. As time passes, it is not just one concept that has to be reexamined microscopically—and revised to meet the needs—it is all of them. The practice of reexploring the same area with different sets of instruments having diverse characteristics, over a period of several years, now involves the interpreter in a complex process of adjustment of characteristics; or it involves the exploration manager in a vexatious and costly decision to abandon some, if not all, of the previous work and reexplore the area once more with the most modern tools.

In the first six chapters the basic requirements of a structural first-look seismic section are established. In the remaining six chapters we explore the additional refinements and modifications to originally crude ideas, which are required for a more detailed attempt at the interpretation of more subtle geological events. It should be no surprise that some of the ideas offer a challenge to understanding. It is not only in the study of history that challenge has been seen as the means by which humans advance.

The last chapter gives an indication of the direction of the most modern advancement. It is at this stage that some readers will really become fully aware of the need for additional tools to cope with the problems of finding stratigraphic oil fields or perturbations in coal seams from the surface. Appendices have been added to some chapters, usually to give more detail than seems to be necessary at first reading.

I am indebted to my present and former colleagues for many stimulating discussions over the years. Many of them had superior abilities to conquer some of the difficult problems; others had keener insight than mine into the nature of the problems. Still others have been inquisitive enough or, at times, demanding enough to require that explanations be clear and concise. In the course of their individual researches they have produced some hitherto unpublished illustrations which I am privileged to use in this book. So, even though I may be guilty of omission, I must express my admiration and my thanks to the following: D. Bahjat, L. H. Berryman, G. L. Brown, W. L. Chapman, J. T. Cherry, J. M. Crawford, E. E. Diltz, W. E. N. Doty, D. E. Dunster, E. L. Erickson, J. A. Fowler, P. L. Goupillaud, G. Harney, B. J. Heath, M. R. Lee, D. E. Miller, C. E. Payton, G. W. Rice, A. J. Scanlan, R. L. Stolt, B. J. Thomas, J. A. Ware, and M. Yancey. These colleagues, with their wide range of expertise, were all necessary in the operations and research departments with which I was associated, and I have derived great pleasure from my association with them. If this work provides help to anyone, the thanks should go to them; if it does not, then the blame is mine.

I am grateful to the Continental Oil Company not only for their very generous permission to allow me to publish this book but also for allowing me to devote a large part of my time during my final year of employment to work on it. If that had not been the case, it would have had to wait for two more years and I doubt that it would even have been finished.

My thanks are due to Mrs. Barbara Watson and to the secretarial services group for coping bravely with the unfamiliar terms and symbols, as well as my persistent requests for retypes and changes. I am also indebted to Mr. Bill Rigsby and members of his drafting staff for many illustrations.

The scientific societies and individual service companies and publishers have been generous with their permission to use some of their illustrations. That kindness is part of the atmosphere of our profession.

This original work was intended as an internal publication of the Continental Oil Company. For more general circulation a large number of revisions and modifications have had to be made. I am indebted to the Phoebe Apperson Hearst Foundation, which has provided the funds making it possible for me to accept an appointment as Visiting Lecturer in the Engineering Geoscience Section of the Material Science and Mineral Engineering Department at the University of California at Berkeley. I have spent part of the first month in the task of making this a more instructive text on reflection seismology.

KENNETH H. WATERS

Berkeley, California
October 1977

Contents

ONE

Introduction

It is the aim of this introductory chapter to set out the goals of seismic exploration in a simple manner, uncomplicated by some of the difficulties dealt with in succeeding chapters. This has turned out to be necessary because many of the later topics are interrelated and it is difficult to talk about the details of one topic without having a general knowledge of all. Thus to some extent the dilemma of ordering the various chapters has been resolved, and for the rest of this introduction we deal with a simplified concept. Later, the reader can expect variations to be introduced to convert this idealistic concept into a realistic one.

The earth consists of a series of layers of rocks. In simple earthquake seismology one can think of layers whose thicknesses are measured in terms of kilometers or hundreds and thousands of kilometers, the layers being built concentrically on a spherical inner core. The outer core, the mantle, and the crust constitute the simplest sequence. For considerations of hydrocarbon finding, however—which is assumed to be our job—there is little need to consider any layer but the crust and then only a small portion of that. Although this simplification is possible, further complications now develop because all oil and gas accumulations are related to the sedimentary rocks in which hydrocarbons are generated. These sedimentary rocks are secondary in nature and occur only in areas of the world where primary rocks have been eroded or where oceans have existed at some time in the past and the salts contained therein or minuscule fauna and flora have deposited themselves on the sea bottom. Most of these locations occur where at some period of the earth's long history a sequence of events established a series of rock deposits of different nature and properties, one upon the other. The rates of deposition have varied widely, as have the times during which a particular form of deposition has taken place. We look at a layered model within these sedimentary basins.

But deposition did not occur at a constant rate all over the basin, nor was the basin itself stationary during its entire history. The large-scale forces that acted on basin sediments as they consolidated caused vertical uplifts, folding due to compressional and shear stresses, and, as rock strengths were sometimes exceeded, breaks of various kinds—faulting—which interrupted the smooth change in the strata characteristics areally.

During the course of these millions of years of evolutionary history the remains of the marine (and, to a lesser extent, terrestrial) fauna were acted on by pressure, heat, and bacteria so that their basic constituents were broken down first to solid prepetroleum materials (e.g., kerogen), then to a combination of liquid hydrocarbons, and, in some cases with higher temperatures, to hydrocarbon gases and carbon dioxide.

Water and its dissolved salts were also contained in the pore spaces of rocks, sometimes tightly bound within rock particles and sometimes, when permeability existed, able to flow in response to differential hydrostatic pressure. The flow of this water out of shales into and through more porous rocks has been largely responsible for the movement and concentration of hydrocarbons within the pores of the rocks.

Various forms of trap exist, some of which are purely hydrostatic in nature (the oil has been pushed by the water into a position that is locally one of minimum potential energy) and some are combinations of hydrostatic and permeability or porosity traps.

From a physical point of view the task is to make use of the variation in the properties of different rocks to find these traps. Moreover, these variations must be mapped by using measurements made on or near the surface. Various physical parameters have been measured in boreholes that pass through rocks, and geologists are familiar with the resulting electrical logs of resistivity and spontaneous potential as well as logs made to indicate changes in density, seismic velocity, and radioactive properties. It is the variation in density and in seismic velocity that is of interest in reflection seismology. If we accept the fact that it is possible to make small shock (seismic) waves propagate through rocks, these waves must travel with a speed dependent on the nature of the rock itself and the possible fluid content.

When two different rocks are juxtaposed a wave initiated at the surface, traveling through one or intercepting the boundary, is partially bounced back, or reflected, and partially transmitted so that it can travel in the second rock. These reflections eventually arrive back at the surface and, by their arrival times, convey some intelligence about how far the waves have traveled and possibly something about the rock boundaries from which they were reflected. Naturally, because there are many boundaries between rocks in the earth, the intelligence conveyed is complex and the interpretation in terms of geology is not easy. Reflection seismology, the sole subject of this book, is the science and art of initiating elastic waves and making surface measurements to be interpreted later to give information on the attitudes of and relationships between rock layers. In addition, whenever possible the lateral changes in rock type and pore fluids must be inferred.

It has been established in more than 40 years of active seismic exploration that investigations have to be done so that areas as small as 1 km^2 (230 acres) can be investigated (oil fields of this size would be economical under the best of circumstances) and knowledge has to be gained for rock layers of a few tens of meters to more than 5 km in depth. Some oil fields may exist because of the oil contained in one porous sandstone with a thickness of as little as 3 m, hence the desired acuity of observation, or seismic resolution, is high.

Although some early oil fields were found by the seismic refraction method (described at length in *Seismic Refraction Prospecting*, published by the Society of Exploration Geophysicists), these methods are not pursued here and our emphasis remains on the use of reflections of artificially created small seismic (elastic) waves.

It should be pointed ut that we are mainly concerned with the problems of obtaining and representing reflection data as free as possible of the adulterating, misleading information artifacts that are collected and sometimes interpreted as though they were bona fide reflections from subsurface layering—and in this book we are able to devote only a little space to the fascinating problems of actual interpretation in terms of geology. In practice this interpretation is rarely unique from the point of view of the physical data alone. By making use of well established geological principles some degree of uniqueness can be achieved. We can therefore expect to deal with the mechanics of the method. It is important to understand these basic principles thoroughly to appreciate later the close interplay between reflection seismology and geology that is needed to make an accurate geological interpretation.

Figure 1.1 is a simple, diagrammatic cross section that illustrates the requirements of the technique. A method must be devised that allows a cross section of the earth to be drawn in sufficient detail to permit probable oil and gas reservoirs, such as A, B, or C, to be found. The sources and receivers of acoustic energy

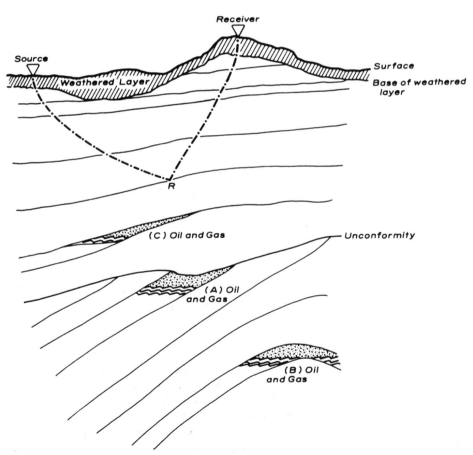

FIGURE 1.1 A diagrammatic cross section shows trapping modes for oil and gas in the pores of rocks.

must reside on or close to the surface. In a manner similar to that used in radar or sonar the method times the energy pulses from onset to echo. In contradistinction to radar and sonar, however, the medium through which the energy travels is heterogeneous, laminated but subject to discontinuities such as the unconformity shown or to abrupt displacements due to faulting. In addition, the weathered layer is a layer of low velocity and density, of variable thickness and constitution, disconformable with the underlying consolidated rocks, and even subject to seasonal fluctuations in properties.

A possible acoustic wave path is that shown from the source to R to the receiver. It is immediately obvious that not just one echo is received at each receiver from a single input pulse because each change in the elastic properties of the rocks should give an echo—just as several radar targets in a series give echoes from a single transmitted electromagnetic pulse. Possibilities of multiple scattering (i.e., the scattered wave from one target being rescattered by other targets) also exist.

The cross section shown in Figure 1.1 in most cases is exaggerated by having the vertical scale larger than the horizontal scale. In other words, the change in the vertical sequence of rocks is not quite so rapid in the horizontal direction as shown in this diagram. Thus most, but not all, prospecting problems can be looked at in the first place as a challenge to measure the positions of rock boundaries vertically beneath the experimental equipment—which can be moved along the surface from point to point—thus gradually building up the cross section from a series of similar measurements. For one of these experimental positions on the surface the problem then reduces to that shown in Figure 1.2, the determination of the coordinates (X_i, Z_i) of the successive reflecting points. This means that the raw reflection times T_i have to be corrected for elevation, for the time taken to pass through the weathered layer, and for the geometric effect introduced because the source and receiver are not at the same place on the surface. Figure 1.3 (also shown as a color plate in the middle of the book) is a seismic cross section. It consists of a great number of vertical seismic traces, each of which shows the variation in received energy with time (the vertical coordinate). This variation in energy level has been converted to a color. The reflection time in seconds is given by the figures on the extreme right.

Because the seismic traces are very closely spaced in the horizontal direction the result is that the variations in energy of the individual traces, as the time of reception changes, are converted to a two-dimensional variation in color that indicates to the observer how the depth of the rock layers varies as the observation point moves along the surface. The numbers along the horizontal axis of this cross section represent the numbers of known points on the surface of the ground, which can be obtained from the land survey at any time. Thus any indication of suitable conditions for oil accumulation (say the pink reflection near R_1) can be related immediately to the position on the surface of the earth and steps can be taken to make appropriate tests by well drilling.

At present monochromatic seismic cross sections are much more common than this polychromatic example. The reasons are discussed later. Moreover, the geological picture presented here is not yet true in all respects. The reasons are pointed out in subsequent chapters. Qualitatively, then, this seismic cross section is the first indication of the geological structure beneath the sequence of experimental stations.

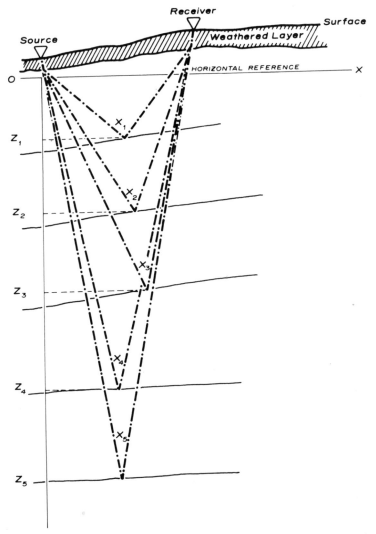

FIGURE 1.2 The coordinates of reflecting points to be determined by reflection seismology.

When reflection seismology is applied to coal technology the emphasis is more on exploitation than on original exploration. The general areas in which coal is present underground are largely known. A knowledge of the *local* conditions of occurrence, however—the thickness, continuity, faulting, or local replacement by other types of rock—can often mean a great deal in ensuring uninterrupted production of coal. The use of seismic reflection methods, modified when necessary to fit this application in an optimum manner, enables dependable maps of the coal seam characteristics to be drawn. After confirmation by a relatively few slim core holes mine layouts can be drawn up before the first shaft is ever sunk. The requirements are as severe as those for finding petroleum products under evermore difficult conditions, but they are not impossible and there is promise that the use of reflection seismology will increase in density of application and worldwide coverage.

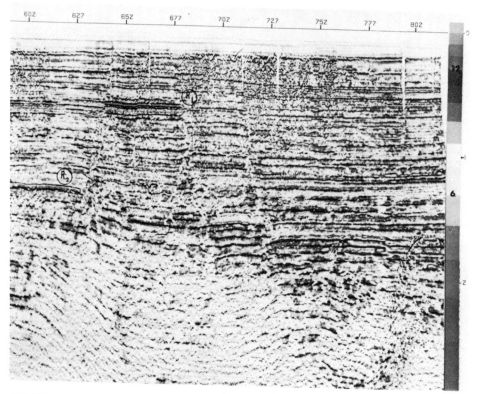

FIGURE 1.3 An example of a finished reflection seismic cross section. (Reproduced from Seis-Chrome (TM) displays with the permission of Seiscom Delta, Inc.) See color section in the middle of the book for color reproduction.

Other applications exist in the field of energy resource exploration. One of the latest has been in defining more exactly the location and extent of geothermal reservoirs. Naturally this application relies on finding anomalies due to the known differences of velocity for seismic waves through rocks filled with water compared with those saturated with superheated steam.

As long as the habitat of mineral energy sources can be associated directly or indirectly with variations in the physical parameters that control propagation of seismic waves there will be an incentive for research and development of the necessary applications.

The art and science of producing optimized seismic cross sections and maps and proceeding from them to more precise geological information on the appropriate energy source location are the subjects of this book.

TWO

Basic Physical Principles:
Waves in Elastic Solids

The science of applied seismology, as it is used in the search for commercial sources of hydrocarbons, owes a debt to the parent science of earthquake seismology in that during the latter part of the nineteenth century it was proved that wave energy travels through and on the surface of the earth. Moreover, such waves were recorded and could be traced from point to point on the surface. These waves, of course, developed from natural causes, usually the accumulation of strain energy at particular localities in the earth until the elastic limit was passed and rupture occurred.

The waves discussed here, which have been put to work for economic purposes, are artificially generated, much smaller in magnitude, and of higher frequency than earthquake waves. Apart from these differences, the physics of wave propagation is the same. The basic principles are discussed in this chapter.

2.1 WAVES IN GENERAL

A disturbance originates in some manner at a particular point A in a liquid medium and propagates outward through the medium. Figure 2.1 shows one representation of such a wave. This makes the assumption that a graph can be drawn that relates one property of the wave to the distance from the source. In this case it is the displacement ξ from the rest position. On the upper line a wavelike form is shown and a particular point $\bigcirc$ on the waveform arrives at B at time t_1. On the lower graph, taken at a different time t_2, we see a similar waveform reduced in amplitude but now having traveled an additional distance d. Because the waveforms are similar, it is possible to pick out a corresponding point on the waveform $\bigcirc^1$. Thus it appears that the point $\bigcirc$ has moved a distance $r_2 - r_1$ in a time $t_2 - t_1$; hence the velocity of this point must be

$$c = \frac{r_2 - r_1}{t_2 - t_1} \tag{2.1}$$

This velocity, associated with a particular feature of the waveform, such as a particular trough or a zero crossing, is called the phase velocity. As long as the wave moves without a change in form, the entire wave packet will move with this

7

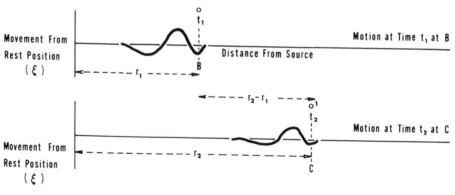

FIGURE 2.1 Wave profiles taken at two different positions, distances r_1 and r_2 from the source.

velocity and the wave is said to be nondispersive. These are the waves we concentrate on for the present.

It is also possible to plot ξ against time while the distance remains stationary (Figure 2.2). It is to be noted that we have drawn the same waveform as in Figure 2.1 but the horizontal scale is now time instead of distance and the waveform is horizontally reversed. In one figure our frame of reference remained fixed in space, whereas in the other it moved with the wave packet. Thus it is possible to represent the shape of the wave packet as a function of both distance and time by the formula

$$\xi(r, t) = A(r)\xi\left(t - \frac{r}{c}\right)$$ (2.2)

It is shown later that the waveform ξ is related to the input stress that caused the disturbance and that $A(r)$ is a factor related to the way in which the medium allows the wave to spread out.

An important fact to note is that it is the abstraction "the wave" that is propagated through the material and not the material itself. Any particular small particle simply vibrates about its own rest position while the wave is passing, returns to its rest position, and remains there afterward. Because motion from the rest position is a necessary characteristic of wave motion, the initial energy, which created the wave, spreads out with it. Thus for certain rather limited special cases it is possible to assign an approximate form to the function $A(r)$. In the body of a homogeneous liquid the wave is generated with total energy E_0, and this is spread over the surface of a spherical wave (surface area $4\pi r^2$). The energy per unit area

FIGURE 2.2 Motion of a point as a function of time.

is thus $E_0/4\pi r^2$. It is shown later that the energy density is proportional to the amplitude squared:

$$\frac{E_0}{4\pi r^2} = k[A(r)]^2$$

$$A(r) = \left(\frac{E_0}{4\pi r^2 k}\right)^{1/2} = \frac{A_0}{r}$$

(2.3)

For a spherically diverging wave (2.2) therefore becomes

$$\xi(r, t) = \frac{A_0}{r}\,\xi\!\left(t - \frac{r}{c}\right)$$

(2.4)

and the amplitudes of all points on the wave packet diminish as the inverse of this distance from the source.

A second case that may be important relates to waves essentially confined to the surface of a homogeneous semi-infinite medium and, in this case, the total energy is confined to a ring of radius r. Then

$$\frac{E_0}{2\pi r} = k[A(r)]^2$$

$$A(r) = \left(\frac{E_0}{2\pi rk}\right)^{1/2} = \frac{A_0}{r^{1/2}}$$

(2.5)

For a wave spreading out over a surface (2.2) therefore becomes

$$\xi(r, t) = \frac{A_0}{r^{1/2}}\,\xi\!\left(t - \frac{t}{c}\right)$$

(2.6)

and the amplitude of the wave packet diminishes inversely in proportion to the square root of the distance. Because the inverse square roots of positive numbers decrease slower than the inverse numbers themselves as the number increases, waves confined to a plane surface (surface waves) decay in amplitude slower with respect to distance than do those propagated within the body of the material (body waves).

2.2 CONSTANT FREQUENCY WAVE COMPONENTS

It is sometimes convenient to consider a source of acoustic energy as capable of emitting waves of any desired constant frequency. The word *frequency*, when used in this book, denotes the number of sinusoidal-type vibrations per second. In this case it is possible to conceive of the measurement of the distance, once steady state is reached, between consecutive similar points of the sinusoidal waveform— a distance called the *wavelength*. To any desired constant frequency there corresponds a wavelength. If the frequency of the wave is f hertz (abbreviated Hz) and the wavelength is denoted by λ it is obvious that a given point on the waveform, produced at a given instant of time, will travel outward from the source a distance of f wavelengths in one second; but by definition the distance traveled by a disturbance in one second is the speed. Therefore we have the relation

$$c = \lambda f \tag{2.7}$$

Although it is not always true that waves of different frequency travel at the same speed, for a wave of frequency f there is always a velocity $c(f)$ that can be used. An *isotropic* medium is one in which the speed $c(f)$ is independent of this direction of travel.

Fortunately, the waves traveling through the body of the earth have a velocity that is *almost* independent of frequency. If this were not so the interpretation of the variations of travel time in geologic terms would be even more difficult than it is.

At the same time that the body waves are propagating the same source generates waves that, because of rapid changes in wave speed near the surface, are channeled along the near surface layers. Measurements of their apparent speeds show that they are strongly dependent on frequency—they are *dispersive*. The consequences of this phenomenon are reviewed shortly.

Up to this point we have dealt with wave propagation largely from the perspective of the time behavior of a wave packet (i.e., some arbitrarily shaped waveform) in a nondispersive medium. For other developments to follow it would be most desirable to be able to use a set of constant frequency signals, work on them individually, and finally merge the results. Of course we are not suggesting anything new.

The process of analyzing wave propagation by examining the constituent single frequencies independently became possible largely because of the brilliant insight of Fourier in the early nineteenth century. It is by no means obvious that a complex waveform can be split into individual harmonically related sinusoidal waves.

We first write down the assumption that for *some* functions of x it is possible to associate a series with $f(x)$:

$$f(x) = \frac{a_0}{2} + a_1 \cos x + a_2 \cos 2x + \cdots + a_n \cos nx$$

$$+ b_1 \sin x + b_2 \sin 2x + \cdots + b_n \sin nx \tag{2.8}$$

$$f(x) = \frac{a_0}{2} + \sum_{1}^{n} a_j \cos jx + \sum_{1}^{n} b_j \sin jx$$

This is simply the mathematical way of writing the Fourier hypothesis. The cosine and sine terms are the single frequency components with amplitudes a_j and b_j, respectively; they increase in frequency as integral multiples of a fundamental frequency. In other words, they are harmonics of the fundamental: (1) $f(x)$ must obey certain simple rules within the interval $0 \le x \le 2\pi$; (2) $f(x)$ and its derivative must be continuous everywhere except at a finite number of points—a condition that must apply to any waveform with which we deal.

Within this region it can be shown that the constants a_j and b_j can be found from

$$a_j = \frac{1}{\pi} \int_0^{2\pi} f(x) \cos jx\, dx$$

$$b_j = \frac{1}{\pi} \int_0^{2\pi} f(x) \sin jx\, dx$$

(2.9)

which are mathematical statements that relate to the degree of similarity between $f(x)$ and $\cos jx$ (or $\sin jx$) over the interval.

An example is given in Figure 2.3 just to fix the ideas clearly in the mind of the reader. First shown is the form of $f(x)$, a square wave with two discontinuities,

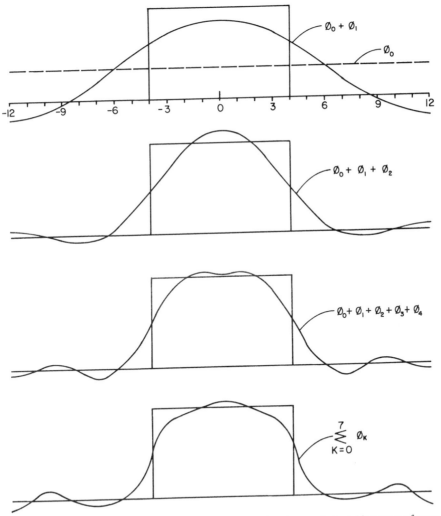

FIGURE 2.3 Successively closer approximations to a box function made by taking sums of successively higher order sinusoidal harmonics (Fourier synthesis).

and then three different degrees of approximation by Fourier series as the number of terms (or constituent harmonics) is increased. Intuition leads to the conclusion that as the number of harmonic sinusoids is increased the Fourier approximation will come closer and closer to the true waveform. This is correct to a point, but there are certain difficulties near the abrupt discontinuities (Gibbs phenomenon); however, they are of little practical importance.

To complete the formal introduction to Fourier analysis and synthesis it is easier in some respects to use the complex exponential form of the series rather than separate cosines and sines.

It is recalled by De Moivre's theorem that

$$\epsilon^{ijx} = \cos jx + i \sin jx ; \qquad i = \sqrt{-1}$$

and

$$f(x) = \sum_{j=-\infty}^{\infty} c_j \epsilon^{ijx} \tag{2.10}$$

whose coefficients (c_j) are given by

$$c_j = \frac{1}{2\pi} \int_0^{2\pi} f(\alpha)\epsilon^{-ij\alpha} \, d\alpha ; \qquad (j = 0, \pm 1, \pm 2, \ldots) \tag{2.11}$$

α is a dummy variable. These coefficients (c_j) are related to (a_j) and (b_j) by the relations

$$
\begin{aligned}
2c_j &= a_j - ib_j ; & j &> 0 \\
2c_0 &= a_0 ; & j &= 0 \\
2c_j &= a_j + ib_{-j} ; & j &< 0
\end{aligned} \tag{2.12}
$$

For more practical usage we obviously have to examine a time signal (sometimes called a *trace*) over a finite length of time (T) for which the data are available. The fundamental frequency of the Fourier analysis then is $1/T$.

The calculations by which the analysis is done are explained in a later chapter; it should be sufficient at this stage to point out the feasibility of the procedure and some of the reasons for it.

Having taken a brief diversion to explain how a complicated waveform can be replaced by a set of simple ones, we can return to the main subject of wave propagation. It should be noted that we are implying the applicability of the principle of superposition; that is, the system is such that the output derived from it for an input of a set of simultaneous sinusoidal waves is equal to the sum of the outputs for each of the sinusoidal waves taken one at a time. This is a *linear* system.

In applied seismology most instruments used in the exploration system are designed to deal with small motions, motions that are part of a linear system. Measurements made on the surface of the earth, however, are affected simultaneously by waves that have been propagated through the body of the earth and by other waves of much larger magnitude that have traveled near the surface. To deal with the small body waves adequately and have the system remain linear some method of reducing the amplitude of the surface waves must be found.

2.3 WAVE FRONTS AND RAYS

As a wave travels outward from the source there is, at any given time, a surface that joins all the points at which motion is just about to start. This surface is called a wave front. In a material with properties independent of position and direction of travel—an isotropic homogeneous material—the wave fronts are a set of concentric spheres with a center at the source. As shown later, however, the form of a wave front is determined by the wave velocity distribution. As shown in Figure 2.4 the energy can be conceived as traveling down a large number of pyramids of infinitesimal cross section. The center line of any one pyramid is regarded as a ray and its direction is normal to the wave front at any time. If the wave front is distorted by a nonuniform velocity distribution the ray is bent so that it always stays perpendicular to the instantaneous wave front. In many situations the ray convention is a convenient one to follow, because it is usually simpler than dealing with the complexities of wave front construction. It is merely a convenience, however, and it leads to difficulties whenever the normal to the wave front cannot be uniquely defined. This matter is examined in more detail later. A special type of wave front is especially useful for ease in understanding additional problems. It is a *plane wave front* and, although not physically realizable, may be regarded as due to a point source of waves an infinite distance away—while still retaining finite amplitudes in the wave along the wave front. It is an artifice, but a useful one.

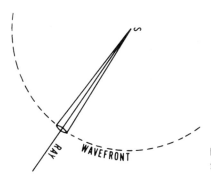

FIGURE 2.4 The relation between the ray and the wave front in an isotropic medium.

2.4 THE EFFECT OF BOUNDARIES BETWEEN TWO DIFFERENT MEDIA ON WAVE PROPAGATION

If we have a wave front *A* at a particular instant of time, as in Figure 2.5, its position at a little later time can be predicted if we regard every point on this wave front as a new (secondary) source of energy and draw wave fronts a little later for each of these new sources. This concept is known as Huyghens' principle and it states that the new wave front *B* is the envelope (the common tangent surface) of all these new spherical wave fronts. In the case of a wave front in a

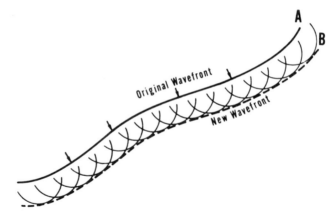

FIGURE 2.5 Use of Huyghens' principle in predicting wave propagation.

homogeneous isotropic material the new surface is parallel to the old; for example, a plane wave front becomes another plane wave front parallel to the first, a spherical wave front generates another spherical wave front with the same center, and so on. The usefulness of Huyghens' principle emerges when, as shown in Figure 2.6, the wave front encounters a plane boundary between two different wave velocities or in even more complex situations.

In Figure 2.6 an incident wave front I, in material 1, with a velocity C_1,

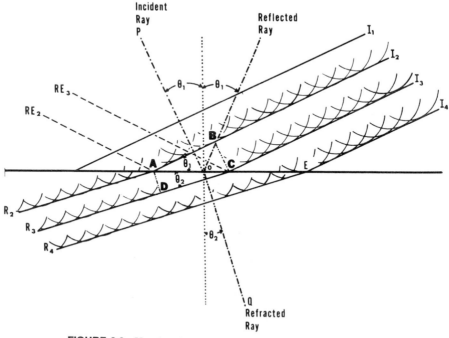

FIGURE 2.6 Huyghens' principle used to determine Snell's Laws.

propagates in a forward direction and at successive small time intervals Δt occupies positions I_2, I_3, I_4, and so on. The spacing between these positions is, of course, $C_1 \Delta t$. As it proceeds a portion of the wave front reaches the boundary (at A, C, E, and so on). If Huyghens' principle is used to generate new wave fronts both forward into material 2 (velocity C_2) and backward into medium 1, it will be seen that a reflection is generated at the boundary, which sends back some of the energy into medium 1, and the angle made with the boundary is equal to the angle made with the incident wave.

The forward wave fronts travel with velocity C_2 which, in the case shown, is lower than C_1, and the angle of advancement of the wave front to the boundary is changed. This is the refracted wave. By consideration of the triangles ACD and ACB, $BC = C_1 \Delta t$, $AD = C_2 \Delta t$, and AC is common; hence

$$\frac{\sin \theta_2}{\sin \theta_1} = \frac{AD/AC}{AB/AC} = \frac{C_2}{C_1} = \mu \tag{2.13}$$

These angles are the same as those between the rays and the normal to the boundary; so Snell's laws, as these relations are called, can be stated in the following form:

> When a wave crosses a boundary between two different media, both reflected waves and refracted waves are generated. The sine of the angle of incidence (between the ray and the normal to the boundary) is related to the sine of the angle of (reflection/refraction) as the ratio of the velocity of the incident wave to the velocity of the (reflected/refracted) wave.

More specification is made after the types of elastic wave have been considered.

This law says nothing about the amplitudes of the waves transmitted (refracted) or reflected, but it does give the directions of the rays in relation to the normal to the boundary. It applies in a very general way to acoustic waves, to electromagnetic waves of all wavelengths, and to the two different kinds of elastic wave that can exist. In other words, it is a property of wave propagation and not of just a particular type of wave.

The laws of reflection and refraction are also derivable from a more general principal of wave propagation known as Fermat's principle, which states that "the ray path between two points in any medium is that which makes the travel time stationary." In other words, a *small* deviation from the correct travel path will not change the travel time appreciably. In Figure 2.7a, for example, the travel path between P and Q, if the medium has a seismic velocity $V(z) = a + bz$, is known to be concave downward. Alternatively, nearby paths from P to Q still differ in travel time only by a second-order amount.

Similarly, Figure 2.7b shows reflection and refraction paths, PRQ and POT, respectively, which involve a curve boundary S. It can be shown that Snell's laws are obeyed at the reflection or refraction points (R or O) on the boundary. Nearby paths ($PR'Q$ and $PO'T$) involve changes of travel time that are second order compared with the true paths.

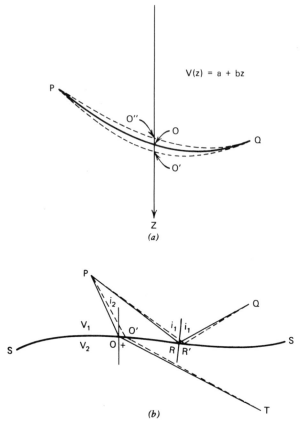

FIGURE 2.7 Fermat's principle illustrated by two different types of ray path: (a) directly between two points in a medium of varying velocity and (b) between two points by reflection from a curved surface and by refraction at the curved boundary between different media.

Therefore the type of path must be specified and one mathematical statement can then be made:

$$\delta \int_{P}^{Q} \frac{ds}{V} = 0 \tag{2.14}$$

where δ stands for "the variation of" and the integral represents the "total time between P and Q." This condition of *stationary* time may lead to a minimum or maximum time or, in some cases, a point of inflection. These three conditions are illustrated in Figure 2.8a,b,c. Note that in the relatively simple cases in which the boundary lies between homogeneous, isotropic materials both reflection and refraction lead to minimum time path.

Although Fermat's principle is conceptually simple and is sometimes the only starting point for complicated wave propagation problems, the calculus of variation techniques are difficult and are not treated further here. For additional details of Fermat's principle reference may be made to standard mathematical texts (e.g., Wylie, 1975).

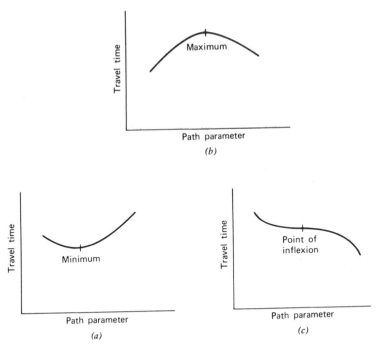

FIGURE 2.8 Three possible cases of travel-time stationarity: (*a*) minimum, (*b*) maximum, (*c*) point of inflexion.

2.5 THE FRESNEL ZONE

Most geophysicists are used to thinking of reflections in terms of the original Newtonian laws, which are illustrated in books by lines drawn from the source to the interface and another directly to the receiver; in other words, rays that have no lateral dimension. We think of the amount of the surface that plays a part in the reflection process as infinitesimally small. The theory of physical optics, however, brings out the fundamentals of image formation by considering illumination of the entire reflecting interface by the source and summing up the scattered waves produced by all parts of the reflecting surface. Because of interference phenomena, only those portions of the surface that produce waves substantially in-phase at the receiver (centered on the Fermat path from source to interface to receiver) contribute substantially to the image. In optics, of course, the wavelengths of visible light lie in the range of 4.0 to 7.0 Å (4 to 7×10^{-7} m), and wave packets that are substantially in-phase at the source must differ by less than one-quarter wavelength or (say) 1.5×10^{-7} m on arrival at the receiver.

There is a surprise for geophysicists, however, when they deal with seismic reflections because the wavelengths may now be one billion (10^9) times as long. A little consideration of a typical reflection shows that a considerable area of the subsurface reflector may be involved in the received image of the source.

Figure 2.9 shows this situation in two dimensions; the reader can readily generalize it in three.

In this figure the source S and the receiver R are separated by a distance $2D$; O

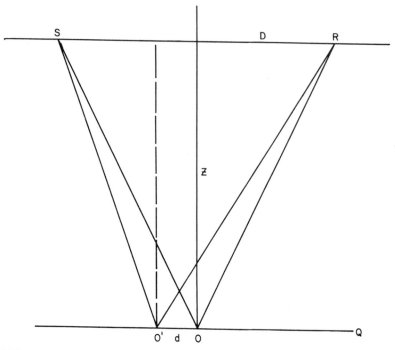

FIGURE 2.9 A diagram of considerations for the determinination of the Fresnel zone for a reflection from a plane boundary.

is the *Fermat* reflecting point; O' is a point of intersection of another ray from S with the assumed horizontal surface Q; $O'R$ represents a scattered ray from this new point of illuminated surface to the receiver.

It can then be easily seen that the path difference $SO'R$ and SOR is

$$\delta P = [Z^2 + (D + d)^2]^{1/2} + [Z^2 + (D - d)^2]^{1/2} - 2(Z^2 + D^2)^{1/2}$$

If we let $D = pZ$ and $d = qZ$, then

$$\frac{\delta P}{Z} = [1 + (p + q)^2]^{1/2} + [1 + (p - q)^2]^{1/2} - 2(1 + p^2)^{1/2} \qquad (2.15)$$

If d is such that δP does not exceed one-quarter of a wavelength of any constituent single frequency of the incident pulse from S, the amplitude of the pulse received will not suffer. Thus we seek a value d such that this criterion is satisfied.

As an example, the following data was used:

$$V = 3000 \text{ m/s}; \ p = 0.5; \ Z = 3000 \text{ m}$$

A series of calculations with a desk calculator shows that for frequencies

$$30 \text{ Hz}: \quad \lambda = 100 \text{ m}; \qquad qZ = 324 \text{ m}$$

$$60 \text{ Hz}: \quad \lambda = 50 \text{ m}; \qquad qZ = 225 \text{ m}$$

All points on both sides of the Fermat path up to a distance of 225 m will therefore contribute positively to the amplitude of a reflected pulse of top frequency 60 Hz and lower frequencies will come from a much wider area, called the Fresnel zone on the reflector. In three dimensions it is roughly elliptical and its size is obviously affected by reflector curvature or roughness. The lesson to be learned is that we must dispossess ourselves of the idea that seismic reflections come from a point or even a small area of the subsurface reflector.

2.6 ELASTIC WAVES

Early in Section 2.1 the medium picked for subsequent discussion of wave propagation was a fluid. This was done to avoid the discussion of elastic waves until the chief properties of wave propagation had been established. There is some precedent for this subterfuge in that most of the exploration geophysics literature talks about *the* seismic velocity or measurements of sonic velocity (singular) in wells. This is not surprising because most seismic experimentation is done with instruments and sources that favor the reception and generation of one type of seismic wave. Modern practices in field seismology, which call for large separations between sources and receivers, bring into question whether other seismic wave types can be ignored. It is our intention to point out the characteristics of elastic waves in general. Whenever it seems appropriate throughout the book dangers that may accrue from making the single-wave-type assumption are discussed. It may be profitable at times, however, to make measurements on the other types of wave, and these occasions are pointed out.

Solids can be distorted by external forces in two different ways. The first form involves the compression of a piece of the material without a change in shape (Figure 2.10a). The second implies a change in shape without a change in volume (Figure 2.10b). Hence the waves are called compressional (P) and distortional (S). The P and S designations, carried over from earthquake geophysics, denote *primus* and *secundus* from their arrival sequence on earthquake records. The S designation is also often taken to be the initial of "shear," which is an alternate name for distortional waves. Because fluids have no shear strength, they cannot support shear waves; hence in these materials only one type of wave (P) has to be considered. It is often called acoustic because these waves constitute sound that we hear.

We deal now in some detail with one type of elastic wave—in one dimension only. The principal features of this wave propagation are derived and the results are extended by analogy to other wave types and more complex types of propagation. The mathematical treatment of elastic waves in general is too advanced to be included here, but several treatments (for those with the necessary mathematical background) are readily available (e.g., see Ewing, Jardetsky, and Press, 1957, or Garland, 1971).

Two physical laws are invoked in the discussion of elastic wave propagation. Newton's fourth law of motion states that a force applied to a body produces an acceleration that is proportional to the force ($F = ma$). The second law required is Hooke's law—applied to perfectly elastic materials—in which it is stated that the strain produced in a portion of an elastic medium is proportional to the stress.

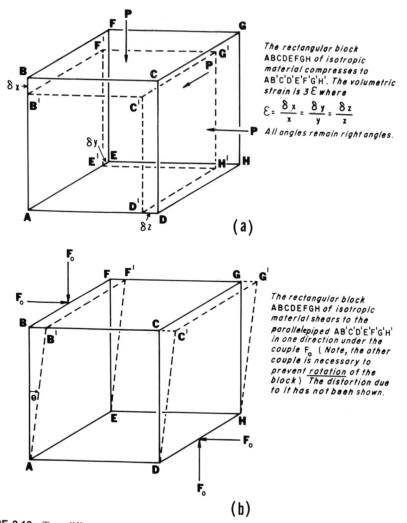

The rectangular block
ABCDEFGH of isotropic
material compresses to
AB'C'D'E'F'G'H'. The volumetric
strain is $3\mathcal{E}$ where

$$\mathcal{E} = \frac{\delta x}{x} = \frac{\delta y}{y} = \frac{\delta z}{z}$$

All angles remain right angles.

(a)

The rectangular block
ABCDEFGH of isotropic
material shears to the
parallelepiped AB'C'D'E'F'G'H'
in one direction under the
couple F_o (Note, the other
couple is necessary to
prevent _rotation_ of the
block) The distortion due
to it has not been shown.

(b)

FIGURE 2.10 Two different kinds of distortion of a rectangular block of material: (a) compressive stress, (b) shear stress.

Strain is defined as the fractional change in a geometrical property of the body (e.g., the length or the volume), and *stress* is defined as the force applied per unit area. Stress is sometimes referred to as *traction*.

We illustrate the use of these laws by considering a rod with a cross-sectional area A and an original length L, which is held at one end and is being stretched by the application of a force F at the other. If the amount of stretch is ΔL the strain is $\Delta L/L$ and the stress is F/A, so the Hooke's law can be written as

$$K \frac{\Delta L}{L} = \frac{F}{A}$$

where the constant K is called a modulus of elasticity. For rods for which the sides are not held the appropriate constant is called Young's modulus.

If we continue with the example of the rod it can be shown that the displacement of one plane of cross section anywhere on the rod in relation to the other cross-sectional planes (e.g., hitting one end with a hammer) immediately causes a stress condition locally that is propagated in both directions from the zone of the original disturbance as a stress or strain wave. The speed of propagation is established in the next two paragraphs as

$$c = \sqrt{\frac{K}{\rho}}$$

Note that only two different properties of the material are involved in the velocity—the proper modulus of elasticity and the density. The wave propagation process is a continuous transfer of energy back and forth from elastic energy (compression of the material locally) to kinetic energy (motion of a small mass of material). A slightly more mathematical discussion follows.

Figure 2.11a shows a tube of constant shape and cross section S with rigid walls and containing the material through which the elastic waves propagates. Under equilibrium conditions two planes perpendicular to the axis are compressed or rarefied by the displacement of both boundaries in the direction of the axis, in general, by unequal amounts. At some instant of time the first boundary is at $x_1 + \xi(x_1)$ and the other plane is at $x_2 + \xi(x_2)$. (*Note*: This is a somewhat naïve way of handling materials by imagining them to be infinitely divisible, but it can be justified if we consider that we are really talking about the *average* displacement of molecules whose *average* position is x_1, x_2, and so on.) Figure 2.11b shows these relationships when the distance apart is the small quantity dx.

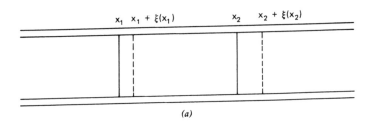

(a)

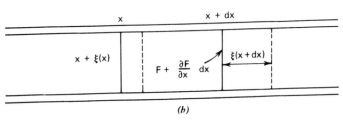

(b)

FIGURE 2.11 Movement of boundaries of a block of fluid in a tube of constant diameter when a wave passes through it.

The law of conservation of matter states that the mass of the material cannot be changed; therefore, regarding the density ρ as constant, the mass is

$$\rho S \, dx$$

where S is the cross-sectional area of the tube.

Now, if there are unequal forces at each end of the segment Newton's law states that the force difference is equal to the mass times the acceleration. Because ξ is the particle displacement from its rest position, $\partial \xi / \partial t$ or $\dot{\xi}$ is the particle velocity and $\partial^2 \xi / \partial t^2$ (or $\ddot{\xi}$) is the acceleration. It is recognized that the partial derivatives are necessary because ξ is a function of time and distance. The force difference is therefore

$$\rho S \, dx \, \ddot{\xi}$$

It is now necessary to consider the force provided by the *elastic* nature of the material itself.

Hooke's law states that

$$\frac{\text{force}}{\text{unit area}} = \frac{F}{S} = K\left(\frac{\text{change in length}}{\text{original length}}\right) = K \frac{\partial \xi}{\partial x}$$

There is a force $F = KS(\partial \xi / \partial x)$ operating on the section at x and an incremented force

$$F + dF = F + \frac{\partial F}{\partial x} dx = KS \frac{\partial \xi}{\partial x} + KS \frac{\partial^2 \xi}{\partial x^2} dx$$

operating on the section at $x + dx$. The net accelerating force acting on the element dx of the material is the difference between the forces acting on the two ends, or

$$KS \frac{\partial^2 \xi}{\partial x^2} dx$$

We can equate the two measures of force to give

$$\rho \frac{\partial^2 \xi}{\partial t^2} = K \frac{\partial^2 \xi}{\partial x^2} \tag{2.16}$$

which is a one-dimensional, partial differential equation, based on simple physical laws, that governs wave motion along the rod of material.

By trial it can be shown that one solution of this second-order partial differential equation is

$$\xi_1(x, t) = A_1 \epsilon^{i(kx - \omega t)} \tag{2.17}$$

because

$$\frac{\partial^2 \xi}{\partial x^2} = -k^2 \xi \quad \text{and} \quad \frac{\partial^2 \xi}{\partial t^2} = -\omega^2 \xi$$

and

$$c = \left(\frac{K}{\rho}\right)^{1/2} = \frac{\omega}{k}$$

In these equations ω is the angular frequency ($=2\pi f$) and k is the wave number. It is related to the wavelength (λ) by

$$k = \frac{2\pi}{\lambda}$$

Now, because the trial solution is independent of the frequency, we can make use of the Fourier series to build a more general solution to obtain

$$\xi(x, t) = \sum_{j=-\infty}^{\infty} A_j \epsilon^{i(k_j x - \omega_j t)} \tag{2.18}$$

which could also be written

$$\xi(x, t) = F_1(x - ct) + F_2(x + ct) \tag{2.19}$$

This general solution represents two waves traveling in the rod, the first in the positive x direction and the second in the opposite direction. It confirms the form of the function of x and t, conjectured near the beginning of this section, but it now adds some important information about the way the wave speed is related to characteristics of the material in which the wave is traveling. So far we have said little about K except that it is a constant of proportionality between stress and strain in Hooke's law. This constant is called modulus of elasticity. Now we can be more specific about which of the moduli this constant K is.

When the rod of material was specified it was contained within a rigid tube that allowed it to expand or contract in one direction only. No motion across the axis could be accommodated. It can easily be seen that these are the same considerations applied to plane waves passing through the body of a material because expansion or contraction of any part of the material is incompatible with conditions for the neighboring material. All the motion is parallel to the direction of wave propagation and results in compression or rarefaction. The waves are called compressional, longitudinal, or P waves (after the waves that arrive first from earthquakes). Their velocity is denoted by α and is given by

$$\alpha = \sqrt{\frac{K + 4/3\mu}{\rho}}$$

where K = bulk modulus of elasticity
     μ = modulus of rigidity
     ρ = density

If these are waves of compression only, why does the modulus of rigidity enter into the velocity? The reason is that during compression, as a plane wave passes through the material, a small portion (originally cubic in shape) becomes a parallelipiped—a change in shape as well as in volume; hence the modulus of rigidity is involved.

2.7 EXTENSION BY ANALOGY

A second mode of motion may occur when the rod of material is twisted but not compressed. We let the amount of twist be $\theta(x, t)$ and, if the same arguments are followed, arrive at the same type of partial differential equation:

$$I \frac{\partial^2 \theta}{\partial t^2} = Q \frac{\partial^2 \theta}{\partial x^2}$$

Here the volume of any small element of the rod has not changed, but the material has been sheared. The quantities I and Q correspond, respectively, to the moment of inertia of the cross section and a constant related to the shear modulus of elasticity. This can be simplified to

$$\rho \frac{\partial^2 \theta}{\partial t^2} = \mu \frac{\partial^2 \theta}{\partial x^2} \tag{2.20}$$

which has the same *type* of solution, but the velocity, denoted by β, is given by

$$\beta = \sqrt{\frac{\mu}{\rho}}$$

Lamb (1934) showed that all plane waves in an isotropic elastic solid propagate with one or the other of these two speeds.

A third condition can develop when the material rod lacks a rigid boundary but is allowed to expand or contract as necessary. Then, for longitudinal waves traveling down the rod Young's modulus E is the relevant modulus of elasticity and the velocity in the rod is

$$C_R = \sqrt{\frac{E}{\rho}}$$

This has little use in seismology per se, but some models are made of rods of materials like rocks and E must be applied.

If we define a quantity σ, called Poisson's ratio, as the ratio

$$\frac{\text{transverse strain}}{\text{longitudinal strain}}$$

four elastic quantities need to be related. A further one, λ, is called Lamé's constant (Lamé also used μ), and all appear in the more formal mathematical theory of elasticity. For those who may need to refer to other literature the quantities are related as follows:

$$K = \lambda + \frac{2\mu}{3} = \frac{E\mu}{3(3\mu - E)} = \lambda\left(\frac{1+\sigma}{3\sigma}\right) = \mu\frac{2(1+\sigma)}{3(1-2\sigma)}$$

$$E = \frac{\mu(3\lambda + 2\mu)}{\lambda + \mu} = 9K\left(\frac{K-\lambda}{3K-\lambda}\right) = \frac{9K\mu}{3K+\mu} = \lambda\frac{(1+\sigma)(1-2\sigma)}{\sigma}$$

$$\sigma = \frac{\lambda}{2(\lambda + \mu)} = \frac{\lambda}{3K - \lambda} = \frac{3K - 2\mu}{2(3K+\mu)} = \frac{1}{2}\left(\frac{\alpha^2 - 2\beta^2}{\alpha^2 - \beta^2}\right)$$

$$\alpha = \sqrt{\frac{\lambda + 2\mu}{\rho}} = \sqrt{\frac{K + 4/3\mu}{\rho}} = \text{compression wave velocity} \qquad (2.21)$$

$$\beta = \sqrt{\frac{\mu}{\rho}} \qquad\qquad = \text{shear wave velocity}$$

$$K = \rho\left(\alpha^2 - \frac{4\beta^2}{3}\right) \qquad\qquad \frac{\alpha}{\beta} = \sqrt{\frac{2 - 2\sigma}{1 - 2\sigma}}$$

$$\lambda = \rho(\alpha^2 - 2\beta^2)$$

Earlier we chose as our example a plane wave that propagates in a single direction. It is necessary, however, for waves that have other geometrical modes of propagation to be considered; for example, those that expand outward from a point source. If the material has properties that do not change from point to point and are not dependent on the direction of wave propagation, the symmetry is such that the waves travel outward as concentric spheres, simply expanding in radius r with time. Intuition should then tell us that by combining arguments made earlier we can write the wave solution in the form

$$\phi_r(r, t) = \frac{1}{r}\left[F(r - \alpha t) + f(r + \alpha t)\right]$$

which represents two compressional waves the first traveling outward from and the second traveling inward toward the source. In most problems, but not all, the inward-traveling waves can be neglected and we have but one term:

$$\phi_r(r, t) = \frac{1}{r}F(r - \alpha t) \qquad\qquad (2.22)$$

Unfortunately, spherically traveling shear waves cannot be treated with quite the same simple type of formula because to generate a shear wave shear traction must be applied to the medium—about an *axis*. This immediately involves considerations of vectors, and the displacement is no longer the same in all directions from the very small source. In the simplest case (a couple about a single axis) the results have cylindrical symmetry about this axis. We have no need for such a development, however. Any readers interested should consult the sources already cited.

This is a suitable time to remark philosophically on the role of the mathematical theory of elasticity in reflection seismology. Formal solutions to wave equations, under even simple boundary conditions, involve very sophisticated mathematical considerations. Fortunately, the solutions found for these cases (which sometimes have only one or two boundaries) have provided guidelines to the types of wave generated, their behavior inside the elastic media and at the boundaries, and ideas related to the distribution of energy between the various waves and the variation in the wave fronts. The importance of this kind of information cannot be overemphasized, and it is given—as the need arises—without proof but usually with a reference for further study.

Field examples are generally so complex that formal solutions are impossible and only approximate answers can be found. To proceed with these approximate solutions, the field model can be simplified by (1) assuming plane parallel layering, (2) stipulating one wave type only, or (3) adopting an iterative type of computer solution that is usually limited to one or two dimensions and integrates the original differential equations (or the difference equation approximations to them). Most of these approximations are used later.

Compressional and shear wave velocities of rocks depend on many parameters. Much information on compressional wave velocities is now available for specific areas in the form of continuous velocity logs taken in drilled holes. Table 2.1 lists typical values only. Shear wave velocities in rocks down to about 70 m depth of burial have been measured by a simple method attributable to Beeston and McEvilly (1977). Measurements in rocks at greater depths have been made by

TABLE 2.1
SEISMIC VELOCITY IN UNCONSOLIDATED SEDIMENTS, CONSOLIDATED SEDIMENTS, AND METAMORPHIC ROCKS

Material	Velocity (km/s)		Velocity (1000 ft/s)		Remarks
	α	β	α	β	
Alluvium	0.5–2.0	—	1.64–6.56	—	Near surface
	3.0–3.5	—	9.84–11.48	—	2000 m depth
Clay	1.1–2.5	—	3.61–8.20	—	
Loam	0.8–1.8	—	2.62–5.91	—	
Loess	0.3–0.6	—	0.98–1.96	—	
Sand					
Loose	0.2–2.0	—	0.66–6.56		
Loose	1.0	0.4	3.28	1.31	Above water table
Loose	1.8	0.5	5.91	1.64	Below water table
Calcareous	0.8	—	2.62	—	
Wet	0.75–1.5	—	2.46–4.92	—	
Weathered layer	0.3–0.9	—	0.98–2.95	—	
Glacial					
Till	0.43–1.04	—	1.41–3.41	—	Unsaturated
Till	1.73		5.67		Saturated
Sand and gravel	0.38–0.50	—	1.25–1.64	—	Unsaturated
Sand and gravel	1.67		5.48		Saturated
Sandstone-shale					
Tertiary	2.1–3.5		6.89–11.48		

TABLE 2.1 (Continued)

Material	Velocity (km/s) α	β	Velocity (1000 ft/s) α	β	Remarks
Cretaceous	2.4–3.9		7.87–12.80		Depth range
Pennsylvanian	2.9–4.4		9.51–14.44		0.3–3.6 km
Ordovician	3.3–4.5		10.83–14.76		0.3–2.1 km
Sandstone	1.4–4.3		4.59–14.11		
Sandstone Conglomerate	2.4		7.87		Australia
Limestone					
Soft	1.7–4.2		5.58–13.78		
Hard	2.8–6.4		9.19–21.00		
Solenhofen	5.97	2.88	19.59	9.45	
U.S. midcontinent	—	2.75		9.02	
and Gulf Coast	3.4–6.1	—	11.15–20.01		
Argillaceous, Texas	6.03	3.03	19.78	9.94	∥ to bedding
Argillaceous, Texas	5.71	3.04	18.73	9.97	⊥ to bedding
Dolomitic, Penn.	5.97	—	19.59		
Cement rock, Penn.	7.07	—	23.20		
Crystalline, Texas, N.M., Okla.	5.67–6.40	—	18.60–21.00	—	
Dense, U.S.S.R.	5.90–7.00	3.03–3.59	19.36–22.97	9.94–11.78	
Salt, cornallite, sylvite	4.4–6.5	—	14.44–21.38		
Caprock, salt, anhydrite, gypsum, limestone	3.5–5.5	—	11.48–18.04		
Anhydrite, midcontinent and Gulf Coast	4.1	—	13.45		
Gypsum	2.0–3.5	—	6.56–11.48		
Chalk, U.S.,	2.1–4.2	1.07 SV	6.89–13.78	3.51 SV	
Germany, France,	2.58	1.13 SH	8.46	3.71 SH	⊥ to bedding
Austin, Texas	3.05	—	10.01		∥ to bedding
Slate, Mass.	4.27	2.86	14.01	9.38	
Hornfels slate	3.5–4.4	—	11.61–14.44	—	
Magnetite ore	5.50	—	18.04	10.81–10.49	$V_P/V_S =$ 1.67–1.72
Marble	3.75–6.94	2.02–3.86	12.30–22.77	6.63–12.66	46 samples
	5.78	3.22	18.96	10.56	Average of 46 samples
Quartzite	6.1	—	20.01	—	
Wet clay, U.S.S.R.	1.50–1.65	—	4.90–5.41	0.36–1.20	$V_P/V_S \sim$ 4.5–13.7
Impermeable argillaceous clay	2.00	0.59	6.56	1.94	
Soil	0.11–0.20	—	0.36–0.66	0.59–1.27	$V_P/V_S \sim$ 1.7–2.0
Tuff	2.16	0.83	7.09	2.72	

Source. These values have been selected from the compilation by Frank Press in the *Handbook of Physical Constants*, rev. ed., Memoir 97, and are printed with permission of the Geological Society of America. Copyright © 1966.

time interval measurement of waves generated at the surface by shear wave vibrators or by explosives when the near shot conditions are not spherically symmetrical. The shear-wave velocities, as well as most of the compressional wave velocities listed, were obtained in most part on small specimens in dimensions of several centimeters by using very high frequency sources and receivers in the laboratory. True shear wave velocities have been obtained in small (AX) dry or water-filled holes by a logging tool described by M. S. King et al. (1974). This tool has not been adapted for holes of a diameter commonly used in oil and gas exploration. As a rather unsatisfactory substitute, energy arriving later than compressional wave energy from a compressional wavelogging tool is often used as a "fracture log." More detail is given and the validity of these measurements is discussed in Chapter 7.

2.8 PARTITION OF ENERGY BETWEEN DIFFERENT WAVES AT BOUNDARIES

Now that it has been established that two different kinds of wave can be propagated in solids we can see that the effect of boundaries on wave propagation becomes much more complex. Not only are reflected and refracted waves generated at the boundaries, but each of them can be compressional or shear. A generalized form of Snell's law still holds as far as the directions of the different waves are concerned, namely (see Figure 2.12),

$$\frac{\sin \phi_i}{\beta_1} = \frac{\sin \theta_i}{\alpha_1} = \frac{\sin \theta_r}{\alpha_1} = \frac{\sin \phi_r}{\beta_1} = \frac{\sin \theta_t}{\alpha_2} = \frac{\sin \phi_t}{\beta_2} \qquad (2.23)$$

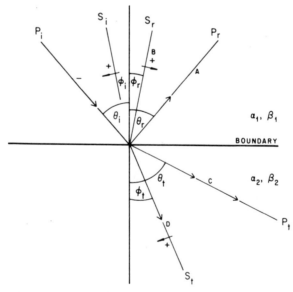

FIGURE 2.12 Angular relations between incident, reflected, and refracted rays when a wave encounters a boundary between two different solid media.

where subscripts 1 and 2 denote the medium and subscripts i, r, and t stand for incident, reflected, and transmitted, respectively. These relations are easily proved by wave-front construction near a plane boundary, as illustrated for a simple case in Figure 2.6.

In Figure 2.12 the direction of particle motion of the shear waves is shown by an arrow with a sign. This establishes that the shear wave, which has particle motion perpendicular to the direction of propagation, has motion in the plane of the diagram. Such waves are given an auxiliary designation SV in that their particle motion lies in a plane perpendicular to the boundary and containing both the normal and the incident ray.

Shear waves with another polarity exist in which particle motion is perpendicular to the plane of the diagram. These are SH waves that, because their particle motion is paralel to the boundary, suffer no change in type on reflection or refraction. Naturally these are the extremes and a random shear wave striking a boundary has both SV and SH components.

At perpendicular incidence, ϕ_i or $\theta_i = 0$, there is no conversion from P to SV or from SV to P, but as the angle of incidence increases the conversion from one wave type to another becomes significant. The Zoeppritz equations determine the amplitudes of the reflected and refracted waves and are given in Appendix 2A. These equations are seldom used in calculations that affect exploration geophysics, in spite of the fact that modern seismic reflection methods use long offsets and involve significant angles of incidence. Normally, the velocity and density contrasts between layers are not large but in special cases (e.g., high reflection coefficients caused by gas in high porosity sands) longer offsets may be a problem.

The reflection coefficients normally used are those for perpendicular incidence. Figure 2.13 is a diagram of the rays associated with an S or P wave at

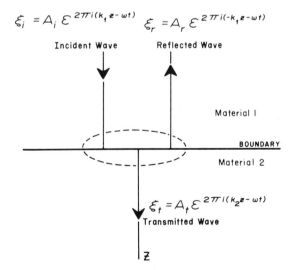

FIGURE 2.13 Showing the incident, reflected, and refracted rays used for calculating the reflection coefficient. All rays are actually along the normal to the surface. The incident and reflected rays are displaced from one another for clarity.

perpendicular incidence. The boundary conditions are simple. The boundary must move by the same amount, whether it is considered the boundary of medium 1 or medium 2. (The displacement must be continuous across the boundary.) Also, the net stress on the boundary must be zero. (The boundary itself has no mass and any stress leads to infinite acceleration.)

A plane wave traveling in the positive z direction can be represented by

$$\xi(z, t) = AF(z - ct)$$

We will not violate any rules if we choose to consider a constant frequency wave (angular frequency ω) represented by

$$\xi(z, t) = A\epsilon^{i,2\pi(kz - \omega t)}$$

This is a commonly used form, and it is shown later that it is possible to combine sinusoidal waves of this type but with a range of frequencies and time relations to one another in order to develop any physically possible form F; k is called the wave number and $\omega/k = c$. Because $c = $ frequency $\times$ wavelength $= f\lambda$ and the angular frequency (radians/s) $= 2\pi f$, we see that $k = 2\pi/\lambda$, a scaled inverse wavelength.

In this diagram the incident and reflected rays have been drawn separately for clarity, but they actually lie along the same normal to the boundary.

The two boundary conditions give rise to two equations:

$$A_i + A_r = A_t$$

and

$$\rho_1 c_1 A_i - \rho_1 c_1 A_r = \rho_2 c_2 A_t$$

by making use of $F = K(\partial \xi / \partial z)$, and from these two it is easy to obtain

$$\frac{A_r}{A_i} = \frac{\rho_1 c_1 - \rho_2 c_2}{\rho_1 c_1 + \rho_2 c_2} = \text{reflection coefficient } R_{12} = \frac{Z_1 - Z_2}{Z_1 + Z_2} \qquad (2.24)$$

where $Z_1 = \rho_1 c_1$ and $Z_2 = \rho_2 c_2$, the acoustic impedances of the two media.

Similarly, the transmission coefficient $A_t/A_i = T_{12}$ is given by

$$T_{12} = \frac{2Z_1}{Z_1 + Z_2} \qquad (2.25)$$

These simple formulas are independent of frequency, which is evident if the two equations of continuity are derived from the expressions for the individual waveforms. If $Z_2 > Z_1$, corresponding to the wave moving from a lower velocity, lower density rock to one with higher parameter values, the reflection coefficient—for *amplitudes*—is negative and the transmission coefficient is positive.

Thus the displacement of the reflected wave is in a direction opposite that of the incident wave. However, the direction of travel of this wave is also reversed, and *in relation to its travel direction* the particle displacement has the same sign as the incident wave.

As long as the medium is an acoustic one (it has no shear strength) it is possible, very simply, to generalize the reflection and transmission coefficients for angles of incidence (i) different from zero.

Then

$$R_i = \frac{Z_1 \cos i - Z_2 \cos r}{Z_1 \cos i + Z_2 \cos r} \tag{2.26}$$

$$T_i = \frac{2Z_1 \cos i}{Z_1 \cos i + Z_2 \cos r} \tag{2.27}$$

where i is the angle of incidence and reflection of the P wave input and r is the angle of refraction of the transmitted P wave. These two angles are, of course, related by Snell's law:

$$\frac{\sin i}{\sin r} = \frac{c_1}{c_2}$$

Even if the material is elastic formulas (2.26) and (2.27) will apply to horizontally polarized shear waves (SH).

Because the particle velocity is a constant times the particle displacement (for a given frequency), it also has the same reflection coefficient and obeys the preceding rules as far as particle velocity is concerned; for the situation described $\dot{\xi}_i = -c\xi_i$ for the downward traveling wave but $\dot{\xi}_r = c\xi_r$ for the upward traveling wave.

In an acoustic wave the pressure obeys the same numerical formula for reflection coefficient but, because it is a number equal to $-\rho c\, \partial\xi/\partial t$, the reflected pressure is actually the negative of that given by (2.24).

In reflection from the surface between water and air the displacement has a reflection coefficient almost equal to 1 (incidence from water to air $Z_1 > Z_2$). Thus the sum of the incident and reflected displacements near such a boundary is nearly twice that of the incident wave. This is also true for particle velocities. The pressure of the reflected wave, however, cancels the pressure of the incident wave. In other words, the air-water surface is a pressure-release surface. For this reason no measurements of excess pressure are made near the surface in marine prospecting—but, provided that noise from water waves can be diminished successfully, particle velocity measurements are double those in the body of the liquid and can be used for detecting acoustic energy.

Because the energy is proportional to $\rho c\dot{\xi}^2$, it can be shown that it is conserved with these reflection and transmission coefficients.

The input energy per unit area (intensity) of the wave front is $\frac{1}{2}\rho_1 c_1 \dot{\xi}_i^2$ (note the analogy to $\frac{1}{2}\,mv^2$ for solid bodies).

The output energy per unit area is the contribution from reflection ($\frac{1}{2}\rho_1 c_1 \dot{\xi}_r^2$) plus the transmitted contribution ($\frac{1}{2}\rho_2 c_2 \dot{\xi}_t^2$):

$$\frac{1}{2}\rho_1 c_1 \dot{\xi}_i^2 \left(\frac{\rho_1 c_1 - \rho_2 c_2}{\rho_1 c_1 + \rho_2 c_2}\right)^2 + \frac{1}{2}\rho_2 c_2 \dot{\xi}_i^2 \left(\frac{2\rho_1 c_1}{\rho_1 c_1 + \rho_2 c_2}\right)^2 = \frac{1}{2}\rho_1 c_1 \dot{\xi}_i^2$$

Thus the principle of conservation of energy is obeyed.

It should be noted that the *amplitude* of a transmission coefficient T can be greater than unity when the reflection coefficient is negative. This does not violate the conservation of energy principle. Note also that in *two-way* passage through a boundary the amplitude is always diminished because the transmission both ways must result in multiplication by $(1 - R)(1 + R) = 1 - R^2$, which is always less than unity. The implications are examined in more detail when transmission and reflection in a many-layered system are considered.

2.9 IMPERFECT ELASTIC MEDIA—ATTENUATION OF SEISMIC WAVES

Observation of propagation of stress waves in solid or fluid media shows that dissipation of strain energy occurs even when the waves have very small amplitude. This dissipation results from imperfection in the structure of the material and the resulting local perturbations in elasticity; for example, small dislocations in crystal structure or minute pores, possibly with some sharply angular boundaries, can cause stress concentrations that change the stress–strain relations locally. Some pores may be totally or partially filled with a mixture of fluids. This dissipation of acoustic energy into heat energy occurs in addition to other forms of loss of acoustic wave amplitude, such as geometrical spreading or scattering.

Fortunately the deviations from Hooke's law are usually small and the approximations of elastic wave theory are sufficiently accurate for most materials, provided the pressure, temperature, and frequency are not too widely varied.

In liquids flow or oscillation is restricted by the viscosity of the liquid and damping is accompanied by heat generation. For solids continuous shearing is possible only when the shearing forces are above the elastic limit of the material. During propagation of elastic waves only very small *average* stresses are involved—well below the elastic limit. Nevertheless, locally there may be sufficient stress concentrations to cause some mineral components, or fluids filling the pores, to flow.

It has been found possible to simulate the effects of viscous damping on wave propagation by making the relevant modulus of elasticity complex (i.e., it consists of a constant term and one dependent on the time derivative); for example, when considering shear waves traveling through a viscous medium the modulus μ is replaced by $(\mu + \eta \partial/\partial t)$ and for a constant frequency shear wave

$$\theta = \theta_0 \epsilon^{i\omega(t - x/c)}$$

The derivative term results in a modulus $\mu + \eta i\omega$, which is complex and *dependent on frequency*. With these assumptions

$$c^2 = \frac{\mu + \eta i\omega}{\rho} = (V + iv)^2$$

$$V \simeq \sqrt{\frac{\mu}{\rho}} \quad \text{and} \quad v = \frac{\eta\omega}{2\rho}$$

As a consequence

$$\theta = \theta_0 \epsilon^{-\omega^2 x/v} \epsilon^{i\omega(t-x/v)}$$

$$= \theta_0 \epsilon^{-\alpha x} \epsilon^{i\omega(t-x/v)} \tag{2.28}$$

where α is the attenuation coefficient and is dependent on the *second* power of the frequency.

Friction caused by one grain sliding against another along the grain boundaries or by imperfections in individual crystals is widely acknowledged to be the most probable mechanism of energy loss. Experiments on seismic wave attenuation in specimens of rock at different humidity levels indicate that adsorption of moisture on the grain boundaries has an important effect. The term "solid friction" is frequently used for nonelastic mechanisms that convert strain energy into heat, thereby damping or attenuating the stress waves in a medium. Table 2.2 lists values of various attenuation parameters for a variety of rocks. The symbol Q is often used in the same way that it is used in electrical engineering to denote the quality of a rock for seismic wave propagation.

TABLE 2.2
INTERNAL FRICTION IN ROCKS

Material	Frequency (kHz)	dE/E	Q	Notes
Basalt	3–4	0.0112	561	20°–900° Average
Diorite (Utah)	10.7	0.035	179	Average of 2
Granite	20–200	0.0202	311	Average of 9
Granite	5.35	0.03	2.09	
Granite, Westerly, R.I.	100–800	0.09807	64.1	Average of 9 separate frequencies
Marble		0.01148	547	7 temps. 20°–1000°C
Quartzite	3–4	0.01603	392	15 temps. 20°–1000°C
Slate		0.02864	219	Average of 12 samples
Caprock	0.12–6.60	0.13375	46.98	Average of 12 samples and frequencies
Dolomite	10.44–15.00	0.03256	192.99	Average of 7 samples
Limestone	—	0.03093	203.12	Average of 12 samples
Chalk	0.002	0.046	136.6	
Hunton limestone	10.55	0.106	59.28	
	2.82	0.088	71.4	
Oolitic limestone	0.002–0.12	0.1350	46.54 ·	Average of 11 frequencies
Shelly limestone	0.002–0.12	0.0915	68.63	Average of 11 frequencies
Solenhofen limestone	0.00389–18.6	0.00925	679.26	Average of 6 frequencies
Amherst sandstone	0.55–3.83	0.2612	24.05	Average of 10 samples
Berea sandstone	20	0.048	130.9	Average of 2 samples
Homewood sandstone	2.68–4.92	0.0909	69.11	Average of 11 samples
Old red sandstone	0.002–0.040	0.0675	93.08	Average of 2 samples
Pierre shale	0.075–0.555	0.36366	17.15	Average of 17 samples
Sylvan shale	3.36–12.8	0.08667	72.5	Average of 3 samples

Source. These values have been selected from the compilation by James J. Bradley and A. Newman Fort, Jr., in the *Handbook of Physical Constants*, rev. ed., Memoir 97, and are printed with permission of the Geological Society of America, Copyright © 1966.

Just as it was found that the viscous damping could be approached by allowing the *elastic modulus* to become complex, it has been found that solid friction can be represented mathematically if the *velocity* is a complex quantity.

In a manner similar to that of the foregoing analysis, we quote directly from Appendix 4A.1, equation (4A.3), in which, for solid friction, it is found that the wave travel can be represented by complex velocity

$$c = V + iv$$

and

$$\zeta = \zeta_0 \epsilon^{-\alpha x} \epsilon^{i\omega(t - x/v)}; \quad \alpha = \frac{\omega v}{V}$$

For the solid friction model, therefore, the attenuation constant is proportional to the *first* power of the frequency. In the remainder of this book it is the preferred model.

In 1962 Futterman showed that a causal wave (i.e., one generated by a sudden release of energy) traveling in a medium that attenuates the energy as the travel time increases, theoretically must be affected by dispersion. This is, of course, because the velocity of wave components is a function of their frequency. It must be noted that in the past it has been difficult to measure this dispersion [e.g., see O'Brien and Lucas (1971)]. Reexaminations of original data taken by McDonal et al. (1958), Wuenschel (1965), and Kjartansson (1979) have shown that there was almost an exact fit with a dispersive, constant Q model.

Futterman calculated that for a linear, frequency-independent Q mechanism and for Q greater than about 10 the velocity $c(f)$ should vary with frequency according to the relation

$$\frac{c_2 - c_1}{c_1} \simeq \frac{1}{\pi Q} \times \ln f_2/f_1$$

As a simple example to illustrate the order of magnitude of the dispersion to be expected in reflection seismic work we can place the spectrum of the transmitted pulse between 15 and 60 Hz and the medium as having a Q of 100; then

$$\frac{c_2 - c_1}{c_1} \simeq \frac{1}{314} \times \ln(4.0) = 4.4 \times 10^{-3}$$

$$\simeq 0.44\%$$

and a reflection pulse may broaden by about $4 \cdot 4$ ms/s of travel. The higher frequencies travel faster.

Recent field measurements by Stewart et al. (1984) and by Stainsby and Worthington (1985) provide additional evidence of the existence of dispersion. Nevertheless, the dispersion is small and its effect on the location of hydrocarbon deposits by seismic reflection work has probably been small. No doubt this can be attributed to the fact that, if present, it is a small near-constant factor in a comparison of arrival times for a reflection at different field locations.

In discussing the gradual loss of amplitude of a seismic wave traveling through an imperfect medium, it should be noted that the change in amplitude is exponential and that each cycle bears the same ratio of amplitude to the preceding one. This ratio is called the logarithmic decrement δ. The quantity $1/Q$ has been designated the specific dissipation constant. Generally speaking, except for unconsolidated, water-saturated sediments, the following is indicated:

1. The specific dissipation constant $1/Q$ is essentially independent of frequency.
2. It appears that $1/Q$ is substantially independent of amplitude for strain below 10^{-4}. The strain in rocks due to the passage of seismic waves used for exploration purposes is usually no more than 10^{-8} except within a few tens of meters of the source.
3. Observations show that dissipation is less for a single crystal than for an aggregate of such crystals.
4. The rate of dissipation decreases with increased pressure.
5. The existing data suggest that dissipation is relatively independent of temperature.

One method of measurement of attenuation is the percentage of stored energy lost *per cycle*. Other measures in common use are

$1/Q$ = specific dissipation constant. It is related to the rate at which the mechanical energy of vibration is converted irreversibly into heat energy and does not depend on the detailed mechanism by which the energy is dissipated.

δ = logarithmic decrement—the natural logarithm of the ratio of amplitudes of two successive maxima or minima in an exponentially decaying free vibration.

a = damping amplitude coefficient in the expression for a free vibration:

$$\epsilon^{-at} \sin 2\pi ft$$

α = attenuation coefficient in the expression for plane waves in an infinite medium:

$$\epsilon^{-\alpha x} \sin 2\pi f\left(t - \frac{x}{c}\right); \quad c = \text{wave velocity}$$

$\Delta f/f$ = relative bandwidth of a resonance curve between the half-power or 0.707 amplitude points for a solid undergoing forced vibrations is a measure of the sharpness of the resonance curve.

$\Delta E/E$ = fraction of strain energy lost per cycle.

Then b, the specific damping capacity, is related to the other parameters by

$$b = \frac{2\pi}{Q} = 2\delta = 2\frac{a}{f} = 2c\frac{\alpha}{f} = 2\pi\frac{\Delta f}{f} = \frac{\Delta E}{E}$$

As examples of the effect of attenuations on amplitudes of seismic data we make use of the formula

$$\alpha = \frac{\pi f}{Qc}$$

to determine how far a plane wave (of frequency 30 Hz) travels before it is reduced to one-tenth of its original amplitude.

In Pierre shale $Q = 17.15$ and

$$\alpha = \frac{3.14 \times 30}{17.15 \times 7000} = 0.000785$$

Then

$$\epsilon^{-\alpha x} = \tfrac{1}{10}$$

$$\ln \tfrac{1}{10} = -2.3 = -0.000785x$$

$$x = 2929 \text{ ft } (892.8 \text{ m})$$

Thus by attenuation alone the amplitude of a plane wave drops by a factor of 10^4 in traveling to a depth of 5858 ft (1785 m) and back.

If the travel material is a limestone of $Q = 200$ and $c = 20,000$ ft/s (6095 m/s),

$$\alpha_{\text{limestone}} = \frac{3.14 \times 30}{200 \times 20,000} = 0.00002355$$

the corresponding x is

$$x_L = 90,766 \text{ ft } (27,664 \text{ m})$$

and the attenuation becomes an insignificant factor in amplitude decay, compared with the spherical spreading.

The foregoing is a simplistic discussion of attenuation and applies only to a single homogeneous rock body. In the practice of exploration seismology the presence of layering has a profound effect on the change in amplitude of a signal as it propagates through the earth layers. This is particularly noticeable when the usual (reflection) observations are made of a wave that has traveled in both directions through the layering. Much more discussion of this is to be found in later chapters.

2.10 BOUNDARY WAVES

The discussion so far has been about waves that travel through an unbounded medium that is homogeneous and isotropic, but because it may be slightly imperfect elastic energy is continuously lost to heat. When, as in all real systems, boundaries are present they tend to guide waves. The first wave to be analyzed was described by Rayleigh and is called, after him, a Rayleigh wave. Rayleigh has shown that a wave is propagated along a free plane boundary with the following characteristics:

1. It has a velocity along the surface that is independent of frequency and is somewhat less than the shear velocity for the body waves in the medium. The actual speed is dependent on the ratio V_P/V_S (or on Poisson's ratio), but for a material of $V_P/V_S = 1.732$ (or Poisson's ratio $= \frac{1}{4}$) the speed of the Rayleigh wave is $0.9194V_S$, where V_P and V_S are, respectively, the velocity of P and S waves in the medium.

2. The motion of a particle on the surface is not linear but an ellipse, described such that, at the top of the ellipse, the particle is traveling toward the source. This motion is called retrograde elliptical motion. The ellipse axes are along the surface and perpendicular to it, and the motion can be regarded as resulting from two linear motions that are 90° out of phase.

3. The amplitude of motion decreases (exponentially) as the depth below the boundary increases. At the surface the vertical motion is about 1.5 times the horizontal motion but vanishes at a depth of 0.192 wavelengths; below that it reverses sign. All these parameters, given for $\sigma = 0.25$, are dependent on the physical characteristics of the rock.

The theoretical ratios of the Rayleigh wave velocity C_R and the compressional and shear velocities α and β of the medium are shown in Figure 2.14, taken from Knopoff (1952).

It is, of course, not possible, because of the invariable presence of a weathered layer, to measure surface motion due to this type of Rayleigh wave under field

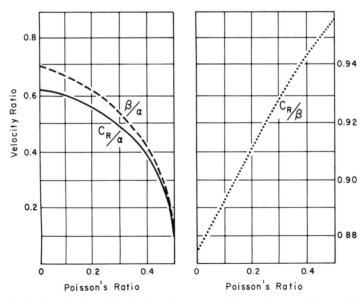

FIGURE 2.14 Ratios of β/α, C_R/α, C_R/β as functions of Poisson's ratio. [*Source*: Knopoff (1952) from Ewing et al., *Elastic Waves in Layered Media*. Copyright © 1957 by McGraw-Hill, Inc. Used with permission of the McGraw-Hill Book Company.]

conditions. There the velocity increases with depth below the surface, and, in addition, the presence of an interface at the base of the weathered layer causes a disturbance of the simple conditions required. Dobrin, Lawrence, and Simon (1951) experimentally measured the particle trajectory variation with depth below the earth's surface for Rayleigh waves from small explosions and found that the motion was retrograde above 40 ft (13 m) and prograde below. The words *retrograde* and *prograde*, as used here, describe the direction of the particle motion of a surface wave. Retrograde refers to particle motion which, at the top of the ellipse, has a direction opposite that of the direction of wave travel, and prograde indicates the opposite particle motion. The crossover depth was 0.136 of a wavelength of the frequency used. These results are valid for the particular area investigated.

Figure 2.15 shows an actual particle trajectory compared with the theoretical particle trajectory and is taken from Howell (1959). The complexity is evident, but it must be remembered that the comparison is again that of an isotropic half-space with a real, complexly layered system. One reason for including this comparison is that it has been suggested that one means of canceling Rayleigh waves may be based on the elliptical retrograde motion in which a fraction of the horizontal motion suitably changed in phase to cancel the vertical motion is used. It seems unlikely that the real Rayleigh waves would yield to such simple treatment. This matter, however, is discussed in more detail later.

Rayleigh waves consist of a mixture of compressional and shear energy, the shear wave motion being *SV*; that is, polarized in a vertical plane through the line of propagation of the wave. There is no corresponding surface wave for horizontally polarized motion (i.e., motion in a vertical plane perpendicular to the direction of wave propagation).

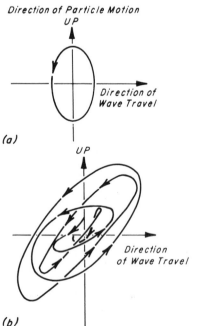

FIGURE 2.15 Particle trajectories from (*a*) a Rayleigh wave on a half-space of isotropic material and (*b*) a real Rayleigh wave. (*Source*: Howell, 1959.)

A guided wave with different characteristics is one controlled alone by the lower boundary of the weathered layer. In reality, there are three different types, some of which can be coproduced if the source is the right type. They are variously called *refracted waves* or *head waves*, but in this case they travel along the boundary with the appropriate velocity (compressional or shear) for the lower medium. Their amplitudes decay exponentially with depth from the boundary and they characteristically leak energy back into the upper layer to be recorded by surface measuring devices.

The paths shown in Figure 2.16 act as though they have been refracted into the second medium from a critically incident ray and are parallel to the boundary. An incident ray is called critical only with respect to one of the refracted (P or SV) rays. It is the ray for which the designated refracted ray has a refraction angle of 90°; in other words, it travels along the boundary. Therefore, because a P wave always travels with higher speed in a medium than an SV wave, the P wave is bent more from the normal than the SV wave and the critical angle for the P wave is smaller than for the SV wave. Plane-wave theory predicts zero amplitude for refracted waves traveling along the boundary. It is only when spherically diverging waves impinge on this boundary that theory shows that some energy is

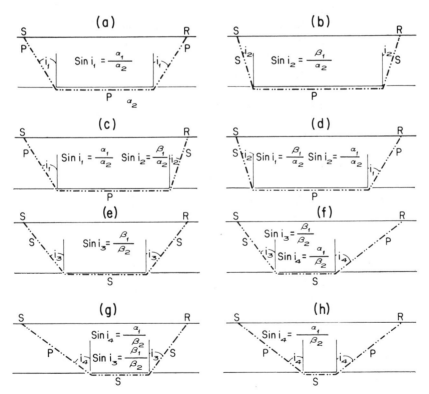

FIGURE 2.16 Possible refraction (or head-wave) paths from the source to the receiver. Medium 1 has velocities α_1, and β_1 and medium 2, velocities α_2 and β_2. Note that if $\alpha_1 > \beta_2$ modes (f), (g), and (h) are not possible. The cases are made more complex by a sloping lower interface or when the source is buried. (*Source*: Ewing et al., *Elastic Waves in Layered Media.* Copyright © 1957, McGraw-Hill, Inc. Used with permission of the McGraw-Hill Book Company.)

diffracted in the horizontal direction. Note that, to be complete, it is necessary to consider waves that change type at the boundary. Thus, for example, a wave can travel from the source to the boundary as SV, continue as P, and be reconverted to SV for upward travel to the receiver. This would be designated SPS, but SSS, PPP, SSP, PSS, PSP, SPP, and PPS are also possible, some of them obviously arriving simultaneously if the depths at the source and at the receiver are equal.

Although Rayleigh waves spread out only on a surface, the amplitude can be expected to diminish as $R^{-1/2}$. Once the refracted waves reach the refracting surface they too spread out over a surface. Contrary to expectation, however, it can be shown (Heelan, 1953a, 1953b) that their amplitudes decay as R^{-2} and they rapidly die out—not rapidly enough, however, that they are not a considerable nuisance when reflection waves are being recorded. This comparatively rapid rate of decay is accounted for by the fact that the refractor continually broadcasts energy into the upper formation.

A refracted wave has an amplitude in the lower medium that decays exponentially with depth below the boundary. The rate of decay depends on wavelength; that is higher frequency waves decay faster than low-frequency waves. As a consequence of this behavior it is possible for energy to enter a second, lower velocity medium below the high-velocity refractor, even though simple ray theory prohibits it. This phenomenon is analogous to the paradox of the escape of alpha-particles from an atomic nucleus even though a potential barrier exists that should prohibit escape of particles with their energy. The wave mechanical treatment does, however, assign a finite probability for alpha-particle escape if the potential wall is thin. Returning to refracted waves, if the high-velocity layer is thick the exponential decay will be so effective that there will be little energy escaping downward from it. Figure 2.17 shows the phenomenon diagrammatically.

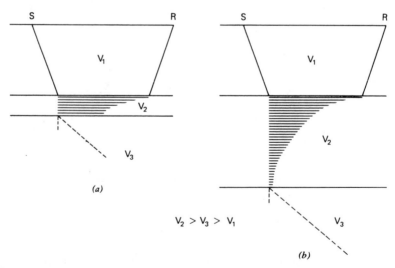

FIGURE 2.17 Diagrammatic representation of the dependence of penetration of refracted wave energy through a high-speed layer: (*a*) shows the high penetration for a thin layer, whereas in (*b*) the energy penetration is greatly reduced. The length of the horizontal shading bars in layer 2 represents the amplitude of the refracted wave. No significance is attached to the vertical starting line.

2.11 SURFACE WAVES THAT RESULT FROM ENERGY TRAPPED IN THE NEAR-SURFACE LAYER—LOVE WAVES

Because of their importance in seismic reflection recording, we show some of the physical characteristics of waves that are one stage more complex than those in the preceding two classes. For a full mathematical treatise on these waves, which are Rayleigh and Love waves formed by energy trapped in a surface layer, the reader is advised to consult Ewing, Jardetsky, and Press (1957). The treatment here is of the physical plausibility of the existence of these waves, together with illustrations actually encountered in exploration practice.

It is easy to see that in a layered situation like that shown in Figure 2.18a rays

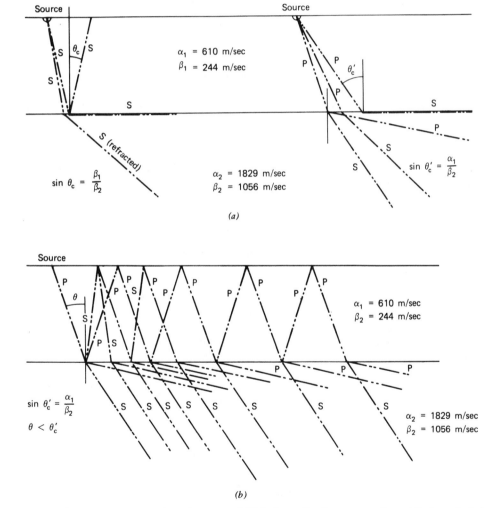

(a)

(b)

FIGURE 2.18 (a) Critically reflected waves and (b) leaky modes (note conversions at the boundaries) are developed by reflections within the layer at angles less than the critical. The latter is not completely developed to preserve some clarity: all reflections from the boundaries should be treated in the same manner, however.

emanating from source S are refracted into the lower half-space; thus a proportion of the energy is lost from the layer if a ray strikes the lower boundary at less than the critical angle. For an incident SH wave, which is not converted at the boundary into any other type, the critical angle is given by

$$\theta_c = \text{arc sin } \frac{\beta_1}{\beta_2}$$

For an incident P wave there is still leakage of shear energy until

$$\theta_c = \text{arc sin } \frac{\alpha_1}{\beta_2}$$

and for an incident SV wave

$$\theta_c = \text{arc sin } \frac{\beta_1}{\beta_2}$$

Because these waves, which are incident at angles less than the relevant critical angle, lose energy, multiple reflections like those shown in Figure 2.18b gradually decay in energy and a true trapped wave does not exist. All waves in the foregoing category are classed as "leaky" modes and, of course, contribute to the energy available for reflection work.

However, at angles beyond the critical angle and at distances from the source such that the wave fronts are appreciably plane an interesting phenomenon of interference, sensitive to frequency takes place. The obvious complication introduced by the conversion from P partially to S at each reflection and from S partially back to P can be deferred if we look first at the simpler SH waves—in which case the trapped waves are called Love waves. These are the large-amplitude surface waves whose presence makes shear wave (SH) reflection records difficult to obtain.

In Figure 2.19 a multiply reflected ray $ABCD$ is drawn for which the angle of incidence θ is greater than the critical angle. It is postulated that we are looking at this ray at a distance from the source such that the wave fronts are essentially plane, and we can disregard amplitude change of the wave by spherical divergence (or by other forms of attenuation). Only one ray has been selected to look at, but it must be borne in mind that there is an infinite number of them, parallel to one another; E then represents one position of a wave front on the ray AB and, as E moves, A sweeps along the surface. It is easy to see that if the wave

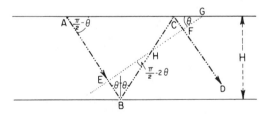

FIGURE 2.19 Conditions for generating a Love wave in a surface layer by multiple reflections of SH waves.

moves through the upper layer with velocity β_1 the wave front will move along the surface with the (faster) phase velocity

$$C_s = \frac{\beta_1}{\cos(\pi/2 - \theta)} = \frac{\beta_1}{\sin \theta}$$

For waves of constant frequency there are points E and F (on the same wave front) that interfere constructively if the distance $EBCF$ is an integral multiple of a wavelength (any phase changes at boundaries must be included in this assessment). For SH waves there is no change in phase at the free surface.

The path length $EBCF$ can therefore be written in terms of the thickness of the layer, the angle θ, and an equivalent length due to the phase change at the lower boundary $L_0(\theta)$:

$$EBCF = \frac{H}{\cos \theta}\left[1 + \sin\left(\frac{\pi}{2} - 2\theta\right)\right] + L_0(\theta)$$

$$n\lambda = \frac{H}{\cos \theta}(1 + \cos 2\theta) + L_0(\theta)$$

$$\frac{2\pi n\beta_1}{\omega} = \frac{H}{\cos \theta}(1 + \cos 2\theta) + L_0(\theta)$$

The value of $L_0(\theta)$ can be found by application of the boundary conditions at the lower surface of the layer; hence various values of θ can be obtained to satisfy this equation for different values of n (the mode number) and ω (the angular frequency).

Because the values of θ vary with ω, it follows that the phase velocity $C_s = \beta_1/\sin \theta$ also changes with the frequency. This is the first characteristic of surface-layer guided waves of all types. It is shown later that this leads to dispersion, that is, broadening of a pulse, so that it becomes a long train of waves after traveling some distance.

A second characteristic is that constant frequency waves set up nodal planes within the layer (i.e., they are standing waves as far as the vertical direction through the layer is concerned). This is obvious from the fact that if the effective length of the paths $EBCF$ is a full wavelength there is some point along this length at which there is a null point. The actual point depends on the phase change at the lower boundary, but the situation is similar to that of an open organ pipe (except that the closed opposite end is replaced by a partial opening). In Figure 2.20 the amplitudes of motion are shown for various modes.

Additional discussion of the Love wave case is necessary to firm up some ideas with respect to the significance of the terms *phase velocity* and *group velocity*. If a pulse travels in a manner that allows it to maintain its shape we can associate with it a velocity that applies to the velocity of motion of any particular part of the pulse (say, a particular peak) and to the velocity of the energy through the medium. For a medium that allows dispersive propagation, however, the different constituent frequencies travel with different speeds. Phase velocity, as used here, really refers to the effective velocity along the ground surface of a particular frequency. If observations are made at points that are sufficiently close together

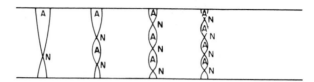

FIGURE 2.20 Possible distributions in a vertical direction of the amplitude of a Love wave. The actual position of the node is dependent on mode frequency and wave number because the phase change at the lower boundary changes with frequency. All modes have antinodes at the free surface and can therefore be generated efficiently by surface sources.

the velocity is simply the change in time of arrival of a particular phase divided into the distance between observation points:

$$c = \frac{d}{\Delta t} = \frac{2\pi f d}{\theta} = \frac{\omega d}{\theta}$$

where θ = phase angle measured for the frequency f.

In a pulse this phase angle as a function of frequency is impossible to see, but in the case of surface waves the dispersion often results in a long train of waves. A portion of this wave train can often be selected and the correlation at two or more sampling points measures the phase velocity. Moreover, when sources of constant frequency (vibrators) are available the comparison of phases at two points is often a convenient method of measuring phase velocity.

Group velocity can be defined as the velocity with which the energy associated with a narrow band of frequencies has traveled. Obviously, if, as in explosion seismology, the time of energy release at the source and the arrival time of the band of frequencies of interest at the receiver are known, the group velocity is just the distance from the source divided by the time of arrival. It is also a function of frequency, the velocity of the media, and the thickness of the formation. The group velocity u is related to the phase velocity c by

$$U = c + k \frac{dc}{dk}$$

where k = wave number = 2π/wavelength λ.

Thus if the slope of the phase velocity–wave number curve is negative the group velocity for a given wave number will be less than the phase velocity. The group velocity sometimes exhibits a minimum, which means that over a certain range of wave numbers the waves travel with the same velocity and arive at the same time. From an energy point of view this is important, because it is a time of maximum surface wave amplitude. It is called, in earthquake seismology, the Airy phase.

Figure 2.21 shows some typical phase and group velocity curves for first- and second-mode Love waves in a selection of velocities and elastic parameters. The vertical scale is a ratio of the velocity concerned to the shear wave velocity in the layer, whereas the horizontal scale is a logarithmic measure of the dimensionless parameter kH.

With the use of a digital computer it is possible to program the amplitude and

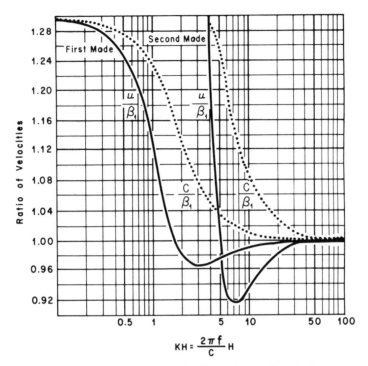

FIGURE 2.21 Phase and group velocity curves for first and second mode Love waves for $\beta_2/\beta_1 =$ 1.297 and $\mu_2/\mu_1 = 2.159$. (*Source*: Ewing et al., *Elastic Waves in Layered* Media. Copyright © 1957 by McGraw-Hill, Inc. Used with permission of the McGraw-Hill Book Company.)

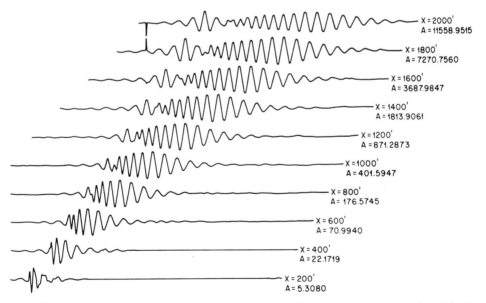

FIGURE 2.22 Love waves generated by a computer program that adds together all possible critically reflected rays. The dispersion with increasing offset is strikingly shown. By this method attenuation in the layer can be included. In this case $\beta_1 = 1400$ ft/s, $\beta_2 = 4000$ ft/s, $\rho_1 = \rho_2 = 2.5$, and $Q = 14$.

phase of each multiple reflection frequency arriving at a particular receiving point. Then, by a summation procedure, the total waveform of the Love wave can be synthesized and compared with that derived from normal mode analysis. The two compare favorably. An example of a sequence of Love waveforms at different distances is given in Figure 2.22. In addition to normal considerations, it is possible in this way to apply attenuation to each ray, both spherical divergence and change into heat. Attenuation with $Q = 20$ or less effectively damps out the high frequencies and makes the true character of the Love wave easier to see.

2.12   SURFACE WAVES THAT RESULT FROM ENERGY TRAPPED IN THE NEAR-SURFACE LAYER—PSEUDO-RAYLEIGH WAVES

The analysis of waves that propagate along the surface of an ideal earth—first made by Rayleigh—shows that these waves maintain the same pulse shape, only diminishing in amplitude with travel distance as a result of cylindrical spreading of the energy. When an imperfect earth is considered, however, the existence of one or more near-surface layers with lower than normal velocities must be taken into account. As far as exploration seismology is concerned, the most important low-velocity layer is the weathered layer, which may include recently deposited alluvium or be due to the alteration of previously competent rock by the action of rain or other agents which leach out soluble compounds and cause partial disintegration of the consolidated rock.

The presence of a low-velocity layer influences the transmission of surface waves profoundly. Because some of the characteristics are changed (e.g., dispersion or broadening of the pulse with distance is now introduced), the surface waves originating from a compressional wave source are called pseudo-Rayleigh waves.

Although the discussion of pseudo-Rayleigh waves follows the same pattern as that of Love waves, the added complication of conversion of waves between P and S at each boundary incidence now exists. Figure 2.23 shows two different path types that give rise to guided waves. The conditions must always be such that the angles of incidence and reflection require all the energy to be contained within the layer. There is no refracted energy because the assumption is that these are plane waves—it does not hold close to the source—but experience has shown that at offsets of five or six times the depth of the layer the guided wave has already become stabilized.

Reflection of SV and P waves from the free surface introduces a change in phase (in distinction from the SH case) that must be taken into account. Given the relevant parameters for the layer and the half-space, dispersion curves can be drawn. There are now, however, more possibilities because the conversion from compression to shear waves (or the reverse) must be considered and the curves become more complex, with many modes and branches similar to those shown in Figure 2.24. In the interpretation of surface waves the phase velocity is measured as a function of frequency and this experimental curve is compared and matched as nearly as possible to one or more theoretical curves. The number of variables makes this a laborious task and there still does not appear to be any way of working directly back from the dispersion curve(s) to a unique two-layer system.

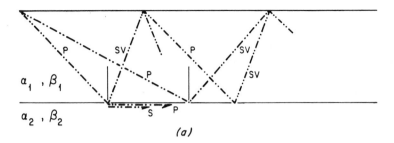

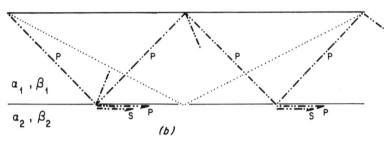

FIGURE 2.23 Some of the many ray paths and conversions that allow the formation of a pseudo-Rayleigh wave in a surface layer: (*a*) conversion from P to SV or SV to P; (*b*) no conversion, All energy is contained in the layer.

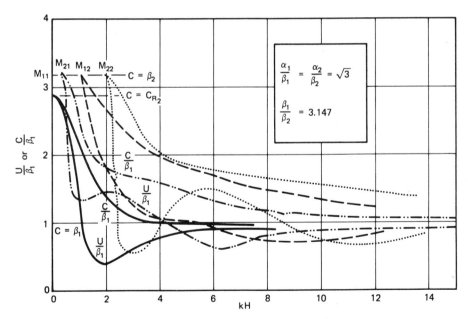

FIGURE 2.24 Dispersion curves for an elastic layer over a simi-infinite solid. The modes M_1 and M_2 correspond to the symmetrical and antisymmetrical modes of a plate, modified by the attachment to a semi-infinite medium. (*Source*: Ewing et al., *Elastic Waves in Layered Media*. Copyright © 1957 by McGraw-Hill, Inc. Used with permission of the McGraw-Hill Book Company.)

In exploration technology the surface waves can be only approximately accounted for by a two-layer system. Except for special cases, the parts of the dispersion curve of diagnostic interest are at very low frequencies $(1.0 \rightarrow 10\,\text{Hz})$. Attempts have been made, without too much success, to use surface wave measurements for weathering calculations. Several facts make this process frustrating. Weathering materials are not constant over the distances needed, nor is the depth of the weathering, because more often than not the weathered layer cannot be regarded even approximately as a single layer.

For some exploration purposes surface waves must be rejected to allow proper visibility of reflected waves. The fact that they generally have frequencies below the band (8 to 60 Hz) used for reflections has been used for filtering—either by geophones which have a low response for these frequencies (the old approach) or by velocity filtering (q.v.) during data processing.

Finally it must be said that the normal mode theory shows that the same kind of standing waves (in a vertical direction) is set up for pseudo-Rayleigh waves as for Love waves but, with the added complication of conversion between wave types, the nodes for different modes, branches, and frequencies are scattered throughout the layer. Any attempt to generate seismic waves when the source is

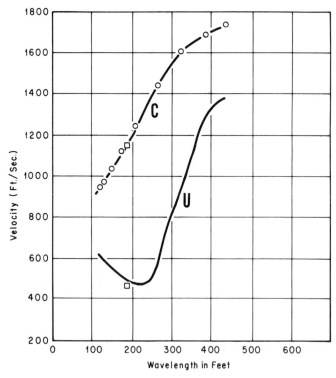

FIGURE 2.25 Group velocity (U) and phase velocity (C) of Rayleigh waves obtained from special exploration seismograms. The square points are data from air shots. (*Source*: Dobin et al., from Ewing et al., *elastic Waves in Layered Media*. Copyright © 1957 by McGraw-Hill, Inc. Used by permission of the McGraw-Hill Book Company.)

in the weathered layer—or on the surface—results in the generation of large-amplitude guided waves as well as body waves used for reflection purposes.

Dobrin, Simon, and Lawrence (1951) applied the theory of Rayleigh wave dispersion by using values for velocities and layer thickness found in borehole surveys. Figure 2.25 shows that fairly good agreement was found between the theoretical and experimental points.

2.13 SUMMARY

If a mechanical disturbance is caused in a body of ideal rocklike material, energy travels from the source throughout the medium in the form of waves. Two types of wave, compressional (P) and shear (S), can exist. They travel with different velocities that correspond to the appropriate elastic constants.

At boundaries between different types of rock a generalized form of Snell's law applies, which regulates the direction of travel, and partial conversion in both reflection and refraction occurs between one wave type and another.

As the wave travels outward from the point source the amplitude of the particle motion diminishes because the energy available is spread out over the entire wave front. It is not required, however, that the energy density be constant over the wave front.

In imperfect rocklike materials an additional loss of amplitude of the wave occurs as a result of the loss of energy due to heat (generated at imperfections in the medium). The specific dissipation constant $1/Q$ appears to be independent of frequency, the wave amplitude decreases exponentially with distance of travel, and the rate of attenuation is roughly proportional to the frequency in most rocklike materials.

In addition to body waves, it is possible for a Rayleigh wave guided by a boundary between the earth and the air to be generated. This form of Rayleigh wave is distinguished by retrograde elliptical particle motion, constancy of pulse shape with distance of travel, exponential decay of amplitude below the surface, amplitude decay with distance as $R^{-1/2}$, and a velocity slightly less than S velocity.

If a low-velocity layer superimposed on a more consolidated rock half-space is present guided waves are possible, the type depending on the mode of initiation of the energy. A Love wave is the result of the interference of different orders of multiple reflections whenever the latter strike the lower boundary at angles greater than the critical angle. This wave is generated by a source of SH motion and generates SH motion on the surface; the energy decays as $R^{-1/2}$ (because it is confined to a layer), but because it is dispersive the amplitudes of individual portions of the wave decay faster than $R^{-1/2}$ (more as R^{-1}). The characteristics of these waves are summarized by the dispersion curves that relate phase and group velocities to the wave number and layer-thickness product.

Pseudo-Rayleigh waves exist in the layer when the excitation produces P or SV waves. The interference of multiply reflected waves (at greater than critical incidence) forms different modes, each of which has characteristic dispersion curves associated with it, giving the phase and group velocities as functions of the mode and branch number, the wavelength, the layer thickness, the P and S velocities, and the densities.

These pseudo-Rayleigh waves or ground roll constitute one of the main difficulties in the recording of seismic reflections because they are of large amplitude compared with the reflections; and because they are usually low-frequency and have low velocities along the surface frequency filtering or other special forms of filtering can be used to reject them from the relation energy.

APPENDIX 2A: ZOEPPRITZ EQUATIONS

According to the notation introduced in Figure 2.12, the equations for determining the amplitudes of reflected and refracted waves (both P and SV) from an incident P wave are the following:

1. Snell's law:

$$\frac{\sin \theta_1}{\alpha_1} = \frac{\sin \phi_1}{\beta_1} = \frac{\sin \theta_2}{\alpha_1} = \frac{\sin \phi_2}{\beta_2} \qquad (2A.1)$$

2. Continuity of tangential displacement:

$$\sin \theta_1 + A \sin \theta_1 + B \cos \phi_1 = C \sin \theta_2 - D \cos \phi_2 \qquad (2A.2)$$

where the P-wave displacement (A and C) is positive in the direction of propagation and the S-wave displacement is positive to the right of the direction of propagation.

3. Continuity of normal displacement:

$$\cos \theta_1 - A \cos \theta_1 + B \sin \phi_1 = C \cos \theta_2 + D \sin \phi_2 \qquad (2A.3)$$

4. Continuity of normal and tangential stress yields two equations:

$$-\sin 2\theta_1 + A \sin 2\theta_1 + \frac{\alpha_1}{\beta_1} B \cos 2\phi_1 = \frac{-\rho_2 \beta_2^2 \alpha_1}{\rho_1 \beta_1^2 \alpha_2} C \sin 2\theta_2 + \frac{\rho_2 \beta_2 \alpha_1}{\rho_1 \beta_1^2} D \cos 2\phi_2 \qquad (2A.4)$$

$$\cos 2\phi_1 + A \cos 2\phi_1 - \frac{\beta_1}{\alpha_1} B \sin 2\phi_1 = \frac{\rho_2 \alpha_2}{\rho_1 \alpha_1} C \cos 2\phi_2 + \frac{\rho_2 \beta_2}{\rho_1 \alpha_1} D \sin 2\phi_2 \qquad (2A.5)$$

where A = amplitude of reflected P wave
C = amplitude of refracted P wave
B = amplitude of reflected SV wave
D = amplitude of refracted SV wave

These equations can be put in a matrix form more suitable for computer solution:

$$\begin{pmatrix} \sin\theta_1 & \cos\phi_1 & -\sin\theta_2 & \cos\phi_2 \\ -\cos\theta_1 & \sin\phi_1 & -\cos\theta_2 & -\sin\phi_2 \\ \sin 2\theta_1 & \dfrac{\alpha_1}{\beta_1}\cos 2\phi_1 & \dfrac{\rho_2\beta_2^2\alpha_1}{\rho_1\beta_1^2\alpha_2}\sin 2\theta_2 & -\dfrac{\rho_2\beta_2\alpha_1}{\rho_1\beta_1^2}\cos 2\phi_2 \\ \cos 2\phi_1 & -\dfrac{\beta_1}{\alpha_1}\sin 2\phi_1 & -\dfrac{\rho_2\alpha_2}{\rho_1\alpha_1}\cos 2\phi_2 & -\dfrac{\rho_2\beta_2}{\rho_1\alpha_1}\sin 2\phi_2 \end{pmatrix} \begin{pmatrix} A \\ B \\ C \\ D \end{pmatrix} = \begin{pmatrix} -\sin\theta_1 \\ -\cos\theta_1 \\ \sin 2\theta_1 \\ -\cos 2\phi_1 \end{pmatrix}$$

$$\mathbf{P} \qquad\qquad \mathbf{Q} \;=\; \mathbf{R}$$

and the solution is

$$\mathbf{Q} = \mathbf{P}^{-1}\mathbf{R} \qquad\qquad (2A.6)$$

All these expressions are relative to an input P-wave amplitude of unity and are given in R. E. Sheriff, *Encyclopedic Dictionary of Exploration Geophysics*, second edition.

A computer program that facilitates the calculation of the reflection and refraction amplitudes from compressional and shear input waves was published by Braile and Young (1976). Beyond the critical angles for P and S waves the respective refracted waves vanish. (These formulas hold only for plane waves.) Therefore they do not allow calculation of head-wave amplitudes which arise from spherical waves incident on a boundary. It is known, however, that spherical waves may be decomposed into plane waves with *both real and complex angles of incidence*. This method was formulated by Sommerfeld, an account of which can be found in Bäth (1968). By including the plane waves with complex angles of incidence (inhomogeneous waves with an amplitude dependent on depth below the boundary) the critically refracted waves may be included in the solution.

Tooley, Spencer, and Sagoci (1965) published curves for the energies associated with the various reflected and refracted waves for all *real* angles of incidence and various material contrasts. The increase in reflection energy near the critical angle was sometimes exploited in seismic surveying—under special circumstances. Beyond the critical angle, however, it is difficult in practice to distinguish these reflections from critically refracted waves. A phase shift is introduced into the reflected wave.

This appendix has been included mainly to show how complex the reflection process becomes for large angle of incidence. The complexity is perhaps illustrated by numerous publications of the Zoeppritz equations that have errors of sign. Ewing, Press, and Jardetsky (1957) obtained the general equations in their book (p. 76) unfortunately with a different set of axes, a different sign convention, and a different notation. It is therefore most important to test any program by taking special cases with well known formal solutions.

Remember that as waves progress through a layered medium these relations apply at every boundary. It is by continued *matrix* multiplication that the "reflectivity" of a laminated medium for a constant frequency plane wave with incidence other than normal is found. Brekovshikh (1960) gives details of this process. For normal incidence a much simpler algebraic iteration is used (Appendix 4A).

Since Muskat and Meres (1940) gave their lengthy tables of reflection coeffi-

cients and their variation as functions of velocities and densities of rocks on ·
opposite sides of a plane boundary there has been a fascination with this problem
of the effect of incident angle on the reflection properties of the boundary. A
complete understanding, as is often the case when many variables are involved
has been thwarted by the complexity of the problem. As a consequence numerous attempts have been made to simplify the Zoeppritz equations; see, for
example, the work of Koefoed (1955), Bortfeld (1961), Ostrander (1984), and
Shuey (1985). As an understanding of the reflection seismic method and its results
improved the desire was, not unnaturally, to derive more information from the
seismic reflections—and their variation with offset. This is often designated the
inverse problem. Carried to its logical conclusion, a successful onslaught might
replace the complexity of the reflection process at many offsets with an enormous
amount of physical data about the sediments—probably equally difficult to
interpret in terms of the discovery of hydrocarbons, but still deemed worth the
effort and cost involved.

REFERENCES

Bäth, M. (1968), "Mathematical Aspects of Seismology," in *Developments in Solid Earth Geophysics*, Vol. 4, Elsevier, Amsterdam.

Beeston, W. E., and McEvilly, T. V. (1977), "Shear wave velocities from downhole measurements," in *Earthquake Engineering and Structural Dynamics*, Vol. 5, Wiley, Chichester, England, pp. 181–190.

Braile, L. W., and Young, G. B. (1976), "A Computer Program for the Application of Zoeppritz's Amplitude Equations and Knott's Energy Equations," *Bulletin of the Seismological Society of America*, Vol. 66, pp. 1881–1885.

Bortfeld, R. (1961), "Approximation to the Reflection and Transmision Coefficients of Plane longitudinal and Transverse Waves," *Geophysical Prospecting*, Vol. 9, 485–502.

Brekhovshikh, L. M. (1960), *Waves in Layered Media*, Academic Press, New York.

Dobrin, M. B., Simon, R. F., and Lawrence, P. L. (1951), "Rayleigh Waves from Small Explosions," *Transactions of the American Geophysical Union*, Vol. 32, pp. 822–832.

Ewing, W. M., Jardetsky, W. S., and Press, F. (1957), *Elastic Waves in Layered Media*, McGraw-Hill, New York.

Futterman, W. I. (1962), "Dispersive Body Waves," *Journal of Geophysical Research*, Vol. 69, 5279–5291.

Garland, G. D. (1971), *Introduction to Geophysics, Mantle, Core and Crust*, W. B. Saunders, Toronto.

Heelan, P. A. (1953a). "Radiation from a Cylindrical Source of Finite Length," *Geophysics*, Vol. 18, pp. 685–696.

Heelan, P. A. (1953b), "On the Theory of Head Waves," *Geophysics*, Vol. 18, pp. 871–893.

Howell, B. (1959), *Introduction to Geophysics*, McGraw-Hill, New York.

Kjartansson, E. (1979), "Constant Q Wave Propagation and Attenuation," *Journal of Geophysical Research*, Vol. 84, 4737–4748.

Knopoff, L. (1952), "On Rayleigh Wave Velocities," *Bulletin of the Seismological Society of America*, Vol. 42, pp. 307–308.

Koefoed, O. (1955), "On the Effect of Poisson's Ratio of Rock Strata on the Reflection Coefficients of Plane Waves," *Geophysical Prospecting*, Vol. 3, 381–387.

Lamb, H. (1934), *The Mathematical Theory of Elasticity*, Cambridge University Press, Cambridge.

McDonal, F. J., Angona, F. A., Mills, R. L., Sengbush, R. L., Van Nostrand, R. G., and White, J. E. (1958), "Attenuation of Shear and Compressional Waves in Pierre Shale," *Geophysics*, Vol. 23, 421–439.

Muskat, M., and Meres, M. W., (1940), Reflection and transmission coefficients for Plane Waves in Elastic Media," *Geophysics*, Vol. 5, 115–155.

O'Brien, P. N. S., and Lucas, A. L. (1971), Velocity Dispersion of Seismic Waves," *Geophysical, Prospecting*, Vol. 19, 1–26.

Ostrander, W. J. (1984), Plane Wave Reflection Coefficients for Gas Sands at Non-normal Angles of Incidence, *Geophysics*, Vol. 49, 1637–1648.

Shuey, R. T. (1985), A Simplification of the Zoeppritz Equations, *Geophysics*, Vol. 50, 609–614.

Stainsby, S. D., and Worthington, M. H. (1985), Q Estimation from Vertical Seismic Profile Data and Anomalous Variations in the Central North Sea, *Geophysics*, Vol. 50, p. 615–626.

Stewart, R. R., Huddleston, P. D., and Kan, T. K. (1984), "Seismic Versus Sonic Traveltimes: A Vertical Seismic Profiling Study," *Geophysics*, Vol. 49, 1153–1168.

Wuenschel, P. G. (1965), "Dispersive Body Waves—An Experimental Study," *Geophysics*, Vol. 30, 539–551.

Wylie, C. Ray (1975), *Advanced Engineering Mathematics*, 4th ed., McGraw-Hill, New York.

THREE

Sources and Receivers

3.1 GENERAL SOURCES OF ELASTIC ENERGY

In nature elastic waves are generated whenever a sudden change in stress occurs in the earth—usually in a form due to strain relief by faulting. The tectonic forces acting on the earth's crust build up strain energy, both compressive and shear, until the elastic limit of the material is exceeded. The change in strain energy implies a change in the gravitational potential energy of the rock masses as well as in the heat energy supplied to the rock in the vicinity of the fault and energy radiated in the form of various kinds of elastic wave.

Although the use of these natural sources of elastic energy supplies some general knowledge of the earth's structure, they are generally unsuitable for more detailed investigations, some of which have to take place beneath the land and others, below the shallow seas that overlay the continental shelves. This is no artifical division, as it turns out because seismic methods of exploration differ considerably for the two environments. We deal first with exploration on land.

Explosions of various types also give rise to radiated seismic waves; the distribution of energy into different types of elastic wave varies, depending on the location of the explosion with respect to significant rock boundaries, and on the type and strength of the rock. The historical (and present-day) use of explosions to generate suitable waves for exploration makes use of the compact chemical energy available in high explosives. Although explosives have been used in the air to generate seismic waves in the earth (Poulter, 1950) for exploration purposes, most of the work has been done with buried, tamped charges. A comprehensive account of the seismic signals generated by explosions has been given by Kisslinger (1963). This reference is recommended for a more detailed account that can be given here. Figure 3.1, which is taken from this work, compares the relative amplitudes of seismic signals (body waves) as a function of height above or depth below the ground–air interface.

In practice, charges are buried by drilling a suitable shothole 10 to 15 cm (4 to 6 in.) in diameter in which an explosive charge of 0.05 to 100 kg ($\frac{1}{8}$ to 200 lb) can be lowered below the weathered layer. These holes range in depths of 10 to 100 m (30 to 300 ft). After the explosive has been loaded the hole is tamped by filling at least the next 3 m (10 ft) with dirt or snow or sometimes by filling the hole completely with water. The charge is detonated electrically. It is usual to record the time of the explosion by an impulse derived from breaking the firing current when the detonator fires. Very high-voltage blasters are used, and the time of

54

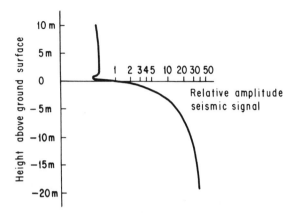

FIGURE 3.1 Relative amplitude of the seismic signal as a function of depth or altitude of the shot, with reference to the ground surface. [*Source*: Kisslinger (1963).]

application of the high voltage and the time of rupture of the bridge wire in the cap are normally simultaneous.

The location of the charge (or any other source) is important to minimize the generation of unwanted waves while simultaneously satisfying the need for the generation and transmission of a spectrum of frequencies as complete as possible

As we pointed out earlier in Chapter 2, it is possible by Fourier analysis to replace the time-series description of the outgoing pulse by the frequency-domain description of amplitude of constant frequency components that make up the pulse. This is the "spectrum of frequencies" or, more commonly, the amplitude spectrum previously referred to. There is additional material on this subject in Chapter 4 and the reader who is interested in a complete treatment is referred to Bracewell (1965).

A simple case of initiation of vibration can be examined to explain the general physical principles involved. Figure 3.2 shows a string held between two rigid

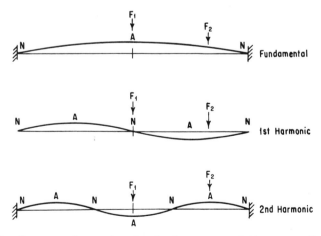

FIGURE 3.2 Standing waves for a string vibrating between two rigid supports. Force F_1 generates only those modes of vibration for which it does not occur at a node; that is, the fundamental and the second harmonic. Force F_2 generates all three modes.

supports. It is well known that this string can be made to vibrate by an impulsive transverse force at some point along its length. Now the string must vibrate in such a manner that the boundary conditions (no motion at either support) are satisfied at all times; and one way of expressing this is to say that the string has a series of (standing-wave) vibrational modes that are integral multiples of a fundamental mode whose wavelength is twice the length of the string. The points marked *A* are antinodes (points of maximum displacement) and those marked *N* are nodes (points of zero displacement) when any of the pure harmonics are excited. It is also known that these harmonics cannot be generated when the string is struck at the corresponding nodal points. When any point is selected at random it is generally possible to excite several modes of vibration but not the mode for which the selected point is a nodal point.

The degree of excitation is again a matter of fitting the initial conditions of the problem, and this requires that the amplitude, velocity, and acceleration of the string at the point in question, as well as at all other points on the string, agree with the sum of the amplitudes, velocities, and accelerations of the individual harmonics and with the impulse supplied.

This general approach can be carried over to three-dimensional models; for example, it has been shown mathematically and by analysis of earthquake records that the earth as a whole has several modes of vibration that can be excited by a sudden rupture and the release of stored elastic energy. The types of vibration are independent of one another, but all are dependent on the near-spherical form of the earth and on the distribution of elastic moduli and density with depth into the earth. These are called the free modes of vibration of the earth.

In the same way near-surface layers of the earth exhibit independent modes of vibration that can be excited by a suitably placed explosion or other source of seismic energy.

In principle, therefore, it is necessary to explode the dynamite, or apply the time-varying force, at a position that is a node of any possible standing wave for the weathering and sufficiently far below the surface that the Rayleigh waves (exponentially diminishing in amplitude from the surface downward) cannot be appreciably excited in the frequency band desired. Because all the standing waves (or normal modes) are frequency-dependent, their nodes occur at different depths and it is impossible to satisfy all the requirements simultaneously. The practice of taking several shots at different depths *below* the base of the weathered layer generates waves of different forms. The major contributory factor is the presence of a reflected wave from the base of the weathering. Although this is a nuisance that tends to introduce unwanted variations in the spectrum of the downward traveling composite pulse, the standing waves in the weathered layer are not initiated to anything like the extent they would be if the shot were in the weathered layer. The near constancy of the reflection coefficient of the base of the weathering then gives generated pulses that, although complex, can be held nearly consistent in waveform.

Aside from the position of the explosion in the earth, the frequency content or form of the waves generated is affected by several other factors. If the assumption is made that the explosion will suddenly increase the ambient pressure in a spherical cavity of radius r and the elastic limits of the material surrounding the cavity will not be exceeded, Sharpe (1942*a*, 1942*b*) showed that the shape of the elastic pulse generated, a highly damped sinusoid, is determined by the radius of the cavity and the elastic constants of the material. It has been suggested that an equivalent cavity can always be found outside which the elastic limits are not exceeded. A high-frequency pulse is generated when the cavity radius is small. Although this explanation carries elements of truth, the actual mechanism is much more complex because almost always crushing of the rock takes place until the shock wave reaches a point at which the behavior of the rock is elastic. This theory does not explain fully the number of relatively low frequencies (5 to 100 Hz) generated in the earth.

The presence of strong reflecting interfaces above the location of the dynamite charge (the base of the weathered layer and, still further away, the ground surface) gives rise to reflected pulses that, after delays, follow the initial impulse. Some attempts have been made to devise explosive sources to mitigate this effect. It is done by making an explosive column that incorporates delays of such magnitude that the effective detonation velocity (initiated at the top of the column) is equal to the P-wave velocity in the sediments surrounding the explosive; for example, dynamite or Primacord with a true detonation velocity of near 6100 m/s may have the effective downward detonation velocity reduced to 2000 m/s, the P-wave velocity of a shale within which the explosive is placed.

The pressure acting on the sediments then causes a wave that gradually increases in amplitude as it is generated in the downward direction but is spread out in time for waves traveling upward and, to a lesser extent, to the side. In this manner the reflections from above, as well as ground roll and other sideways traveling waves, are reduced in average frequency and effectiveness. Winding the Primacord spirally around a wooden pole allows the effective detonation velocity to be controlled by the pitch of the spiral (Martner and Silverman, 1962).

There is no question that the use of explosives in drilled holes is an inconvenience and that this method is at an economic disadvantage compared with the more modern practice of surface sources. This inconvenience and expense are, however, mitigated by the following advantages:

1. Relative freedom from surface waves
2. The possibility of measuring directly the time required for the wave to pass through the weathered layer

To quote Kisslinger (1963) in his "Summary and Conclusions: The State of the Art":

> The study of explosion-generated seismic waves is still primarily an empirical science. In the absence of a complete theory of the processes that result in the radiated wave, progress in further understanding of the subject is dependent on well designed experiments. Unfortunately, the complex properties of earth materials and geologic structure in which full-scale tests must be carried out make the design of experiments difficult.

3.2 DIRECTIONALITY

In the discussion of explosive and other types of source we talk about the energy (or amplitude) of the wave not only as a function of distance but also of direction.

The general plane wave solution of the wave equation in Cartesian (x, y, z) coordinates is

$$\xi(x, y, z) = F[lx + my + nz) - ct] \qquad (3.1)$$

where l, m, and n = cosines of the ray path with the x, y, and z coordinate axes. For the present purpose it is more convenient to express the solution to the wave equation in spherical coordinates:

$$\xi(\theta, \phi, R) = \frac{1}{R} P(\theta) Q(\phi) F(R - ct) \qquad (3.2)$$

where $P(\theta)$ = a function of the azimuth θ only
   $Q(\phi)$ = a function of the vertical ϕ only
   $F(R - ct)$ = a function of R and t only

For an explosion in the body of an infinite isotropic medium the wave has no way of showing a preference in direction; therefore *by symmetry* we put $P(\theta)$ and $Q(\phi)$ equal to constants. This leaves the wave amplitude ξ as a function of distance only.

A single spherically symmetrical source in an infinite medium is, however, an abstraction, and in any real situation one or more boundaries are near it. A relatively simple condition is shown in Figure 3.3, where the spherically symmetrical source is buried at a small distance h below the free surface. If the

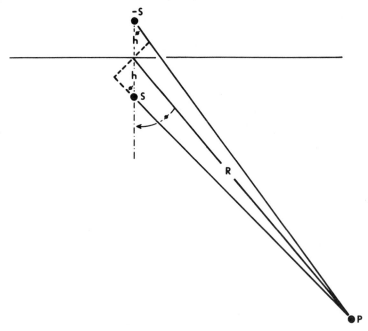

FIGURE 3.3 Geometry for the calculation of amplitude as a function of the vertical angle ϕ for a source below the free surface (acoustic medium).

measurement of amplitude is made at a point P, where the distance R is much larger than h, we can make use of the method of images to satisfy the (free) boundary conditions at the surface. If only P waves are considered (the entire medium is a fluid) a *negative* image of the same strength S as the original source is assumed to be present at a height h above the boundary. By so doing the surface boundary conditions are satisfied. Because the real and virtual sources lie on the vertical axis, this source combination produces waves that are independent of the azimuth. The variation with the vertical angle ϕ is calculated as follows:

The wave amplitude at a distance R from any single source *of frequency* ω is given by

$$\xi_R = \frac{A}{R} \left[\cos \omega(t - t') - \cos \omega(t + t') \right]$$

where t' = time difference due to path difference $h \cos \phi$

$$\omega t' = \frac{\omega h \cos \phi}{c} = \frac{2\pi h}{\lambda} \cos \phi$$

Therefore

$$\xi_R = \frac{A}{R} (\sin \omega t) \sin\left(\frac{2\pi h}{\lambda} \cos \phi \right) \tag{3.3}$$

The amplitude at distance vibrates 90° out of phase with the individual source standard phase (accounting for travel time) but varies with ϕ. The manner of variation is much controlled by the factor $2\pi h/\lambda$; that is, the ratio of the depth of the source to the wavelength of the sound in the medium. Figure 3.4 shows three cases:

1. If the depth below the surface is small compared with the wavelength (which is usual in seismic exploration) the variation is nearly proportional to $\cos \phi$. (In polar diagrams the length of the line from the origin is proportional to the amplitude at that angle from the vertical.)
2. For $h = \frac{1}{2}\lambda$ the variation is determined by the angle and phase difference between the two sources for the frequency concerned ($f = c/2h$).
3. For $h = \lambda$ the variation with angle becomes more rapid, going through two complete half-cycles. Note now that the net amplitude is negative over part of the range of angles.

Even for this simple combination of constant-frequency source and image the directivity pattern can be a complicated function of the vertical angle. More complex source patterns are dealt with later in this chapter.

Although these spherically symmetrical sources generate no shear waves by themselves, the combination of source and image, when considered together, can yield SV waves when the combined effect is measured close to the source. It is necessary first to draw the lines of equal phase (wave fronts) for the combination and then to calculate the amplitude and direction to determine whether two components, one along and one perpendicular to the ray, are necessary to explain

$\dfrac{h}{\lambda} = .01$

(a) Maximum amplitude is down (= .0627)
Variation near as cos ∅. All values positive.

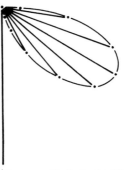

$\dfrac{h}{\lambda} = 0.5$

(b) Maximum amplitude is ∅ = 60° (1.00)
All values positive.

$\dfrac{h}{\lambda} = 1.0$

(c) Maximum amplitude is − .992
Values cycle from negative to positive.

FIGURE 3.4 Directivity functions for a constant frequency P-wave source located at different distances below a free boundary of an acoustic medium; λ is the wavelength of the signal and h is the source depth.

60

the motion. Of course, in an elastic medium a particular type of wave impinging on a boundary always undergoes partial conversion of type. This is, qualitatively, the way the Rayleigh wave is set up. The generation of Rayleigh waves from an explosive, buried shot has been the subject of numerous investigations by Dix (1955) and others.

3.3 IMPULSIVE SURFACE SOURCES

The inconvenience and expense of drilling holes, the inconsistency of the seismic pulse generated, and, to some extent, the continued existence of areas in which no reflection records could be obtained led to experimentation with other types of seismic source. Although some appeared temporarily to offer an advantage over others, during the course of time only a few types remained. Poulter (1950) showed that reflection records could sometimes be obtained by arrays of explosive charges fired in the air a few meters above the surface of the ground. Generally, these arrays were about 100 m in diameter, with 7 to 13 charges of $2\frac{1}{2}$ kg (5 lb) of dynamite mounted on poles 2 to 3 m (6 to 10 ft) high. Interconnections between the charges were usually made by Primacord, although separate electric blasting caps were sometimes substituted. Although used for oil prospecting for a short time, the disadvantages outweighed any (dubious) advantages and the method succumbed to progress.

The use of arrays of small (2- to 5-kg) explosive charges in multiple holes a few feet deep gained some acceptance, largely because of the advantages of arrays over single charges in diminishing the size of surface waves and increasing the downward traveling energy. The drilling and tamping of 2- to 5-m holes, however, were a deterrent. Whereas this method gradually died out, the concept of arrays or patterns of sources and receivers showed promise, provided that a suitable unit source could be found.

The advent of the weight-drop method ushered in the era of surface sources. In this method a large mass that weighed 1500 to 2000 kg (3000 to 4500 lb) was dropped from a height of about 2 to 3 m (6 to 10 ft) onto the earth, the time of initiation of the seismic wave being derived from an accelerometer contained in the concrete block. The weight could be raised easily, the position of the truck, changed quickly, and the generation of source patterns, facilitated.

If the mass always fell flat on the earth and the latter had consistent elastic and density characteristics, an impulsive pressure on the earth which was responsible for producing both surface and body waves would be generated. A discussion later in this chapter shows that the amount of energy diverted into surface waves was probably 10 times that for compressional waves. In practice, in addition to the perturbation in wave shape caused by imperfect falls and by roughness of the ground surface, it was found that the form of the impulse generated was controlled by the damped resonant system of the large mass and the spring constant of the earth surrounding the impact area.

A typical pulse shape generated was given by Neitzel (1958) (see Figure 3.5). The difficulties experienced were due largely to inconsistencies in the soil and near-surface parameters but also to the predominance of low frequencies (5 to 20 Hz) generated. As shown in the examples given by Neitzel, the pulse shape

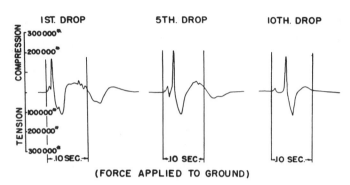

FIGURE 3.5 Force-time curves at the bottom of a concrete mass. The approximate value for the first peak is 90,700 kg (200,000 lb.) [*Source*: Neitzel (1958). Reprinted with permission from *Geophysics*.]

changes even when the weight is dropped many times in a single location. The result is a varation in the effective initiation time of the low-frequency spectrum. In the later process of summation of the results of multiple drops only the low-frequency reflections survive and even their phase (or initiation time) is in question by an undesirable amount. Because of the relative simplicity of the equipment and the fact that lower frequencies often have an advantage in prospecting for deeper geological structures (lack of attenuation and the tendency for geological formations to be thicker in older sediments), this method now appears to be losing ground to newer methods. The advantage (or necessity) of using arrays of surface source and geophones was nevertheless firmly established.

The Dinoseis® (a registered trademark of ARCO Inc.) method substitutes a controlled explosion of a propane-air mixture to create a transient pressure on the surface of the earth. The explosion is contained in a chamber with one flexible side pressed against the earth by the weight of the vehicle. A fixed amount of energy is generated each time, and the method is sufficiently mobile to allow patterns or arrays of source to be used easily. If the surface soil or rock had consistent properties, consistent pulses resulted; however, this again is not realistic, and the method suffers from some of the disadvantages of the weight drop method. Its ease of use in generating arrays of sources has gained it some popularity for land use. It does not involve the problem of interaction of the mass of the concrete block dropped with the spring constant of the earth because the diaphragm of the Dinoseis method has virtually zero mass; hence the resonant frequency is high—well outside the normal seismic frequency range. Thus the prevalence of very low-frequency generation in the spectrum of the weight drop pulse is not carried over to the Dinoseis pulse. This method is therefore more suitable for the generation of higher frequency reflections and, of some consequence, has less of a tendency to generate ground roll.

3.4 THE VIBROSEIS® SYSTEM

Although Vibroseis® (a registered trademark of CONOCO, Inc.) is a surface-source method, it has the advantage of exerting some control over the frequency spectrum of the energy injected into the earth. By the feedback system, using a monitor signal of the motion of the baseplate resting on the earth, the force applied to the earth over a long signal of many seconds duration can be varied to accommodate phase changes brought about by inconsistencies in earth materials. All this control results from the nature of the method itself, which is described by Crawford, Doty and Lee (1960).

The source now generates a prolonged signal and causes a stress on the surface which follows some preselected form. The increased length of signal serves two purposes:

1. It makes the input signal easier to control by analog electronic circuitry.
2. The total power emitted by the source is increased in proportion to the length of time the signal is emitted.

Before discussing the type of source needed it is necessary to describe a unique signal-processing step called *correlation* which is used after reception of the signal to perform the operation of converting the facsimiles of the input signal into shorter pulses while maintaining a measure of the time these output signals have been delayed in their path through the earth.

Facsimiles of the source output are reflected at every interface in the reflection path and there is some modification of the amplitude spectrum by attenuation in the earth. The receivers on the surface, however, receive a total signal that is the sum of all the reflected signals—different in amplitude and delayed in time. It is not recognizable as a reflection record at this stage because of the complexity of interference of the overlapping signals of strength a_i and delay time τ_i:

$$F_2(t) = \sum_{i=1}^{N} a_i F_1(t - \tau_i) \tag{3.4}$$

The second part of the Vibroseis system involves the process of correlation of the received signal F_2 with the original (control) signal F_1; these two signals have a common reference zero time. Correlation can be regarded as a method of seeking out the control signal components of the received signal and replacing them at the proper delay time and with the proper reflection amplitude by a much shorter pulse. Correlation can therefore also be described as a method of filtering the received signal trace with a matched filter.

Physically, the process consists of the following steps:

1. The two signals are regarded as being laid down side by side and the control signal (F_1) is movable by a delay (τ) with respect to the received signal F_2.
2. Over that part of both signals where they overlap in time a continuous cross multiplication and addition is performed to obtain a measure (as explained earlier for Fourier analysis) of the degree of likeness of the two signals at this delay.
3. This process is performed for all delays and the cross correlation function (of τ) is obtained. Because the overlap length is nearly always the length of the control signal (T), there is little need to normalize these cross-correlation $[\phi_{12}(\tau)]$ values.

Mathematically, we can write

$$\phi_{12}(\tau) = \frac{1}{T} \int_0^T F_1(t) \cdot F_2(t + \tau)\, dt \tag{3.5}$$

If this formula is applied to the presumed form of the received signal (3.4), the result is

$$\phi_{12}(\tau) = \frac{1}{T} \int_0^T F_1(t) \cdot \sum_{i=1}^N a_i F_1(t - \tau_i)\, dt \tag{3.6}$$

$$= \sum_1^N a_i \cdot \phi_{11}(\tau_i) \tag{3.7}$$

where $\phi_{11}(\tau)$ is called the autocorrelation of $F_1(t)$ and is a symmetrical (*zero-phase*) pulse with an amplitude spectrum that is the square of the amplitude spectrum of $F_1(t)$. It is important to note that the autocorrelation pulse has its maximum value at the delay time (τ) and that choosing ("picking") this value is much easier than trying to locate a rather nebulous first arrival of a normal reflection on an impulsive-type seismogram.

Instead of doing the correlation process in the time (delay) domain, as already explained, it is just as easy to convert both signals (by Fourier analysis) into their (complex) spectra and to do the correlation in the frequency domain.

It can be shown that the mathematical formulation of correlation in the frequency domain is as follows:

1. The formula corresponding to (3.4) is

$$G_2(\omega) = \sum_1^N a_j G_1(\omega) \epsilon^{-i\omega\tau_j} \tag{3.8}$$

2. The formula corresponding to (3.6)

$$\gamma_{12}(\omega) = \sum_1^N a_j G_1(\omega) \cdot G_1^*(\omega) \epsilon^{-i\omega\tau_j} \tag{3.9}$$

3. The (complex) spectrum $\gamma_{12}(\omega)$ of the cross-correlation function has to be subjected to Fourier synthesis to recover the time domain representation of the cross-correlated seismic trace.

Thus we have disclosed the general principles behind the Vibroseis method, invented by Crawford, Doty, and Lee (1960). These principles are simple enough—it is the implementation in the field and processing lab that requires further explanation. Much of the mystique of the Vibroseis system was associated with the early need for analog-recording practices, which include methods of preserving the time relation between the reference signal and the geophone group signals. The control signal had to be laid down on one track of a magnetic recording sheet and then picked up and radioed to the vibrators as the geophone signals were being laid down on the same recording sheet—often with the use of narrow-head recording to facilitate later compositing. Further, methods had to be devised to perform analog correlations between the reference signal and the field traces (Anstey, 1964). All these ingenious pieces of equipment and the field vibrators had to work together precisely and reliably. Their completion represented a *tour de force* in experimental geophysics.

The onset of digital processing and analog-digital conversion (or straight digital recording) made many of these ingenious methods obsolete and rendered the Vibroseis method more accessible to the reliability and precision of digital correlation, filtering, and stacking.

It has just been established that the shape of the autocorrelation pulses that make up a Vibroseis record (sometimes called a correlogram) is dependent *only* on the amplitude spectrum of the signal input to the earth. Because the phase remains unspecified in this input signal, it indicates that any signal with a given amplitude spectrum suffices; in fact, this is true to a limited extent. Later we talk about the use of this concept (pseudorandom sweeps) to design a system that allows more than one source to operate and produce independent data, although received by the same recording setup. For the moment, however, we confine ourselves to a straightforward single system. What determines the choice of an input time signal?

First we require that the output signal, after correlation, be as clean as possible in two senses:

1. The near-in shape $\phi_{11}(t)$ for $t < T_0$, where T_0 is about one period of the midfrequency needed, must be as clean.
2. The pulse must die away rapidly without ghosts or local increases in amplitude when $t > T_0$.

Figure 3.6 illustrates the situation. Here we are looking at one side of an autocorrelation function—the other side is a mirror image about zero time—the

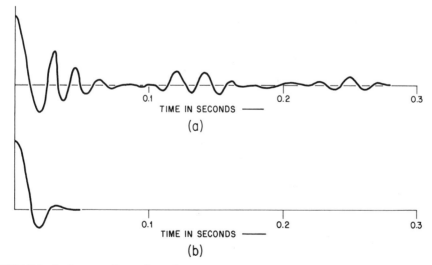

FIGURE 3.6 In these two illustrations of autocorrelation functions (*a*) represents an undesirable type that has a ringy near-in shape and shows local increases in amplitude at 0.13 and 0.25 s; (*b*) represents a more desirable, fast-decaying pulse with a near-zero background for $t > 0.05$ s.

frequency spectrum of which is unspecified. A fast-decaying near-in shape is obviously desirable because it indicates a reflection; a very "ringy" shape can easily hide several reflections and obviously reduces the resolution obtainable with the Vibroseis method. The second requirement is related to the same problem: that a given autocorrelation wave shape occur at each reflection time. We know in practice that reflections generally decrease in amplitude with time; therefore it may be necessary, during the processing step, to increase the gain of the system to make later reflections visible. But, if this is done all the ghosts that occur on a single autocorrelation pulse will be amplified and will appear to be legitimate, strong reflections. However, they portray only a much shallower structure and therefore constitute noise.

These requirements become stringent for the input wave shape—the control signal—and have been shown to allow only signals that

1. Are wide-band
2. Have no sudden changes in amplitude
3. Have *instantaneous* frequencies that do not repeat
4. Have *instantaneous* frequencies that are smoothly varying
5. Have, predominantly, a flat-amplitude spectrum

To some extent these qualifications overlap; for example, a constant-amplitude time signal that dwells temporarily (or even changes the rate of change in instantaneous frequency) causes the amplitude spectrum not to be flat.

Some qualifications conflict; for example, the desire for a wide-band, flat-amplitude spectrum within certain frequency limits causes a sudden amplitude change at the ends of the spectrum. This results in ringing at these two frequencies and, as a consequence, a poor autocorrelation pulse both near-in and at later times because these particular frequencies do not die out quickly. These phenomena are well known in electrical engineering practice and have been discussed at length by Edelmann (1966) and Krey (1969).

As a consequence of these requirements, a standard, linear-swept frequency signal has been adopted for the control signal for use in normal situations. It has the form

$$F(t) = A(t) \sin 2\pi \left(at + \frac{bt^2}{2} \right) (0 < t < T) \tag{3.10}$$

This signal (looking at the sine function first) consists of a sinusoidal function of gradually (and uniformly) increasing frequency—beginning at $\omega_0 = 2\pi a (f_0 = a)$ and ending at $\omega_f = 2\pi(a + bT)$. Thus a sweep that passes from 10 to 100 Hz in 10 s has values of $a = 10$ and $b = 9.0$.

The purpose of the function $A(t)$ is to have a multiplying factor in the sinusoidal function to prevent the sweep from starting or stopping abruptly in amplitude. The taper at each end is usually linear over a few cycles at each end of the sweep but is sometimes sinusoidal or a more complex function (see Figure 3.7).

Suggestions have been made that filtered random noise can be used as a control signal for the Vibroseis method. Random noise is chosen because, with a

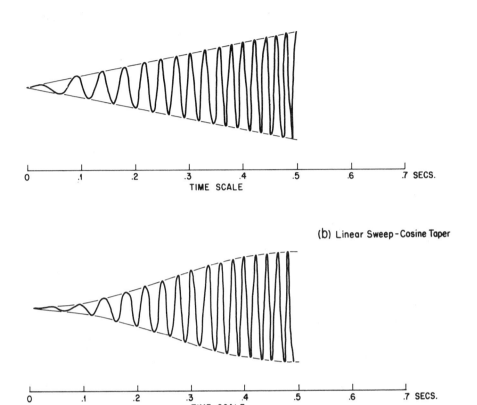

FIGURE 3.7 Two forms of taper for sinusoidal sweep signals. The linear taper (*a*) is usually considered adequate and is featured in some automatic sweep generators now widely used. Type (*b*) produces a quieter autocorrelation signal at times >0.05 s.

suitable distribution for the amplitudes of the noise trace, the spectrum can be made flat over a wide band of frequencies. This is obviously a good starting point in view of requirement 5. The original bandwidth, however, is controlled by the sample time and the spectrum can be flat for bandwidths much greater than required—or greater than *can be accepted* by the mechanical vibrator system. Therefore the random noise trace is filtered with a zero-phase pulse designed to leave the spectrum amplitude alone inside the required band and to cut it off (with a taper, if necessary or desired) outside this band. It is particularly necessary, as shown later, to remove low frequencies outside the design range of the vibrator.

We can call this filtered version a pseudorandom sweep. It is now found that the filtered version no longer has a constant-amplitude envelope but has random variation.

For practical reasons this type of sweep is not desired because the vibrators are usually required to operate at maximum amplitude at all times, so that the signal entering the earth will have maximum energy.

For theoretical (or processing and interpretation) reasons, moreover, the sweep is undesirable because local repetitions of instantaneous frequency cause the overall form of the autocorrelation pulse to deteriorate by the formation of ghosts—where two different parts of the sweep correlate to a greater extent than is desirable.

Having selected the signal needed, there remains the problem of introducing a primary signal into the ground. The conventional method (see Crawford, Doty, and Lee, 1960) has been to vibrate a large plate (called the baseplate) which is kept in contact with the ground during the time of generation of the signal. If the vibratory signal desired has a range of force of $2A$ kg the baseplate must be held down on the ground with a force of at least A kg; otherwise it would leave the ground and hammering would result. The weight of the vehicle is used as the hold-down weight and the vehicle is isolated from the vibration by air bags or similar "soft" springs. With the weight of the truck on it, the system corresponds to a low-frequency-damped resonant system. The period is kept well below the lowest frequency to be generated by the vibrator. Large, heavy vehicles are therefore necessary to have a high enough mass to hold down the vibrator and to carry the weight of the diesel power supplies, vibrator units, and auxiliary equipment. With the continual demand for increasing the low-frequency characteristics of vibrators a compromise is becoming necessary between the weight of vehicles, their maneuverability, and road weight limits.

The actual mechanisms that produce the vibrating force must be designed with special care to suit the main purpose. Basically, a servohydraulic actuator follows the design illustrated in Figure 3.8. A central cylinder is bored through a large steel mass that can be sealed by sleeves with O-ring seals. A piston is formed by steel ridges on a central steel bar in which are drilled several holes that can (1) convey oil under pressure to two enclosures (30 and 32) or (2) exhaust oil to a low-pressure accumulator (not shown). Enclosures 30 and 32 are alternately connected to a high-pressure, servocontrolled hydraulic system. Also, 30 and 32 alternately connected to the return hydraulic line.

Under the action of these hydraulic pressures in the enclosures the mass oscillates up and down about some neutral point, which is obtained and regulated by bleeding off part of the high constant-pressure hydraulic fluid through hole 52 into chamber 46.

If a force exists on the steel mass 16 an equal and opposite force is applied to the baseplate 18, which is in contact with the earth and, in turn, induces seismic energy into the earth. The baseplate is held down by the weight of the vehicle by decoupling members (low-pass frequency filters) such as air bags or springs. The hold-down weight must be greater than the peak dynamic force supplied by the oil pressure in the cylinder-piston assembly. Peak forces up to 16,300 kg (36,000 lb) have been obtained. The rigid baseplate and piston assembly have a mass that is coupled to the impedance of the earth. For the seismic frequency range the earth acts approximately as a spring, although the action is more complicated because of damping (by radiation of energy and heat loss) and because of the nonlinear behavior of the baseplate–earth system. As pointed out earlier when the weight drop method was being considered, the earth consolidates under the action of successive impacts and presents a varying load to the force being impressed on it. In the case of the vibrator the behavior of the earth under the baseplate is more

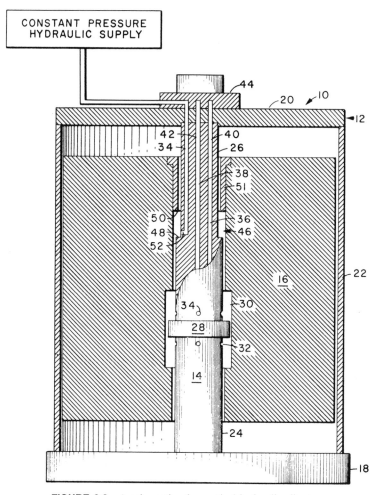

FIGURE 3.8 A schematic of a vertical hydraulic vibrator.

complex because the zone is not in contact with the air between successive force peaks and the heat generated in crushing can change the nature of the soil. From the point of view of signal shape the effect of the nonlinear behavior is to introduce harmonics (see Figure 3.9) into the low-frequency signal generated, sometimes almost to the extinction of the desired fundamental.

The vibrator mechanism is attached to the lift system through the isolating air bags, and it can be raised or lowered onto the ground by hydraulic cylinders capable of lifting the entire truck off the ground so that its weight is fully on the baseplate. Actually, it is desirable to leave a small proportion of the weight on the front wheels to stabilize the truck. The lift system is therefore positioned slightly aft of the center of gravity of the truck and mounted components. Raising and lowering have to be done in a few seconds to allow quick moving. In the raised position the vibrator mechanism is well above the ground to achieve maximum clearance in rough terrain.

These features are made evident in Plates 1, 2, and 3, which illustrate,

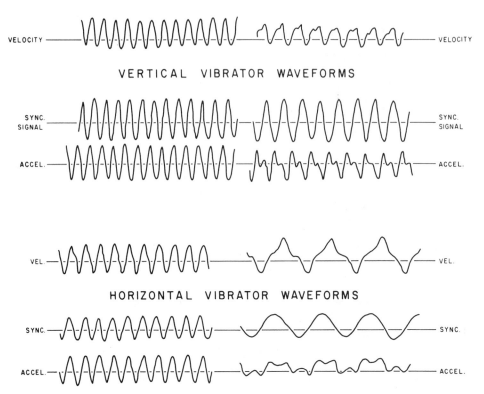

FIGURE 3.9 The vertical and horizontal vibrator baseplate motions that correspond to a given synchronizing electric signal. High frequencies are on the left, low frequencies on the right. Development of the double frequency components as the frequency is lowered is evident.

respectively, a new, commercially produced general-purpose (Mertz #26/219) servohydraulic vibrator capable of generating SH waves as well as P waves, a large low-frequency research vibrator (CONOCO Inc.), and a commercially produced horizontal vibrator that is now in use for the production of shearwave reflections (CONOCO M-13). The comparative specifications of these vibrators are given in Table 3.1. Shear-wave vibrators are built according to the same general engineering principles but have a different method of introducing motion into the earth.

Developments in land-based seismic sources have included a vibrator and an impulsive source based on a surface-effect vehicle (or Hovercraft). As we all know, it rides on a cushion of air provided continuously by an air compressor. A flexible skirt forms a partial seal against air leakage near the earth and the vehicle platform rises to a height automatically determined by the available air flow, air pressure, and the quality of the seal against the ground surface.

R. A. Broding and D. E. Miller suggested independently that this vehicle could be made into a seismic source by (1) vibrating the platform with a reaction-type vibrator or (2) modulating the air stream. Both modes of operation do, in fact, work, but a larger force can be obtained more conveniently by using the second method. Figure 3.10 is a schematic of a vibrator. The action of the modulators raises a combination of dc and ac air pressure inside the skirt—the

PLATE 1 Universal vibrator #26/219 which may be field-adapted for vertical or horizontal vibration. (*Source*: Courtesy Mertz, Inc.)

PLATE 2 Low-frequency research vibrator (CONOCO, Inc.). Peak force 36,000 lb; low frequency limit 2.3 Hz.

PLATE 3 A Model 13 shearwave vibrator. (*Source*: Courtesy Mertz, Inc.)

former holding the vehicle near an equilibrium position, the latter giving an alternating pressure on the ground. The alternating pressure is relatively small [about 0.035 to 0.07 kg/cm^2 (0.5 to 1.0 lb/in.^{2})], but by using a large (3.04 × 6.08 m) vehicle the total force [about 0.65 to 1.30 × 10^4 kg (20,000 lb)] can be made as large as that of a conventional field vibrator. Several advantages can accrue from this arrangement:

1. Coupling can be made to a rocky surface
2. Easily damaged ground (permafrost) can be traversed without environmental damage
3. Pressure can be provided on the ground that does not exceed the elastic limit of the soil
4. Work can be carried across tidal flats and other marginally accessible areas

A vehicle of this design has been built and tested. In addition, a similar concept has been used in which the air is suddenly released inside the air cushion to produce a sudden increase in pressure—which, of course, decays rapidly because of the leak at the ground surface.

An air-cushion vehicle is resonant at a low frequency determined by the mass of the platform and the volume of air in the cushion. It is, of course, damped because of the air leak at the ground surface and is therefore predominantly a low-frequency vibrator with an output for constant peak input pressure variation that decays by a factor of 2 each time the frequency is doubled (an octave). Expressed in the usual way, the rate of decay is 6 dB per octave. [The *decibel change* is given by $20 \log_{10}(A_2/A_1)$, where A_2 and A_1 are the *amplitudes* of the measured signals.]

TABLE 3.1

COMPARATIVE SPECIFICATIONS OF SELECTED VIBRATORS

Characteristic	Mertz M26/219 V	Mertz M26/219 H	CONOCO LFV	CONOCO No. 13 Shear
Vibrator mounting	Custom hydraulic drive	Custom hydraulic drive	Truck 6 × 4	Custom 8 × 4 hydraulic drive
Piston Area (cm²) (in.²)	127.9 (19.82)	64.5 (10.0)	78.64 (12.22)	66.45 (10.3)
Peak force (kg) (lb)	27,216 (60,000)	13,880 (30,600)	16,636 (36,600)	14,045 (30,900)
Usable stroke (cm) (in.)	5.08 (2.00)	17.8 (7.00)	22.86 (9.0)	25.4 (10.0)
Actuator weight (kg) (lb)	3,900 (8,600)	2,722 (6,000)	6,910 (15,200)	3,273 (7,200)
Displacement	7.0	4.0	2.3	2.9
High frequency Limit (Hz)	250	100	—	—
Vibrator pump model	Sundstrand 27	Sundstrand 27	Vickers PVB 90	Vickers PVB 90
Vehicle wheelbase (m) (in.)	5.944 (234)	5.944 (234)	5.03 (198)	4.98 (196)
Total vehicle weight (kg) (lb)	27,216 (60,000)	27,216 (60,000)	22,841 (50,250)	24,136 (53,100)
Pad weight (kg) (lb)	27,216 (60,000)	27,216 (60,000)	22,159 (48,750)	21,880 (48,135)
Servovalve	Moog 79C266	Moog 79C266	—	—
Pad area (cm²) (in.²)	— —	— —	25,084 3,888	20,000 V max / 3,100 V max / 6,729 H max / 1,043 H max

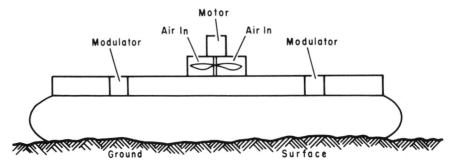

FIGURE 3.10 Schematic of a vibrator based on a Hovercraft. The two air modulators are sychronized with a sweep signal. A shutter is open to allow airflow only during the linear increase of frequency and closed for a return to the original starting frequency.

3.5 EARTH RESPONSE TO A DRIVING FORCE DISTRIBUTED OVER AN AREA OF THE SURFACE

The earliest treatment of the production of elastic waves due to prescribed, periodic stresses on the surface of a semi-infinite isotropic solid has been provided in two important papers by Miller and Pursey (1954, 1956) entitled "The Field and Radiation Impedance of Mechanical Radiators on the Free Surface of a Semi-infinite Isotropic Solid" and "On the Partition of Energy between Elastic Waves in a Semi-infinite Solid." Only a few results are quoted here. Different types of motion on the surface were considered:

1. An infinitely long strip of finite width vibrating in a direction normal to the surface of the medium
2. An infinitely long strip of finite width vibrating tangentially to the surface and normally to the axis of the strip
3. A circular disk of finite radius vibrating normally to the surface of the medium (a model for a vertical vibrator)
4. A torsional radiator in the form of a disk of finite radius performing rotational oscillations about its center (a model for a torsional vibrator)

The directivity functions in Figure 3.11 are taken directly from the first paper (1954), but note that the motions are given as functions of R (distance from the origin) and θ (the vertical angle used earlier). It should be noted that the radial motion U_R is almost proportional to $\cos \theta$ and varies only a small amount with a change in Poisson's ratio. For most work in sedimentary sections $\sigma = \frac{1}{3}$ (or $V_p/V_s = 2$) fits the field data closely.

The SV motion (perpendicular to the direction of travel and in a vertical plane) is a much more complex function. The directivity function has two loops with a zero in between. Actually the phase changes by 180° between the two loops, thus making the zero a natural consequence of the (sudden) phase change. There is no SV motion down the axis (by symmetry this must be the case) and the strongest SV motion is developed at an angle near 30° from the vertical axis.

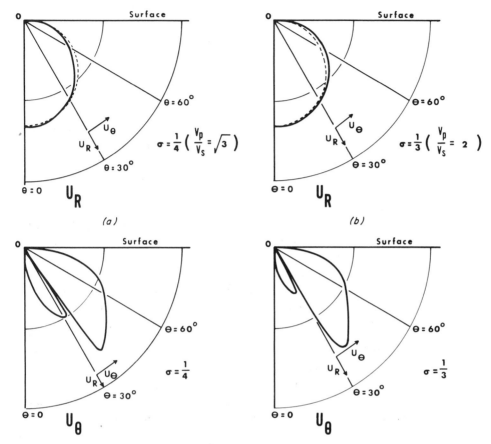

FIGURE 3.11 Directivity functions for a disk that vibrates perpendicular to the ground surface. Two examples are shown: (*a*) Poisson's ratio = 0.25, $V_s/V_p = 0.577$, and (*b*) $\sigma = 0.33$, $V_s/V_p = 0.5$. [*Source*: Miller and Pursey (1954).]

The distribution of the available energy from the vibrating source (for the particular case $\sigma = \tfrac{1}{4}$) has been evaluated in Miller and Pursey (1956) as

$$\text{compressional energy} = W_c = 0.333 \, \frac{\pi^3 \nu^2 a^4 P_0^2}{\rho V_c^3}$$

$$\text{shear energy} = W_{SR} = 1.246 \, \frac{\pi^3 \nu^2 a^4 P_0^2}{\rho V_c^3}$$

$$\text{surface wave (Rayleigh) energy} = W_{Su} = 3.257 \, \frac{\pi^3 \nu^2 a^4 P_0^2}{\rho V_c^3}$$

where ν = frequency
$\quad a$ = radius of disk
$\quad P_0$ = stress under disk (peak)
$\quad \rho$ = medium density
$\quad V_c$ = medium compressional wave velocity

Thus the distribution of energy is

compressional energy	6.9%
shear energy	25.8%
surface wave energy	67.3%

In practice the body-wave energy figures appear to be optimistic. The presence of a surface layer causes much of the body-wave energy radiated at high angles to be multiply reflected inside the surface layer and thus becomes additional surface wave energy. However, the effect of the surface layer on the mechanical efficiency of the source does not appear to have been investigated, and the foregoing conclusions may not be valid. The effect of a layered medium, including, of course, a near-surface low-velocity layer, was recently published by Luco (1974, 1976). No general conclusions are stated here and reference should be made to the original papers.

As with the explosive sources, we deal later with the pattern effect of multiple-surface sources. This pattern effect is obtained whether the vibrators are used simultaneously or sequentially and the results are summed afterward. There may, however, be some question concerning the mutual effect between simultaneously operated mechanical sources situated close together so that the motion due to one source changes the radiation impedance (or efficiency) of another. Physically, this nonlinear effect must be present. Results from finite elastic modeling show that the surface motion from one vibrator dies off so rapidly with distance that for all practical distances (say, greater than 5 m) one vibrator cannot "feel" the effect of another in any significant amount. This matter is also discussed (mathematically) in Miller and Pursey (1956). Experimental evidence has been obtained that in a real situation the influence of one vibrator on the efficiency of another nearby is negligible. Multiple, simultaneously operating vibrators are used in field practice, but for considerations of signal/noise ratio and economics rather than any gain in efficiency due to interaction between vibrators. For a given random noise power four vibrators operating simultaneously increase the signal/noise level four times over the action of a single vibrator. If four single-vibrator traces are taken at different times, however, the noise samples must be assumed to be different and the signal/noise ratio increases only as the square root of the number of trials, or 2:1 in this case. Thus it is always more beneficial to increase the signal strength than to try to increase the number of samples by the same proportion. In fact, this increase in sampling brings diminishing returns as the number is increased and eventually becomes uneconomical.

The early work of Miller and Pursey (1954, 1956) was based on the assumption that a constant stress is maintained over the area of contact of the vibrator with the earth. This assumption was not realistic for most vibrators (except perhaps for the air-cushion vibrator), but it did allow the displacement of the surface to be variable, and resulted in a simpler set of boundary conditions, and it did allow calculation of the distribution of energy within the earth for many types of surface motion.

Since that time, largely because of the interest of civil engineers in the same problem, Bycroft (1956), Lysmer (1965), and Luco (1974) have made numerical calculations of the elastodynamic radiation from surface sources that are *rigid*. This condition results in constancy of displacement, whereas the stress under the source must be variable.

To compare these results Farrell (1979) adopted the idea of radiation impedance (Z_R), defined as the (complex) ratio of the mean normal stress $\langle T_{xx} \rangle$ to the mean normal particle *velocity* $\langle \dot{x} \rangle$, each averaged over the stressed region. To a good approximation it has been found that

$$Z_R \approx B\rho\beta(1 - i/ka)$$

or even better by

$$Z_R = B\rho\beta\left(R - \frac{iI}{ka}\right) \tag{3.11}$$

where R and I are coefficients of the real and imaginary terms of the radiation impedance. They correspond to the *radiative* and *storage* parts of the impedance function and can be frequency-dependent. In these formulas ρ = density, β = shear wave velocity, k = wave number = ω/β, and the product ka is the usual nondimensional frequency, which is always an important part of radiation calculations.

Lysmer (1965) has shown that if B is picked in such a way that I tends to unity as ka tends to zero this scaling suppresses most of the dependence of I *and* R on Poisson's ratio. In this case (constant stress)

$$B = \frac{3\pi}{8(1 - \sigma)} \left.\begin{array}{l} \\ \end{array}\right\} \quad \begin{array}{l} = 1.57 \text{ for } \sigma = \frac{1}{4} \\ = 1.77 \text{ for } \sigma = \frac{1}{3} \end{array}$$

For the rigid punch problem (constant displacement)

$$B = \frac{4}{\pi(1 - \sigma)} \left.\begin{array}{l} \\ \end{array}\right\} \quad \begin{array}{l} = 1.70 \text{ for } \sigma = \frac{1}{4} \\ = 1.90 \text{ for } \sigma = \frac{1}{3} \end{array}$$

Two important points emerge. First, within practical limits, it is immaterial whether we consider the baseplate to be flexible or rigid as far as production of waves in the earth is concerned. Second, the impedance the earth offers to the *vertical* vibrator depends largely on the *shear wave* velocity as well as the frequency and a baseplate representative dimension.

At high frequencies (e.g., $ka \to \infty$) the baseplate *must* launch plane compressional waves, and it is known that the radiation impedance must be equal to $\rho\alpha$ (where α is the compressional wave velocity). The two different assumptions then yield

$$\text{constant stress limit } R = \frac{8\sqrt{2}}{3\pi} \frac{(1 - \sigma)^{3/2}}{(1 - 2\sigma)^{1/2}}$$
$$ka \to \infty$$

$$R = 1.10 \quad \text{for } \sigma = \frac{1}{4}$$

$$r = 1.07 \quad \text{for } \sigma = \frac{1}{3}$$

$$\text{constant displacement limit } R = \frac{\pi\sqrt{2}}{4}\,\frac{(1-\sigma)^{3/2}}{(1-2\sigma)^{1/2}}$$
$$ka \to \infty$$
$$R = 1.02 \quad \text{for } \sigma = \tfrac{1}{4}$$

$$R = 0.99 \quad \text{for } \sigma = \tfrac{1}{3}$$

and we find that it is relatively unimportant at high frequencies, whether the constant displacement or constant stress conditions apply. Even further, the impedance is relatively independent of Poisson's ratio.

If some numbers typical of oil exploration practice are used, with $a = 1.0$ m, $\beta = 600$ ms^{-1}, $\omega = 2\pi \times 30$, it turns out that $ka = 0.3$ and justifies us in adopting Lysmer's choice of $R = 0.85$, $I = 1.0$. It also means that the storage impedance term has a value of 3.33, whereas the radiative part is only 0.85—a rather poor radiative efficiency.

All of the foregoing justifies the treatment of the earth itself as a lumped mechanical spring–dashpot system, and it can easily be seen that the spring constant of the earth

$$k_e = \pi a B \rho \beta^2 I \tag{3.12}$$

which increases linearly with the baseplate radius and the damping coefficient (radiative term) is

$$\lambda_e = \pi a^2 B \rho \beta R \tag{3.13}$$

which increases as the square of the baseplate radius. Clearly (under these conditions before considering the internal mechanics of the vibrator), the radiative term increases in proportion to the radius of the area of contact. The air-coupled vibrator is excellent in this regard.

We now superimpose the typical mechanical construction of a normal vibrator on this radiation impedance framework. Figure 3.12 shows schematically the arrangement of a baseplate of mass m_b resting on the earth, represented by its spring constant (k_e) and damping coefficient (λ_e). A force (f), a spring (k_r) and a damping coefficient (λ_r) represent the coupling between a reaction mass (m_r) and the baseplate mass (m_b). This is all the system to be considered in the mathematical treatment, but in addition there is shown (dotted connections) the mass of the truck (m_t), which is placed on the baseplate by the soft-spring (air-bag) system previously mentioned. In almost all analyses this "dead weight," which changes considerably the operating point on the stress–strain relationship of the soil, decouples the truck from the movement of the baseplate and prevents all the extra mass (m_t) from having to be accelerated.

Farrell (1979) has shown that the effect of a drive force (f), which acts between the reaction mass and the baseplate, gives rise to motion governed by the following two differential equations; dot superscripts denote time derivatives:

$$m_b\ddot{x}_b = -(k_e + k_r)x_b - (\lambda_e + \lambda_r)\dot{x}_b + k_r x_r + \lambda_r \dot{x}_r - f$$
$$m_r\ddot{x}_r = -k_r(x_r - x_b) - \lambda_r(\dot{x}_r - \dot{x}_b) + f \tag{3.14}$$

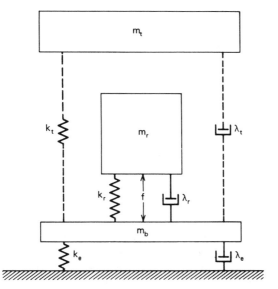

FIGURE 3.12 Schematic of relationship between mechanical parameters of the earth, the vibrator baseplate, the reaction mass, and the truck. Dotted connecting lines signify that the truck elements were not considered in the formal system analysis.

These equations are solvable by Laplace or Fourier transform techniques. The results only are given in terms of X_b, X_r (transformed displacements) and F the transformed force:

$$X_b = \frac{F}{m_b} \frac{N_b(s)}{D(s)}$$

$$X_r = \frac{F}{m_r} \frac{N_r(s)}{D(s)} \qquad (3.15)$$

$$X_d = \frac{F'_m}{m_b} \frac{N_d(s)}{D(s)}$$

where

$$m' = \frac{m_r + m_b}{m_r}$$

$$X_d = X_r - X_b$$

$$s = i\omega$$

and

$$N_b(s) = s^2 \qquad (3.16)$$

$$N_r(s) = s^2 + \frac{\lambda_e s}{m_b} + \frac{k_e}{m_b} \qquad (3.17)$$

$$N_d(s) = s^2 + \frac{\lambda_e}{m_r + m_b}\, s + \frac{k_e}{m_r + m_b} \tag{3.18}$$

$$D(s) = s^4 + \frac{\lambda_e}{m_b}\left(1 + \frac{\lambda_r}{\lambda_e}\, m'\right)s^3 + \frac{k_e}{m_b}\left(1 + \frac{k_r}{k_e}\, m' + \frac{\lambda_e \lambda_r}{k_e m_r}\right)s^2$$
$$+ \left(\frac{\lambda_r k_e + \lambda_e k_r}{m_r m_b}\right)s + \frac{k_e k_r}{m_b m_r} \tag{3.19}$$

We note that the reaction mass support elements k_r, λ_r have no effect on the numerator (N) polynomials and that the reaction mass (m_r) affects just the N_d, the displacement difference polynomial. These three elements do, however, affect the denominator polynomial $D(s)$. It may be shown that this denominator polynomial $D(s)$ has one quadratic factor nearly equal to $N_r(s)$ and another quadratic factor that is *almost* independent of k_e and λ_e.

Thus the system is about the equivalent of two slightly coupled resonant systems: first, the reaction mass and base-plate coupling and, second, the baseplate–earth coupling.

Further insight into the nature of the coupling between the vibrator and the earth is obtained by a more complex root-locus examination of the zeros of $D(s)$. Interested readers should consult the paper by Farrell (1979) in which this analysis is done in detail. The type of vibrator (e.g., electromagnetic or hydraulic) is important in this regard because the spring constants between the reaction mass and the baseplate are so different—being, respectively, a comparatively weak spring holding up the reaction mass and a spring due to the compliance of the hydraulic fluid.

It is not possible to pursue this matter because it rapidly becomes specialist material. A more cogent reason, probably, is the fact that because the earth materials vary so rapidly as the vibrator moves along any line the vibrator characteristics give rise to undesirable phase and amplitude changes that make the vibrator action inconsistent.

We shall see, shortly, how the inconsistent phase behavior is reduced to a negligible amount by the use of feedback; but first the practical design method should be studied.

The usual starting points in the design specifications (Brown and Moxley, 1964) are the maximum peak force and low-frequency cutoff values. At high frequencies the system is limited in force because of the pressure available from the power supply; for example, the usual dc hydraulic fluid pressure is 3000 lb/in.[2] (232.6 kg/cm[2]), and there is normally a drop of 500 lb/in.[2] (38.8 kg/cm[2]) across the servovalve, leaving 2500 lb/in.[2] (193.8 kg/cm[2]) to be applied to the piston area. If this were 5.0 in.[2] (32.3 cm[2]) the total peak force (one direction) available would be 12,500 lb (6260 kg). The low-frequency limit is usually the total stroke length. For an actuator whose displacement is given by

$$y = A \cos \omega t \qquad y_{\max} = A \tag{3.20}$$

the maximum velocity is ωA and the maximum acceleration is $\omega^2 A = 4\pi^2 f^2 A$. Because the force available is M times the acceleration,

$$\text{maximum force} = 4\pi^2 M f^2 A \tag{3.21}$$

In the midband of frequencies the maximum velocity of the actuator is limited by the servovalve flow or the pump flow. It is not entirely necessary to provide for a pump flow that is sufficient at all times of the swept frequency signal because the demand at high frequencies can be supplied by an oil accumulator that is replenished during the idle periods between sweeps.

Although fluids are normally considered incompressible, the high dynamic response demanded of the system necessitates consideration of the compressibility of the fluid, which acts as a spring between the hydraulic servovalve and the piston actuator. (Of course, the compliance of any coupling tubing also has to be taken into account.)

The entire system is controlled by the two-stage hydraulic servovalve, controlled by an electric torque motor (see Figure 3.13). Five watts of electric power results in a deflection of ±0.015 in. (0.038 cm), or a midposition force of ±13 lb (5.91 kg). This controls a pilot spool which, in turn, ports high-pressure fluid to the main spool. The motion of the main spool is monitored by a linear transformer that supplies a signal to be used for feedback control of the torque motor.

An additional feedback control is used for the main poston to make it constant *displacement* below a given frequency—at a point at which there is no danger of the driven mass exceeding the stroke limit. Bumpers are usually placed inside the framework to prevent damage, should the stoke be temporarily exceeded. Over the entire midrange the servovibrator tends to act as a constant-*velocity* system and, beyond the high-frequency break, as a constant-*acceleration* system. Fluid compression sets in soon and produces an additional 12 dB per octave drop in output (see Figure 3.14).

Figure 3.15 is a diagram of the salient features of the restrictions on vibrator output. The solid lines show the restrictions of the primary equipment, whereas the dotted lines show how they are affected by closing the servoloops.

Although the design of the hydraulic servovibrator itself is complicated enough, there are still other considerations. Earlier experiments showed that the earth acts mainly as a spring, even though there is a mass of earth to be driven.

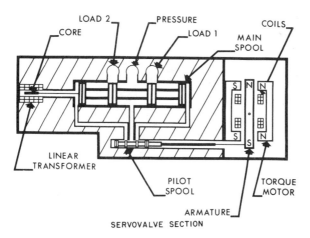

FIGURE 3.13 Schematic of a two-stage hydraulic servovalve with torque motor control and a linear transformer for feedback signals.

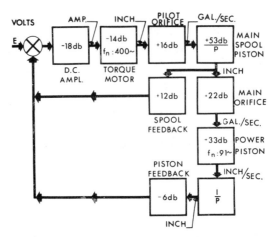

FIGURE 3.14 A typical servosystem block diagram shows gains at each stage and feedback controls.

This mass usually about that of the baseplate but varies with baseplate area and soil characteristics. Because seismic surveys, of necessity, cross different soils with different spring constants, the inconsistency of phase is intolerable in a system whose purpose is to measure times accurately. Between the early high-frequency reflections and deeper lower frequency signals a change in time interval of as much as 0.035 s has been known to exist. This is due largely to phase inconsistencies between the baseplate motion and the control signal.

Of course, baseplate motion can be monitored, recorded, and used as the reference signal with which the remote earth motions are correlated, but an important consideration is the possibility of using only one reference signal—that

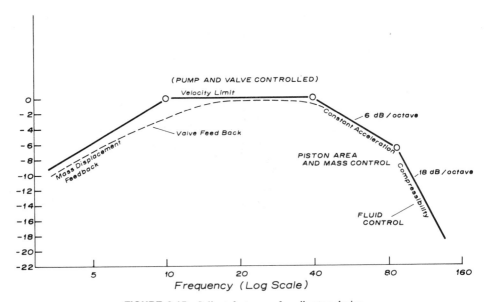

FIGURE 3.15 Salient features of a vibrator design.

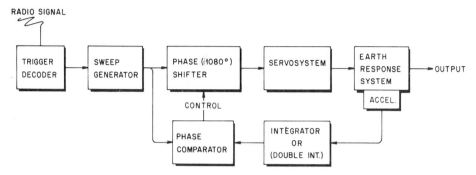

FIGURE 3.16 Block diagram of a phase compensation system.

originally sent out by the sweep generators. This unique reference signal makes data processing much simpler, and for this reason a phase compensation system was developed.

A phase compensation system is shown, in block diagram form, in Figure 3.16. The modern field concept is to have a digitally controlled sweep generator in each vibrator truck as well as in the recording truck. A trigger set off in the recording truck sends a coded radio signal to the vibrators, where it is decoded, and the individual sweep generators are triggered simultaneously. This operational principle prevents a noisy radio signal from feeding into the vibrator(s). The sweep generators usually have controls to allow a change in initial frequency, final frequency, and the linear tapers at the two ends of the sweep.

The control signal is now split between the phase compensator and the phase shifter, the latter having the capability of shifting by a maximum of 1080°. The phase-shifted signal then goes to the servosystem shown in Figure 3.14 and its output controls the motion of the baseplate against the earth. The earth acts as an added baseplate mass and a spring. Finally, an accelerometer, designed for strong motion and integrated to give baseplate velocity (or sometimes doubly integrated to give baseplate position), delivers a signal to one side of the phase comparator which compares it with the reference signal and issues a (rectified) control signal to change the operating point of the phase shifter in a direction that reduces the discrepancy. Note that it does not work by applying a voltage feedback of the same frequency as the signal it is controlling. Instead, it controls the phase added to the signal by a rectified signal from the comparator. When this rectified signal acts on the phase-shifting network the *difference* in phase between the *control signal* and the *output velocity* is kept constant. A system of this type locks in more readily at high rather than at low frequencies; therefore the tendency is to use downsweeps with phase compensators, even though this is dangerous from the point of view of harmonic distortion (considered later). Note that no attempt is made to keep the amplitude spectrum constant from one vibrator position to the next. This type of consistency of output must be achieved by later digital processing.

All of these considerations apply equally to horizontal (shear wave) vibrators. The accelerometers are mounted on the baseplate to measure horizontal acceleration. With the large force available a good deal of attention must be given to the

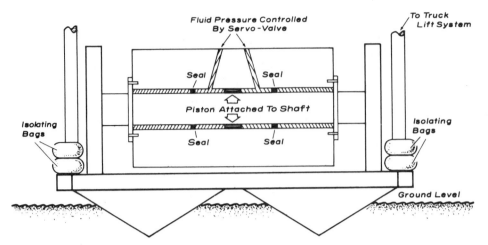

FIGURE 3.17 Side schematic of a shear vibrator baseplate and actuator.

problem of coupling the horizontal vibrator to the ground. Unless the coupling is continually good, some frequencies are inadequately conveyed to the ground. Several different systems have been tried, but the most satisfactory found so far consists of large pyramids welded (inverted) to the light baseplate so that as the weight of the vibrator truck is placed on the baseplate the pyramids are pushed into the ground. Thus a sideways force is counteracted by the truck's weight, and even if there is a tendency for the soil to consolidate the combination of pyramid and truck weight keeps the sides of the pyramids in contact with the soil (see Figure 3.17). The damage to the surface, particularly in intensive patterns, may be severe but can usually be repaired.

Light baseplates are, of course, desirable because the baseplate mass has to be accelerated by the hydraulic forces without producing any radiated power. However, considerations of strength usually limit the reduction in baseplate mass to about 2000 lb (~1000 kg).

3.6   USE OF MORE THAN ONE VIBRATOR

Field experiments have shown that the law of linear superposition holds closely in the case of multiple vibrators operating simultaneously. In other words, the use of N vibrators, all synchronized with the control signal, produces a signal in the earth, vertically under the group, which is N times that of a single vibrator. At other points in the earth the N vibrators produce a signal that is the vector sum of each vibrator signal, thereby taking into account different travel times.

The mutual interaction effect, by which the action of one vibrator affects the transfer impedance between the other vibrators and the ground, must exist but must also be very small.

For all practical purposes, therefore, multiple vibrators can be used and disposed on the surface as the geophysicist wishes (see Figure 3.18). In many cases the vibrators stay within a few feet of one another, either abreast or in line,

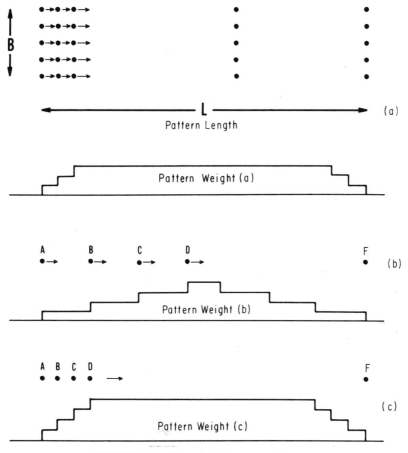

FIGURE 3.18 Vibrator deployment schemes.

for one or more sweeps; they then move up to give energy at the next location in the pattern required. In open country (e.g., pasture or desert) it may be desirable to use vibrators abreast to give a lateral pattern that will reduce interference coming from the side of the recording line. When the trails are narrow this is a deterrent and in-line patterns have to be used. Some companies dispose the available vibrators to allow them, after moving several times, to generate a tapered pattern.

Because the principle of superposition holds, the positions in a pattern can be occupied sequentially or simultaneously—whichever is expeditious. For a given random noise power four vibrators operating simultaneously increase the signal/noise level four times over the action of a single vibrator. If, however, four single-vibrator traces are taken at different times the noise samples must be assumed to be different and the signal/noise ratio increases only as the square root of the number of trials, or $2:1$ in this case. Thus it is always more beneficial to increase the signal strength than to try to increase the number of samples by the same proportion. In fact, this increase in sampling brings diminishing returns as the number is increased and eventually becomes uneconomical.

3.7 THE USE OF ARRAYS OF SOURCES

In this discussion of arrays or patterns of sources the sources themselves are considered omnidirectional; that is, each radiates equally in all directions. The effect of the arrays can then be compared to the action of diffraction gratings (or crystal lattices in x-ray diffraction) as a means of obtaining high-intensity beams in certain directions. In seismic reflection prospecting arrays are used to reduce the effects of waves traveling in a near-horizontal direction while at the same time amplifying the effects of almost vertical traveling waves. As shown later, not all reflected waves can be considered as traveling vertically because large horizontal offsets between source (patterns) and receiver (patterns) cause off-vertical rays to be considered. For simplicity, however, we assume that the discrimination required is between interfering horizontally traveling waves and vertically traveling reflected waves. The effects are obtained by the same physical mechanism; namely, constructive or destructive interference between the component waves produced by the individual sources that constitute the array. Figure 3.19 illustrates this array, and we investigate the strength of the composite signal as it is transmitted at an angle θ to the vertical. For the present, all sources are assumed to have the same strength and to be oscillating in synchronism. Although it is possible to consider each source as producing a signal of some arbitrary shape, we constrain them to oscillate at a constant angular frequency ω. Therefore the problem is resolved into finding a method of summing the contributions from a set of equally spaced, omnidirectional sources within the array.

If the individual sources have their own directivity function (as surface sources and near-surface-buried sources have) the final directivity function is given by the product of the individual source directivity function (taken as a class) and the array directivity function. This statement, naturally, is known as the directivity product theorem.

The method of obtaining the array directivity function is illustrated by considering a line of N discrete sources separated by equal distances d and examining the directivity in the plane perpendicular to the earth's surface and containing the line of sources. In Figure 3.19 we see that once the waves from the individual sources reach the line OO' (perpendicular to the required direction of propag-

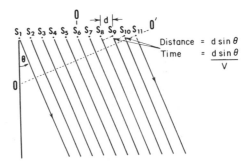

FIGURE 3.19 A diagram showing an array of equidistant sources $S_1, S_2, S_3, \ldots, S_{11}$ that contributes to the illumination in a direction θ to the vertical. In the general case neither the spacings nor the intensities of the sources are equal.

ation) they will have equal time paths to a point at infinity. Therefore the directivity arises because of time differences in reaching OO' from all sources. We can see that the difference in distance of successive source to the line OO' is $d \sin \theta$ and that the time difference is $d \sin \theta / V$, where V is the velocity of waves in the material. Because we are dealing with a single-frequency wave of angular frequency ω (period $2\pi/\omega$), the waveform from one source is delayed, compared with that from its right-hand neighbor, by a fraction of the period equal to $\omega d \sin \theta / 2\pi V$. If the component from one source has the form

$$A_0 \cos \omega t$$

the next one will have the delayed signal

$$A_0 \cos(\omega t - \phi)$$

where $\phi = \omega d \sin \theta / V$.

A summation of the contributions gives the array directivity function as

$$A(\omega, \theta) = A_0 \sum_{K=0}^{N-1} \cos(\omega t - K\phi)$$

This directivity function has the same form on both sides of the vertical axis and the formula

$$|A(\omega, \theta)| = A_0 \frac{\sin N\phi/2}{\sin \phi/2} \tag{3.22}$$

An example for $N = 18$ over an array length of $17d$ is shown in Figure 3.20, in which the directivity function is plotted with the source interval or apparent wavelength along the array as the independent variable. The main features to be noted are the following:

1. The large maximum for vertical transmission (the apparent wavelength is infinite).
2. A series of zeros and subsidiary maxima corresponding to angles for which $\sin 9\phi$ is zero or unity. In seismic prospecting the initial maximum and the first zero are principally of importance.

If the source strengths are unequal the contribution of each source is given by $m_k \cos(\omega t - \phi_k)$, and this can be represented, as in Figure 3.21, as a vector sum:

$$R_N = \sum_{k=0}^{N-1} m_k \cos(\omega t - \phi_k)$$

The same general method applies when the *distance* between the individual sources is no longer constant.

If it can be assumed that a vibrator is consistently of the same strength; no matter how the earth material changes its characteristics, these nonuniform sampling methods may have some advantages. One claimed advantage makes use

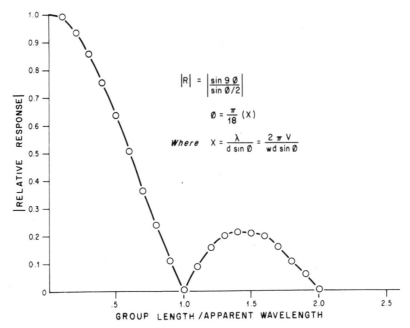

$$|R| = \left|\frac{\sin 9\,\emptyset}{\sin \emptyset/2}\right|$$

$$\emptyset = \frac{\pi}{18}\,(X)$$

$$\text{Where} \quad X = \frac{\lambda}{d\sin\emptyset} = \frac{2\,\pi\,V}{wd\sin\emptyset}$$

FIGURE 3.20 A directivity function of an 18-source array.

of different sampling with receivers and sources to achieve freedom from aliasing (i.e., inadequate sampling for small wavelengths on the surface) and to gain otherwise unobtainable pattern characteristics (Muir and Morrison, 1973). In the field, however, consistency of vibrator action is hard to obtain, and the advantages of these claims are hard to substantiate. Similar problems with suggested weighted patterns of receivers are considered later.

The question of aliasing is nevertheless an important one, and considerable

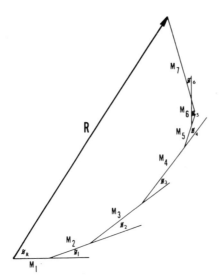

FIGURE 3.21 The vector addition of contributions from weighted elements of an array for a particular angle, frequency, spacing, and wave velocity. In this diagram M_k are the weights of the sources and $\phi_i = (\omega d_j \sin\theta)/V$.

care must be taken to ensure that the waves being rejected (generally surface and refracted waves) are sampled by sources and/or receivers and that more than two, preferably four or more, geophones or sources occupy the shortest wavelength expected. Rayleigh waves in the weathered layer may have phase velocities of 300 m/s, and, if the highest frequency of interest is 60 Hz, the wavelength will be 5 m and sources or receivers should occupy positions only 1.25 m apart along the pattern length. It is shown that the directivity of the pattern itself is such that a null occurs when the pattern length is equal to one or more complete apparent wavelengths. If the additional cosine directivity function for vertical vibrators on the surface is disregarded it will be necessary to choose a pattern length long enough to discriminate significantly against the *longest* wavelength in the interference to be rejected. This is, of course, the highest phase velocity, divided by the lowest frequency. It should not, however, be mandatory that the source pattern alone reject this lowest frequency interference, and economics usually dictate that some optimum rejection system be found by selecting arrays of both sources and detectors.

3.8 MARINE SEISMIC SOURCES

In seismic exploration confined to water-covered areas like the continental shelves the early source was always a form of unconfined explosion. Environmental damage was recognized here, however, perhaps earlier than the damage done on land, because the killing of fish was an immediate economic threat to the livelihood of fishermen. Therefore immediate pressure was applied to devise means of creating seismic waves, in water, that did not have peak pressures high enough to cause damage to marine fauna. Moreover, an additional reason was based on the so-called bubble effect. As long as the bubble of gas formed immediately after the explosion is spherical, a compression of the water continues, as a result of inertia, for a short time after the average pressure within the bubble becomes equal to the water pressure. For a 50-lb charge of TNT the outward expansion is brought to a halt after 200 ms and bubble contraction starts. An implosion results because the water converges into the limited spherical volume with higher and higher velocity until suddenly there is no space left. The pressure rapidly increases, overshoots, and the expansion starts again. This process continues, with less and less amplitude (energy is lost each time and some heat is lost to the water). Eventually the bubble breaks the water surface, the spherical symmetry is broken, and the process stops (see Figure 3.22).

As far as the production of seismic waves is concerned, a series of nonuniformly spaced pulses is obtained which decays in amplitude but, at the same time, gives rise to a complex source pulse (or signature) that is far from predictable. To avoid this complex signature the explosive charges were used inefficiently at depths so low that the first bubble broke the surface.

Another predictable consequence of the difficulties of using dynamite was taken other marine sources were developed and tested. There were analogs of the weight drop and Dinoseis and Vibroseis methods, but none has become popular, although all are sometimes still adopted for special purposes.

Because of the long swept frequency signal normally used and the fact that the

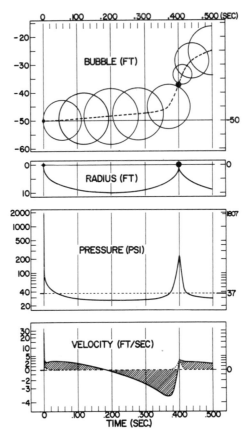

FIGURE 3.22 Pressure and velocity signatures for the explosion of 22 kg (50 lb) of TNT taken at a distance of 10 m (33 ft). The depth below the water surface was 16 m (50 ft). [*Source*: Seismic-Energy Sources. United Bendix (1968).]

water medium allows continuous motion of boat, sources, and receivers, the Vibroseis system is faced with the fact that a considerable length of subsurface is covered for reflections during the duration of the sweep. If the sweep length were, for example, 15 s long and the boat traveled at 6 knots (11 km/h) the boat actually would move 45.7 m while the sweep is taking place—and still further while waiting for deep reflections to return in full. It is only then that a new sweep can be started, and this consideration (as opposed to land Vibroseis practice) allows only about one sweep per horizontal position.

If that were the case the power injected to the medium during the single sweep from (say) four marine vibrators would have to be large enough to produce an acceptable signal/noise ratio for the deepest reflections. Each vibrator has to put out a large amplitude signal and even with the largest acceptable units *cavitation* sets in at the higher frequencies. This condition, and the fact that rather complex mechanisms (multiple units) have to be immersed at a depth of 13 m in seawater and dragged along by the boat, caused Vibroseis operations to be subject to breakdowns and uncertainties at sea. Although still in use for special purposes, the Vibroseis method has failed to become truly operational.

The requirements for a marine seismic source are the following:

1. The ability to generate a discrete powerful pulse or a signal that can be subjected to later compression in time such as in the Vibroseis system.
2. A rechargeable or repeatable system that can be used in a sequence of operations at short intervals of time (10 s or so).
3. A relatively simple system that will operate consistently, be trouble-free, and have a long life between overhauls.
4. A system that can be used at a constant depth below the water surface and results in a minimum drag on the vessel carrying it.
5. A system that does not injure marine life.
6. A system that minimizes the bubble effect.

The three repetitive impulse methods that have become operational are the Aquapulse system, the air-gun method, and the water-gun method. Their *partial* defeat of the bubble effect was obtained, however, by different methods.

For its energy the Aquapulse source uses a propane–oxygen mixture that is exploded inside a steel, mesh-supported, cylindrical rubber sleeve. After detonation the sleeve expands from its initial supported cylindrical form but retains its cylindrical symmetry, which does not allow the strong oscillating gas-bubble phenomenon to take place. After the initial sudden expansion the sleeve returns to normal size and shape relatively slowly, the explosion-generated gases are expelled, a new charge enters, and the system is ready for another cycle. The cycle time is about 8 s, which is adequate for most marine exploration.

Another source, the air gun (Figure 3.23a), uses compressed air that is released suddenly into the water to create a bubble and a sound wave. This action is controlled by a freely operating spindle with an axial hole that (1) allows compressed air from a suitable source to enter a chamber until equilibrium is reached; a subsidiary side tube is closed; (2) the subsidiary side tube which is open by a valve operated electrically, allows an excess force to operate on the under side of the spool, forcing it wide open, and allowing the compressed air from the sealed chamber to escape through ports into the sea; (3) when the air pressure in the chamber falls the spool returns to its former position and the whole system is ready for another cycle. This system, shown diagrammatically in Figure 3.23a, is extremely simple and minimally subject to breakdown.

The problem, of course, is that the bubble formation is not controlled and the seismic signal generated is not a single, discrete pulse. The frequency of oscillation is controlled largely by the chamber volume (Giles and Johnson (1973)). At least a partial answer to these oscillations lies in the observation that by varying the volume in a number of air-gun units, all of which can be triggered simultaneously, the first pulse from these units can be made to add together, whereas the later oscillations, at different frequencies, tend to cancel.

Finally, a different system called the water gun (Figure 3.23b) was developed by Societé pour le Development de la Researche Apliquée (Sodera), aided by Shell. This system, too, obtains its energy from compressed air. The operating principle, however, is entirely different, even though a free-moving shuttle—this time with no axial hole—is used. An upper chamber (the firing chamber) filled with compressed air is sealed by the upper piston of the shuttle, which is held in position by hydrostatic pressure. Water from the sea is allowed to fill the lower chamber.

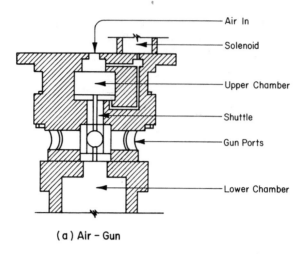

(a) Air – Gun

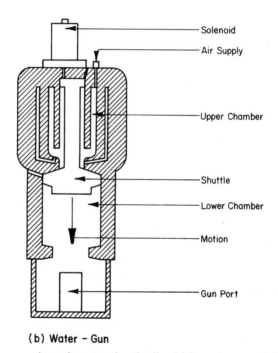

(b) Water – Gun

FIGURE 3.23 A comparison of construction details of (a) an air gun and (b) a water gun.

A solenoid, when energized, forces the free shuttle to move slightly downward; when it "feels" the full air pressure on its upper surface it shoots down. This causes water to be expelled from the lower chamber through side ports ("gun ports") and a number of cavities in the water result. Some energy from the water expulsion has been radiated but most is retained as potential energy of these bubbles, which now collapse in a single implosion. The full theory of this rather complex action has not been developed, but approximate theory and experiment have shown that the pulse transferred to the water (*near field pressure signature*) is quite different from that given by a single air gun. A full, readable account of two versions of this water gun, which includes near and far field pressure signatures, has been given by Safar (1984). Figure 3.24 is a diagrammatic comparison of near field pulses that are emitted from the air gun and water gun. Time periods are obviously dependent on the dimensions of the two instruments chosen. It is evident, however, that the single water gun does not produce the sequence of bubble pulses that emanates from an air gun. This fact becomes important in later data processing.

Although marine sources were originally used as fixed devices close to the boat, tests have shown that advantages can be gained by deploying them (air gun sets or water guns) in a limited array laterally with respect to a central towed cable. The lateral spread is sometimes provided by towing each source unit at different distances outboard from a long boom supported by the ship or by paravanes ("lifting bodies," according to the U.S. Navy) which are diverted from the wake of the ship once the ship is underway. By suitable adjustments of the paravane the amount of sideways diversion can be controlled.

It is also possible to create a rectangular pattern of sources by towing several sets behind each paravane. The objective, of course, is to create a directional effect. Careful consideration of the array directivity pattern is required:

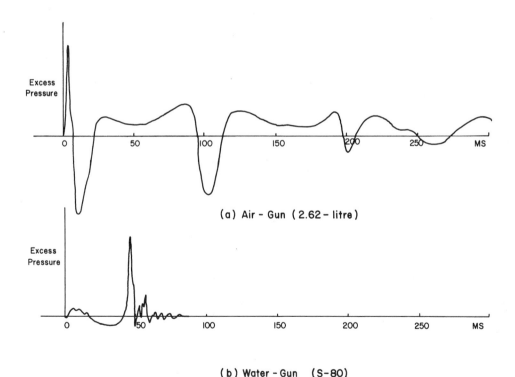

(a) Air - Gun (2.62 - litre)

(b) Water - Gun (S-80)

FIGURE 3.24 A comparison of near-field signatures of two marine seismic sources immersed in seawater at a normal towing depth. No attempt has been made to relate the excess pressure scales.

1. The wavelengths in water lie in the 150- to 22-m range.
2. The total spread of sources in the water is of the same order as the maximum wavelength.
3. The number of source units spread laterally is small (usually 2 to 4).
4. Over the area covered by a survey, the regional dip or the local dip, of strategic strata may vary considerably.
5. The course of the survey traverses may change with respect to the strike of the formations being mapped.

Most of the successful sources have been air, water, or gas guns for deep seismic prospecting and the sparker or Boomer for shallower, higher frequency work. Figure 3.25 is a comparison chart that relates the strength and other characteristics of these various sources.

Rayleigh showed (1917) that the period of spherical bubble oscillation is related to other parameters by the equation

$$T = 1.83 A_m \sqrt{\frac{\rho}{P_0}}$$

where T = period of bubble oscillation (s)
A_m = maximum radius of the bubble (cm)
ρ = specific gravity of the fluid (g/cm^3)
P_0 = ambient absolute hydrostatic pressure (dyn/cm^2)

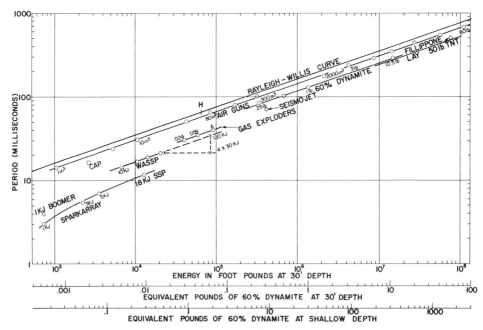

FIGURE 3.25 Rayleigh-Willis diagram. Comparison of different source strengths versus period of bubble oscillation. [*Source*: Seismic-Energy Sources, United Bendix (1968).]

Willis (1941) combined this relationship with the formula for the potential energy Q of a bubble of radius A_m:

$$Q = \tfrac{4}{3}\pi A_m^3 P_0$$

which yielded

$$t = 1.14\rho^{1/2}P_0^{-5/6}(KQ)^{1/3} \qquad (K = 1 \text{ when cgs units are used})$$

known as the Rayleigh-Willis formula. The diagram shows this relation on a log-log scale.

3.9 TYPES OF LAND RECEIVER

Although geophones of other types have been designed to be used on land, by far the greatest number in field use are of the moving-coil variety (see Figure 3.26). A radial magnetic field between a center pole piece and a permeable magnetic case passes through a cylindrical coil suspended within the field. Any motion of the coil in the magnetic field then generates a voltage in the coil, which is proportional to the velocity of the coil with respect to the field. The mass of the coil is supported by some type of spring, and the coil itself is usually constrained to move only vertically by the action of diaphragms. The latter are made from thin metal and are also used for electrical connections to the coil. They exert a negligible restoring force on the vertical movement of the coil. By selection of the masses of the coil form and wire and the sizes of the supporting spring different resonant frequencies can be obtained. Usually the coil and form are designed in such a way that when connected across the proper load resistor the geophone is nearly critically damped. Typical resonant frequencies in use now are 4.5 and

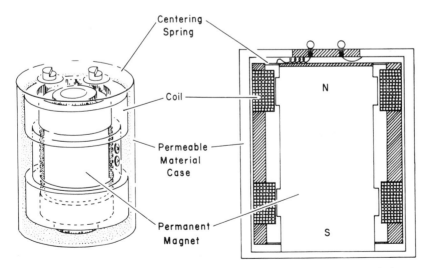

FIGURE 3.26 A schematic of a moving-coil geophone element (not to scale, electrical connections are diagrammatic).

7.5 Hz. The geophones are equipped with spikes to ensure proper planting in the·
earth—good coupling—and are usually arranged in multiple groups or strings.
Thus a typical unit may be a string of 15 geophones separated by 7 m (20 ft) to
give a string of length 105 m (345 ft). The geophones are connected in series or
parallel to match the input impedance of the amplifiers.

To a first approximation, the geophone moves with the surface of the earth. If
a seismic wave is perpendicularly incident on the surface the surface motion will
be double that of the wave amplitude because of the necessity of including the
effect of the reflected wave.

For this type of geophone the output voltage is proportional to the rate of
change in the flux through the coil or roughly proportional, over part of the
frequency range, to the velocity of the earth's surface. A mathematical analysis of
the motion of the geophone on the surface of an elastic earth has been given by
Wolf (1944). The geophone, however, is a mechanical–electrical transducer with
the mass, spring-coil system linked to the electrical parameters of the output
circuit through the magnetic field linkage with the coil. As a consequence the
geophone system, considered by itself, has a resonant frequency and damping,
and the output is not strictly proportional to geophone velocity except over a
limited bandwidth. The frequencies for which proportionality exists are those
above the resonant frequency. An additional complication due to phase distortion
occurs because at the resonant frequency (with 70% critical damping) the phase is
$\pi/2$ radians different from the phase at frequencies well above resonance. For
frequencies well below the resonant frequency there is a total phase shift of π
radians.

Like many other factors that affect applied seismology, the choice of
geophones is a compromise. Low-resonant-frequency geophones are not so
rugged as high-resonant-frequency units and are usually heavier and larger. For
most modern interpretations of seismic results it is desirable that equipment
effects, particularly distortion of the waveform, be minimized, and this means that
preferably the geophones should have a resonant frequency below the frequencies
of interest in the reflection band. To complicate matters still further the surface
waves are usually strongest at the lower end of the frequency spectrum and are
not required. To accept them in any geophone means that the geophones and
their associated amplifiers must have a wide dynamic range over which their
linearity (proportionality of output to input) is good. With modern amplifiers and
digital recording systems the tendency is to accept all types of wave without
filtering by geophones and then to remove the unwanted events in postrecording
processing. This is possible only if the system records all amplitudes linearly. A
detailed account of seismic instruments has been given by Anstey, Evenden, and
Stone in *Seismic Instruments* (1970).

The sensitivities of geophones (Wolf, 1942) must be such that a string of them
will produce a signal from earth motion that is substantially above the intrinsic
electrical noise level of the amplifiers. This establishes a minimum electrical
output voltage level for the minimum seismic signal motion desired. In practice,
the limiting sensitivity is not that of the Brownian movement of the mass.
Reflection signal levels at the lower limit become obscured by ground-noise
motion caused by wind and other types of microseismic noise. Motions of 10^{-8} cm
can be detected and voltages supplied to amplifier inputs are in the microvolt-to-
millivolt range.

In the best geophones sensitivity to motions other than those along the major axis of alignment has been reduced to a few percent. The high-frequency limit of moving-coil geophones is usually set by parasitic oscillations, whereas the low-frequency limit is a mechanical one of total allowable motion before harmonic distortion becomes unacceptable. In most field geophones the allowable motion (limited by mechanical stops) is of the order of 2 mm peak to peak, although stronger motion, moving-coil geophones are available for special purposes. Geophones based on other electromechanical principles such as the magnetostrictive or piezoelectric, have been made, but they are not used much for land work.

3.10 MARINE RECEIVERS

In marine seismic prospecting the requirements are entirely different and have been met by making up streamers of geophone elements. These elements are contained in a fluid-filled (kerosene or silicone liquid) flexible tube. They consist of piezoelectric (ceramic) devices in which the output voltage is proportional to the stress (hydrostatic excess pressure) caused by the seismic wave. It has been found convenient to use many elements in parallel to achieve a reasonable impedance match to the cable-input transformer system and, at the same time, some noise cancellation. The seismic signal has to pass through the plastic tubing and through the liquid filling the tube to act on the piezoelectric elements. For this reason materials are chosen so that the acoustic impedance of the material of the tubing and the fluid is close to the acoustic impedance of seawater. For other purposes special rubbers and silicone liquids which have desirable acoustic impedances have been manufactured, but for commercial seismic streamer cables plastics and kerosene are often the choice. A strain-member steel rope runs through the entire cable to reduce the possibility of breakage caused by the usual frictional and unusual (snagging) tensions developed when the streamer is pulled through the water.

The piezoelectric units are responsive to bending produced by acoustic pressure. Similar bending results from acceleration of the cable through the water and, to avoid noise due to this acceleration, modern units installed in marine cables are acceleration-canceling units. Two bender or diaphragm units are placed back to back and both respond positively to acoustic pressure, but one has a response to acceleration negative to that of the other. When connected in series the acceleration-induced noise voltages cancel, whereas the acoustic pressure voltages add. Because these ceramic (piezoelectric) elements are responsive to acoustically produced excess pressure, the streamer cannot be located close to the surface of the water.

For low-frequency operation it is usual to drag the cable through the water at a depth of approximately 13 m. Because the output characteristics of the cable are affected by its depth, special depth controllers (such as the pressure-controlled Condep® devices) are placed on the cable at intervals of a few hundred meters to maintain it at a constant preset depth. Because of the reflection of the seismic waves at the water surface, a pressure geophone at depth has a variable frequency characteristic due to the effective presence of a negative image at an equal distance above the water surface. Assuming that the sea surface is smooth, the received signal is

$$A\{\cos \omega t - \cos[\omega(t + \delta)]\} = -2A \sin\left[\omega\left(t - \frac{\delta}{2}\right)\right] \sin \frac{\delta\omega}{2}$$

for a constant frequency ω; δ is a delay due to a travel time of $2D/V$, D is the depth of the cable in the water, and V is the velocity of sound in water (1500 m/s approximately). The signal suffers a phase change dependent on the frequency (a received waveform is distorted). The amplitude is also a function of frequency. For low frequencies the output is *proportional* to the frequency. In deep water it is common to set the depth of operation of the cable at 12 to 13 m.

The change in output for both amplitude and phase with frequency is only one of the changes in characteristics of the received signal caused by source-receiver characteristics. The cumulative effect is a filter and its effect must be removed before the seismic trace can in any way be regarded as a near approximation of the signal caused by the layering characteristics of the earth.

To distinguish between sources (which are almost always used at full output) and receivers it is possible to use weighted receiver patterns. Because the marine environment is so consistent, these patterns can be used with greater fidelity to their calculated characteristics than in land operations.

As with all equipment and instruments used in reflection seismology, improvements are continually being made that will contribute to the reliability, safety, and, hopefully, increased fidelity of reproduction of the earth's layered system. In recent years much attention has been paid to cable positioning with respect to the towing ship not only in depth below the water surface but also in horizontal position at a number of points along the length of the cable. For this purpose compasses fitted with electrical transducers are positioned within the cable at regular intervals and their signals are carried to the ship's recording equipment along with the hydrophone group signals that register the incoming reflection wave pressures. In a later chapter the use of all of these data is discussed.

Although it is only of academic interest to the geophysicist, the communication of hydrophone (geophone) and other instrumental signals along cables is changing gradually to carrier modulation methods and to light signals carried down optical fibers. Each system has its own advantages and there is by no means unanimity in regard to the best method.

The main objective of cable design must be to reduce noise in the transducers (hydrophones) due to cable motion through the water or to accelerations of the cable caused by motions of the towing ship or the depth-keeping devices and the tail buoy.

3.11 MOTION OF A VERTICAL GEOPHONE AS A RESPONSE TO COMPRESSIONAL AND SHEAR WAVES INCIDENT ON THE SURFACE

For angles of incidence other than perpendicular the action of a vertical geophone on the surface of the earth is not so simple as that already discussed. Compressional waves, with which we are mainly concerned, give rise to compressional and shear waves on reflection. Figure 3.27 shows the geometrical considerations for incident compressional and shear (SV) waves (after Ewing, Jardetsky, and Press, 1957). A full discussion is provided in this reference (p. 24) and only the main results are given here.

If the velocities of the compressional and shear waves are given by α and β, respectively, and c is the phase velocity of a wave along the surface for compressional waves,

$$\sqrt{\frac{c^2}{\alpha^2} - 1} = \tan e$$

and, for shear waves,

$$\sqrt{\frac{c^2}{\beta^2} - 1} = \tan f$$

If Poisson's ratio $\sigma = \frac{1}{4}$, then $\alpha^2 = 3\beta^2$ and $\cos^2 e = 3 \cos^2 f$; under these conditions

$$\frac{A_2}{A_1} = \frac{4 \tan e \tan f - (1 + 3 \tan^2 e)^2}{4 \tan e \tan f + (1 + 3 \tan^2 e)^2}$$

$$\frac{B_2}{A_1} = \frac{-4 \tan e (1 + 3 \tan^2 e)}{4 \tan e \tan f + (1 + 3 \tan^2 e)^2}$$

It can be seen that the shear wave amplitude B_2 is zero only for cases of

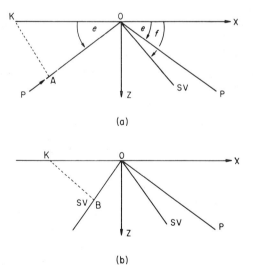

(a)

(b)

FIGURE 3.27 The reflection of P and SV waves at the free surface of an elastic solid. (*Source*: Ewing et al., *Elastic Waves in Layered Media*. Copyright © 1957, McGraw-Hill, Inc. Used with permission of the McGraw-Hill Book Company.)

grazing incidence of the P wave and for perpendicular incidence. For all other angles part of the P-wave energy is converted to SV-wave energy. A useful concept is the angle of emergence $\bar{e}$ which is defined as the angle given by

$$\bar{e} = \tan^{-1} \frac{A_v}{A_h}$$

where A_v is the vertical motion and A_h, the horizontal motion of the point O. It can also be shown that

$$\tan \bar{e} = \frac{1 + e \tan^2 e}{2 \tan f} = \frac{\tan^2 f - 1}{2 \tan f} = -\cot 2f$$

Thus, measured at the surface by the vertical and horizontal motions A_v and A_h, the compressional wave appears to arrive at an angle different from the true angle. To determine how much the effect can be for an extreme reflection seismology case we take a P wave arriving below the base of the weathered layer at an angle of incidence of 45°. Then, if $\alpha = 2000$ ft/s (610 m/s), $\beta = 1154$ ft/sec (352 m/s), and the P-wave velocity for the subweathering is 6000 ft/s (1828 m/s), it can be shown that, given $\sigma = \frac{1}{4}$,

$$e = 76.37°$$
$$f = 82.185°$$
$$\bar{e} = 74.37°$$

and the difference is only 2°.

The relations may be used for rare cases in which the angle of emergence at the surface is not nearly vertical.

3.12   THE USE OF ARRAYS OF GEOPHONES

When surface sources are used the large amplitudes of waves propagated along the surface or in the near surface region (waves we call interference to differentiate them from random noise signals) make it almost mandatory that arrays of detectors as well as arrays of sources be used. The amplitude of these interference signals must be reduced before the total received signal is amplified and recorded on magnetic tape. In the older analog tape recording (FM or AM) the range of signals that could be recorded linearly (i.e., retrievable with the same ratio of amplitudes and the same frequency spectrum as the original signal) was about 100 : 1 (40 dB). In present-day digital recording this range has been extended considerably, but the advisability of reducing the interference/reflection signal ratio still remains. Because of the cost of the individual sources, source patterns are occupied sequentially and the effect of the source array is realized only in later summation. It is therefore the task of the detector, or geophone, pattern to reduce the surface wave/reflection signal ratio before energy due to a single source is recorded. Geophones are much cheaper than sources and can be prewired in arrays or strings.

Once these strings of geophones are made up, however, they must be usable at any point in the reflection field setup because it would be much too difficult to find a string for a particular location among the hundreds available. We have already noted that the amplitude of surface waves decreases, at least as $r^{-1/2}$, where r is the distance from the source. The greatest part of this decrease occurs in the first few hundred meters from the source. Our interest springs from a need, when designing an array of geophones, to have equal contributions from each geophone or a knowledge of its likely contribution to the overall array response. Arrays are made up by equal contributions from all geophones and, if this is done, the array manufactured must be used at distances no less than several hundreds of meters from the source so that the wave amplitude over the entire string can be sensibly constant. For conventional P-wave recording in the 5- to 50-Hz frequency range a minimum offset of about 300 m (1000 ft) is acceptable.

In this discussion of receiver arrays the presence of a weathered layer is ignored, as, in fact, it was for sources. Later this omission is discussed in terms of the equivalent input signal for the exploration system.

The physical principles involved in geophone array design are the same as those for sources. The receivers respond to the same plane wave signal but receive it with a different phase, depending on the angle of incidence. The vector addition of signals was illustrated in Figure 3.21, and we note that unequally weighted contributions can be obtained. Equally weighted geophones in any array have the same directivity function as the same array of sources, if the question of the directivity function of the individual elements is disregarded. Thus Figure 3.20 describes the response of an 18-element receiver array.

Vertical geophones have a response that is approximately $\cos \theta$, where θ is again the angle of incidence of the seismic wave. Horizontal geophones used for SH recording, however, record the full amplitude of the wave independent of its angle of incidence. In the case of SH strings therefore only the array directivity is available for discrimination on a direction basis.

In the discussion that follows we pursue the question from a theoretical point of view, whether or not arrays of receivers that reject surface waves even better than the constant-amplitude arrays customarily used can be designed. Because we make use of Fourier transforms—introduced only briefly in Chapter 2—it would be advisable for the reader not versed in this technique to read the early part of Chapter 4, in which they are developed further. Reference to standard texts, such as R. Bracewell, *The Fourier Transform and Its Applications* (1965), will provide a more thorough treatment.

It is instructive to consider the concept (shown in Figure 3.28) of a continuous linear geophone system in which the response function is represented by

$$A(\omega) = \int_{-L/2}^{L/2} \sigma(l)e^{i\phi} \, dl$$

where $e^{i\phi}$ = delay factor

$$\phi = \omega l \sin \theta / V$$

and $\sigma(l)$ is the weight of an output element at the point l. Putting $c = (\sin \theta / V)(1/2\pi)$,

$$A(\omega) = \int_{-L/2}^{L/2} \sigma(l)e^{2\pi i \omega l c} \, dl$$

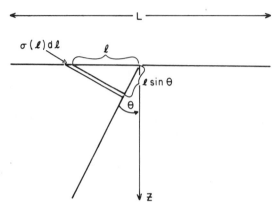

FIGURE 3.28   The concept of a continuous linear geophone system.

This is a Fourier integral, and procedures exist to determine $A(\omega)$ if $\sigma(l)$ is known or $\sigma(l)$ if $A(\omega)$ is known.

The $A(\omega)$ obtained gives the output of the continuous system (for a given angle θ) as the angular frequency (ω) is varied.

When designing field methods or using arrays, however, it is often more convenient to reject signals of a given frequency that approach the surface at a known set of angles. It can be seen that this amounts to the ability to reject certain apparent wavelengths (along the surface) or to reject the corresponding apparent wave numbers k_{app}. The subscript is omitted but understood in

$$\phi = \frac{\omega l \sin \theta}{v} = 2\pi k l$$

Hence

$$A(k) = \int_{-L/2}^{L/2} \sigma(l) e^{2\pi i k l}\, dl$$

and the Fourier inverse is

$$\sigma(l) = Q \int_{-\infty}^{\infty} A(k) e^{-2\pi i k l}\, dk$$

where $Q = $ constant.

The fact that the response of all the arrays discussed so far varies with frequency or wavelength, even within the acceptance band, is a source of concern. It is much better to have an array that does not disturb the form of the incoming pulse.

An interesting example of the Fourier transform procedure was suggested by Goupillaud (1955). When infinite Fourier transforms are used it is known that the Fourier transform of a bandpass in frequency (or k) is equivalent to a $(\sin kx)/kx$ pulse in time $(x = t)$ or length $(x = 1)$. This suggests that for a continuous geophone system to accept all k between $-k_c$ and $+k_c$ (cutoff wave numbers) the weighting function for the system should be of the form $(\sin ql)/ql$. In any finite system (as indicated in Figure 3.29) it is not possible to use the form $(\sin ql)/ql$ without truncation; hence the bandpass in k is not rectangular but approaches it, depending on the ratio of the total length L to the cutoff wavelength $\lambda_c = 2\pi/k_c$. It is noted here that this pattern or array involves some negative weighting. A further approximation is required because of the need for discrete numbers of geophones in the array.

Weighting schemes other than the constant or the $(\sin kx)/kx$ have been suggested (Holtzman, 1963), but all rely on the constancy of sampling achieved by the individual geophones. In field practice the number of possible recording channels has increased dramatically to the order of 1000. This makes possible the recording of individual geophones so that their outputs may be edited or weighted for recombination later. Earlier experiments with permanently weighted geophones in an array showed clearly the lack of consistency of the sampling, and schemes of weighting like one half-cycle of a sine wave yielded an expectation of no better performance in rejecting unwanted interference than a linear pattern only one-half the length. With the availability of extremely fast, reasonable-cost, off-line computers for input data examination and modification (e.g., change of amplitude or phase spectra under program control), the matter is now one that is

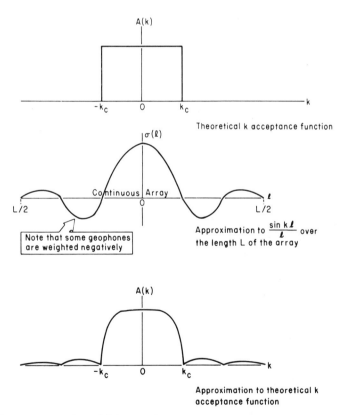

FIGURE 3.29　An attempt, by weighting geophone outputs, to develop an acceptance function that is constant over a given range of wave numbers and zero outside that range. Note the negative weighting of some geophones.

controlled by economics. Of course, it is not *necessary* to sum the individual source contributions in the field. If suitable facilities exist they can be examined and consolidated after controlled modification. A large volume of tape handling which has to be weighed against the advantages shown by experience is involved.

Because the marine environment is so constant, it is perfectly feasible to adopt weighting procedures for the recording array. There is an almost constant transfer impedance between the wave field in the water and the transducers, and the geometry of the transducers with respect to one another is fixed. But source distances change with boat speed, and some care (and extra information) would be necessary in combining individual source contributions to given an expectation of better results than simply adding the contributions.

3.13 NOISE, NOISE CANCELLATION, MULTIPLE GEOPHONES AND SOURCES, AND TWO-DIMENSIONAL ARRAYS

In this entire chapter so far the materials through which waves are to be propagated and the conditions of observation have been assumed to be perfect. Under these conditions the paths taken from the source to the receiver are those predicted geometrically, and the plane of propagation of the various waves generated and received is a vertical plane through the source and receiver. It was therefore legitimate to discuss only linear arrays that could be oriented on the surface to lie in this vertical plane. Furthermore, the conditions of observation were perfect, for the geophones picked up only those signals (wanted or unwanted) that were generated by the sources provided. The unwanted signal is classed as interference and is differentiated from noise that comes from extraneous sources beyond the control of the experimenter. Discontinuities in the materials (cracks, sudden lithological changes), particularly in the near-surface layers, are often so intense and random in nature that the seismic waves from the artificial sources used can become incoherent and can approach the geophones of the linear arrays in directions other than along the length of the array. Even if the interference were coherent the array would not have been designed to reject it. Furthermore, if it is coherent over the array the design of the array will have little to do with its rejection.

Noise, however, can be coherent over the dimensions of an array, and some cancellation can be achieved if the wave number (in the direction of the array) is within the reject band. This, however, is not certain and in general, fortuitous. Noise is generated by local and distant earthquakes, by artificial sources such as traffic, railroads, and aircraft, by the wind swaying trees, bushes, and grasses, and by the fluctuating pressures in the air caused by wind variation due to the topography (Frantti, 1963; Junger, 1964; Gupta, 1965).

In the marine environment interference often derives from scattering by objects, both natural and artificial, floating in the water. Submarines and whales and large fishes, alone or in schools, scatter sound waves because of the air trapped in their internal cavities. Other scatterers are ships and buoys on the surface, irregularities of the (hard) sea bottom, and gas bubbles seeping up from underground reservoirs. Interference also comes indirectly from the underside of waves; therefore the sea state is a major factor, as would be expected, in the success of marine surveys.

Sea waves also produce areally distributed pressure waves in the water, thus setting up incoherent noise.

Additional problems therefore must be solved in some areas:

1. Incoherent noise
2. Coherent energy coming from outside the source-receiver vertical plane

The first of these problems has the characteristic that, once the geophone of source samples lie outside a given distance, the noise samples are uncorrelated; that is, samples taken at different locations or at different times at the same location tend to add randomly and the average amplitude increases only as the square root of the number of samples taken. Signals that are coherent, however, add linearly with respect to the number of samples. Thus the technique for handling random noise is to use many geophones at separations greater than the coherence radius and to add them together. The coherence radius can be regarded as the average distance between points in the area for which the correlation coefficient between noise samples is 0.5. It would do no good to increase the number of geophones indefinitely, regardless of separation, because then the samples would be correlated and would tend to add as the signal does. The statistical characteristics of the noise (spatially and temporally) must therefore be known.

The technique of reducing noise must be linked further to signal strength because an increase in the signal strength of N is equivalent to adding together N^2 samples as far as the signal/noise ratio improvement is concerned. If the noise is spatially random but results from incoherent scattering of the signal f(in other words, it is interference) no good will come from increasing the source strength.

The second problem, that of coherent noise from points not in the source-receiver vertical plane, must be solved by using samples in the plane of the surface rather than in a single line in the plane. If there is a preferred direction from which the noise originates the two-dimensional pattern can be optimized to give the highest signal/noise ratio. The design of weighted two-dimensional arrays has been treated by Parr and Mayne (1955), although not from the point of view of two-dimensional Fourier analysis, which appears to be simpler now that computer methods are available. In exploration practice these arrays are usually radial or star patterns of seven units upward.

Much larger compound fixed patterns (Figure 3.30) (Capon et al., 1968) have been used in earthquake seismology to achieve higher angular resolution and noise cancellation. Modern digital-processing techniques allow beam steering and optimization of the noise cancellation performance (Capon, Greenfield, and Kolker, 1967; Burg, 1964).

In modern exploration practice two-dimensional patterns are rarely used because of the noise cancellation afforded by the common depth point (CDP) method to be described later. In particularly stubborn areas, however, they are still required. Rectangular patterns of several hundred meters on a side are more usual than star or radial patterns, and both are used only as a last resort because economic considerations weigh heavily against them. Two patterns are shown in Figure 3.31.

In some methods of exploration, with P or S waves, it becomes important to reject surface-wave interference, regardless of its azimuth of travel. This means

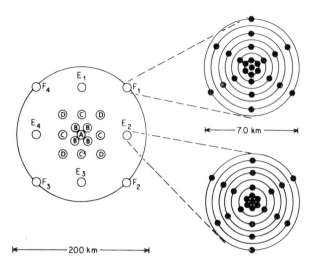

FIGURE 3.30 A complex compound array used in earthquake seismology at the Long Aperture Array, Montana. [*Source*: Capon et al. (1969), reprinted with permission from *Geophysics*.]

that a circular array must be used. For P waves a uniform density of vertical geophones over the area of the circle is used, whereas for shear waves each element must consist of a pair of orthogonal horizontal geophones. These elements would then be composited to yield a signal of the correct magnitude. The azimuth of travel of the wave could be determined if both horizontal outputs were recorded and processed.

An interesting and disappointing aspect of uniform circular arrays is their output given by

$$S(k, a) = \frac{3\pi}{2} \cdot \sigma a^2 \cdot \frac{J_1(ka)}{ka} = \frac{3\pi}{2} \cdot \sigma a^2 \cdot \frac{J_1(a/\lambda)}{a/\lambda}$$

where a = the radius of the circular array
σ = the output per unit area for a unit input
J_1 = a Bessel function of the first order and first kind
k = the horizontal wave number
λ = the horizontal wavelength

As ka increases the first zero of $J_1(k, a)$ is when $ka = 3.83$. Hence to cancel completely a horizontally traveling wave of wavelength λ the radius of the array must be 3.83λ. For a *linear* array the same cancellation can be achieved by a uniform array length $l = \lambda$.

Uniform, continuous coverage is, of course, not possible at this time, but, assuming that four samples are necessary in any wavelength to avoid aliasing, a 10-Hz ground roll with a phase velocity of 280 m/s would have a wavelength of 28 m and one geophone per 7.0 m would be needed. The circular array would have a radius of 107 m and approximately 740 geophones would be needed. The linear array would be 28 m in length and would require four geophones—a high price to pay for versatility.

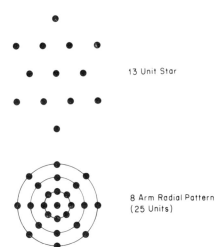

13 Unit Star

8 Arm Radial Pattern
(25 Units)

FIGURE 3.31 Different types of two-dimensional array used in exploration seismology.

3.14 HARMONIC DISTORTION

All vertical vibrators and, to a lesser extent, horizontal vibrators react with the earth in a nonlinear fashion when high-amplitude (usually low-frequency) signals are required. Seriff and Kim (1970) have discussed this phenomenon at length. Essentially, the harmonics of the low-frequency end of the sweep correlate with the proper fundamental frequency in the control signal and a ghost is produced. The time of arrival of this ghost can easily be determined. As an example, we take a sweep signal that passes from 10 to 45 Hz in 7 s. The rate of change in frequency is thus $35/7 = 5$ Hz/s. Now, if the second harmonic of 10 Hz is to correlate with 20 Hz in the control signal the delay will be $(20-10)/5 = 2$ s. Thus each event on the Vibroseis correlated record that has a 10-Hz component will have a distortion ghost *starting* 2 s away from the true event. The second harmonics are usually more prominent at the low-frequency end of the sweep but there is a range of frequencies that will be affected. Slightly higher frequencies (say, 12 Hz) would have a ghost at 12/5 or 2.4 s. this means that the ghost is dispersed in relation to the original sweep, and correlation against the control signal will show the ghosts as dispersed events instead of autocorrelation functions. Because the highest amplitude signals arriving at the receivers are generally first arrivals (in spite of precautions to the contrary), the ghosts will proclaim their presence by having the same time shift ("move-out") across the record as the first arrivals—ghosts from other events will be present but less prominent.

It is now up to the geophysicist to design the sweep in such a way that the main body of the seismic cross section will be free from ghosts. Theoretically, of course, this can be done by using upsweeps because the ghosts will then appear before zero record time. Early-phase compensation circuitry did not lock in easily on an upsweep, but modern circuit developments have removed this difficulty. Even so, many users of Vibroseis equipment still prefer downsweeps and alter the length of the swept frequency signal to permit the ghost to occur at a time greater than any reflections of interest. In the preceding example, if it had been known that all reflections of interest occurred before 4 s, the first ghost (that of the first arrivals) could have been made to appear at 4+ s simply by making the sweep 15 s

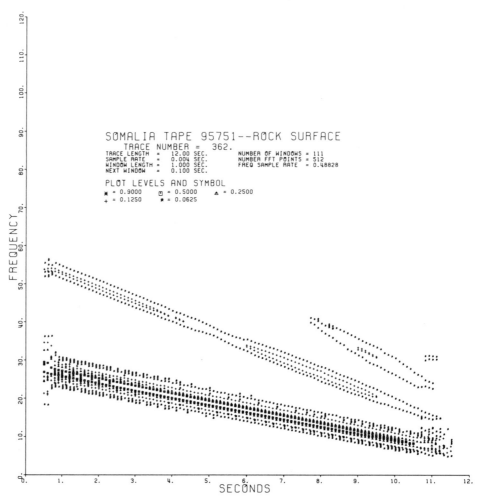

FIGURE 3.32   A computer analysis of the motion of the vibrator baseplate. Note that some double and triple frequency motion is present because of the nonlinear nature of the load.

long. The rate of change in frequency is now $35/15 = 2.333$ Hz/s and the harmonic ghost of 10 Hz occurs at $(20 - 10)/2.333 = 4.286$ s. With digital sweep generators any reasonable sweep length is easy to obtain.

B. J. Heath (personal communication) designed a digital program to examine and plot the spectral components within a short, sliding window so that a graphical picture of baseplate distortion is relatively easy to obtain. Figure 3.32 is an example. It is shown that in addition to a second harmonic a small amount of third harmonic is present. the symbols allow an estimate of the intensity to be made in 6-dB steps down from the maximum amplitude.

3.15   SPECIAL CONSIDERATION OF VIBROSEIS FIELD METHODS

The Vibroseis method, as normally used, is a surface-source technique; therefore it generates large-amplitude surface and near-surface waves. In contrast to impulsive sources, the long swept frequency signal ensures that the geophone array will receive surface waves at the same time it receives reflections.

One primary need in Vibroseis is a recording system that is linear over a wide amplitude range, although attempts are made in the field to minimize the amplitudes of the interfering waves. This is the primary function of the pattern of geophones because the source pattern is often occupied sequentially.

The geophysicist must decide if it is possible to sacrifice some amplitude of shallow reflections with respect to deep reflections; any linear pattern of geophones will to some extent reject energy from shallow reflections if it is made to reject near-surface interference. Figure 3.33 makes this clear.

We take as an example a frequency range of 20 to 60 Hz, a presumed pattern length of 400 ft, and the velocities and geometry of Figure 3.33; Figure 3.34 shows the positions of the acceptance curve for the receiver pattern occupied by the deep reflector, the shallow reflector, and the refractor. It turns out that the Rayleigh wave occurs near $X/\lambda = 10$ to 30, and that is very high rejection. The refractor coverage lies within the second and third lobes of the acceptance curve and, because these lobes are 180° out of phase, causes distortion of the refraction pulse. Note that with these parameters the shallow reflector occurs over the steep part of the first and second lobes (including the first null). It is therefore reduced in size, predominantly because of high-frequency rejection. It is obvious that

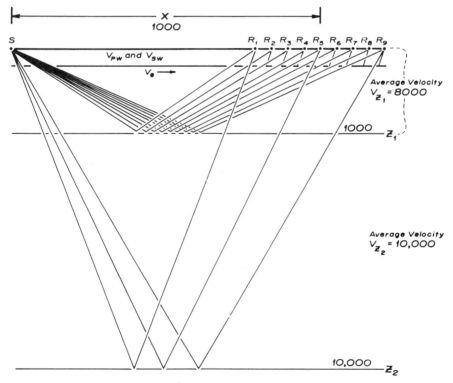

FIGURE 3.33 Geometrical considerations for designing the geophone (or source) array. Phase velocities for the various waves are (*a*) refractive, V_e; (*b*) Rayleigh waves, $V_R \approx 0.9\,V_S$; (*c*) shallow reflection $(2V_{z_1}/X)\sqrt{X^2/4 + Z_1^2}$; (*d*) deep reflection $(2V_{z_2}/X)\sqrt{X^2/4 + Z_2^2}$.

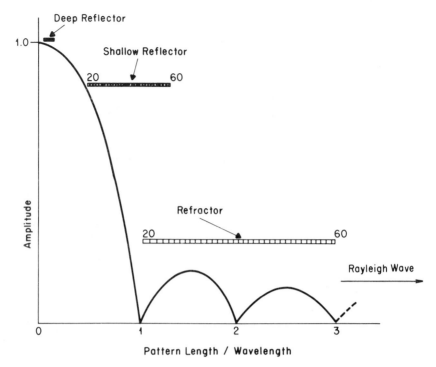

FIGURE 3.34 The relative positions on the array acceptance curve for a deep reflector, shallow reflector, and refracted events. The Rayleigh wave is off the diagram to the right ($X/\lambda = 10$ to 30). Beware of aliasing problems.

other compromises must be possible and each area has to be treated on the basis of its own characteristics.

Two further points can be made about this method of rejection by patterns. First, the foregoing simple explanation rests on the assumption of a continuous detector or a high sampling rate that affects the reflections very little but it can be important for waves, such as refraction and Rayleigh waves, that have small wavelengths. The cancellation shown by the curve is valid only if even the smallest wavelength is sampled with no less than two (preferably more) samples.

In a high-frequency (60-Hz) Rayleigh wave the wavelength is only 13.3 ft; therefore the geophones have to be no more than 6 ft apart. This requires a large number of geophones per pattern (approximately 70).

The second point to be made is that all the geophones in the pattern must accept the wave with equal amplitude. If the patterns are started too close to the source the near geophones will be subjected to much higher interference amplitudes than the far ones and the theoretical pattern effect will not be obtained. This is one reason why Vibroseis field work is often done with a relatively large (1000 ft) minimum offset.

Using a geophone pattern of this type the dynamic range of amplitudes dealt with by the recorder is reduced, and the interference is no longer intolerably larger than the reflection amplitudes during any given sweep injection. Of course, a similar pattern can be used sequentially for the sources or, in some cases, a different pattern that places zeros (nulls) where the geophone pattern has maximum acceptance of the interference can be devised. The proper choices are determined by experiment. The final output at the geophone pattern, after compositing of the results due to all the sweeps, is the product of the two separate acceptance functions and results in a high signal/interference ratio for the reflection work.

The third reason for a relatively large initial offset is to prevent too great a change in reflection amplitude with time. As the offset is increased the rate of change in travel time (hence spherical divergence) decreases. Inasmuch as the primary pulse for Vibroseis work is an autocorrelation function controlled by the reference signal, it is important that the side lobes of this autocorrelation pulse be maintained at a low level compared with other primary peaks occurring at the same time. Theoretically the rate of decay of the autocorrelation function can be made 100:1 or better in 0.1 s, but in practice the rate of change in gain needed for proper visibility of reflections should be kept small. In shot reflection work gain changes from 0.1 to several seconds can be of the order of 10^4, but in Vibroseis work it should not be necessary to increase the gain over the same interval by more than a factor of 20.

If Vibroseis reflection results fail to show an adequate signal/noise ratio, particularly for deeper reflections, the most effective way to increase the available energy is obviously to increase the number of vibrators. There is, however, an economic, and often a logistic, limit to this procedure. Because we are trying to improve the signal and reduce the random noise, an increase in the time of injection of energy is another possibility. The sweeps can be made longer and/or more of them used for each reflection source position. The signal increases linearly with the time of operation of the vibrators, whereas random noise amplitudes increase as the square root of the recording time.

As a consequence, a net gain equal to the square root of the time increase is realized. An extreme example has been described by Fowler and Waters (1975) to provide sufficient energy for investigation of deep crustal reflections. In this case 240 separate sweeps were composited to achieve a theoretical increase in the signal/noise ratio over one sweep of 15.5. In most circumstances this would be an uneconomical procedure, but smaller uses of the time increase are often possible in normal reflection practice.

3.16 DESIGN CRITERIA FOR FIELD WORK AND INITIAL FIELD EXPERIMENTATION

The versatility of the Vibroseis system in obtaining reflections of high or low frequency at great and shallow depths and with a diversity of ground surface conditions means that it is possible to tailor the operating conditions to suit the job. Obviously the exactness of fit depends on how much is known about the work to be done—chiefly about the surface and subsurface conditions in the area.

The depth of the primary reflection objective and some velocities are possibly the most important factors. Essentially, they determine the frequency range of the sweep, the maximum and minimum offsets, and the pattern sizes that can be allowed. The experimental work needed immediately is a determination of the surface-wave and refractor velocities that determine pattern lengths. This is usually done first with a single spread of bunched geophones and with source positions at multiples of the spread length (Figure 3.35a). This work should be done with a very wide frequency band because later filtering can be used to determine the frequency characteristics of the various events recorded. It is rarely the case that a reflection is seen clearly on this interference record. The refraction velocity for the top of the unweathered layer and any subsequent higher velocity refractors can be easily measured, however, and the time offset cross section shows clearly those zones that are free from high-amplitude surface waves. The latter are dispersive and the range of velocities can also be measured.

It is desirable, of course, to place the geophone spread at an offset distance at which the interference amplitude is low for the zone of the most important reflection times. On these initial interference records the proper placement may not be clear, but the decision is made easier by repeating the interference spread with different lengths of geophone patterns. Figure 3.35b is an example in the same area as Figure 3.35a and the improvement is clearly evident. It must be remembered that an equal amount of improvement can be expected on the final reflection records as a result of the effect of the source patterns.

In some particularly difficult areas usually associated with a hard rock surface, rough topography, or surface cracking the elimination of in-line interference may not be sufficient to show the reflections clearly. The interference spreads should then be repeated with geophone *nests* that have lateral coverage. In some cases lateral source coverage is necessary also to reduce scattering from inhomogeneities in the near-surface formations.

All these increments in field effort necessarily reduce the amount of effective work that can be done per day and therefore must be used with discretion. There is, nevertheless, no substitute for an adequate signal/interference ratio in the final reflection record. The examples given are illustrative only, and each area legitimately demands its own interference investigation.

Although it is always desirable to obtain reflections by using a bandwidth as wide as possible, there may be conditions that justify a small reduction in bandwidth if, by so doing, other goals are achieved. One goal may be to dispense with long arrays of sources and receivers to record reflections from both shallow and deep horizons. Evidently this reintroduces the specter of large Rayleigh waves. Sometimes, depending on near-surface conditions, this interference can be almost entirely eliminated by raising the lower sweep frequency limit. An interesting example is displayed in Figure 3.36. It refers to one of the gold mining areas of South Africa in which in the last two or three years Vibroseis surveys have been made to delineate structure in the ancient basins where gold occurs. On the left is a correlated field record in the 10- to 48-Hz band, whereas on the right is a record taken with a swept frequency band of 20- to 70-Hz. The '*swamping*' action of the ground roll from about 0.5 to 1.0 s on the lower frequency record has been almost completely eliminated by increasing the lower limit from 10 to 20 Hz. Reflection energy is easily visible at 0.8 to 0.9 s on the right-hand record. Much shallower reflections can be obtained once CDP records are obtained from combinations of these "100%" raw records. I am indebted to Mr. Emil Weder for providing this illustration.

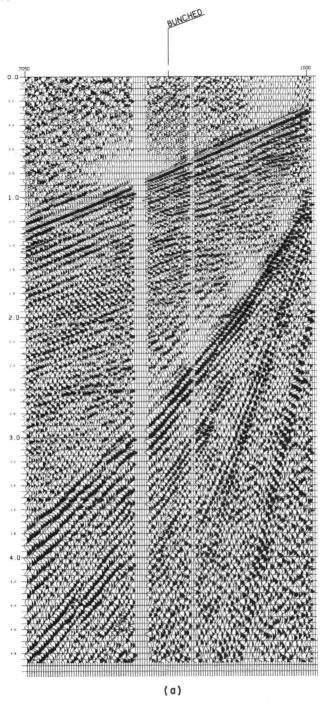

(a)

FIGURE 3.35 (*a*) An interference record that uses bunched geophones (no array).

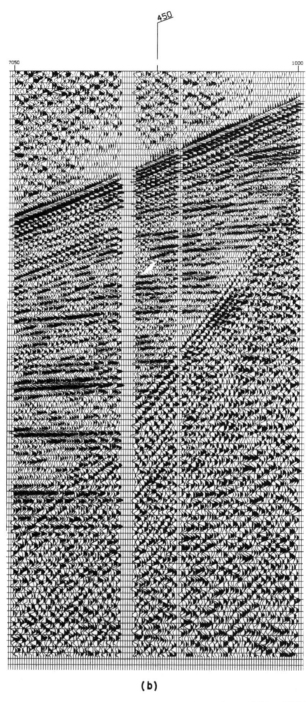

FIGURE 3.35 (*b*) The same interference when an array of geophones 137 m (450 ft) long was used.

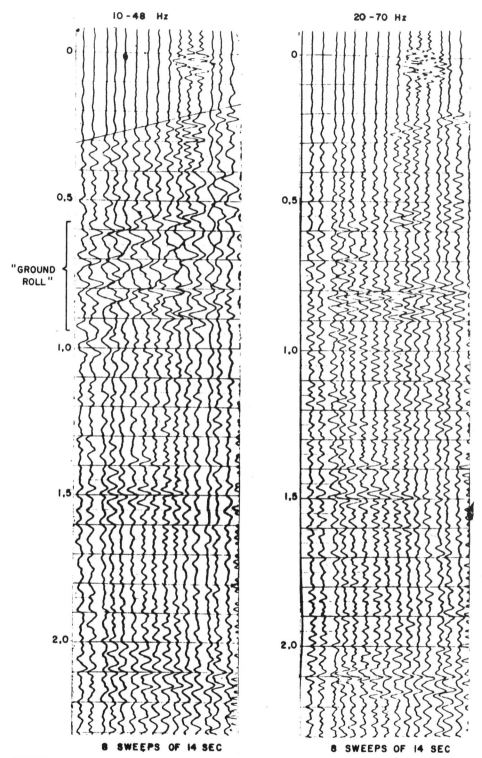

FIGURE 3.36 An example of the frequency sensitivity of ground roll (pseudo-Rayleigh waves) in some areas. On the left the bandwidth of 10–48 Hz allows strong '*ground roll*' from 0.5 to 1.0 s. This is removed by increasing the lower cutoff to 20 Hz. [*Sources*: The copyright in this matter is retained by Rand Mines (Mining & Services) Limited.]

3.17 PSEUDORANDOM SWEEPS AND THEIR USES

An allusion has been made to the possibility of using swept frequency control signals that are not linear increases in frequency with time. The total bandwidth of seismic signals is not great for most areas. In some cases, by design or otherwise, two field crews, or a composite field crew, may need to have two sets of vibrators working at the same time, yet have distinguishable results. This problem has been incompletely investigated. Random noise, filtered to the seismic bandwidth, has the major disadvantage that the vibrators are not used at maximum efficiency. This part of the problem can be solved by a method in which a standard linear sweep is decomposed into individual, numbered, positive, and negative half-cycles. These numbered half-cycles are then reassembled under the control of a random sequence generator, with the provision that positive and negative half-cycles alternate. It is, of course, possible to do this (for a sweep signal that may consist of 500 or more half-cycles) in an almost infinite number of ways; some will satisfy other considerations and others will not. It is necessary to filter the reassembled sweep with a zero-phase bandpass filter to eliminate the slope changes that inevitably occur when half-cycles are assembled in a random manner. A (trivial) set of examples is given in Figure 3.37.

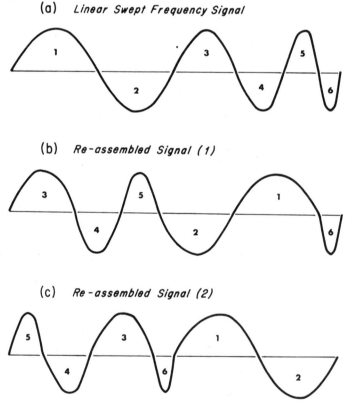

FIGURE 3.37 Pseudorandom sweeps (*b*) and (*c*) are derived from linear sweep (*a*). In the field the number of half-cycles can be two orders of magnitude greater than the six shown here.

In this case, however, every sweep has to be correlated with the geophone output before compositing and this adds to the difficulty with present field recording systems. The advent of new, extremely fast modules that can be interfaced with standard computers promises to make possible new recording systems that can correlate in the field and store only the correlated or composited data. Single-trace, real-time correlators already exist and have been used in the field for special applications.

An entirely different approach to control signals has been taken by Cunningham (1979). His proposed signal consists of a sequence of single sine-wave elements, $2^N - 1$ in all. The constant central frequency is f_c, chosen near the center of the exploration bandwidth required. Whether the sine-wave elements are positive or negative is determined by a random sequence, there being one more positive than negative numbers in the $2^N - 1$ sequence.

A small adjustment is then made to the amplitude of the negative elements compared with the positive elements. *Statistically*, the autocorrelation function of this class of signal for any given N can then be shown to be zero at times greater than $1/f_c$ from the central peak. A further interesting property is the low spectral amplitude at the central frequency (due to having only one more positive than negative cycles in a total of $2^N - 1$).

It is quite evident, however, that the *ensemble* or *average* properties of this class mean nothing unless this is the way they are used; that is, there must be a large number of members of a particular class, the field results, cross-correlated with the *particular* signal, and the results, averaged. This could, of course, be done by recording the field results and the control signal for every sweep by correlating each set and then averaging.

A number of signals have been generated, however, for $N = 5, 6$; *individually* they possess poor autocorrelation functions and do not die away within (say) 0.100 s to 1/100 of the peak value—a property easily attainable with tapered linear sweeps. Two problems also appear to exist with all sweeps of this type:

1. The existence of cusps in the control-signal time function will need a wide bandwidth in the response function of the vibrator. In practice this may not be a severe restriction. Field tests shown by Cunningham (1979) with sweeps in which $N = 7$, 8, or 9 are convincing that substantially the same reflection records are obtained as with long aperiodic sweeps.
2. The '*phase locking*' of the baseplate response to the control signal, although relatively simple with the aperiodic sweep, may be quite difficult with this new signal. The use of a vibrator without phase compensation can cause serious changes in time intervals on the correlated records, depending on the surface material on which the vibrator is working.

3.18 SUMMARY

Explosive sources of seismic energy have now largely been replaced by repeatable, consistent sources that operate on the surface of the earth or in the water. Different types are required for the two different environments. The newer sources are always of less strength than is true of the explosive sources but, because of their mobility, can be used with ease in multiple units, simultaneously or sequentially. The results of these separate unit sources can then be summed to give a result equivalent to that obtained from a single stronger source. In the presence of random noise the use of N separate sequential sources results in an increase in the signal/random noise ratio of $\sqrt{N}$. In a like manner signals from detectors with different noise environments also add like $\sqrt{N}$ as far as the noise is concerned.

Patterns of sources and detectors (geophones) can give directivity to source-created energy or equivalent directivity in the reception of the receivers. These pattern directivities are multiplicative, as is the directivity of the individual source (detector), with that of the pattern itself.

The calculated responses of patterns cannot in practice be fully realized, largely because of the inadequacy of the model used in the calculation. In land work transfer characteristics between individual transducers (sources or receivers) are not consistent enough for full array characteristics to be realized. In marine work, however, these transfer characteristics *are* consistent. In this case it is the effect of the water surface, due to waves, that causes a problem.

The Vibroseis system makes stringent demands on the form of the extended signal that is induced in the earth. Various modifications have been proposed for special purposes, but the best basic signal still remains that in which the instantaneous frequency changes linearly with time. The amplitude of this control signal is usually flat, except for some form of tapering near the ends, the objective being to prevent *ringing* at the two end frequencies.

All surface sources work on the earth, which must be regarded as a spring with some damping because of the radiation of energy and the nonlinear characteristics of the near-surface rocks. A damped resonant system is formed by the mass of the baseplate with the earth spring. For this reason the baseplate must be light and as near rigid as possible.

One of the most important features of the Vibroseis system is the degree to which it is possible to use feedback as a means of making the output of the system a facsimile of the input control trace by which phase changes due to surface material variation can be avoided. The practical consequence is that random-reflection time shifts are avoided.

In marine work the Vibroseis system, due to constant motion while the swept frequency signal is being generated and to complexities of operation at sea, has not been used for production work. It has been replaced largely by impulsive *guns* of various kinds that are designed to avoid the *bubble effect*. This is accomplished by using sets of units with different resonant frequencies or by making their symmetry so far from spherical that bubbles cannot persist.

Arrays of sources in marine work may be towed behind laterally extended booms attached rigidly to the ship, or use may be made of the capabilities of paravanes that swing outward from the ship's course.

REFERENCES

Anstey, N. A. (1964), "Correlation Techniques—A Review," *Geophysical Prospecting*, Vol. 12, No. 4, p. 355.

Anstey, N. A., Evenden, B. S., and Stone, D. R. (1970), *Seismic Instruments*, Vols. I and II. Gebrüder Borntraeger, Berlin, Germany.

Bracewell, R. (1965), *The Fourier Transform and Its Applications*, McGraw-Hill, New York.

Brown, G. L., and Moxley, S. D. (1964), *IEEE International Convention Record*, Vol. 12, Part 8.

Burg, J. P. (1964), "Three Dimensional Filtering with an Array of Seismometers," *Geophysics*, Vol. 29, pp. 693–713.

Bycroft, G. N. (1956), "Forced Vibrations of a Rigid Circular Plate on a Semi-infinite Elastic Space and on an Elastic Stratum," *Philosophical Transactions of the Royal Society of London, Ser. A*, Vol. 248, pp. 327–368.

Capon, J., Greenfield, R. J., and Kolker, R. J. (1967), "Multidimensional Maximum-Likelihood Processing of a Large Aperture Seismic Array," *Proceedings of the IEEE*, Vol. 55, pp. 192–211.

Capon, J., Greenfield, R. J., Kolker, R. J., and Lacross, R. T. (1968), "Short Period Signal Processing Results for the Large Aperture Seismic Array," *Geophysics*, Vol. 33, p. 452.

Cole, R. H. (1948), *Underwater Explosions*, Princeton University Press, Princeton, N.J.

Crawford, J. M., Doty, W. E. N. D., and Lee, M. R. (1960), "Continuous Signal Seismograph," *Geophysics*, Vol. 25, No. 1, pp. 95–105.

Cunningham, A. B. (1979), "Some Alternate Vibrator Signals," *Geophysics*, Vol. 44, pp. 1901–1921.

Dix, C. H. (1955), "The Mechanism of Generation of Long Waves from Explosions," *Geophysics*, Vol. 20, No. 1, pp. 87–108.

Edelmann, H. (1966), "New Filtering Methods with 'Vibroseis,'" *Geophysical Prospecting*, Vol. 14, No. 49, pp. 455–469.

Ewing, W. M., Jardetsky, W. S., and Press, F. (1957), *Elastic Waves in Layered Media*, McGraw-Hill, New York.

Farrell, W. E. (1979), "Linear Analysis of the Interaction between a Small Electromagnetic Vibrator and an Elastic Half-space," Personal communication.

Fowler, J. C., and Waters, K. H. (1975), "Deep Crustal Reflection Recording Using 'Vibroseis' Methods—A Feasibility Study," *Geophysics*, Vol. 40, No. 3, pp. 399–410.

Frantti, G. E. (1963), "The Nature of High Frequency Earth Noise Spectra," *Geophysics*, Vol. 28, p. 547.

Giles, B. F., and Johnson, R. C. (1973), "System Approach to Air-gun Array Design," *Geophysical Prospecting*, Vol. 21, pp. 77–101.

Goupillaud, P. L. (1954, 1955, 1956), personal communications.

Gupta, I. J. (1965), "Standing Wave Phenomena in Short Period Seismic Noise," *Geophysics*, Vol. 30, p. 1179.

Holtzman, M. (1963), "Chebyshev Optimized Geophone Arrays," *Geophysics*, Vol. 28, p. 145.

Junger, A. (1964), "Signal to Noise Ratio and Record Quality," *Geophysics* Vol. 29, p. 922.

Kisslinger, C. (1963), "The Generation of the Primary Seismic Signal by a Contained Explosion," Vesiac State of the Art Report, Universtiy of Michigan, AD 403708.

Krey, T. (1969), "Remarks on the Signal to Noise Ratio in the Vibroseis System," *Geophysical Prospecting*, Vol. 17, No. 3, pp. 206–218.

Luco, J. E. (1974), "Impedance Functions for a Rigid Foundation on a Layered Medium," *Nuclear Engineering and Design*, Vol. 31, pp. 204–217.

Luco, J. E. (1976), "Vibrations of a Rigid Disc on a Layered Visco-elastic Medium," *Nuclear Engineering and Design*, Vol. 36, pp. 325–340.

Lysmer, J. (1965), "Vertical Motion of Rigid Footings," Ph.D. Thesis, University of Michigan.

Martner, S. T., and Silverman, D. (1962), "Broomstick Distributed Charge," *Geophysics*, Vol. 27, pp. 1007–1015.

Merzner, P. (1965), "Abstracts of E.A.E.G. Meeting," *Geophysical Prospecting*, Vol. 13, No. 1, p. 140.

Miller, G. F., and Pursey, H. (1954), "The Field and Radiation Impedance of Mechanical Radiators on the Free Surface of a Semi-Infinite Isotropic Solid," *Proceedings of the Royal Society, A*, Vol. 223, p. 521.

Miller, G. F., and Pursey, H. (1956), "On the Partition of Energy between Elastic Waves in a Semi-Infinite Solid," *Proceedings of the Royal Society*, A, Vol. 225, p. 55.

Molotova, L. V. (1963), "Velocity Ratio of Longitudinal and Transverse Wave in Terrigenous Rocks," *Bulletin (Izvestiya) of the Academy of Science of the U.S.S.R.*, Geophys. Series, No. 12, p. 1769.

Muir, F., and Morrison, J. P. (1973), "Anti-Aliasing of Spatial Frequencies by Geophone and Source Placement," U.S. Patent No. 3,719,924.

Parr, J. O., and Mayne, W. H. (1955), "A New Method of Pattern Shooting," *Geophysics*, Vol. 20, p. 539.

Poulter, T. C. (1950), "The Poulter Seismic Method of Geophysical Exploration," *Geophysics*, Vol. 15, p. 181.

Safar, M. H. (1976), "The Radiation of Acoustic Waves from an Air-gun," *Geophysical Prospecting*, Vol. 24, pp. 756–772.

Safar, M. H. (1976), "Efficient Design of Air-gun Arrays," *Geophysical Prospecting*, Vol. 24, pp. 773–787.

Safar, M. H. (1984), "On the S80 and P400 Water guns: A Performance Comparison," *First Break*, Vol. 2, pp. 20–24.

Seriff, A. J., and Kim, W. H. (1970), "The Effect of Harmonic Distortion in the Use of Vibrators," *Geophysics*, Vol. 35, No. 3, p. 234.

Sharpe, J. A. (1942a), "The Production of Elastic Waves by Explosion Pressures (I)." *Geophysics*, Vol. 7, p. 144.

Sharpe, J. A. (1942b), "The Production of Elastic Waves by Explosion Pressures (II)," *Geophysics*, Vol. 7, p. 311.

Wolf, A. (1942), "The Limiting Sensitivity of Seismic Detectors," *Geophysics*, Vol. 7, p. 115.

Wolf, A. (1944), "The Equation of Motion of a Geophone on the Surface of an Elastic Body," *Geophysics*, Vol. 9, pp. 29–35.

Ziolkowski, A. (1970), "A Method for Calculating the Output Pressure Waveform from an Air-gun," *Geophys. J.R. Astr. Soc.*, Vol. 21, pp. 137–161.

Ziolkowski, A. (1971), "Design of a Marine Seismic Reflection Profiling System Using Air-gun as a Sound Source," *Geophys. J.R. Astr. Soc.*, Vol. 23, pp. 499–530.

FOUR

Description of Wave Trains and the Characteristics of the Reflection Process

4.1 TIME AND FREQUENCY DOMAIN CONCEPTS

The concept of time of arrival of an event and the delay time after the instant of initiation are familiar ones related to our everyday experience that one thing happens after another. It is orthodox to plot these arrivals as a graph of their amplitude, particle velocity, and pressure as a function of time in which the time line is horizontal (the independent variable) and the event parameter is the dependent variable. Thus Figure 4.1, the time domain description of a series of reflected seismic wave amplitudes arriving at a geophone, is a conventional, graphic means of illustration.

From the point of view of any extended analysis (apart from a simple time difference measurement made between two selected points) a time domain description is sometimes inconvenient. An electrical or communications engineer, for example, needs to know how best to design amplifiers, filters, and other processing equipment. Geophysicists are interested in the effects on interpretation of geology due to filtering of the signals—by passage through the layered earth and by instrumentation used in creating and receiving the seismic waves. Fortunately, as is well known in electrical engineering, there are other ways of describing time series. The best known is obtained by Fourier analysis—a concept that was introduced in Chapter 2 and shown to be feasible even for time series with a limited number of discontinuities. The name for this procedure is *frequency domain* description. As described earlier, it consists of building up a series of sinusoidal signals of known amplitudes and phases, the frequencies being related to a fundamental frequency in a harmonic manner; that is, the frequencies used for the sinusoids are integral multiples of the fundamental frequency.

The first step is to decide the length of time (a finite time, for the present) over which the description is needed. If this is T the fundamental frequency f_0 is established as $1/T$. In seismic signals, for example, we can be satisfied to take a seismic trace (time series) 6.0 s long simply because any reflections that take longer than 6 s to arrive come from greater than current drilling depths. The fundamental frequency would be $1/6$ Hz. The Fourier theorem says that it is possible to describe any single-valued time series (with only a finite number of discontinuities) within the interval T by adding up a series of sinusoidal signals which have frequencies $f_0, 2f_0, 3f_0, \ldots, Nf_0$, and a constant. The degree of

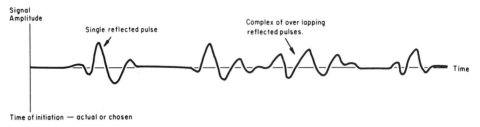

FIGURE. 4.1 A time-domain illustration of a series of reflected pulses arriving at a receiver.

approximation to the time function increases as N increases. The amplitude and phase associated with each of these sinusoidal signals must be determined by Fourier analysis, which is described shortly.

Even at this stage we must note that the choice of a time interval (T) describes the fundamental frequency (f_0) and the interval between the sinusoidal frequencies. Although there *could* be an infinite number of constituent sinusoids, it is likely that at some stage the closeness of fit would be exact enough and the set of constituents, terminated.

In the reverse process of Fourier synthesis a number of constituents must be added together—obviously an impossible task unless corresponding points are specified in the time domain in which the additions are to take place. This leads to the idea of *sampling* the previously continuous time series at a set of points equally spaced in the period T that has been chosen.

Alteration of the amplitudes and phases of the constituent frequency sinusoids (on Fourier synthesis) produces a different time series which can be regarded as a fundamental definition of "filtering." To filter a time series it is necessary to use Fourier analysis to gain knowledge of the amplitude and phase of the harmonic frequencies, alter them in a manner prescribed by the filter action, and then use Fourier synthesis to go back to the filtered trace.

In mathematical terms

$$S(t) = a_0 + \sum_1^N a_j \cos 2\pi jf_0 t + \sum_1^N b_j \sin 2\pi jf_0 t \qquad (4.1)$$

is a statement of the Fourier synthesis procedure in which it is assumed that the coefficients $a_0, a_1, a_2, \ldots, a_N$ and $b_1, b_2, \ldots, b_N$, as well as f_0, are known.

If $F(t)$ and f_0 are known, however, the coefficients $a_0, a_1, a_2, \ldots, a_N$ and $b_1, b_2, \ldots, b_N$ can be found from the formulas

$$a_0 = \frac{1}{T} \int_0^T F(t) \, dt$$

$$a_j = \frac{1}{T} \int_0^T F(t) \cos 2\pi jf_0 t \, dt \qquad (4.2)$$

$$b_j = \frac{1}{T} \int_0^T F(t) \sin 2\pi jf_0 t \, dt$$

It is noted that these processes really describe the calculation of a correlation coefficient between the known $F(t)$ and each of the cosine and sine functions that correspond to the individual harmonics.

Each harmonic is independent of another harmonic, and this is the power of the frequency domain description because in a filtering operation the various frequencies can be dealt with separately.

If we form the complex number C_j

$$C_j = a_j + ib_j \qquad (\text{where } i = \sqrt{-1}) \tag{4.3}$$

these relations become

$$a_j + ib_j = C_j = \frac{1}{T}\int_0^T F(t)e^{ijf_0t}\, dt \tag{4.4}$$

Because

$$\cos 2\pi jf_0(t + T) = \cos 2\pi jf_0 t$$

and

$$\sin 2\pi jf_0(t + T) = \sin 2\pi jf_0 t \qquad (\text{for all values of } j)$$

we can see that Fourier analysis and synthesis really describe a signal that is periodic and repeats itself for every T interval.

One way of looking at this is that the trace being examined (or synthesized) is in the form of a circle of circumference T and that, as t increases, the process moves continually around the circle. There is a reference point on the circle at which the beginning of the time trace joins on to the end. This may result in a discontinuity, but the value of the trace at the point is $\frac{1}{2}[F(0) + F(T)]$. See Figure 4.2.

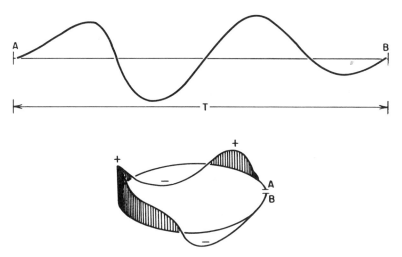

FIGURE. 4.2 A diagram that shows that the seismic trace AB can be joined, head to tail, to give a circular trace of circumference T. Filtering operations can cause some energy to cross the join AB.

To describe the signal $S(t)$ in the frequency domain involves two different graphs in which amplitude is plotted against discrete values of frequency (separated by f_0). One graph is for a_j, the other for b_j (sometimes called the real and imaginary parts, respectively). (See Figure 4.3a.)

We can also define quantities called the amplitude

$$A_j = \sqrt{a_j^2 + b_j^2} \tag{4.5}$$

and the phase

$$\phi_j = \tan^{-1} \frac{b_j}{a_j} \tag{4.6}$$

and by plotting them, the amplitude and phase spectra, we have an equivalent representation (Figure 4.3b). In both illustrations $N = 34$ and the time interval T

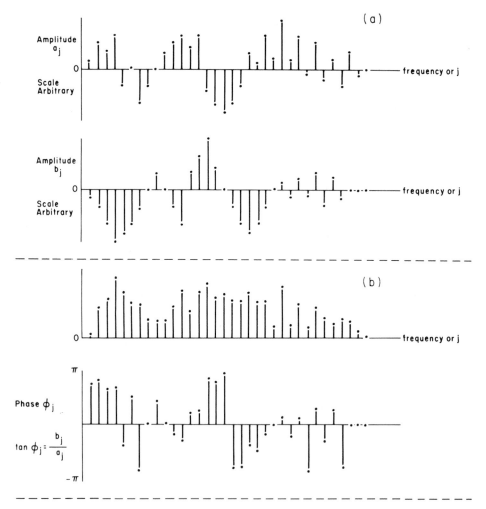

FIGURE 4.3   A representation of a time function over the interval $0 < t < 1/f_0$: (a) real and imaginary components for each frequency; (b) amplitude and phase spectra.

must be divided into $2N$ intervals. The $2N + 1$ points to be obtained are equidistant samples throughout the time series. This is in agreement with the fact that N cosine coefficients, N sine coefficients, and one constant are to be determined.

It is obvious that dividing the time interval t into $2N$ intervals can lead to inadequate sampling because much more sudden (high-frequency) variations may be present in the trace than can be adequately sampled by a given sampling rate.

4.2 SAMPLING CONSIDERATIONS

We wish to consider how often equidistant samples must be taken to describe a given time function adequately. We have already seen that it is possible to describe a function in the frequency domain as a series of sinusoidal functions of different amplitudes and phases and that these amplitudes and phases are independent of one another as we go from one frequency to the next. It seems intuitively obvious that the gradual variations in a function can be described by widely spaced samples, whereas rapid variations require high-frequency sampling.

Figure 4.4 shows the same frequency sinusoidal wave sampled with a sampling

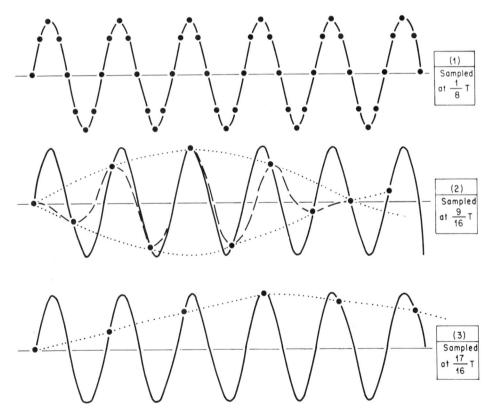

FIGURE 4.4 How insufficient sampling of a signal (less than two per cycle) introduces new, or aliased, frequencies that were not present in the original continuous signal.

interval much less than one-half the period, again with the sampling interval greater than one-half the period, and finally when the sampling is greater than the period of the sinusoid. It is shown that artificial low frequencies are created by inadequate sampling, a process known as *aliasing*, which is dangerous in the computation of Fourier spectra. Hence a cardinal rule is that there must be at least two sample points per cycle for the highest frequencies present in the signal to be analyzed—not the highest frequencies of *interest* but the highest frequencies actually present. Any frequency in the range $(0 < f < f_c)$ is aliased with

$$(2f_c \pm f), \qquad (4f_c \pm f), \ldots, (2nf_c \pm f)$$

This can be shown quite easily because the digital sample points of a cosine wave of frequency f, digitized at an interval of $1/2f_c$ are

$$\cos(2\pi f \cdot m/2f_c) = \cos \frac{m\pi f}{f_c}$$

But $\cos(2\pi(2nf_c \pm f)t$ sampled at the same time intervals will have values

$$\cos 2\pi(2nf_c \pm f)m/2f_c = \cos(2nm\pi \pm m\pi f/f_c)$$

$$= \cos\left(\pm \frac{m\pi f}{f_c}\right)$$

because the addition of an integral number of 2π to the cosine argument does not change its value.

These extra aliased amplitudes beyond f_c are reflected back into the amplitude spectrum ("folded") about the value of f_c and make the affected values incorrect.

The same process actually occurs in a sampling of the waves traveling on the earth's surface, where two geophones *at least* are needed for the shortest wavelengths present. The proof is left to the reader.

Actually two points per cycle is the absolute minimum sampling rate; for example, these points may occur at nodes or, with the same probability, at other *constant value* points on the waveform. For practical purposes four sample points per cycle should be used for all frequencies of interest and at least two for the highest frequencies present. This must be remembered when sampling rapidly varying logs. In seismic recording practice electrical antialiasing filters are provided before the sampling (or digitization) step to ensure that frequencies above an established maximum are severely attenuated by analog filtering. As an example, suppose that it has been established that digital samples are needed every 0.004 s; this establishes that the highest frequency that can be sampled adequately is 125 Hz (period 0.008 s). Analog filters are then included in the circuit which rejects frequencies above (say) $62\frac{1}{2}$ Hz severely. The frequency that corresponds to the two samples per cycle is called the Nyquist or folding frequency.

This phenomenon is discussed at greater length when the general subject of data processing is taken up. For the present we use it only to examine its effect on Fourier analysis and synthesis.

In going back to the Fourier series analysis of (4.2), it is seen that we have selected N separate frequencies, separated by f_0. Actually, $2N + 1$ separate amplitudes must be computed and this means that $2N + 1$ pieces of independent data must be supplied. It is for this reason that the period over which the Fourier analysis is to take place is divided into $2N$ intervals, given rise to $2N + 1$ sample points.

Because only a finite number of points has been specified, it is not possible to compute uniquely the integrals given in (4.2). They must now be replaced by summation formulas:

$$a_j = \frac{1}{N} \sum_{k=0}^{2N} S_k \cos \frac{2\pi jk}{N}$$

$$b_j = \frac{1}{N} \sum_{k=0}^{2N} S_k \sin \frac{2\pi jk}{N}$$

(4.7)

These formulas hold for all j except 0 and N, when they become

$$a_0 = \frac{1}{2N} \sum_{k=0}^{2N} S_k$$

$$a_N = \frac{1}{2N} \sum_{k=0}^{2N} S_k \cos \pi k$$

(4.8)

$$b_N = \frac{1}{2N} \sum_{k=0}^{2N} S_k \sin \pi k$$

4.3 FOURIER INTEGRALS—CONTINUOUS-FREQUENCY DISTRIBUTIONS

Although the Fourier series represents the only practical way to analyze a time function, it has been shown that by restricting the length of the function to be represented a tacit assumption is made that the function is periodic, with the chosen time interval as the fundamental period. In practice, if we require the transformation of a pulse in the time domain into its corresponding frequency domain spectra, there is little problem because points with zero amplitudes can be added before and/or after the pulse duration; the interval between frequency components f_0 can then be made as small as required. So also can the highest frequency be made as high as required by taking the sampling frequency high enough.

There is, however, a formal extension of the Fourier series representation to the continuous Fourier integral representation in which the time and frequency domain functions are piecewise continuous functions of the time or frequency. These integral forms are often used, because the mathematical manipulations are easier to perform. Stratton (1941) gives a readable account of this extension. In the complex notation normally used this transform pair is

$$f(t) = \frac{1}{\sqrt{2\pi}} \int_{-\infty}^{\infty} g(\omega) e^{+i\omega t} d\omega$$

$$g(\omega) = \frac{1}{\sqrt{2\pi}} \int_{-\infty}^{\infty} f(t) e^{-i\omega t} dt$$

(4.9)

4.4   THE CONCEPT OF AN IMPULSE—OR THE DELTA FUNCTION

It is useful, on many occasions, to determine first what information can be gathered from the seismic method if no constraints are placed on the frequencies that can be transmitted into the earth and received by the geophones. The limitations on this knowledge, due to frequency restrictions of whatever form, can be determined as desired.

For this purpose the concept of an impulse is useful. It is a curious function because it has no amplitude except at one point in time. It can be approached through several more familiar concepts.

1. It can be regarded as the limit of a rectangular pulse in time in which the function has a constant amplitude over a range of time, if the product of the amplitude and the time duration are kept constant as the time duration shrinks to an infinitesimal value. See Figure 4.5a. In mathematical terms (for impulse at time $t = \tau$)

$$\int_{-\infty}^{\infty} \delta(t - \tau)\, dt = 1$$

$$\delta(t - \tau) = 0 \qquad (t \neq \tau)$$
$$\delta(t - \tau) = \infty \qquad (t = \tau)$$

(4.10)

2. It can be regarded as the limit of the function

$$f(t) = \frac{e^{-(t-\tau)^2/h^2}}{h\sqrt{2\pi}} \qquad \text{(the error function)}$$

(4.11)

as $h \to 0$. See Figure 4.5b.

3. It can be looked at as the pulse that has infinitesimally small but *constant amplitudes of all frequencies*. The location of the impulse is controlled by the slope of the phase function. Under these conditions

$$g(\omega) = \frac{1}{\sqrt{2\pi}} \int_{-\infty}^{\infty} f(t) e^{-i\omega t}\, dt$$

(4.12)

and

$$f(t) = \frac{1}{\sqrt{2\pi}} \int_{-\omega_0}^{\omega_0} g(\omega) e^{i\omega t}\, d\omega$$

(4.13)

Because $g(\omega) = A$ constant and $f(t)$ has to be real,

$$f(t) = \frac{A}{\sqrt{2\pi}} \operatorname{Re} \int_{-\omega_0}^{\omega_0} e^{i\omega t}\, d\omega = \frac{A}{\sqrt{2\pi}} \operatorname{Re}\left[\frac{e^{i\omega t}}{it}\right]_{-\omega_0}^{\omega_0}$$

$$= A\, \frac{2}{\sqrt{\pi}}\, \frac{\sin \omega_0 t}{t}$$

(4.14)

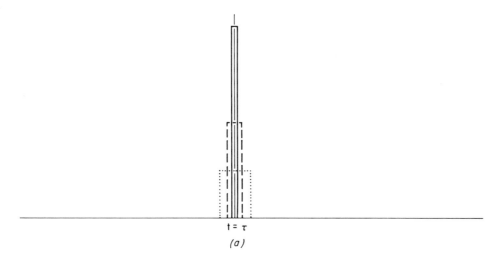

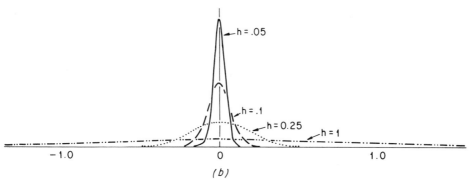

FIGURE 4.5 Approximations to the delta function: (*a*) successive approximations of rectangular pulses with a constant area but successively decreasing width; (*b*) successive approximations based on the error functions as the parameter *h* approaches zero.

This is a $(\sin x)/x$ type function (Figure 4.6) that, as ω_0 approaches infinity, gradually approaches an impulse at zero time and is near zero elsewhere. Thus we can regard the delta function $\delta(t)$ as one that has equal amounts of all frequencies present. This is a useful property because the delta function can be used to find out (for example) the characteristics of the reflection process of the earth—the impulse response—if the input pulse contains all frequencies in equal proportions. Deviations from this assumption can be corrected later by filtering; that is, the adjustment of the amplitudes and phase of all frequencies to conform to what is known about the input pulse.

The duality in description of a time function (i.e., in the time or frequency domain) appears throughout this book. Sometimes it is convenient to use one representation and at other times, the other, but we switch between the two, using Fourier analysis or synthesis without further question.

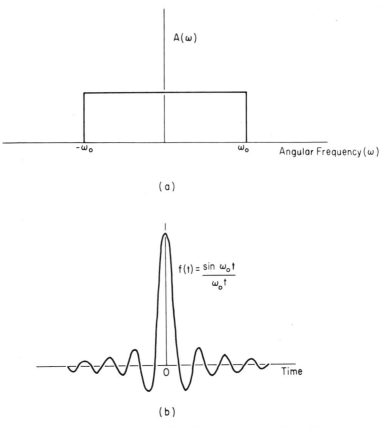

FIGURE 4.6 An approximation to the delta function that shows a pulse with a constant amplitude-spectrum up to the cutoff frequency ω_0.

4.5 THE z-TRANSFORM

It is essential to realize that the time series that results from a Fourier synthesis consists of a series of δ-functions, equally spaced but with differing amplitudes. Suppose that the amplitudes of the time series, taken at intervals Δ from some time zero, are given by

$$S \equiv \{a_0, a_1, a_2, \ldots, a_{n-1}, a_n\}$$

and that we define a quantity $z = \epsilon^{-i\omega\Delta}$ we can see that z is a shift operator because it acts on a δ-function at time zero to make a δ-function at time Δ.

$$z\,\delta(t) = z \int_{-\infty}^{\infty} \epsilon^{i\omega t}\,dt = \int_{-\infty}^{\infty} \epsilon^{i\omega(t-\Delta)}\,dt$$

$$= \delta(t - \Delta)$$

Thus the trace S, as defined here, can now be written

$$S \equiv a_0 z^0 + a_1 z^1 + a_2 z^2 + \cdots + a_{n-1} z^{n-1} + a_n z^n$$

At first it appears that all we have done is to change the symbolism, but, in fact, the transformation to a polynomial in z makes possible an easy method of filtering. This stems from the fact that these z-transforms have an identity similar to the frequency domain and to perform filtering we simply multiply the two z-transforms concerned.

Obviously, if z represents a delay then $1/z$ or z^{-1} represents an advance in time. Accordingly, a symmetrical pulse is represented by an expression of the form

$$\cdots + a_2 z^{-2} + a_1 z^{-1} + a_0 z^0 + a_1 z^1 + a_2 z^2 + \cdots$$

In the remainder of the book examples are given to demonstrate the versatility and ease of use of the z-transform. In Section 4.6 we compare and contrast three different ways of performing a filter operation; the use of the z-transform is one of them.

4.6 GENERAL FILTERING

Filtering is one of the most often used operations in the processing of seismic data. We deal first with a single sampled seismic trace—the time series picked up by a single receiver unit or a composite made from several other original traces.

If it were possible to introduce energy into the earth in the form of a single impulse (or δ-function) the reflections would arrive as delayed impulses due to the various travel times involved. This would be true whether the reflections were multiple reflections or primaries. This sequence of impulses is called the *impulse response* of the earth. If some pulse other than a δ-function had been used we could invoke the requirement of linearity of the system to replace each impulse in the impulse response with the sampled new pulse. This operation can be done in one of three different ways. The first, which has been described in outline, is to Fourier-analyze the impulse response, modify the coefficients of the Fourier series to reflect the fact that the delta function has been replaced by a pulse whose Fourier analysis is also known and then perform a Fourier synthesis on the result. Symbolically, we can describe the entire process by

$$F^{-1}[P(\omega)G(\omega)] \qquad\qquad (4.15)$$

where

$$G(\omega) = F[I(t)] \qquad\qquad (4.16)$$

and

$$I(t) = \text{impulse response}$$

$$P(\omega) = \text{known spectra composition of the desired pulse}$$

Note that the spectra of the impulse response and desired pulse are multiplied together.

There is another way in the time domain. Let the desired pulse $P'(t)$ be sampled at the same sampling interval as the impulse response. Physically, then, the filtering is done (as shown in Figure 4.7) by replacing each delta function with the properly scaled pulse and adding up the result of the overlapping pulses at each sample time. The values of the resultant signal samples $S(t)$ are then

$$S_1 = a_1 p_1$$

$$S_2 = a_1 p_2 + a_2 p_1 \qquad\qquad S_i = \sum_{j=1}^{i} a_j p_{i-j+1} \qquad (4.17)$$

$$S_3 = a_1 p_3 + a_2 p_2 + a_3 p_1$$

This process is known as *convolution*, and symbolically we write

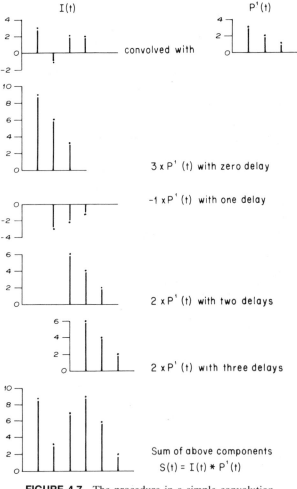

FIGURE 4.7 The procedure in a simple convolution.

$$S(t) = I(t) * P'(t) \qquad (4.18)$$

A third way is by means of the z-transform. The z-transforms of the pulse and impulse response are obtained from the sample values. The two z-transforms are then multiplied by using the frequency domain characteristic of the z-transform:

$$I(z) = a_0 z^0 + a_1 z^1 + a_2 z^2 + \cdots + a_n z^n$$

$$P'(z) = p_0 z^0 + p_1 z^1 + p_2 z^2 + \cdots + p_m z^m$$

$$S(z) = I(z) \cdot P'(z) \qquad S_i = \sum_{j=0}^{i} a_j P_{i-j}$$

If the example shown in Figure 4.7 is treated in this manner

$$
\begin{array}{ll}
I(z) = 3 - z + 2z^2 + 2z^3 \\
P'(z) = 3 + 2z + z^2 \\
\hline
3 \times I(z) & 9 - 3z + 6z^2 + 6z^3 \\
2z \times I(z) & + 6z - 2z^2 + 4z^3 + 4z^4 \\
z^2 \times I(z) & + 3z^2 - z^3 + 2z^4 + 2z^5 \\
\hline
& 9 + 3z + 7z^2 + 9z^3 + 6z^4 + 2z^5
\end{array}
$$

Note that the scale and shift operations done in convolution are implicit in this z-transform multiplication.

These three methods of linear filtering are equivalent. Convolution in the time domain obtains the same result as transform multiplication, which is equivalent to multiplication in the frequency domain.

4.7 ALTERNATIVE WAYS OF DESCRIBING RECORD SECTIONS

In the last few sections several different ways of representing a seismic trace have been described. Except in abstraction, these methods deal with a set of numbers derived from amplitudes of a single seismic trace of equal intervals of time. Time. (or, rather, delay time) is the only independent variable.

In seismic prospecting, however, the usual output is a set of seismic traces that in effect is derived by performing the same experiment at a sequence of locations along the line of tranverse. It is devoutly hoped that these experiments *are consistent* because, without this assurance, experimental perturbations may (later) be identified as geologic variations. Much care and expense is devoted to removing as many inconsistencies as possible from this plurality of experiments. Thus a new unit of information must be dealt with—the seismic cross section. Ideally, the seismic cross section consists of a series of (sampled) seismic traces taken at equidistant points along a line on the earth's surface. The value of the amplitude of the jth seismic trace at a sample time of $(i \, \Delta t)$ may be labeled $A(i \, \Delta t, j \, \Delta x)$ (where Δt and Δx are known and understood), and we can regard the entire set of points as a rectangular matrix of impressive dimensions. A 1500×800 matrix would not be unusual. If this set of numbers is displayed as a function of x and t the result can be interpreted (visually) to show a (sometimes distorted) picture of the manner in which the earth's geologic makeup varies with distance along the depth below the seismic traverse.

Clearly, there may be some demands to display the information in a different way. The rectangular matrix may be regarded as a set of columns or rows. Both columns and rows consist of a large number of equally spaced amplitudes. It has already been explained how each *column* (the seismic trace) can be represented by an alternative (the complex spectrum provided by Fourier analysis). In that representation the amplitudes and phases of a number of constituent harmonic frequencies are obtained.

There is, of course, an equal probability that the *rows* might be treated in the same manner. A spectrum in the spatial frequency domain can be obtained (the corresponding parameter to frequency is wave number).

A more common transformation, however, is the dual transformation from $F(x, t)$ to $G(k, \omega)$—the two-dimensional Fourier transform analysis. The inverse transformation from $G(k, \omega)$ to $F(x, t)$ also exists. It turns out that the two-dimensional transform may be done by a combination of two successive one-dimensional transforms:

1. Each column is transformed from t to

$$F(x, t) \rightleftharpoons F'(x, \omega)$$

2. The *rows* of $F'(x, \omega)$ are analyzed to transform x to k:

$$F'(x, \omega) \rightleftharpoons G(k, \omega)$$

Each of these transforms is reversible, as indicated by the arrows. Without going into details at this stage, filtering of the cross section can take place by *multiplying* the complex amplitudes of two displays in the temporal and spatial frequency domain, point by corresponding point. Naturally, a 2-D Fourier synthesis is required before the filtered results can be compared with the original seismic section.

Sampling considerations apply doubly to cross sections because there is a need to be concerned about the adequacy of sampling in *both* the x and t directions. Obviously two folding or Nyquist frequencies are involved (for which there are only two sample points per period (in the vertical direction) or the wavelength (horizontally)).

Because two dimensions are involved, it is possible to select matrix values such that i and j form a special relationship. Because it has been shown that

$$\frac{\omega}{k} = \frac{2\pi f}{2\pi/\lambda} = f\lambda = c \quad \text{(the velocity)}$$

we can plot a straight line through the coordinate origin which is a particular velocity on the (k, ω) representation. This facility allows division of the (k, ω)

plot into regions in which the velocity is greater than, or less than, a particular value. As we show in Chapter 6, a special filtering procedure can be devised.

Another transform available for special purposes is called the p-τ-transform. In exploration seismology it has been given the name '*slant stack*' because in reality it is the plane wave decomposition of the original cross section. It can be regarded as a set of '*linear events*' whose slope is p and whose intercept (on the t-axis) is τ. The transformation, which is capable of being inverted, accomplishes a number of desirable functions:

1. It removes the problems of travel-time triplications; that is, reflections on the same record as refractions from the same stratum.
2. The transformation is an objective procedure, independent of the interpreter's difficulties.
3. Reflections which in the (x, t) domain are hyperbolas become ellipses in the (τ, p) domain.
4. Refraction events, linear in the (x, t) domain, become points in the (τ, p) domain.
5. Multiple reflection times are exact multiples of the primary time.

In view of these features it is possible to construct new filtering methods. These features are examined in more detail in Chapter 6.

4.8 SYNTHETIC SEISMIC RECORDS FROM WELL LOG CHARACTERISTICS

Until recent years, the knowledge that sound waves could be generated, reflected from geological discontinuities or strata, and recorded at the surface by suitable instruments was all that was required to sustain a very active seismic exploration program. The reflections (although their characteristics had not been related in any way to the lithological sequence) were near enough alike that they could be correlated by eye, and the relative time of occurrence at sequential subsurface points could be determined. After suitable corrections, which are detailed later, a structural map could often be made. This was the era of the search for *structural* hydrocarbon traps, and this era has by no means ended. In more intensively investigated areas, however, this approach is rapidly becoming less fruitful.

It was nevertheless inevitable that some notice would be given to the form or character of seismic recordings, particularly after it became possible to make continuous measurements of seismic, compressional wave velocity and the density of formations with instruments lowered into boreholes. Well logging can provide these continuously measured parameters in a form similar to those shown in Figure 4.8. These logs have a ragged, or high-frequency, character, and it is evident that the problem of sampling them is a difficult one if aliasing is to be avoided. For the moment it is supposed that any digitization is done with sufficiently close sampling that aliasing will cause no problem.

When it was recognized that reflection character could be used to give additional information about the earth, data processing—an entirely new field—

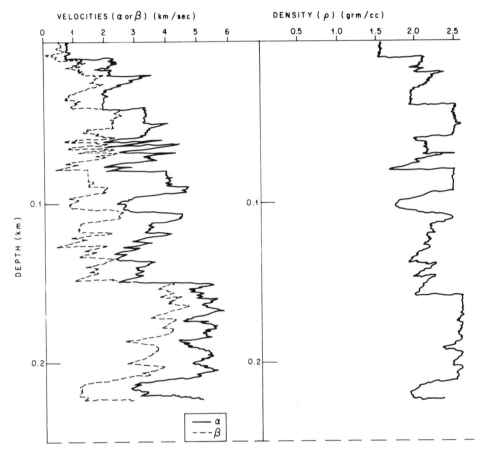

FIGURE 4.8 Log characteristics needed as input for reflectivity determination. At the present time only the P-wave interval velocity (α) and the density (ρ) can be obtained in continuous form.

was developed to maintain consistency in the treatment of field data and to organize this treatment to remove some of the distortions caused by known processes.

It is our purpose to examine the ways in which seismic reflection sequences are formed and to learn how to match their characteristics with the known parameters of the earth's layering.

Although other approximations are possible, the simplest representation of the system of rocks beneath the earth's surface is a series of plane, parallel layers, each having its own characteristic—but constant within the layer—parameters of velocity and density. Compressional and shear velocities should be used to characterize the rock, but at the present time shear velocities are not measured accurately. Because we are talking about compressional waves, the lack of shear parameter information is not important, but if the shear velocities had been available the treatment of shear reflection output would have paralleled this development.

The approximation we are talking about is shown in Figure 4.9. Berryman, Goupillaud, and Waters (1958) considered a more complex case in which the

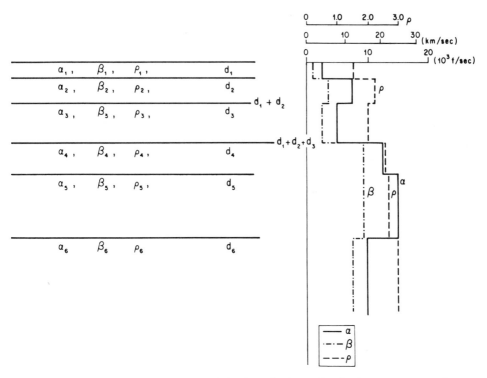

FIGURE 4.9 The layered system assumed for investigations of reflection response, α, P-wave velocity; β, shearwave velocity; ρ, density.

velocity and/or density can be a linear function of depth within each layer. Our treatment assumes that they are constants.

There are two different ways of tackling the problem of the reflection response. One in the frequency domain has been described by Berryman, Goupillaud, and Waters (1958). The expressions for up and down propagating waves of constant frequency in each layer are written down, and the necessary boundary conditions are fitted to all the amplitudes in relation to one another. These amplitudes are functions of frequency, and it is necessary later to use Fourier synthesis on the amplitudes in the uppermost layer to obtain the reflection output in the time domain. In Appendix 4A the details of the method are given, and complex velocities are used to take attenuation into account.

The other method in the time domain is more revealing of the effects of multiple reflection and transmission losses and the relation of the reflection character to the form of the velocity and density logs. The examination in the time domain is facilitated by arranging the layer thicknesses in integral multiples of the sample interval. If more detail is required the fundamental sample interval is made smaller.

The geological sequence of rocks in which the vertical dimension is travel time and all layers have a constant "time thickness" can then be drawn. Quantizing the time is a computational feature and not a physical requirement. It is paralleled by the need (in the frequency domain) to compute the complex wave amplitudes for a large but finite number of frequencies when the frequency domain method is used. This, too, is an approximation demanded by computational needs.

To complete the initial picture we consider only those waves that propagate vertically, perpendicular to the layering, although for clarity the various ray paths are drawn, as in Figure 4.10, at nonnormal incidence. Because the rays are really perpendicular to the boundaries, the diagrams do not show any difference in angle for reflected and refracted rays.

A pioneering attempt to derive synthetic seismograms from a sequence of velocities, as determined by well logging, was made by Peterson, Fillipone, and Coker (1955). If this log is continuous we can consider two points infinitesimally separated by δt for with the velocities are V and $V + \delta V$. When density differences are neglected the reflection coefficient

$$\delta R = \frac{V + \delta V - V}{V + \delta V + V} = \frac{\delta V}{2V} = \frac{1}{2}\,\delta(\ln V) \qquad (4.19)$$

neglecting second-order quantities.

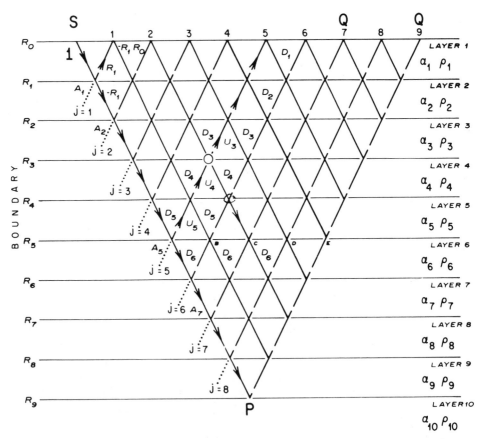

FIGURE 4.10 All possible ray paths in a constant travel-time layered model are shown. Any path that starts from S, travels continually to the right, and returns to the surface is a possible path. The output at points $1, 2, \ldots, n$ on the surface is the sum of all constant $(2j \cdot \Delta t)$ paths that emerge at the jth position.

Peterson et al. neglected multiple reflections and transmission losses and derived a primary reflection trace simply by converting the velocity log into a reflection coefficient log and using a nonlinear logarithmic transformation. This was followed by a simple filtering scheme that essentially placed a particular-shaped pulse (which could be designated) at each reflection coefficient at the correct amplitude. This was linear filtering accomplished electrically. In spite of its success in pointing out a basic relationship the idea was not physically sound. Nevertheless it was a useful approximation from which further developments have come. The model described here is physically correct and takes into account the following facts:

1. If an interface between two rocks causes a reflection for a wave traveling in one direction it must give rise to a reflection for a wave traveling in the opposite direction. The reflection coefficients are opposite in sign but of the same magnitude.
2. In crossing a boundary, a transmitted wave has an amplitude different from that of an incident wave. If the down-going reflection coefficient is R the amplitude of the downward transmitted wave will be $1 - R$. As a corollary, the upward reflection coefficient will be $-R$ and the amplitude ratio of the transmitted to the incident upgoing wave will be $1 - (-R) = 1 + R$.

The transmission (both ways) through a boundary thus gives rise to multiplication of the original amplitude by $1 - R^2$, which is always less than unity. This model has been studied in a more formal, mathematical manner by Wuenschel (1960), and his paper forms a useful adjunct to the description given here. It is not necessary, however, to have an exact physical picture of the process within the confines of the model.

In our system the input waveform seems to be neglected because we use only amplitudes, but there is an implicit assumption that the input is really a delta function. Because the reflection wave shape is the same as the incident wave shape when reflecting from a single boundary, it is sufficient to record amplitudes as a function of a series of integers; the times of arrival are always even integral multiples of the one-way travel time through the layer Δt.

After the log has been properly sampled over the interval logged additional problems develop because of the lack of information in the near surface and the need to ensure that the travel times to certain depths agree with any measured times available. These measured times may be taken from actual check shots made at the time of the well survey or they may be based on statistically derived velocities from reflection data. An estimate must be made of the thickness and velocity applicable to the weathered layer. A smooth velocity-depth function is then derived which (1) agrees with the probable subweathering velocity; (2) blends smoothly into the known interval velocities near the top of the logged interval; and (3) integrates to give the correct arrival time of a wave, starting from the base of the weathering and ending at a known checkpoint. There are no standard prescribed ways of doing this, but a velocity that is a quadratic function of depth is usually assumed and the coefficients are adjusted to fit the known information.

With this accomplished it is then possible to adjust to the known travel times by multiplying the velocity by the time interval and adding these values until the check-shot depth is bracketed. An interpolation is made to fit the exact depth and the calculated time is compared with the measured time. An adjustment can then be made to all velocities to achieve exact time agreement.

The same process is continued from one check shot to the next deeper one until all the check-shot times are satisfied. It is necessary to be critical of the check-shot times themselves and to edit those that yield unlikely interval velocity averages within the check-shot interval. When all this is finished the velocities are tabulated for each layer to form part of the basis for the synthetic record calculation. Density values are usually entered as read; however, care must still be taken to avoid aliasing and they must be given at the proper time values instead of the depth values as originally read.

The computational procedure is the following:

1. The sequential velocity and density values α_i and ρ_i are stored in a computer.
2. The sequential reflection coefficients $R_1, R_2, \ldots, R_N$ are calculated from the velocities and densities:

$$R_i = \frac{\alpha_{i+1}\rho_{i+1} - \alpha_i\rho_i}{\alpha_{i+1}\rho_{i+1} + \alpha_i\rho_i} \tag{4.20}$$

The reflection coefficient R_0 for the air–earth interface must be assigned. Although this value should be close to unity for perfect earth materials, the attenuation for elastic waves in the weathering can be taken into account approximately by reducing the value of R_0. If this is not done experience shows that the synthetic record takes on a ringy character; that is, it contains long trains of waves of near-constant frequency because of reverberations in the weathered layer.

3. With reference to Figure 4.10, the most general situation is found at, say, point O, where both downgoing and upgoing incident waves and upgoing and downgoing reflected or transmitted waves are present. For an upward, right-trending wave path the index of arrival at the surface is j (in this specific case $j = 5$). The time of arrival is j times the two-way time through any layer. For a given boundary the index is i [between the ith and the $(i+1)$th layers]. Recursion equations can be written as follows:

$$U_{ij} = D_{ij}R_i + U_{(i+1)j}(1 + R_i)$$

$$D_{(i+1)j} = D_{ij}(1 - R_i) - U_{(i+1)j}R_i \tag{4.21}$$

$$D_{(i+1)j} = D_{(i+1)(j+1)}$$

4. To start off, the condition at point A_1 is considered, and here $D_{11} = 1$ (the initial pulse amplitude) and $U_{21} = 0$ (which would correspond to a wave generated before zero time):

$$U_{11} = R_1 \cdot 1 + 0 = R_1$$

$$D_{21} = (1 - R_1) \cdot 1 - 0 = (1 - R_1)$$

Any value of U_{1j} calculated is stored in a location that is representative of its arrival time (in terms of $2\Delta t$). Strictly speaking, it should be necessary to multiply U_{1j} by $1 + R_0$ to arrive at the particle velocity observed at the surface, but this is a constant factor and for most purposes it can be omitted when playing out the reflection amplitude arrivals U_{1j}. It is necessary, however, to calculate and store in an appropriate location $D_{21} = -R_1 R_0$.

5. The process now continues by moving to A_2 and making iterative calculations for all points on the right uptrending diagonal path ($j = 2$). Note that all values of $D_{(i+1)j}$ will be needed for calculations when j is incremented to $j + 1$ but all values of U_{ij} may use that same storage as they replace one another as calculated.

6. We continue to calculate the output by incrementing the time and continuing the same basic procedure. By this method any wave that starts at S and finishes at a surface point Q_j gives rise to an output Q_j which includes all possible reflections and takes into account any change in amplitude due to transmission through the boundaries. The output consists of a sequence of numbers that represent the amplitudes of the signal arriving at integral multiples of $2\Delta t$. We can identify the signal by

$$\{a_1, a_2, a_3, \ldots, a_j, \ldots, a_N\}$$

where $2\Delta t$ delay between each pulse is implied. If the input to the geological section had been a delta function the output from the layered section would be a sequence of delta functions called the impulse response.

The final step from the unreal concept of transmission of an infinite range of frequencies to the real world of a restricted bandwidth now remains. In this chapter we have not tried to deal with attenuation in the earth in which the pulses that have traveled the farthest are restricted the most in bandwidth (high-frequency loss). Rather, at this stage a simple filtering process or convolution is used.

4.9 CHARACTERISTICS OF THE REFLECTION PROCESS

We have described a means by which well-logged parameters can be converted into synthetic, or expected, seismograms and can now discuss some experience obtained from wholesale use of this process and a comparison with actual seismograms taken in the immediate vicinity of the logged well.

First, however, it should be noted that it is possible to produce synthetic seismograms, without multiples, by essentially the same process as that used by Peterson, Fillipone, and Coker (1955). The digital computer program is arranged in such a way that if the option "without multiples" is desired the work of computing the impulse response of the layered system is bypassed and the desired result is achieved simply by filtering the sequence of reflection coefficients

$$\{R_1, R_2, R_3, R_4, \ldots, R_j, \ldots, R_N\} \tag{4.22}$$

by convolution with the desired pulse.

If transmission coefficients, but not intrabed multiples, are desired the appropriate sequence to be convolved with the filter pulse is

$$\left\{ R_1, R_2(1 - R_1^2), R_3(1 - R_1^2)(1 - R_2^2), \ldots, R_j \prod_{k=1}^{j-1} (1 - R_k^2), \ldots \right\} \quad (4.23)$$

where $\prod_{k=1}^{j-1} (1 - R_k^2)$ stands for a continued product of all the individual transmission factors $1 - R_k^2$ for the allowable values of k.

Some discussion of the desired pulse is necessary because it changes, depending on the type of source used for the field records with which the synthetic record is compared. There are two basic types:

1. The impulsive type
2. Vibroseis® pulses

These pulses are fundamentally different, inasmuch as the impulse type gives rise to an indication of motion only after the cause; that is, after an explosion or similar event has taken place. This basic waveform is difficult, if not impossible, to determine except by direct measurement. It must take into account the response characteristics of the recording instruments as well as the source characteristics. The latter are particularly difficult to determine on land because they can vary with the source placement, the receiver location, and the characteristics of the near-surface layers. Measurements of marine sources have been made in deep water where the hydrophones cannot record anything but the direct pulse. These pulses are shown in Figure 4.11.

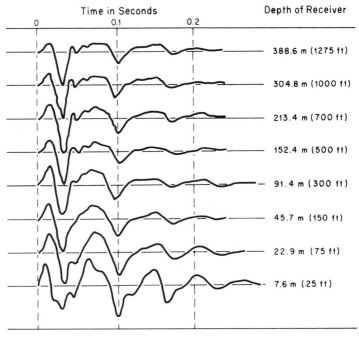

FIGURE 4.11 Deep-water tests of a pulse for a combination of four Aquapulse guns 12.2 m (40 ft) deep.

In regard to Vibroseis it has already been stated that the source generates a surface motion that is constrained as nearly as possible to be a swept frequency signal. In theory, every reflecting boundary then returns to the surface a replica of this signal, multiplied by the reflection coefficient, attenuated by transmission coefficients, and delayed in time by the two-way travel time from the source to the reflector and back to the receiver.

The Vibroseis method then consists of cross correlating the received signal with the input signal with the result that each reflecting horizon is represented by a band-limited, zero-phase (symmetrical) signal in which the peak occurs at the proper travel time. Thus the Vibroseis seismogram apparently displays earth movement before, as well as after, the effective cause. There are, of course, effects of the near-surface layering (such as multiple reflections in the weathered zone at the source and the receiver) that must be taken into account in establishing the effective pulse to be used, but, in comparing field records with synthetic records, many of these effects have (to the amount known) been included. A Vibroseis basic pulse is shown in Figure 4.12, which is the autocorrelation function of a signal with a constant-amplitude spectrum between 5 and 20 Hz. It is possible to alter this basic waveform when the Vibroseis system is used by changing the form of the swept frequency control signal. The autocorrelation pulse shape is a function of the amplitude spectrum only. Theoretically, this allows an infinite number of ways of constructing a time series because the phase spectrum has not been specified. It should be noted, however, that the pulse given in Figure 4.12 does not take into account the characteristics of the weathered layer. It is presumed that this layer is dealt with adequately in the calculation of the reflection coefficients and thus is included in the impulse response of the earth.

A display of synthetic traces calculated for a shallow well (using velocity only for calculating the reflection coefficients) is given in Figure 4.13. Synthetic records

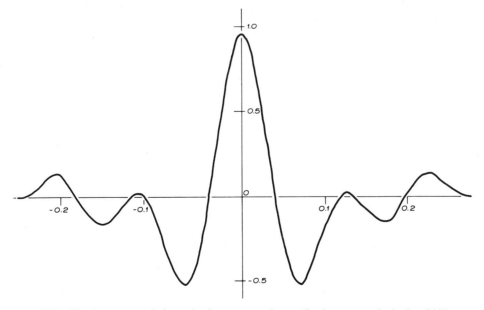

FIGURE 4.12   An autocorrelation pulse for a rectangular amplitude spectrum in the 5-to-20 Hz range.

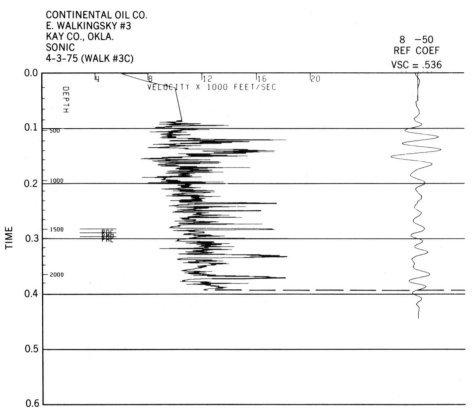

CONTINENTAL OIL CO.
E. WALKINGSKY #3
KAY CO., OKLA.
SONIC
4-3-75 (WALK #3C)

8 −50
REF COEF
VSC = .536

VELOCITY X 1000 FEET/SEC

DEPTH

TIME

FIGURE 4.13 A series of synthetic traces calculated from a velocity log in the CONOCO, E Walkingsky No. 3 well, Kay County, Oklahoma. The filter range and multiple characteristics are shown above each trace. The relative plotting scales are also given.

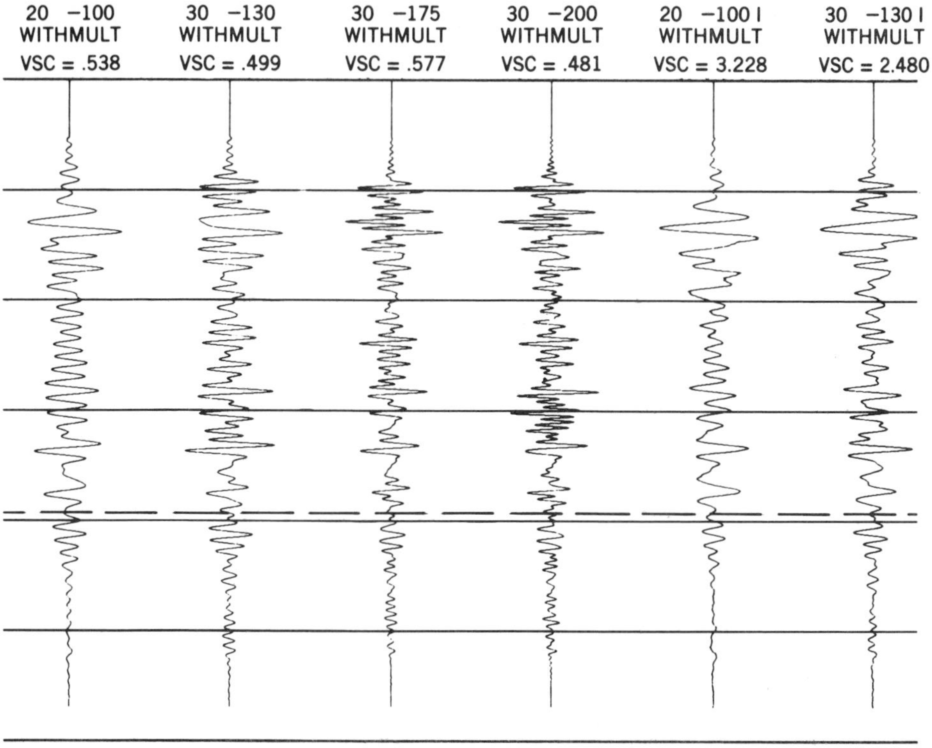

| 20 −100
WITHMULT
VSC = .538 | 30 −130
WITHMULT
VSC = .499 | 30 −175
WITHMULT
VSC = .577 | 30 −200
WITHMULT
VSC = .481 | 20 −100 I
WITHMULT
VSC = 3.228 | 30 −130 I
WITHMULT
VSC = 2.480 |

FIGURE 4.13 Continued.

145

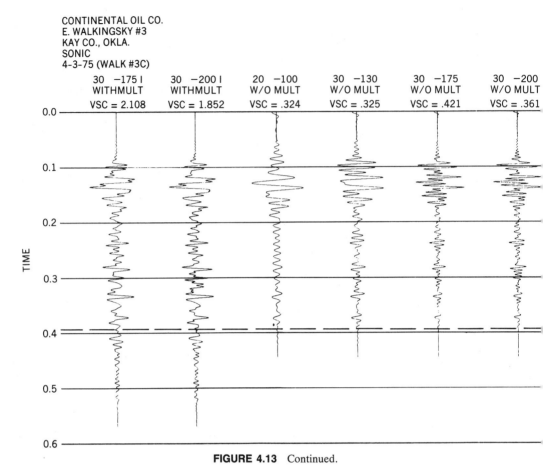

FIGURE 4.13 Continued.

146

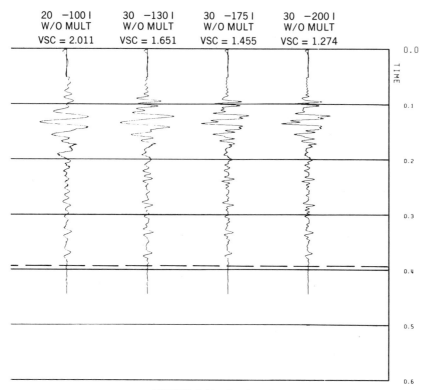

FIGURE 4.13 Continued.

made with and without multiples are included. It serves to illustrate some of the characteristics of the reflection process. This well penetrated the Permian and part of the Pennsylvanian section of Oklahoma, a section that consisted generally of thin sands and shales with occasional thin limestones. Reflection coefficients were quite large and close together. The only thick, high-velocity layers were concentrated in the first 1000 ft (305 m) and, for the lower frequency synthetic records (the spectrum is given above the trace), the low-frequency seismic reflection energy is concentrated there. The general impression is that there is a correspondence between the filtered velocity (or acoustic impedance) log and the filtered reflection coefficients. A closer examination also reveals time shifts; therefore the peaks of the reflection coefficients occur at times when the velocity is changing most rapidly.

Reflection records without multiples follow closely the reflection coefficient traces filtered with the appropriate filter. The only difference is that the rate of decay is faster than that in the reflection coefficients alone and is due to transmission losses.

There is a surprise awaiting us when we compare the synthetic record with multiples because the rate of decay is reversed; that is, a synthetic with multiples decays less per unit time than one without multiples. Both have the transmission losses included in the calculation. This apparent paradox is explained by consideration of the change in the downward traveling pulse as it penetrates deeper into the geological section. The form of this pulse can be examined relatively easily by using the same synthetic record program.

Referring again to Figure 4.10 and observing a particular line of constant j (say $j = 5$), we can see that D_{65} consists only of energy that has traveled directly downward from the surface: D_{55}, however, as illustrated in Figure 4.14a, has energy that *at some point of its path* has traveled up through one layer and then down again. In other words, D_{55} contains all the energy that has reverberated *once* in some layer. Similarly, if we look at D_{45} we will see only the energy that has reverberated twice (in any combination of layers above). Of course, if we follow a line of constant j all these energies are composited. However if we deal with points A, B, C, D, and so on, all at the same level (a reflector), the downward values D_6 ($j = 5, 6, 7, 8, \ldots$) give the successive amplitudes of the resultant pulse traveling down through that particular reflector. These values are easy to segregate from the mass of values calculated during the synthetic record computation.

Thus, as shown in Figure 4.14b, the downward traveling pulse gradually accumulates trailing energy which not only delays the reflection from a given reflector but can also cause it to lose some high frequencies. The deeper the reflector, the more these *peg-leg* reverberations tend to accumulate after the theoretical time of arrival. This process of accumulating peg-leg energy occurs not only on the downward path to a reflector but also on the upward path from the reflector to the surface.

It is not immediately obvious that the peg-leg reflections should have predominantly the same sign as the initial pulse. Consideration of several acoustic impedance situations, however (Figure 4.15), shows how this occurs in cases of geological interest. In case *a* in which the acoustic impedance log is erratic—consisting of alternating hard and soft layers and alternating signs for the

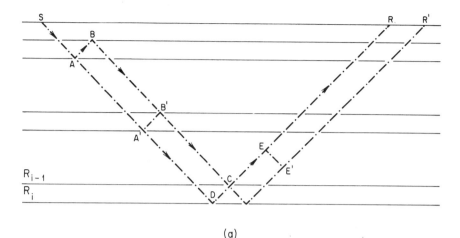

(a)

There is a single primary reflection SAA'DCR. *There are many possible "peg leg"*
multiples of the first order which are reflected along paths such as SABB'CR *or*
SA'B'CR. *These occur at the same time as the primary shown. On the other hand,*
multiple reflections, including reflection from the R_i *boundary, will arrive with*
one incremental delay behind the primary. Note that these multiples can also
occur on the upward path, such as SAA'DEE'R.

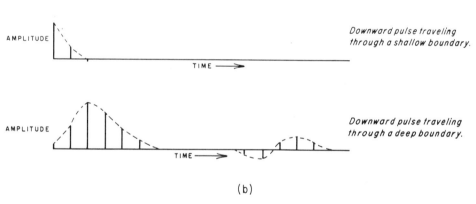

(b)

FIGURE 4.14 The formation of peg-leg multiples and the effect of these reverberations on a
downward traveling pulse.

reflection coefficients—all the first-order multiples are positive, as is the primary.
Although the multiples have amplitudes that involve the product of small
reflection coefficients, with enough layers the amplitude of the composited
multiples can exceed that of the primary. It must be remembered that this process
also occurs on the upward path of the reflected ray.

In case *b* in which the velocity increases at each step it is found that the
first-order multiple has a sign *opposite* that of the primary pulse. Thus, if it
becomes large enough by the superposition of events from many layers, the initial
reflection traveling upward could have the wrong polarity. Again the upward path

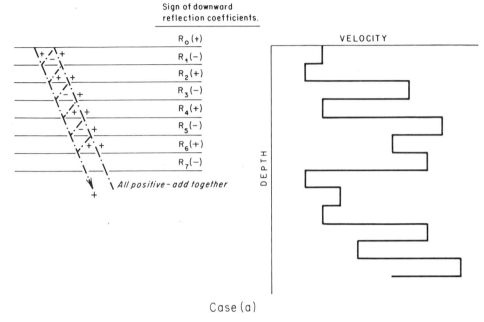

Case (a)

*All first order multiples have positive sign and add. Thus, even through the individual
reflection coefficients may be small, and the multiples (two reflection coefficients
multiplied together) individually much smaller, eventually the summed multiples grow
bigger than the transmitted primary.*

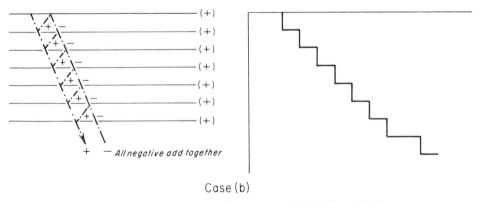

Case (b)

*Reflection coefficients all positive. A negative peg-leg multiple builds up behind a
diminishing positive primary.*

FIGURE 4.15 The addition properties of reverberations: (*a*) reflection coefficients of alternating
sign; (*b*) reflection coefficients that are positive.

is as effective in multiple production as the downward path. The polarity of the
reflection is reversed and the main energy is carried by the (delayed) multiple
event.

The case of continually decreasing velocity can be argued in the same manner
as that of continually increasing velocity and leads to the same conclusion.

O'Doherty and Anstey (1971) have pointed out two types of contributor to the
intrabed multiples, ramplike changes in velocity (in which the reflection coeffi-

cients must be small because of the limited velocity range), and alternating velocity changes that give rise to alternating reflection coefficient sequences with relatively large reflection coefficients.

Information on the size of intrabed multiples of delays greater than one is given by the autocorrelation of the impulse response. It varies from one velocity log to another.

In general, the amplitude for all reverberations of all orders can be obtained best by synthetic calculation. The gradual change in form of the effective downward traveling pulse and its transformation on the upward path may at times be of great importance to the ability of the reflection method to define lithological change. A complex pulse from a simple interface produces a complex reflection, and, if the shallow section changes appreciably, this deeper reflection, besides being complex, will be inconsistent in character.

One result of great importance, however, has been deduced; that is, the internal reverberations help to sustain the amplitudes of the reflections compared with primary reflections alone but at the cost of a small, but measurable, time delay. This effect is essentially one of scattering, dispersing the original sharp impulse, and providing a second—sometimes severe—form of attenuation for higher frequencies. This may be the final constraint on our ability to achieve high resolution with a seismograph when dealing with deep reflectors. It is not an accident, then, that reflections tend as a rule to become lower and lower in frequency as the depth (or time of occurrence on the reflection record) increases.

4.10 MORE GENERAL SYNTHETIC SEISMOGRAMS

The synthetic seismograms described to this point are limited to the following:

1. The representation of the medium is by a series of horizontal plane parallel layers, each of which is specified by one elastic constant and the density.
2. The layer thicknesses are constant in terms of the chosen wave-type travel time. (This restriction is removed in the frequency domain treatment in Appendix 4A.)
3. The incident waves have wave fronts that are plane and parallel to the layering. For this reason there is no interchange of energy between wave types at the boundaries.

This formulation, although proved to be invaluable in some seismic interpretation problems (see Section 11.5), is far from actual field practice, in which, for example, the seismic traces are made with energy from near-point sources, the media need two elastic constants and density for minimum specification (in addition, the attenuation constants may be specified), and the layer thicknesses are arbitrary. Furthermore, when the rays from the point source are not normal to the layer boundaries conversion between wave types must exist. Our physical intuition tells us that this conversion must be important, especially when large reflection coefficients exist. The presence of shear wave reflections on an offset seismic trace, in addition to possible multiple reflections, no longer makes it possible—as some interpreters do—to assume a one-to-one relationship between time of occurrence of a reflection and the length of its travel path. Now we see that change of energy from one type to another could give rise to a mixed velocity path, unrelated to the velocity structure of paths of nearby reflections.

Although this effect of conversion from one wave type to another has always been known to exist theoretically, it has been argued that it cannot be of much practical consequence because relatively short offsets have been used; hence near-normal incidence. Since the introduction of longer and longer surface offsets between sources and receivers, there has been a nagging worry that the longer offset seismograms may be contributing unwanted events to the composite seismograms. These effects are described in more detail later. Fortunately, however, two methods of constructing an offset source-receiver synthetic seismogram for the horizontally plane layered earth have been developed to the point of practical use. The *reflectivity method* was described basically by Fuchs (1968a, 1968b) and developed by Fuchs and Müller (1971) and Kind (1976). Finally, it has been used by Fertig and Müller (1978) to calculate some instructive examples.

An alternative method has been used by Kennett (1974, 1975, 1978), based on a similar but two-dimensional theory. His method also permits a small modification if a decision is made to include attenuation or to limit the number of multiple reflections.

Reflectivity Method

A point source of spherical waves is located on the surface of a plane parallel-layered system (Figure 4.16), specified by a wider parameter list (α_i, β_i, ρ_i, h_i) for each layer than for systems described earlier. The results of this methodology describe multiple reflections of any order but *head* waves are excluded.

Despite the mathematical complexities, it is possible to give a physical description of the method. Briefly, the spherical waves developed by the source at constant frequency are represented by an equivalent set of plane waves, according to a method developed by Sommerfeld. A complete set used angles of incidence on the layered system that are both real and complex. Complex angles of

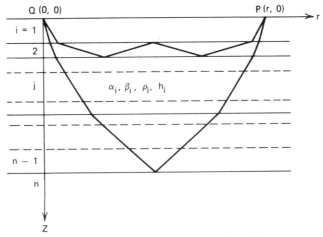

FIGURE 4.16 The general model used by J. Fertig and G. Muller for computing synthetic seismograms. The physical parameters of the layers are α, P velocity; β, S velocity; ρ, density; and h, layer thickness.

incidence correspond to inhomogeneous waves that decay with depth in a layer and represent critically refracted waves. In this application, however, only those in the set that have real angles of incidence and that lie between certain calculable limits (those that can physically start at the source and be received by the geophone) are used.

Each plane wave is dealt with in turn by the theory of the reflections of a plane wave from a layered elastic system—a theory developed essentially from work by Thompson and Haskell.

The main body of the computer work is done in the frequency domain and the response for each of the constituent plane waves is obtained by multiplying the complex spectrum of the input pulse by the complex transfer function due to the layered system and then summing for all real angles of incidence. After the response at the geophone has been evaluated (on the free surface) the inverse FFT is performed to bring the response back into the time domain. Both horizontal and vertical displacements of the geophone(s) may be found.

The increase in computer work (and cost) that results from this extension from a one-dimensional to a three-dimensional system is enormous, and as a consequence examples of its use have been restricted in quantity. One of the models chosen by Fertig and Müller, however, is instructive and answers immediately some of the questions that have arisen because of the limitations of a one-dimensional model. Figure 4.17 shows a model that consists of two thin coal seams at a general depth of 1050 m (3445 ft) with characteristics

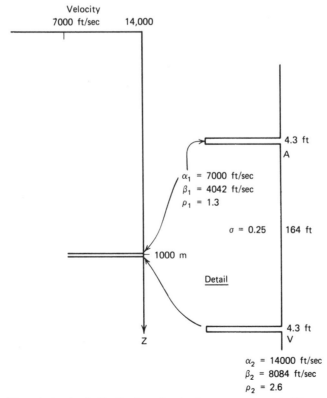

FIGURE 4.17   The velocity-depth distribution of a simple two-seam model. [*Source*: Modified from J. Fertig and G. Muller (1978), *Geophysical Prospecting*, Vol. 26, pp. 868–883.]

$$\alpha_1 = 2133.6 \,\text{m/s}, \quad \beta_1 = 1232.0 \,\text{m/s}, \quad \rho_1 = 1300.0 \,\text{kg/m}^3, \quad h_1 = 1.3 \,\text{m},$$
$$\quad\; 7000 \,\text{ft/s}, \qquad\qquad 4042 \,\text{ft/s} \qquad\qquad\qquad\qquad\qquad\qquad 4.3 \,\text{ft}$$

in a rock medium the characteristics of which are in turn

$$\alpha_2 = \;\; 4267.3 \,\text{m/s}, \quad \beta_2 = 2464.0 \,\text{m/s}, \quad \rho_2 = 2600 \,\text{kg/m}^3.$$
$$\quad\;\; 14000.0 \,\text{ft/s}, \qquad\qquad 8084.0 \,\text{ft/s}$$

As the first illustration of the difference shown between treating this model as an elastic or acoustic (P) system Figure 4.18a shows the reflected P pulses for plane

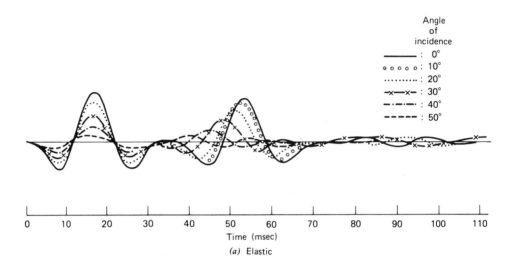

(a) Elastic

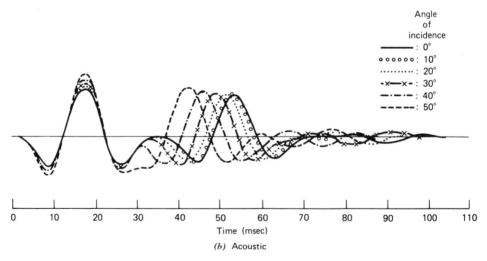

(b) Acoustic

FIGURE 4.18 (a) The reflected P pulses for plane waves of different angles of incidence from the *elastic* model in Figure 4.17; (b) the reflected P pulse for plane waves of different angles of incidence from a *liquid* model. [*Source*: J. Fertig and G. Muller (1978), *Geophysical Prospecting*, Vol. 26, pp. 868–883.]

waves for an elastic model at different angles of incidence, whereas Figure 4.18*b* shows the P pulses for an acoustic model. The waveforms composed of derivatives of the input waveform, as in reflections from thin beds, are aligned to compensate for arrival-time differences at different angles of incidence. The decrease in distance between the pulses as the angle of incidence increases is due to the fact that the curves are measured taken on a line perpendicular to the emergent reflection rays. It can be shown that this time difference

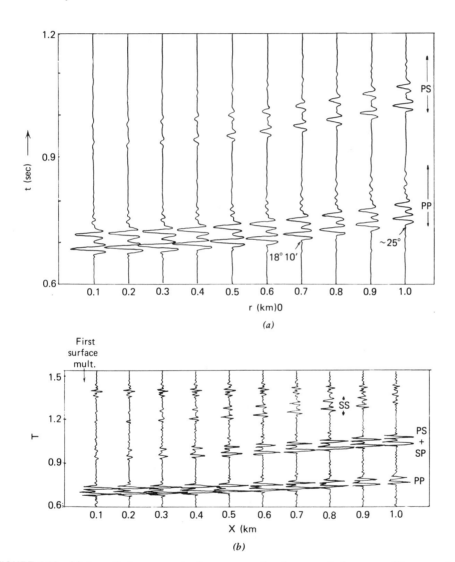

FIGURE 4.19   (*a*) A vertical component seismogram section for an explosive source (P waves only) at the top of the model of Figure 4.17; the dominant phases are PP and PS. The approximate angles of incidence for the PP reflections are given at two offset distances. (*b*) A vertical component seismogram section for a vertical impulse applied to the surface. Because S waves are initiated, SS and SP phases are present. A compressed time-scale relative to Figure 4.19*a* has been used. [*Source*: Both seismogram sections (*a*) and (*b*) were produced originally be J. Fertig and G. Muller and are reproduced with their permission.]

$$\Delta t = \frac{2(d_2 - d_1)}{\alpha_1} \cos \phi$$

where d_1 and d_2 = the coal seam depths
ϕ = the angle of incidence
α_1 = the country rock velocity

Hence as ϕ increases the time difference decreases. Note that the *elastic* and *acoustic* arrivals are identical for 0° incidence, but as the angle of incidence increases changes occur. Particularly important is the large decrease in amplitude in the elastic model for all later events as the angle of incidence increases.

Figure 4.19 shows the arrivals as a function of offset distance. In Figure 4.19a the arrivals are due to a point source of P waves at the top of the model of Figure 4.17. After the PP pulses (and some between-seam multiples) occur there are some P–S reflections, low in amplitude at small offsets and growing in importance as the offset increases and becoming equal in amplitude to the PP events as the offset approaches 1 km.

Figure 4.19b shows the results for the same model if the source is a vertical force on the surface. (This is referred to as Lamb's problem when only an isotropic half-space is involved.) Note now that the PS + SP events are larger than the PP (the amount of shear wave energy generated by this type of source is about four times greater than the P energy). Small SS reflections and some surface multiples are also present. Large "numerical" pulses originating in the calculation and appearing on the original diagrams have been edited out.

We must note that the relative sizes of the PP and PS or SS events might have been drastically altered if attenuation had been included. The attenuation for S waves is four times greater than for P waves of the same frequency and for the same travel distance. For this reason the convolution of the impulse response for a nonattenuating layered system with a single pulse is not physically correct. It would be more nearly correct if both P and S wave velocities were made complex to include attenuation. This method predicted seismic records under a variety of circumstances close to reality. Much more work remains to be done to uncover all the effects of variation of offset distance, attenuation, and other parameters.

Kennett's Method

In a number of papers Kennett (1974, 1975, 1978) has established a method that resembles the reflectivity method in a number of ways. The calculation scheme includes plane wave superposition and Fourier synthesis. The main *reflectivity* of the layered system is calculated by the *usual* iterative system, resembling in some ways the Thomson-Haskell treatment but going into more detail about the relation of the individual and composite reflectivity matrices. Therefore it becomes relatively easy to leave out reverberations within layers, to leave in only one (peg-leg multiples), or to include reverberations of all orders. Those readers interested in the details of the derivations are advised to consult Kennett's (1978) latest paper.

Kennett specified a line source instead of an explosive spherical source to preserve the two-dimensionality of the problem needed to obtain the result by the use of a two-dimensional Fourier synthesis and to translate results from the domain of the frequency wave number to the time-offset distance domain. It may be worth noting at this point that a revolutionary method of computing Fourier transforms that makes use of the Cooley-Tukey algorithm (1965) and reduces the labor of such calculations considerably has been developed. This is particularly true when the number of points is large and can be expressed as an integral power of two (2^N). This method is so common that it is known as the fast Fourier transform, or by the initials FFT for Fourier analysis and FFT^{-1} for Fourier synthesis.

General Purpose Ray-Tracing Programs

There exists also a need for a method of creating a synthetic seismic cross section, to be made from a two-dimensional generalized geologic model in which the boundaries are no longer parallel, in which unconformities and geological faults can be specified. Several examples exist in the industry, some proprietary and others that are available for purchase or for hire to make contract synthetic seismic sections. They generally have the following specifications:

1. They are acoustic models needing only one velocity and density for each rock segment.
2. Arrival times for each boundary reflection, at a number of specified source and receiver surface locations, can be found by ray-tracing methods. In these methods reflection coefficients are usually constant for a particular boundary, Snell's laws are obeyed, and transmission coefficients may be used.
3. Multiple reflections are usually ignored but multiples within one or two layers may be permitted.
4. In some model programs spherical divergence can be simulated.
5. In some model programs diffractions from terminating boundaries can be simulated.
6. The boundaries may be curvilinear, specified by a series of coordinates—beginning, intermediate, and end—to which cubic splines are fitted to permit curvatures and locations of intermediate points to be calculated.

Some examples obtained with one of these programs appear in Chapter 8 (Figure 8.17) and in Chapter 11 (Figures 11.11, 11.15, and 11.16). Additional details are provided in Section 11.6. The most important point to notice is that much of the sophistication, including elastic modeling and the use of attenuation, has to be given up once the parallel-layered model is abandoned.

Finite Difference and Finite Element Models

In addition to the modeling programs already discussed, for nearly 20 years a number of computer programs have been based on an iterative application of fundamental equations of state and on the laws of motion applied to small elements of a body in which a shock or elastic wave is passing. The interaction between these elements provides the necessary link that makes a study of wave propagation possible.

These programs (codes) were developed for dealing with phenomena associated with nuclear bombs in which many different states of matter (plasma, gaseous, plastic, fractured, and elastic) may be involved. State-of-the-art computers were needed to deal with the immense number of calculations required.

The original TENSOR code was simplified by Cherry and Hurdlow (1966) and could be applied more readily to the typical seismic problem. Even at present, however, applications are usually those that explore problems of wave propagation in both P and S waves when the boundaries concerned between materials are not necessarily plane and horizontal.

Developments in the field have been rapid, both for earthquakes and for exploration purposes. Boore (1972) and Alterman and Loewenthal (1972) have established a firm mathematical basis for finite difference approximations. A more limited code, in the sense of being suitable only for linear elastic conditions, is available from Systems, Science and Software, of La Jolla, California. It is called SAGE (seismic analysis for geologic environments) and allows investigation of two-dimensional Cartesian or axisymmetrical models. Both basis systems are shown in Figure 4.20. In the plane of the diagram all grids are rectangles, but the size can vary separately in each direction. Surface loading or interior pressure loading is possible. The duration, several pulse forms, and whether the loading is pressure- or velocity-controlled can be specified. Each grid can, if necessary, have its own elastic parameters and densities given separately, although at present only simple models are being investigated. The types of permissible motion at the boundaries of the model can also be named; that is, the cells can be allowed to move in a particular direction; for example, vertical motion only along the radial model axis or motion in any direction in the plane for the free surface and other boundaries. As Boore (1972) showed, there are potential instabilities associated with the finite difference model, and some (small) damping factors can be specified that effectively damp out the high-frequency oscillations that can be falsely generated.

There are two interrelated, troublesome aspects of the elastic modeling system. One is that even in the two-dimensional system computer time consumption can rapidly become prohibitive. Associated with this expense are the large arrays needed. If they are to be held in core memory the sizes are too large, whereas if the parameters are held on disk memories the transfer times in and out of the computer become a controlling factor.

At first sight there appears to be no need for overwhelmingly large arrays. The stipulations, however, first of the method requirements and then of the results needed, soon cause their size to increase. An example may make this clear. It is necessary that at least seven grids be present in the shortest compressional wavelength. If conventional frequencies are used and the upper frequency is 50 Hz, whereas the lowest P-wave velocity is 2000 ft/s (609.6 m/s), the grid size may not exceed 5 ft (1.5 m).

We must then determine that the model is large enough to ensure that the reflections from any of the edges do not return to interfere with the desired results; for example, if the results are needed out to 0.2 s and the major portion of the medium has a P-wave velocity of 8000 ft/s (2438 m/s) the model should be at least 1600 ft (488 m) long and deep, and, allowing a safety factor of 1.5, the

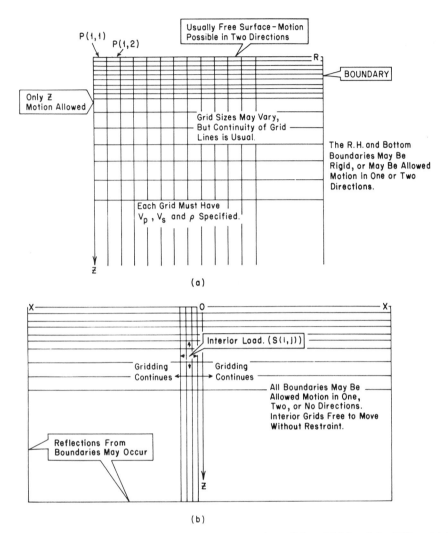

FIGURE 4.20 Illustrations of the basic arrangements for the radial model (*a*) and the *X-Z* model (*b*) of the SAGE elastic modeling program.

number of grids in each direction should be 480. Five parameters are stored for each grid, making the total storage requirement 1,152,000.

The cycle time (rate of progression of the waves in the time domain) must not exceed three-eights of the time taken for a compressional wave to travel through the fastest grid. In our sample case the transit time is 0.625 ms and a cycle time of 0.0002 s would be wise. Thus for 0.2 s of data 1000 cycles would be required. Even on very fast machines this sample problem would take several hours of computation time.

Nevertheless, these models simulate the elastic wave equation and can answer some hitherto unanswerable questions. They can be used to simulate actual seismic experiments, if done in three dimensions, and much valuable information can be obtained from the more restricted two-dimensional or axisymmetrical systems (Cherry, 1973; Kelly et al., 1976). To some extent, if the restrictions are

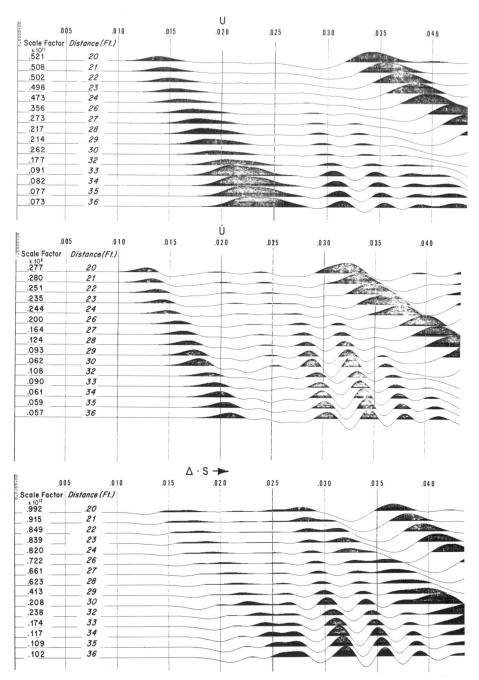

FIGURE 4.21 The results of an elastic modeling program described in the text. (*Source*: Courtesy CONOCO, Inc.)

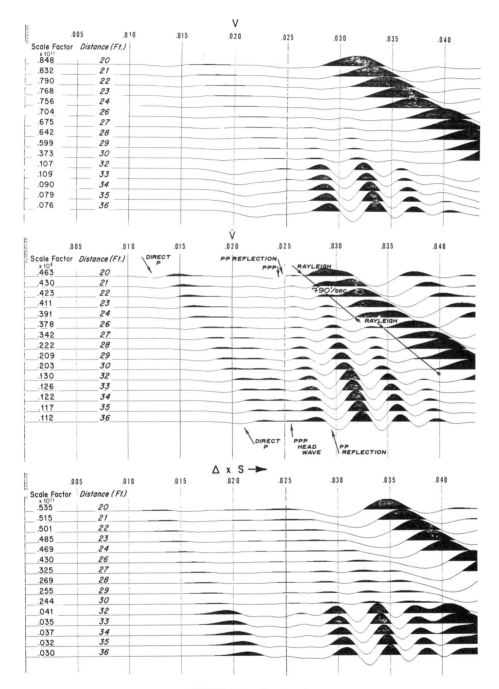

FIGURE 4.21 Continued.

borne in mind, they may replace some forms of experimental field test. Although the ultimate test for any new equipment, or new idea, is a field test, accurate testing by computer simulation offers an alternative that makes the expensive field test necessary only when the idea has crystallized, all unsuspected "bugs" have been removed, and it shows definite promise.

This tool may seem esoteric to the practicing field geophysicist and remote from the realities of exploration, but consideration of the following two unsolved problems may be evidence that research of this nature can bear fruit. The first problem is that of the influence of the conversion of compressional waves to shear waves, or vice versa, in the long-offset traces now commonly used in the CDP reflection method. It is not logical to assume that the extension to longer and longer offsets will continually increase the accuracy of velocity determinations or the quality of the reflection P-wave data. At some stage, as the angle of incidence of the P wave increases, the amount of conversion from one wave type to another increases, and, of course, refractions or head waves of both types (P and SV waves) are possible. Thus it is a matter of practical importance to decide what the limit of the spread length should be to optimize the desirable factors compared with the deleterious effects.

A second possibility is the use of finite difference programs to investigate in detail the influence of a weathered zone on the character of the reflection record. As a start to such an investigation I have investigated the arrivals from a surface source placed on top of a normal weathered zone. Figure 4.21 shows the results of the input of a 200-Hz pulse at the surface of a radial model that incorporates a weathered zone 20 ft (6.1 m) thick, velocities $V_p = 2000$ ft/s (609.6 m/s) and $V_s = 800$ ft/s (244 m/s), and specific gravity 1.6 superimposed on a half-space with parameters $V_p = 8000$ ft/s (2438 m/s), $V_s = 4000$ ft/s (1219 m/s), and specific gravity = 2.2.

The output is taken from groups of geophones at every foot on the surface from intervals 20 to 24 ft (6.1 to 7.3 m), 26 to 30 ft (7.9 to 9.1 m), and 32 to 36 ft (9.8 to 11.0 m). Six different quantities are displayed: the horizontal and vertical displacements (U and V), the horizontal and vertical particle velocities ($\dot{U}$ and $\dot{V}$), and the compressional and shear strains ($\nabla \cdot \mathbf{S}$ and $\nabla \times \mathbf{S}$). The time of the different theoretical arrivals [theoretical impulse time plus one-half the pulse width (0.0025 s)] is shown only on the $\dot{V}$ record. This detailed information helps to solve the problems of taking measurements close to a surface source.

All the records were played out in a program that scales the largest event and its amplitude is just equal to the trace spacing. The program also makes available scale factors to allow absolute amplitudes and their rates of decay with distance to be calculated.

To reduce the complexity of computer codes, and consequently the computer cost of solutions, some finite difference programs have been written that restrict wave propagation to the acoustic type. These programs undoubtedly have a place in research and development, but with recent advances in computer technology in which extremely fast programmable arithmetic modules interfaced to a small general-purpose computer have been produced, the outlook is already rosier for substantial, relatively low-cost use of finite difference programs.

As long as the problem for solution is one in which all strains can be assumed to occur within the elastic limits of the materials there is an alternative to the finite difference system—called the finite element system. In the latter the small blocks of material, which are stressed and the strains are computed, are replaced by a system of masses and springs. The spring-mass system is computed so that wave propagation through the system is completely analogous to the *elastic* wave propagation through the finite difference representation. Because it is simpler, the finite element representation permits faster computation; nevertheless many of the same limitations apply. There are restrictions on element spacing, cycle time, and, of course, boundary conditions for the edges of the model. Recently transparent boundary conditions were formulated by which no reflections emanate from the model external boundaries. This feature makes it appear as through a wave passes through these boundaries and is lost from the computational matrix. For this reason the boundaries no longer have to be in a position such that reflections from them do not arrive during the crucial part of an experiment. They simply do not arrive at all and computations can be much shorter.

4.11 SUMMARY AND CONCLUSIONS

Continuous seismic record traces can be described in two different ways. One, the time domain method, simply gives the variation of amplitude as a function of time. The other, the frequency domain method, selects a basic time interval and establishes the amplitudes and phases of constituent sinusoidal waves that are integral multiples of the fundamental frequency.

A seismic trace can be filtered by Fourier transformation into the frequency domain by changing the amplitudes and phases of the constituent harmonic frequencies in some predetermined manner and inversing Fourier transforming back into a time domain trace. *Practically*, this involves the use of sampled values of the original seismic trace.

Seismic signals can be prepared for digital computer operations by sampling at fixed time intervals. These intervals must be carefully selected to obtain at least two samples of the highest frequency present in the record to be sampled. Failure to do so produces errors in the representation of the trace and errors due to *aliasing* of high frequencies into lower frequencies. The highest frequency that can theoretically be represented by a given sample rate is called the Nyquist, whose period is twice the sampling rate.

Once the sampling has been done a third method of representation called the z-transform makes use of impulses (or δ-functions) at the sampling points; z represents the delay between sample points and the seismic trace is expressed at a polynomial in z with coefficients equal to the trace amplitudes at the sampling points.

The most useful characteristic of the *impulse* is that it contains equal amounts of all frequencies. Hence, if the impulse response of a circuit or process is found, the response to other waveforms can be obtained by filtering in the frequency domain or (more conveniently) the z domain or by convolution.

In two dimensions several descriptions are possible. The first is in the time–distance domain—a simple result of experiment in which the various reflection/refraction/scattering event times are recorded as a function of offset distance from source to receiver.

In a manner completely analogous to the one-dimensional procedure the seismic cross section is convertible into an $(\omega\text{-}k)$ representation. This 2-D Fourier transform is also reversible and two-dimensional filtering is possible in both domains.

Still further, the p-τ domain describes the process of plane wave analysis by which the complete section is thought of as consisting of linear events with slopes (p) and zero intercept times (τ). Because events that occupy a particular region of the X-t plane will reside in a totally new region when mapped on to the p-τ plane, some novel methods of filtering are possible in the new representation; for example, a band-limited refracted event that occupies a linear band in the normal X-t plot reduces almost to a point in the p-τ plane (Diebold and Stoffa (1981), Stoffa et al. (1981), Stoffa et al. (1982)).

If a simplistic model of the earth, which consists of a large number of layers with constant travel-time thickness, is selected, it is possible to use a digital computer to calculate the impulse response of a system for waves traveling perpendicular to the layering. Use is made only of the appropriate velocity and density of the layers derived from well logs. Contrasted with the simple (but physically impossible) synthetic record pioneered by R.A. Peterson, these synthetic traces are realistic and include the effects of transmission losses and multiple reflections.

These multiples, principally 'peg-leg' multiples, explain many of the experimentally derived characteristics of seismic records; for example, the scattering or reverberation accounts for anomalous attenuation, some delay of reflections, and a gradual decrease in average reflection frequency with depth. In some cases of geological interest deeper reflections may be altered complexly in form as a result of the scattering in the multiple process.

If the notion of plane parallel layers is retained two different methods are available for computing offset reflection traces. One, in the frequency domain, is called the *reflectivity method* and is due to developments of the Thomson-Haskell method by Fuchs, Fertig, Muller, and Kind. The second, largely due to B.L.N. Kennett, is a matrix method that allows some flexibility in the amount and type of reverberation to be included.

Once the plane parallel layered-earth is abandoned recourse must be made to ray-tracing programs in media in which velocities only are specified. Multiple reflections are usually ignored but refracted events and diffractions are sometimes permitted.

Finally, much more complex finite difference and finite element programs are available in 2-D and 3-D forms that permit the consideration of different equations of state for the materials and that automatically compute P and S waves traveling through the media. These programs are expensive to run on large, fast computers, and it is to be hoped that some progress will be made in their implementation on less expensive machines equipped with special, fast calculation modules.

APPENDIX 4A: SYNTHETIC RECORDS WITH COMPLEX VELOCITIES

As mentioned earlier in Chapter 4, the frequency domain provides a convenient method of computing synthetic seismograms, but details were omitted there because the time domain approach gives an easy physical picture of the multiple process. When, however, frequency-dependent attenuation must be included the frequency domain approach is simpler, although it involves the introduction of complex arithmetic computer subroutines.

This appendix contains a detailed derivation of the frequency domain approach and the use of an iterative formula, starting from the bottom layer and working up to the layer just below the surface. This is done for every frequency of interest, and the response of the uppermost layer is related to the stress or to the initial velocity applied to the surface.

Fourier synthesis finally takes all the coefficients (responses) and adds them together to obtain a time domain output.

4A.1 THEORY

The plane wave equation is

$$\frac{\partial^2 \zeta}{\partial t^2} = C^2 \frac{\partial^2 \zeta}{\partial x^2} \qquad (4A.1)$$

where the velocity is complex and can be represented by

$$C = V + vi \qquad (4A.2)$$

Assume a solution of (4A.1) of the form

$$\zeta = \zeta_0 \epsilon^{i\omega[t-(x/c)]} \qquad \text{(for waves advancing in the positive } x \text{ direction)}$$
$$= \zeta \epsilon^{i\omega\{t-[V-vi/(V^2+v^2)]x\}}$$

where $v^2 \ll V^2$

$$\zeta = \zeta_0 \epsilon^{i\omega[t-(x/V)]} \epsilon^{(-\omega vx/V^2)}$$
$$= \zeta_0 \epsilon^{-\alpha x} \epsilon^{i\omega[t-(x/V)]} \qquad \alpha = \frac{\omega v}{V^2} \qquad (4A.3)$$

This is a wave that attenuates with distance as the first power of the frequency— the solid friction model—and

$$\frac{1}{Q} = \frac{2v}{V}$$

We now take two consecutive layers in the earth and in each we have upgoing A^- and downgoing A^+ waves. Because these waves attenuate as they pass through the layer, we write the continuity equation with the convention that the amplitudes A_m^- and A_m^+ are measured at the extreme uppermost point of the layer.

As Figure 4A.1 shows, the downgoing waves attenuate to the right, the upgoing waves to the left.

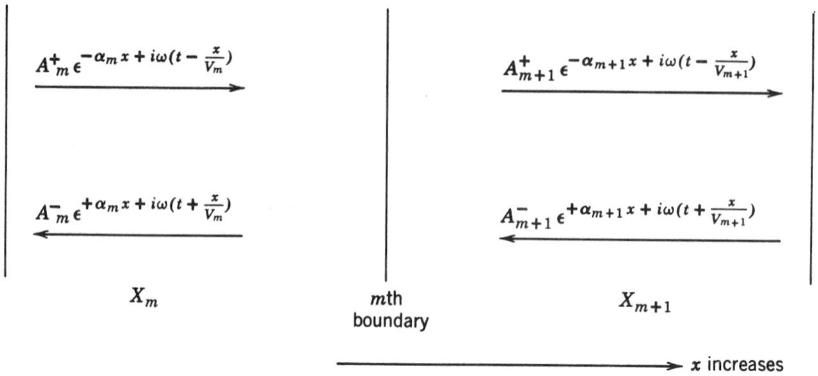

FIGURE 4A.1 Up- and downgoing waves in two consecutive layers.

Continuity of displacement at the mth boundary gives

$$A_m^+ \epsilon^{-\alpha_m x_m - (i\omega x_m/V_m)} + A_m^- \epsilon^{\alpha_m x_m + (i\omega x_m/V_m)} = A_{m+1}^+ + A_{m+1}^- \qquad (4A.4)$$

Continuity of stress at the mth boundary gives

$$C_m^2 \rho_m \left(\frac{\partial \zeta_m}{\partial x}\right)_{x=X_m} = C_{m+1}^2 \rho_{m+1} \left(\frac{\partial \zeta_{m+1}}{\partial x}\right)_{x=0} \qquad (4A.5)$$

Note that ζ_m is the particle displacement due to the combination of upgoing and downgoing waves.

$$C_m^2 \rho_m \left[A_m^+ \left(-\alpha_m - \frac{i\omega}{V_m}\right) \epsilon^{-\alpha_m X_m - (i\omega X_m/V_m)} + A_m^- \left(\alpha_m + \frac{i\omega}{V_m}\right) \epsilon^{\alpha_m X_m + (i\omega X_m/V_m)} \right]$$

$$= C_{m+1}^2 \rho_{m+1} \left[A_{m+1}^+ \left(-\alpha_{m+1} - \frac{i\omega}{V_{m+1}}\right) + A_{m+1}^- \left(\alpha_{m+1} + \frac{i\omega}{V_{m+1}}\right) \right] \qquad (4A.6)$$

For brevity we write

$$q = \epsilon^{-\alpha_m X_m - (i\omega X_m/V_m)}$$

and

$$K_m = \frac{\alpha_{m+1} + (i\omega/V_{m+1})}{\alpha_m + (i\omega/V_m)} \frac{C_{m+1}^2 \rho_{m+1}}{C_m^2 \rho_m}$$

Equations (4A.4) and (4A.6) reduce to

$$A_m^+ q + \frac{A_m^-}{q} = A_{m+1}^+ + A_{m+1}^- \qquad (4A.7)$$

$$-A_m^+ q + \frac{A_m^-}{q} = K_m(-A_{m+1}^+ + A_{m+1}^-) \qquad (4A.8)$$

Divide (4A.7) by (4A.8) and let

$$R_m = \frac{A_m^-}{A_m^+} \qquad R_{m+1} = \frac{A_{m+1}^-}{A_{m+1}^+}$$

Then

$$R_m = q^2 \left(\frac{R_{m+1} + r_m}{1 + R_{m+1} r_m} \right) \tag{4A.9}$$

where

$$r_m = \frac{1 - K_m}{1 + K_m}$$

We now make a further simplification of K_m, neglecting v^2 with respect to V^2 and substituting $\alpha = \omega v / V^2$:

$$K_m = \frac{C_{m+1}^2 \rho_{m+1}}{C_m^2 \rho_m} \frac{\alpha_{m+1} + (i\omega/V_{m+1})}{\alpha_m + (i\omega/V_m)}$$

and, after some tedious algebra,

$$K_m = \left[\frac{V_{m+1} + iv_{m+1} - (2v^2/V_{m+1})i}{V_m + iv_m - (2v_m^2/V_{m+1})i} \right] \frac{\rho_{m+1}}{\rho_m}$$

Again neglecting v^2/V compared with v,

$$K_m = \frac{(V_{m+1} + iv_{m+1})\rho_{m+1}}{(V_m + iv_m)\rho_m}$$

Therefore K_m is the ratio of the acoustic (complex) inpedances across the boundary, and

$$r_m = \frac{1 - K_m}{1 + K_m}$$

is the complex reflection coefficient associated with the mth boundary.

Thus we obtain the general iteration formula

$$R_m = \left(\frac{R_{m+1} + r_m}{1 + R_{m+1} r_m} \right) q^2 \tag{4A.10}$$

There now remains the problem of relating the motion in the layers to the motion of the surface induced by the source (see Figure 4A.2).

Because the upper boundary is free,

$$-A_0^+ \left(\frac{i\omega}{V_0} \right) + A_0^- \left(\frac{i\omega}{V_0} \right) + 1 = 0 \tag{4A.11}$$

if the stress on the boundary due to the soruce is 1 and

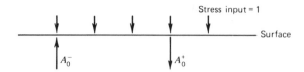

FIGURE 4A.2 The relation between surface stress input and the up- and downgoing waves in the near-surface layer.

$$\frac{A_0^-}{A_0^+} = R_0 \qquad \text{(determined from the iteration procedure)}$$

$$(4A.12)$$

$$A_0^- = \frac{V_0 R_0}{i\omega}\left(\frac{1}{R_0 - 1}\right)$$

For Vibroseis-type sources we compare (correlate) the input velocity with the output velocity; hence we need to convert from a stress input to a velocity input constant with frequency.

$$\left(\frac{\partial \zeta}{\partial t}\right)_{x=0} = i\omega A_0^- + i\omega A_0^+$$

$$A_0^+ = \frac{A_0^-}{R_0}$$

$$(4A.13)$$

$$\therefore i\omega(A_0^+ + A_0^-) = V_0 R_0\left(\frac{1}{R_0 - 1}\right)\left(1 + \frac{1}{R_0}\right)$$

$$\text{velocity input} = V_0\left(\frac{R_0 + 1}{R_0 - 1}\right)$$

Thus the upward displacement A_0^- due to a *unit velocity* input is [combining (4A.12) and (4A.13)]

$$A_0^- = \frac{V_0 R_0}{i\omega}\left(\frac{1}{R_0 - 1}\right)\left(\frac{R_0 - 1}{R_0 + 1}\right)\frac{1}{V_0} = \frac{R_0}{i\omega}\left(\frac{1}{R_0 + 1}\right)$$

Finally, the upward velocity $\dot{A}_0^-$ is $i\omega A_0^-$:

$$\dot{A}_0^- = \frac{R_0}{R_0 + 1} \qquad\qquad (4A.14)$$

4A.2 SUMMARY

1. We have described a method of making synthetic records with multiples and attenuation by using a complex velocity $C = V + iv$.
2. V is the velocity measured on the continuous-velocity logs and v must be estimated in some manner.
3. The density ρ is the conventional density measured on density logs or computed from porosity and particle density.

4. α_m is an attenuation constant for the mth layer, given by

$$\alpha_m = \frac{\omega v_m}{V_m^2}$$

5. This constant is related to the Q by the formula

$$\frac{1}{Q} = \frac{2v}{V}$$

6. R_m is the ratio of the amplitude of the up-traveling wave to that of the down-traveling wave at the upper boundary of a layer:

$$R_m = \frac{A_m^-}{A_m^+}$$

7. r is the complex reflection coefficient of the mth boundary (at the base of the mth layer):

$$r_m = \frac{1 - K_m}{1 + K_m}$$

8. K_m is the ratio of the complex acoustic impedance of a layer $m + 1$ to that of a layer m:

$$K_m = \left(\frac{V_{m+1} + iv_{m+1}}{V_m + iv_m}\right)\frac{\rho_{m+1}}{\rho_m}$$

9. A general iteration formula (for layers of unequal thickness) is

$$R_m = \frac{R_{m+1} + r_m}{1 + R_{m+1}r_m} \epsilon^{-2[\alpha_m x_m + (x_m/V_m)i\omega]}$$

10. If the layers are of equal time thickness $\tau = x_m/V_m$:

$$r_m = \epsilon^{-2\pi i\omega}\epsilon^{(-2v_m/V_m)\omega\tau}\frac{(R_{m+1} + r_m)}{(1 + R_{m+1}r_m)}$$

where the factor $\epsilon^{-2i\omega\tau}$ is a complex constant for a given frequency.
11. When the layer stops, that is, when the last layer is assumed to be infinite (no reflection comes from deeper than a given depth), $A_{m+1}^- = 0$ and R_m can be calculated.
12. The iteration continues until R_0 is obtained.
13. The final output is $R_0/(R_0 + 1)$.
14. Because these quantities are complex and dependent on frequency, the values of $R_0/(R_0 + 1)$ are computed for all relevant frequencies and a Fourier synthesis is performed to obtain the time domain reflection output. The usual precautions with Fourier syntheses are needed to ensure no overlap of head and tail and to give the requisite signal length.

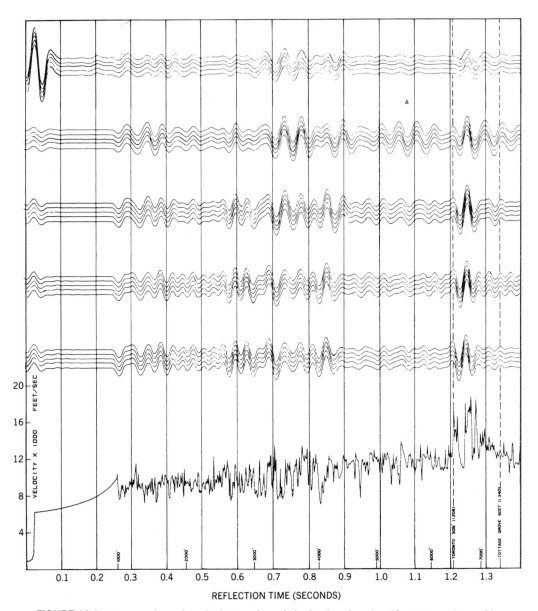

FIGURE 4A.3 A comparison of synthetic records made in the time domain *without attenuation* and in the frequency domain *with attenuation*. (The latter is the top set of traces.) Velocity data taken from the Humble No. 1 Harris well, Grady County, Oklahoma.

170

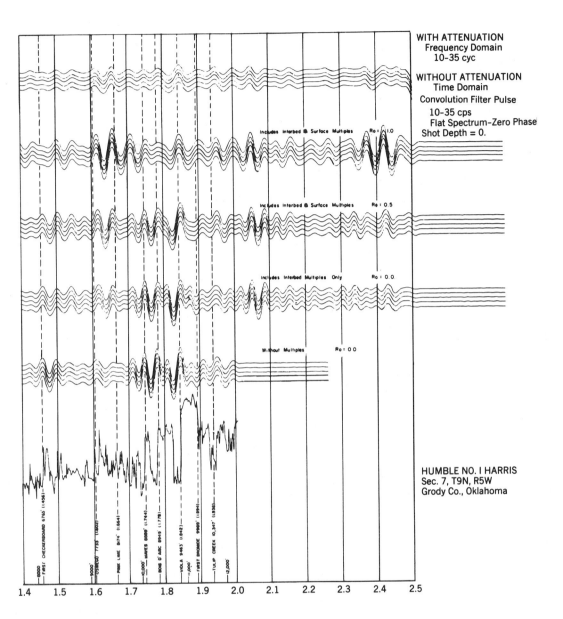

WITH ATTENUATION
Frequency Domain
10-35 cyc

WITHOUT ATTENUATION
Time Domain
Convolution Filter Pulse
10-35 cps
Flat Spectrum–Zero Phase
Shot Depth = 0.

Includes Interbed & Surface Multiples   Ro = 1.0

Includes Interbed & Surface Multiples   Ro = 0.5

Includes Interbed Multiples Only   Ro = 0.0.

Without Multiples   Ro = 0.0

HUMBLE NO. I HARRIS
Sec. 7, T9N, R5W
Grody Co., Oklahoma

1.4   1.5   1.6   1.7   1.8   1.9   2.0   2.1   2.2   2.3   2.4   2.5

171

4A.3 EXAMPLES

An example is given in Figure 4A.3 in which a synthetic record with attenuation is compared with synthetic records made in the time domain without attenuation but otherwise the same.

Two different versions of a second example, including attenuation, have been made; the model is essentially that of Figure 4.17, except that the depth to the first coal seam is 1000 ft (305 m). Figure 4A.4 illustrates this example. The parameters are the following:

Layer Number	1	2	3	4	5	6
Velocity (ft/s)	14,000	7,000	14,000	7,000	14,000	—
Density (kg/m^3)	2,600	1,300	2,600	1,300	2,600	—
Thickness	1,000	4.3	164.0	4.3	00	—
Q		50	25	50	25	100	—

The synthetic record was filtered in the frequency domain by multiplying the real and imaginary spectra by a window that had a *hanning* taper of 2 to 12 Hz at the low-frequency end and 80 to 140 Hz on the high-frequency end. After the inverse FFT this has the effect of replacing a δ-function with a symmetrical damped pulse having a bandwidth of about 7 to 110 Hz.

The first group of large pulses contains the two primary reflections from the two coal seams (it is shown that these are approximately differentiated with

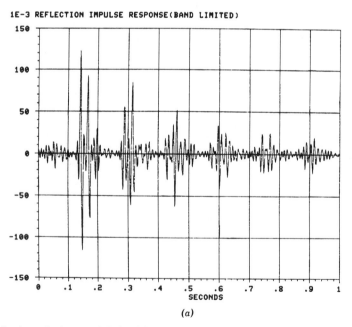

(a)

FIGURE 4A.4 A synthetic record derived from the layered system in Figure 4.17. Attenuation was included. Six sets of multiple reflections were obtained between the coal seams and the surface. There was no weathered layer.

respect to the input) and is followed by a smaller event, which is the first reverberation between the coal seams.

The second aggregate can also be analyzed as consisting of pulses that have traveled *twice* from the surface to the individual seams and return. In these simple analyses the coal seams have no thickness.

Please note that because these records were made in the frequency domain, followed by an inverse Fourier transform, a small high-order, multiple group of pulses occurring at greater than the *allowed* time has been circulated around to the front of the record trace; in other words, *aliased*. This could have been avoided in a computer with a larger memory simply by specifying the frequency interval as one-half the present one, thereby providing a record time twice as long.

To illustrate how sensitive the synthetic records are to a relatively small change in the constitution of the layered system Figure 4A.5 shows the synthetic record with a weathered layer 30 ft (9.1 m) thick added at the surface. The thickness of the second layer has been adjusted to allow the first group of reflections to occur at the same travel time. No other changes were made. The Q of the weathered layer was made 0.2 (a very low value) to reduce *ringing*. It must be remembered that the reflection coefficients at the surface and base of the weathered layer are large (1.0 and 0.85, respectively), but the Q of 0.2 represents a two-way loss—even at 10 Hz—of about 0.01 and the effective reflection coefficient of the surface is only 0.01. The explanation for the intense reverberation lies in the effect of the weathered layer on both source and receiver "ends" of the reflection path; these effects are compounded.

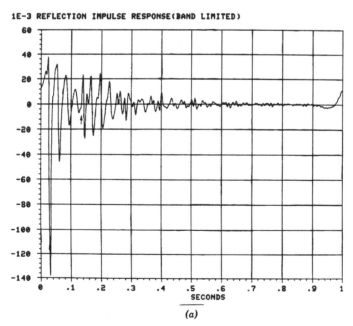

FIGURE 4A.5 The synthetic record produced by the deep-layered system in Figure 4.17, with a weathered layer with a Q of 0.2 added. Adjustment of the depth was made to give the same primary reflection time at about 0.13 s, indicated by an arrow.

REFERENCES

Alterman, Z., and Loewenthal, D. (1972), "Computer-Generated Seismoforms," *Methods in Computational Physics*, Vol. 12, pp. 35–164.

Berryman, L. H., Goupillaud, P. L., and Waters, K. H. (1958), "Reflections from Multiple Transition Layers—Theoretical Results," *Geophysics*, Vol. 23, pp. 223–243.

Boore, D. M. (1972), "Finite Difference Methods for Seismic Wave Propagation in Heterogeneous Materials," *Methods in Computational Physics*, Vol. 11, pp. 1–37.

Cherry, J. T., and Hurdlow, W. R. (1966), "Numerical Simulation of Seismic Disturbances," *Geophysics*, Vol. 31, No. 1, pp. 33–49.

Cherry, J. T., (1973), "Computational Simulation of Stress Wave Propagation in One and Two Dimensions"; paper presented at the 43rd Annual International Meeting of the Society of Exploration Geophysicists, Mexico City.

Diebold, J. B., and Stoffa, P. L., (1981), "The Travel Time Equation, tau-p Mapping, and Inversion of Common Midpoint Data," *Geophysics*, Vol. 46, pp. 238–254.

Fertig, J., and Müller, G. (1978), "Computations of Synthetic Seismograms for Coal Seams with the Reflectivity Method," *Geophysical Prospecting*, Vol. 26, pp. 868–883.

Fuchs, K. (1968a), "Das Reflexions—und Transmissions vermögen eines geschichteten Mediums unit beliebiger Tiefen—Verteilung der elastischen Moduln und der Dichte für schrägen Einfall ebener Wellen," *Zeitschift für Geophysik*, Vol. 34, pp. 389–413.

Fuchs, K. (1968b), "The Reflection of Spherical Waves from Transition Zones with Arbitrary Depth-Dependent Elastic Moduli and Density," *Journal of Physics of the Earth*, Vol. 16 (Special Issue), pp. 27–41.

Fuchs, K., and Müller, G. (1971), "Computation of Synthetic Seismograms with the Reflectivity Method and Comparison with Observations," *Geophysical Journal of the Royal Astronomical Society*, Vol. 23, pp. 417–433.

Kelly, K. R., Ward, R. W., Treitel, S., and Alford, R. M. (1976), "Synthetic Seismograms. A Finite Difference Approach," *Geophysics*, Vol. 41, pp. 2–27.

Kennett, B. L. N. (1974), "Reflections, Rays and Reverberations," *Bulletin of the Seismological Society of America*, Vol. 64, pp. 1685–1696.

Kennett, B. L. N. (1975), "The Effect of Attenuation on Seismograms," *Bulletin of the Seismological Society of America*, Vol. 65, pp. 1643–1651.

Kennett, B. L. N. (1978), "Theoretical Reflection Seismograms for Elastic Media," *Geophysical Prospecting*, Vol. 27, pp. 301–321.

Kind, R. (1976), "Computation of Reflection Coefficient for Layered Media," *Journal of Geophysics*, Vol. 42, pp. 191–200.

O'Doherty, R. F., and Anstey, N. A. (1971), "Reflections on Amplitudes," *Geophysical Prospecting*, Vol. 19, pp. 430–458.

Peterson, R. A., Fillipone, W. R., and Coker, F. B. (1955), "The Synthesis of Seismograms from Well Log Data," *Geophysics*, Vol. 26, p. 138.

Stoffa, P. L., Buhl, P., Diebold, J. B., and Wenzel, F. (1981), "Direct Mapping of Seismic Data to the Domain of Intercept and Ray Parameter. A Plane Wave Decomposition," *Geophysics*, Vol. 46, pp. 255–267.

Stoffa, P. L., Diebold, J. B., and Buhl, P. (1982), "Velocity Analysis for Wide Aperture Seismic Data," *Geophysical Prospecting*, Vol. 30, pp. 25–57.

Stratton, J.A. (1941), *Electromagnetic Theory*, 1st ed., McGraw-Hill, New York, p. 284.

Wuenschel, P. E. (1960), "Seismogram Synthesis Including Multiples and Transmission Coefficients," *Geophysics*, Vol. 25, pp. 106–129.

FIVE

Seismic Data-Gathering Methods

5.1 INTRODUCTION

Because it is known from experience that oil fields occur in random sizes and at random locations in basins about which little is known, it appears reasonable to seek information on the subsurface at points uniformly distributed on the surface. The earliest attempts to gather geological information by using a reflection seismograph were made in this manner with a technique known as isolated correlation. This technique presupposed that the character of the seismic trace was caused only by a succession of reflections—as described in the discussion of synthetic records. Of course, with a single shot point and receiver, which produced a single seismic trace, there was no way to check on the occurrence of other types of event on this trace. Therefore a single shot point was recorded at a small number of single geophones separated by a few tens of meters with minimum offset from the shot point of no more than 35 m (100 ft).

The several traces then allowed assessment by eye of the occurrence of some types of unusual event; for example, the Rayleigh wave and the refracted (head) wave could be distinguished by the rate of change in time of occurrence as the distance of the geophone from the shot point increased—a finite rate for near-surface disturbances and an extremely small rate for reflections (or indistinguishable multiple reflections). Nevertheless, this information was gathered at section corners or at other quasi-equidistant points to achieve a grid coverage of the area. The recording apparatus was simple and easy to lay out, but the rate of coverage was limited by its density, the difficulty of drilling, and sometimes by the rate at which adequate land surveying and permitting the landowners could be achieved. (Before seismic work could be done permission by landowners had to be obtained and maps had to be made of seismic locations.) The ground roll could be reduced by proper placement of the charge in the hole and by increasing the low cutoff frequency of the geophones and amplifiers—to 25 Hz at times. This was the era of seismography as an art. Figure 5.1 illustrates the method.

The idea of uniform area coverage was good, but there were too many deleterious factors to deal with to allow it to remain the sole means of obtaining reconnaissance and detailed information. Changes in seismic character with shot depth and medium, rapid changes in subsurface geology, and the presence of noise made correlation by eye between successive recording locations too difficult and ambiguous.

Most of the difficulty was, correctly or incorrectly, assigned to the problem of

175

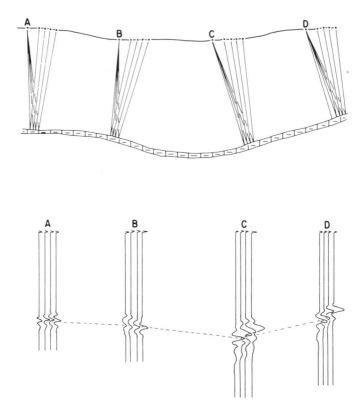

FIGURE 5.1   A scheme of an isolated correlation method and its interpretation in a simplistic case.

rapid geological variation, and the next attempts, reversed profiling or continuous profiling, used the available geophones and amplifiers—which tended to increase in number to 20, 24, or even 40 channels—but still usually only one geophone per channel to cover spread distances between holes of about 400 m (1320 ft). The method emphasized, as shown in Figure 5.2a, the provision of seismic traces equally spaced on the surface, the idea being that it is easier to make 12 small correlation decisions than possibly one larger one. With this concentration of holes between which cables and geophones had to be laid the idea of equal coverage by area was abandoned.

The continuous-profiling method and the methods of correcting near the far traces (from a given shot point) so that a time tie for each selected reflection could be made on offset traces shot from opposite directions have been well covered in the literature. Corrected times, or depths if a velocity survey was available, were computed individually for the near and far traces and time sections consisting of bars to join corresponding reflection points were made. This method, with minor modifications of overall field layouts, continued until almost 1962. The advent of surface sources, however, had given considerable emphasis to the need for groups of geophones connected to provide an average ground displacement over a substantial surface distance—sometimes a few hundreds of meters. As described earlier, these arrays of sources and receivers caused a considerable differential gain in reflection amplitude, compared with that of

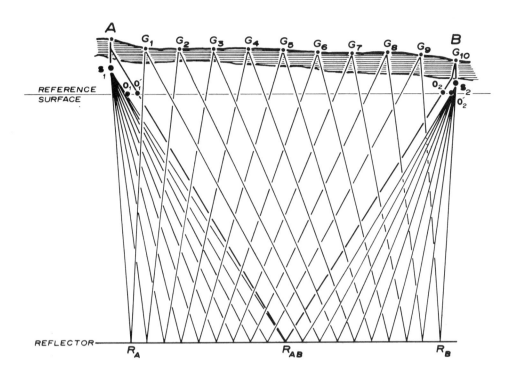

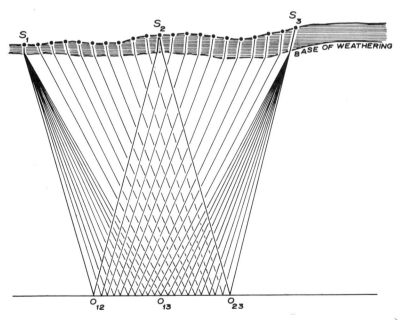

FIGURE 5.2 (*a*) The continuous, reversed profile reflection method with subweathering sources. It is necessary to correct the arrival times to a reference surface before time ties can be effected. (*b*) The time-tying method when offset geophone spreads are used, as in Vibroseis®. The surface-to-surface time $S_1 O_{12} S_2$ is correlated between traces to the time $S_1 O_{13} S_3$, which is equal to the reverse time $S_3 O_{13} S_1$ on the next record but one. Further correlations then carry the reflection to $S_3 O_{23} S_2$.

surface waves. In addition, the number of geophones averaged, aided by accumulated knowledge of the necessity for firm planting or spiking of the geophones into the ground, caused a considerable drop in the random noise level caused by wind, traffic, and other microseismic noise. Offset continuous profiles (Figure 5.2*b*) were commonly used.

Other seismic spreads, such as isolated dip spreads, were used in areas in which reflection continuity was poor both along a single line and with components in two directions at right angles. In these cases phantom horizons had to be drawn on the cross sections. These phantom horizons started at an arbitrary point and were drawn parallel to dips nearby (in time and space). The lack of reflection continuity, although it could obviously occur in clastic sections or in highly faulted areas, was not so prevalent as at first supposed and as later reflection seismograph methods testified.

5.2   MODERN LINEAR SEISMIC REFLECTION METHODS

Dependence on large arrays of source points and receivers can be carried only to the point at which these multiple elements become large enough that an appreciable amount of the subsurface is used as a reflector. Beyond this point the averaging of subsurface energy tends to modify the structure and impairs the resolution. These uses, however, had emphasized the vital point that noise can still be a controlling factor in interpreting subsurface geology from geophysical data.

One method of avoiding this problem is to use reflection traces, with different offsets, all of which refer to the same reflecting point on the subsurface. Of course, the geometrical effect of increased arrival time of reflections as the offset is varied has to be corrected by using an adequate velocity function to remove the normal moveout (NMO).

As early as 1938 Green suggested the use of multiple offsets centered about a common depth point as a means of eliminating the effect of dip on velocity determinations. It was Mayne (1962), however, who proposed the modern CDP method. This technique found acceptance largely because of the increased need to find oil fields in difficult areas and because surface sources could be utilized easily. The economics of seismic surveying had reached a point at which the increased effort and cost were affordable. Another consideration was that recording on magnetic tape had also reached a point at which the transcriptions from one tape to another—necessary for analog correction and reshuffling of seismic traces—could be performed without greatly increasing the signal-to-noise ratio.

Nevertheless these analog procedures for correction and reordering data were cumbersome and subject to human error during the tape-handling process. As a result it was only when digital processing of seismic data established itself that the CDP method took hold. It was the fortunate combination of a viable idea with a technological development that resulted in one method of vastly improving data quality.

As far as seismic cross sections are concerned the CDP method is now almost standard. It is therefore necessary to discuss the data-gathering method in detail, including the type of information that can be generated. Data-processing techniques, however, are deferred to a later chapter.

The CDP technique described in this section is designed basically for use in oil and gas exploration in which the depths of interest usually fall between, say, 3000 ft (1000 m) and 30,000 ft (10,000 m). Over this long span of depth attenuation and scattering of the seismic waves reduce the usable range of frequencies to 5 to 80 Hz in the shallowest section and 5 to 35 Hz in the deepest.

Thus in the use of arrays of sources and receivers considerable care must be taken to avoid large changes of elevation over the arrays. These changes could lead to time delays through the near surface comparable to a period of predominant frequency of the reflection energy and to cancellation or undesirable changes in the character of the reflected signal. This should be as much a matter of concern to the field manager as the effect of arrays that are incorrect in length or the effects of persistent noise. It is entirely worthwhile for the field manager to keep diagrammatic data for each line, up to date, with pertinent comments, available to himself and to clients and office personnel. In this manner perturbations on normal practice (which are necessary in nearly all field operations) can have their effects readily assessed. Figure 5.3 is a portion of a diagram in which the source points are shown as a series of small circles along a horizontal straight line, numbered at each tenth position. Small dots along lines at 45° (up to the right and down to the left) show the traces to be derived from each source. Lines joining these dots down to the right are due to a single geophone group whose number is obtained by projecting the line until it reaches the numbered source line. A count of the number of dots *vertically* through any position reveals the number of the fold coverage for the point *on the subsurface*. The comments are self-explanatory and refer to field problems. Note how this line started off with single-ended spreads (with a five-interval minimum offset and 48 geophone groups) but encountered a rather wide zone of *no source permit*. There would have been a serious drop in stack if it had not been possible to continue the

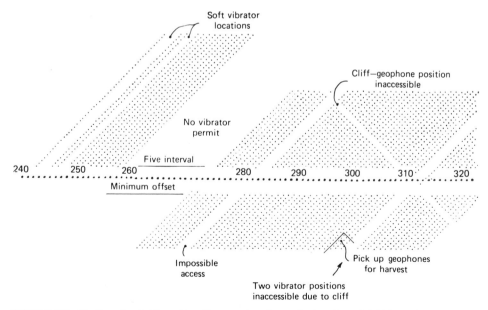

FIGURE 5.3 Portion of a field *stacking* diagram that shows the location of field hazards and permit problems with respect to source and receiver CDP positions.

right-hand portion of the line with 20 geophone groups (offset five groups) *behind* the sources. A 400-ft cliff provided impossible access for one geophone and two vibrator groups. The usefulness of the type of diagram extends far beyond the field. Examination of the raw field records will reveal possible recording faults that must be eliminated at the processing stage.

Normally, therefore, the frequencies and wavelengths used present no special difficulties in the design of arrays nor in laying out the groups of detectors (or surveying the source locations) on anything but extremely rough terrain; but if there is a requirement for higher than normal frequencies special precautions have to be taken.

Figure 5.4 illustrates the different ray paths that, after appropriate time corrections are stacked together to form one composite trace. It is noted here that these traces have different sources and receivers as well as different paths through the upper formations. Thus it would be expected that random events would (on stacking several traces together) add up only as the square root of the number of traces, whereas the reflection amplitude, common to all traces at the same time, would add up linearly as the number of traces. Therefore an increase in the signal/noise ratio as the square root of the number of traces is to be expected. Moreover, in any geological section in which the velocity increases with depth the geometrical time increases of long-period multiples are not corrected by the standard NMO correction for a primary reflection at the same time and, given the proper disposition of stacked traces, tend to cancel (see Figure 5.5). This cancellation, however, is a sensitive process that needs seismic traces that have a constant time difference for the multiples. These traces are not spaced uniformly along the surface. A brute force stack of all traces probably does not cancel multiple reflections as well as the addition of certain selected traces. In addition,

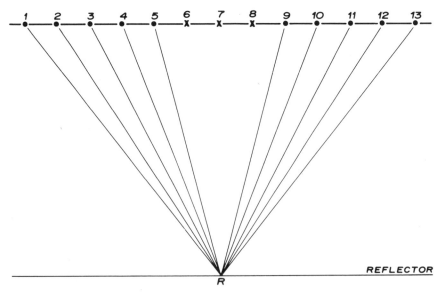

FIGURE 5.4 Different ray paths in a simplistic model, which are added together after a geometrical time correction, to give a fivefold stacked trace (applicable to position 7 on the surface).

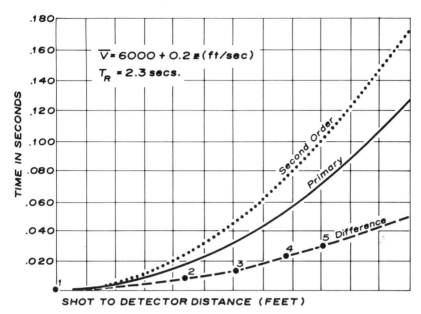

FIGURE 5.5 The NMO for primary and multiple reflections (and their differences) as a function of the offset distance. Positions 1, 2, 3, 4, and 5 have equal differences and traces can be stacked to give a cancellation of waves of period 0.035 s. [*Source*: Mayne (1962). Reprinted with permission from *Geophysics*.]

because this effect is exactly the same as the use of arrays on the surface, the sequence of multiple reflections over the traces used (after NMO correction) must cover a time interval of at least one period of the predominant frequency in the reflected energy. Sometimes, with small increases in velocity with depth, this requirement means the use of extremely long offsets—with the consequent danger of error in NMO removal due to lack of adequate knowledge of the reflection ray paths. Weathering and elevation corrections are also extremely important (Mayne, 1967; Marr and Zagst, 1967).

The layout in the field to accomplish the CDP procedure in the most efficient way is simple, flexible, and ingenious. Source points and receiver points, which are at least partly shared, are laid out equidistantly along the line of profile (see Figure 5.6). The cable is divided into sections that can be plugged together. Sections can be removed from the beginning of the spread and plugged in at the end to provide a system that proceeds efficiently and with a minimum of labor. The recording truck is equipped to handle two sets of geophone inputs, with a switch to select the appropriate traces from those available as the shot points (or other source positions) are occupied. Thus the recording vehicle moves only after one complete spread is used up. The number of traces has now increased considerably over the original 20 to 24; 48 to 96 trace units are not uncommon and this makes possible 24-fold, or higher, stacking when poor record quality demands it.

Whenever the terrain is neither flat nor open enough to allow reasonably straight traverses it is necessary to resort to making use of trails, which give rise to *crooked lines*. The chief difficulties result from two factors:

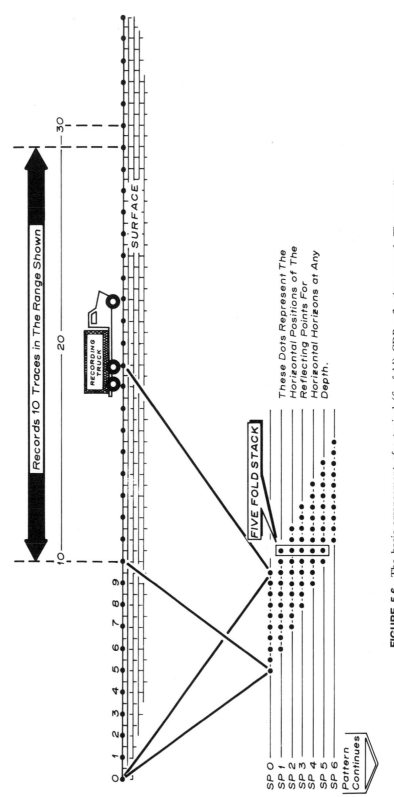

FIGURE 5.6 The basic arrangement of a typical (fivefold) CDP reflection spread. The recording truck contains a switch that allows selection of the correct sequence of geophones as the shotpoint is moved up. As convenient, portions of cable and geophones are moved from the left end and connected to the right end of the cable.

182

1. The arrays of receivers are not always in line with the source point(s) and cannot be expected to yield optimum cancellation of near surface waves. The effective length of an array is its actual length multiplied by the cosine of the angle between the mean array vector and the line joining the center of the array to the source point. In the usual crooked line the cosine of this angle does not often differ much from unity. If the field manager foresees that the array may be *foreshortened* excessively array coverage may be increased.

2. Distances between groups, for cable-laying purposes, are usually measured along the trail, thus giving incorrect straight line distances from source to receiver groups. In this case the surveying must be done accurately in two dimensions and *airline* distances must be calculated for use in later processing.

If a shear wave survey is being made there is an additional effect. Now the azimuth of polarization of the source may not be the same as the receivers. Again, this is an effect dependent on the cosine of the difference of azimuth angles and should not need consideration except in severe cases. The proper remedy uses a pair of orthogonal horizontal geophones and records each separately, leaving the combination of the two signals as a computer operation.

Crooked lines do, on occasion, offer the chance to gather some rough estimates of dip at right angles to the main line of traverse. This possibility is explored in more detail in Chapter 6.

5.3 MODERN MARINE LINEAR REFLECTION METHODS

In marine environments, of course, the situation is even simpler because the recording boat, which carries the source(s) and tows the cable, needs no division of the cable. It is arranged instead that the sources operate in synchronism with the motion of the boat and cable. New source energy is introduced at the appropriate points, which the receivers *will later occupy* as they are towed. In principle, of course, there is nothing new. In practice there are some differences. Although a marine cable has the advantage that the offsets of receivers are always constant and can always be occupied, the effects of cross currents can be a severe problem because the cable does not lie along the course pursued by the towing vessel (see Figure 5.7). In this case the offsets are known but reflection points are not common. An even more disastrous situation occurs when the cross current is not constant in speed and direction over the length of the cable.

Although these problems were known (from observations of the tail-buoy attached to the end of the cable), few active measures have been devised to realign the cable in the proper direction and position. Waters (1977) suggested the use of one or more servocontrolled variable lift paravanes to keep the cable on the line of traverse rather than dead astern.

These phenomena, however, *have* prompted passive attempts to obtain better quality data by incorporating instrumentation in the cable by which the cable position can be accurately measured and appropriate corrections made in the data-processing sequence.

The positioning in relation to the water surface is equally important, a problem that has been solved by the use of a Condep® controller. This device allows the

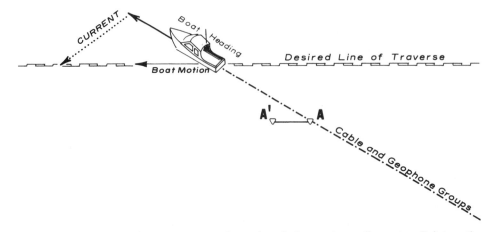

FIGURE 5.7 The effect of cross currents on the motion of a boat and recording system. Points on the cable, such as **A**, move parallel to the desired traverse (e.g., to **A′**). Different offset depth points travel along different subsurface lines and *do not stack*.

cable to pass through it and then positions it below the water surface by using water pressure to actuate a vane. The vane, as the cable passes through the water, raises or lowers the cable unless it is already at the correct (preset) depth. This device, simple in concept and action, has proved to be effective and is used almost universally.

5.4 MODERN LINEAR SEISMIC REFLECTION METHODS FOR COAL EXPLOITATION

As explained in Chapter 1 and in more detail in Ziolkowski (1979), seismic methods are now used increasingly for coal exploitation. The rewards of successful, high precision seismic reflection work lie not in the realm of the discovery of more coal but in the detailed working out of the location of problem areas in which coal exploitation could be hazardous economically and physically. In other words, ways are to be found in highly detailed seismic information to obviate any *false starts* sinking shafts or otherwise initiating a mine. Those geological anomalous features such as faults or *wants* (where coal may have been replaced by sandstone lenses) certainly cause difficulties in maintaining the output from a highly mechanized modern mine and are often associated with roof conditions that can be dangerous to miners.

It is hoped that most, if not all, of these geological anomalies will be visible on the seismic cross sections. While adhering to the basic CDP format a striking change of scale is necessary; for example, most new oil fields are found at 1000 m (3200 ft) to 10,000 m (32,000 ft), whereas the coal fields that are at present economically feasible to work lie at depths of 0 to 1000 m. Our interest lies in the range of possibly 100 m to 1000 m because surface stripping can deal with most of the near-surface coal deposits.

To deal with this reduced depth range and still maintain the maximum offset from source to receiver of the same order as the depth consideration must be given to spread lengths of 1000 m or less which, with 24 trace instruments, require trace spacings of about 40 m. If these traces were made from single-source and single-geophone surface locations with the same theoretical midpoints the depth measuring points on the subsurface (coal) would be 20 m apart. If, on the other hand, surface arrays were used for sources *and* receivers these subsurface locations would be spread over the reflector as a weighted pattern consisting of the convolution of the source and receiver array patterns concentrated in length by a factor of two. At first sight this diffusion of the information location caused by instrumentation may appear to the contrary to the requirement for detailed information. Physically, however, the scattered waves from anomalies of the coal seam must be regarded as diffractions that spread out laterally as the spherical wave diverges and returns to the surface. Data processing (including migration) can restore some of the lost detail due to wave divergence.

Ziolkowski and Lerwill (1979) in some well designed experiments have shown results that can be achieved by taking one extreme viewpoint in designing a seismic system and using small dynamite charges in holes and single well planted geophones on the surface or hydrophones in water-filled holes. Yet surface waves do cause reflection-data deterioration, even under the almost perfect conditions. The near-constant conditions of high humidity and small temperature variation must have produced almost perfect seismic conditions in their experimental sites in England.

In the United States coal seams are rarely worked at depths greater than 400 m, the spread lengths are limited to about 400 m, and even greater theoretical density of reflection points is achievable. In practice the surface terrain and the weathering conditions preclude the use of single-source and single-receiver locations. Surface sources in small arrays and surface receivers in similar arrays can still produce indisputable evidence of subsurface anomalies (see Figure 8.21). Two parameters may be controlled by using the Vibroseis system. First, the input spectrum can be selected to have a lower cutoff frequency of 40 Hz. Second, the number of geophones per group can be regulated to give adequate sampling for the smallest wavelengths encountered. For both shear and compressional waves the smallest wavelength is 300/170 m, or about 2 m, which is sampled adequately by 1-m geophone positions. The longest surface waves have a wavelength of 7.5 m for which an array of 30 m is more than adequate.

There seems to be no reason why high-frequency shear waves could not be used in these surveys. Apart from the disadvantage of a higher attenuation rate, they would offer several advantages; for example, a high reflection coefficient for existing air-filled or *water-filled* mines, greater resolving power for the same frequency range, and lack of conversion at horizontal boundaries to compressional waves.

Of course, as in all surface source and receiver methods the problem of making weathering and elevation corrections remains. Although it is true that arrays of sources and receivers average the energy received from different paths through the near-surface, low-velocity layer, a single geophone can give a poor estimate of the signal received, compared with an average of 20, unless the surface conditions are extremely uniform. Under arid conditions even more locations may be necessary to reduce randomly scattered interference or noise. The main problem is to avoid large elevation (and weathering) changes between individual source or receiver elements.

Because the total reflection times are much smaller for coal prospecting, the

NMO will be correspondingly smaller for traces with similar geometry. The same care is needed to remove the excess time for slant reflections because higher frequencies are being used. This aspect of coal exploitation work is covered in more detail in Chapter 6.

5.5 METHODS OF OBTAINING TWO-DIMENSIONAL CDP COVERAGE

It is axiomatic that the subsurface lithology varies in three dimensions, and a conventional CDP linear spread provides a cross section that shows only apparent two-dimensional variation. For structural geological measurements in which the third dimension (at right angles to the line of cross section) may give rise to only slow variation the problem is not important, but for stratigraphic trap problems, in which the changes in sedimentation may be rapid (channel sands, offshore sand bars, reefs, etc.) there is a need to search for variation in three dimensions. This has led to the use of methods by which three-dimensional variation can be examined.

First, it must be realized that a series of parallel CDP profiles can be used if terrain and cost are not deterrents. If however, for reasons of surface conditions the quality of subsurface records is dependent on the location of the sources and receivers every effort should be made to maintain, as far as possible, communality of source and receiver positions. The next step therefore could be two parallel CDP lines recorded simultaneously because a row of depth points (Figure 5.8) can be obtained between and beneath the two lines (Davis, 1975). These center points are recorded from both sets of shot points and, as a consequence, have a double stack compared with those below the surface spreads. The method is difficult to stabilize at sea. On land CDP lines are usually constrained by available roads and trails; hence the positions of the depth points and the distance apart of the lines of reflecting points are constrained. We return to the original recording

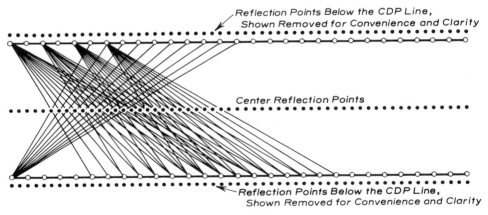

FIGURE 5.8 The method of formation of CDP reflecting points between two simultaneously recorded lines. The number of folds of stack is arbitrary, depending on the source–receiver relation and the number of traces recorded.

philosophy that the points of information about the geology should be equally spaced in two orthogonal directions.

To satisfy this need the idea of recording points (geophones) on lines at right angles to the line of sources was conceived. Actually the recording lines can be at any angle to the line of sources, but when they are a mile or two long, and the line of sources is a continuous line several miles long, a right angle is most convenient, particularly in sectionized country. Figure 5.9 illustrates this concept and shows how a swath of reflecting points is generated. Note that the number of geophone stations per line and the number of source locations (or intervals) between the geophone lines are arbitrary. The number of offset distances that can be stacked together is related to several factors:

1. The total number of geophone groups that can be recorded simultaneously.
2. The number of geophone groups per line.
3. The maximum and minimum offset distances that can be used.
4. Receiver separations.
5. Source separations.

Economic and logistic factors contribute to this decision. It is intuitively obvious that this method of laying out the sources and receivers—and communication of the collected information to the recording truck—must be more time-consuming and/or require more manpower than a conventional CDP profile; a great deal more information is being gathered, however. Probably the factors that have the greatest control of the use of the method are economics and the need for advanced communication between the geophone groups and the recording vehicle.

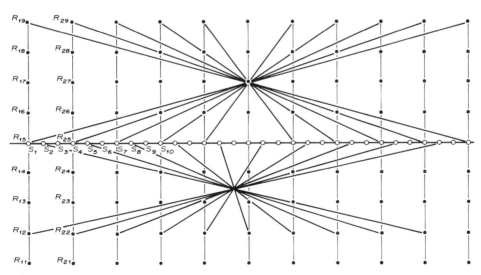

FIGURE 5.9 The seismic swath arrangement. Sources stay along the central horizontal line. Receivers occupy lines at right angles. If 99 traces can be recorded simultaneously they will yield a 10-fold stack or more.

Additional field problems are the following:

1. If linear groups of geophones are laid out at each group location and a linear set of sources is used they cannot be colinear for all pairs of sources and receivers used. Consideration should be given to the use of radial patterns of geophones.
2. The vertical and horizontal locations of geophones must be known—just as in linear methods of surveying—but are more time consuming to obtain. New geodetic surveying methods are needed to do this quickly and accurately.
3. The permit problem may be much more difficult in some areas.

The difficulties are not discussed further but they do represent problems that will have to be solved before CDP information in two dimensions, with the same quality as the linear CDP method, becomes a routine procedure.

The Wide Line Profiling® method of CGG, Michon (1972), treats the need for three-dimensional information from a slightly different point of view. The main objective is to provide knowledge of the component of dip at right angles to the main line of profile. This is useful to differentiate between arrivals recorded from different locations and directions but has not been emphasized as a method of tracing sedimentary trends within are wide swath of data. The field method differs little from that already described and is shown in Figure 5.10.

To generalize two-dimensional seismic surveying even further, there are, of course, two-dimensional analogs of the crooked line profile, and in very rough terrain the area can be reached only by heavy surface sources along a few intersecting trails. The more portable geophone stations can be located more easily inside the source loops (see Figure 5.11), provided cable linkage or recording by digital radio communication is possible. Communication and surveying problems are sometimes a nightmare but can be solved if sufficient economic

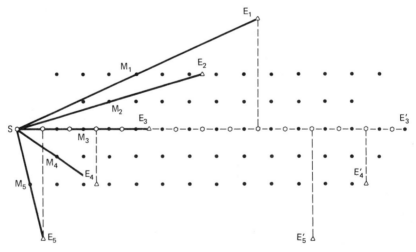

FIGURE 5.10 The method of Wide Line Profiling: E_i represents a shot point, S, a receiver, and M, a corresponding reflecting point. (*Source*: Michon. Reprinted with permission from the *Oil and Gas Journal*, November 27, 1972.)

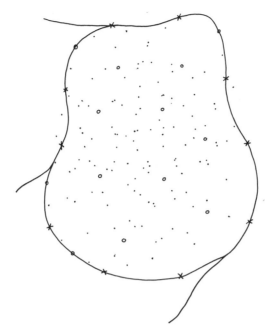

FIGURE 5.11 Two-dimensional surveying in rough terrain. Small circles show portable geophone nest locations, crosses are source locations on roads, and the dots show locations of all the reflecting points from this sparse source–receiver system.

incentive exists. Naturally, the reflecting points on the subsurface are irregularly spaced but suitable computer programs can be devised to calculate their (x, y) coordinates and to group them accordingly to preselected rules.

For P-wave work the use of linear geophone or source arrays is pointless. Circular or disk-type arrays would be acceptable but are difficult to lay out in rough terrain. For shear-wave work it is necessary to use horizontal geophones in orthogonal pairs and to combine the traces obtained in the computer.

All these methods have been used on land but sometimes their implementation in marine surveys is more difficult. Hedberg (1971) proposed the use of sources whose positions can be controlled in two dimensions by paravanes to give three-dimensional information in marine surveys. Other suggestions based on the use of paravanes have been made. Figure 5.12 shows Hedberg's scheme. In theory, of course, there is no reason why these methods should not be successful. The difficulties occur largely because of currents and tides in the ocean and the characteristics of paravanes—lift versus drag, and so on. Furthermore, if the swath method could be carried out it would be difficult to navigate this system in relatively crowded waters; for example, around already existing drilling platforms.

It is paradoxical that in the last few years 3-D surveys have been made more commonly at sea than on land. Although no new efficient method has been developed, quality control of positioning, combined with real time display for the navigator of the ship, allows accurate navigation of a preset pattern of survey traverses and documentation for final control during the data-processing phase. Thus a rectangular pattern of common midpoints (depth points) is obtained by a

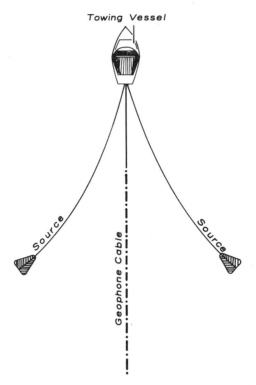

FIGURE 5.12 A marine three-dimensional method suggested by Hedberg (1971). It has since been indicated that the source paravanes can be steerable and servo-controlled.

series of meticulously controlled survey lines, parallel to one another. In some cases use is made of two sources and/or two cables deployed simultaneously.

5.6. EXTERNAL FACTORS THAT AFFECT THE SUCCESS OF SEISMIC SURVEYS

In common with all other physical experimentation, the measurements made in geophysical exploration are contaminated by unavoidable errors. Seismic methods, of course, include recording the motions of the earth over relatively long periods of time, motions derived from stresses purposely introduced and random stresses induced by factors over which we have no control. The continuous variations caused by these random stresses must be regarded as noise—an unwanted component of the recorded motion at any point. It is to be expected that the noise component is uncorrelated statistically with the signal contributed by the source. In general, the noise field varies with time and position, but because we expect to be dealing with wavelengths of the order of tens of meters and periods of tens of milliseconds it is obvious that there can be areas within which the noise can be regarded as coherent. Samples taken within this area will correlate to a high degree.

There are many contributors to the noise observed on a seismic trace and it is convenient to divide them into two categories:

1. Natural (rarely completely controllable).
2. Artificial (controllable to some degree).

Natural Noise

The natural phenomena that contribute to noise are chiefly wind, waves (in marine work), microseisms, and animals.

These factors, in addition to sometimes having a direct action, give rise to a seismic wavefield, both time- and space-varying. The nature of the wavefield depends largely on the near-surface conditions; for example, the areas of coherency in places where surface cracks are abundant are much smaller than where the near-surface is spatially consistent in its constitution. The presence of trees and/or small bushes can, by their response to the wind and seismic wave generation by their root systems, cause local variations in the wavefield.

Direct action of the wind on the geophones or their attached cables can often be prohibitively large and can *sometimes* be alleviated by burying the geophones after they have been carefully and firmly *planted* in the earth. Note well that a loosely planted geophone cannot be tolerated. It is beneficial to a successful survey and to personnel training to require constant monitoring of geophone planting by high-level field personnel. Drying of the soil, melting of ice and snow, blowing away of loose sand, the actions of curious onlookers (both people and animals)—all can wreak havoc with a carefully set out group of detectors. Burial of the geophones is usually helpful when wind speeds are below 30 kph (20 mph) and when no stalks, leaves, or other parts of brush or grass touch the instruments. When the wind exceeds this limit, however, the motion of the air over surface irregularities—gullies, hills, houses, etc.—causes differential pressures to be applied to the earth's surface and seismic waves to be produced, in which case the survey should probably be suspended until the wind dies down (possibly in the evening or more probably in a matter of days). In an emergency it may be necessary to increase the source strength to achieve an acceptable signal/noise ratio.

Similar but different effects occur in marine prospecting. Fortunately the detectors are deployed at some depth (usually 10 to 13 m) below the surface, where the hydrophones are not so sensitive to wave action. Nevertheless, the sea state eventually dictates an end to successful recording because of pressure variation and the effect of the boat's motion on the cable, which induces waves in the cable itself. It is not so easy in the normally smooth-flowing sequence of events of a marine survey to overpower the noise by increasing the signal strength. That can be done only on a long-term basis by increasing the number and power of the sources available.

Generally, the most prevalent methods of increasing the signal/noise ratio of seismic records are (1) to increase the power of the source, (2) to average out the noise by increasing the number of independent samples, and (3) to move receivers to less noisy locations, if available.

Artificial Noise (Power Lines)

A problem that has plagued the seismic prospector since the very early days is that of *electrical* pickup from power lines present in the area being surveyed. These power lines operate at a number of frequencies (60 Hz in America, 50 Hz in Europe for main power and $16\frac{2}{3}$ Hz for electric railway power) and their harmonic currents are also induced in the earth.

When these power lines are in operation resistive and capacitative leakage from the seismic cables to the earth produce voltages at the input of the seismic amplifiers that can be much greater than the voltages produced by the geophones, thus ruining any chance of detection of seismic reflections. Various schemes have been evolved for alleviating this effect (Ensing, 1983). We must be aware of the need for at least a preliminary elimination in the field, even if there are facilities for a final *clean-up* of the unwanted frequencies in the later data-processing stage (qv).

Some countermeasures that can be taken in the field and their problems are the following:

1. Suppression by means of a balancing (or *hum-bucking*) circuit at the input of the seismic amplifier. These circuits usually use two potentiometers for each amplifier to balance out the in-phase and out-of-phase components of the picked-up noise. The problem here is that the load on the electrical power line is rarely constant. It takes several seconds for a skilled operator to balance the input for each amplifier, and with multiple amplifiers there is a high probability that the power-line interference will change before all balancing is finished.

Ensing (1983) has described this method in detail and has included automatic balancing methods that are rapid and effective.

2. Suppression by common mode attenuation is possible with a special input circuit to modern amplifiers. These are the differential amplifiers. When signals are balanced about a center tap ground it is possible to reduce the voltages common to both sides of the input by as much as 120 db (amplitude ratio—10^6). Nevertheless when connected to an unbalanced input line, caused by capacitative or resistive leakage, this capability cannot be used successfully. Methods of balancing the input line have been described in detail by Smithers and Pater (1982). When this common mode attenuator is used the full common mode rejection ratio of the amplifier may be used for rejection of the power-line interference.

The design of the common mode attenuator must be regarded as specialist material and is not discussed further here. Laboratory and field tests have demonstrated that common mode signals of any frequency and of amplitude 100 mV can be reduced to an effective 1 μV of interference.

3. Suppression in the field can be accomplished when Vibroseis sources are used by the following technique:

The full seismic source input for any location can be made to consist of an even number of sweeps and the plan is to operate one-half positively and one-half negatively.

This can be accomplished in two different ways:

1. By having a reversing switch on each vibrator, which is operated alternately, *and* at the same time allowing the initiation of the sweep to take place
   only when the (separately) monitored power-line interference reaches a peak.
   Then, because the input to the compositing unit (the adder of separate
   operations) consists of power-line pickup *plus* earth response for one sweep
   and power-line pickup *minus* earth response for the next consecutive sweep,
   the subtraction of alternative inputs to the compositing units will yield twice
   the seismic earth response. Full power-line pickup elimination occurs only
   when the power line operates in a constant manner (i.e., there are no power
   fluctuations). The problem, of course, is to *ensure* that all reversals required
   are actually made.
2. A better method results from leaving the vibrators alone but triggering the
   sweeps on alternative peaks and troughs of the power-line pickup. The
   successive (all positive) inputs to the compositing unit are earth response *plus*
   power-line interference and, alternately, earth response *minus* power-line
   interference.

Note that if either of these methods is accomplished successfully it works for all
power-line frequencies and harmonics. It is also to be noted that these alter-
nations *must* accomplish *some* reduction even if it is not the complete elimination
of pickup that is desired.

3. The use of a narrow-band reject filter is often regarded as a third possibility. It
   should then be realized, however, that one reject filter is necessary for every
   frequency concerned and that analog null filters have large phase effects on all
   other frequencies recorded. They do, of course, also filter out their own
   characteristic frequencies from the *seismic* input. In data processing zero-phase
   null filters can be constructed and the interested reader should refer to the
   proper section in Chapter 6.

In summary, attempts should be made a reduce the power-line interference in the
field—if the seismic traverses *have to* be placed near power lines. Avoidance is the
preferred procedure, but if that is not possible the signal/noise should be
improved by (1) keeping leakage of cables and geophones to ground to a
minimum, (2) balancing out (hum-bucking) automatically whenever possible, (3)
adopting the foregoing field procedure, and (4) adopting the use of special
common mode attenuators with high-quality CMRR amplifiers.

Artificial Noise (Traffic and Airborne Noise)

Traffic noise, people walking near the geophone spread and other ground
disturbances, and the noise of aircraft are the bane of an observer's life because
these noises are picked up while useful seismic information is also being recorded.
Some field carelessness has been caused by the reputation of the Vibroseis system
"that it is not sensitive to field noises, since they do not correlate with the vibrator
signals." Although, in principle, there is a modicum of truth in this belief, it must
be remembered that traffic noise and the motion of people and animals may cause

seismic waves that, in the location of some geophones, are much larger than the reflection signals being recorded. It is well known that recording along roads that normally carry heavy vehicular traffic is better done at night when the traffic density is lowest.

Other than that partial remedy, the use of multiple geophones and an increase in source strength and time—more heavier vibrators and more and longer sweeps—are the only recourses available to the production of a larger signal/noise ratio.

5.7 THE MEASUREMENTS OF AUXILIARY SEISMIC INFORMATION

In addition to the generation and reception of seismic reflections, other data are needed to make the necessary corrections for the topography and varying thicknesses of the low-velocity layer. Although attempts are now being made to provide weathering corrections for surface source use from the reflection data by methods to be described later, these techniques have disadvantages that sometimes can be alleviated by making direct measurements.

It is worthwhile at several points in an area to have complete knowledge of near-surface velocities, both P and S wave. Beeston and McEvilly (1977) have shown how this can be done in a 6-in.-diameter hole, cased with 3-in.-diameter plastic pipe, and back-filled around the pipe with pea gravel. Care must be taken to ensure that this material settles slowly without bridging or development of large voids. Three component orthogonal 4.5-Hz geophones are enclosed in a waterproofed sonde that can be clamped to the casing. Filtering is used to remove unwanted noise usually by allowing a recording band of 30 to 150 Hz. Once set this band remains unchanged to avoid possible phase shifts between records.

The source for P waves is a hammer blow on a metal plate about 1 m from the top of the hole. S waves can be generated by horizontal (wooden) hammer blows on the end of a wooden plank 8 ft long, held down on the ground by the front wheels of a light truck. Apparently light to moderate blows that do not cause strains above the elastic limit of the soil are better than heavy blows. The latter generate unwanted P waves and the S waves generated tend to be incoherent.

Sample record groups for down-hole surveys 10 ft apart are shown in Figure 5.13 and a composite time-depth graph, in Figure 5.14. I am indebted to Professor McEvilly for drawing my attention to this work and for giving his permission to use these figures. The quality of these results, made from the carefully designed experiments, speak for themselves. Of course, not all holes in an area need to be surveyed with this degree of detail. Most of the time corrections are based on simple up-hole times.

For source shots of dynamite the procedure is a simple one that consists of positioning a geophone (cluster) nearby the hole and recording its output. The uphole time, the depth of the shot, and measurements of the weathering and subweathering velocities provide all the information necessary. Figure 5.15a shows this diagramatically and also shows a method of obtaining the necessary data. Alternatively (Figure 5.15b), use can be made of the fact that the times of the first arrivals (first breaks) are those of waves whose travel paths to successive

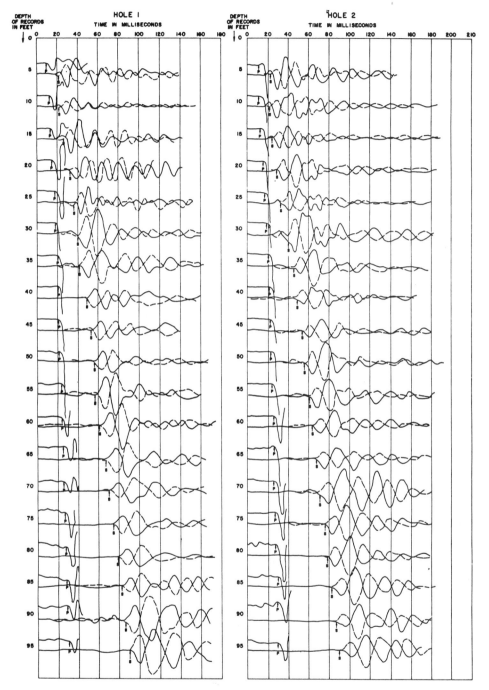

FIGURE 5.13 Two record groups, showing P-wave and S-wave arrivals for down-hole surveys 10 ft apart. [*Source*: H. E. Beeston and T. V. McEvilly (1977) with permission from John Wiley & Sons, Ltd.]

195

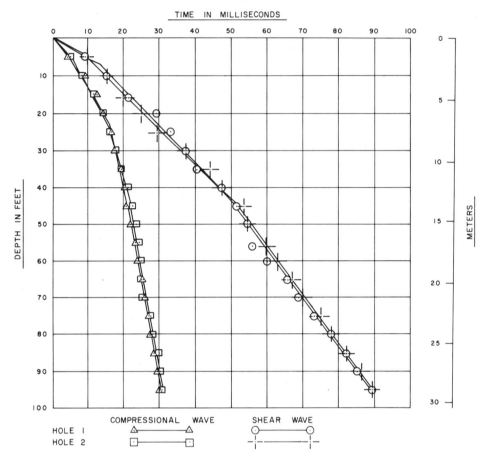

FIGURE 5.14 A composite time-depth graph for two holes 10 ft apart. [*Source*: H. E. Beeston and T. V. McEvilly (1977) with permission from John Wiley & Sons, Ltd.]

geophones are predominantly horizontal through the subweathering region and then almost vertical through the weathered zone. The differences in weathering time can be calculated if allowance is made for horizontal travel distance differences. By comparison with up-hole time information these times can be made absolute.

For surface sources, however, a different weathering-time procedure is necessary (Figure 5.16). A succession of shot points is shot into receivers at the next two successive shot points and the arrival times are recorded. This procedure is repeated down the line. For any three successive points, A, B, and C, the times t_{AC}, t_{AB}, and t_{CB} will (eventually) be known and the weathering time at B is

$$t_{WB} = \frac{t_{AB} + t_{CB} - t_{AC}}{2}$$

This is a simple procedure and works well when the weathered layer can be regarded as a simple two-layer case. When, however, there is an appreciable

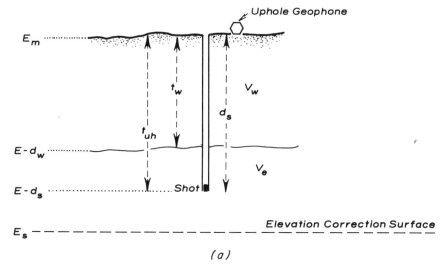

(a)

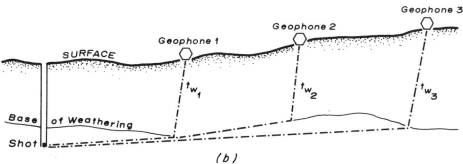

(b)

FIGURE 5.15 (*a*) Determination of corrections for a shot below the weathered layer. (*b*) Determination of differential weathering information by refraction arrivals.

gradient in velocity below the base of the weathering the horizontal travel times are not additive and the method fails.

In some areas, in conjunction with a down-hole survey similar to that of Beeston and McEvilly (1977), it might be desirable to conduct a more extensive refraction survey. The description of procedure that follows provides minimal information for conducting a survey and obtaining preliminary information about

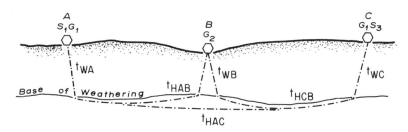

FIGURE 5.16 The ABC refraction weathering method.

the near-surface velocities. Readers are advised, however, to supplement this description by referring to relevant material in *Seismic Refraction Prospecting* (1967) edited by Albert W. Musgrave and published by the Society of Exploration Geophysicists.

Ideally the field setup is a set of single geophones spaced at equal distances along a straight line on the surface. If it is required that SH-wave refractions, as well as P-wave refractions, be used (each with its appropriate source) the geophones needed must be a combination of vertical and horizontal instruments, properly oriented. Standard three-components geophones are available for vertical and two horizontal components. As a rule of thumb, the total spread length should be four to five times the depth to be investigated. If a straight line is not available because of terrain difficulties surveying must supply airline distances and even more care is necessary to keep the horizontal geophones oriented parallel to one another and to the source.

The sources are used at each end and a record is made of all geophone signals. Some modern refraction instruments will accept the records from successive source pulses for composition, thereby achieving signal/noise improvement. With vibrators available, the Vibroseis® method may be used with as many sweeps as desired to record the refraction arrivals. Of course, cross correlation between the control and field traces is a requisite before interpretation. Times of arrival of the first energy are plotted on a time-distance graph similar to Figure 5.17.

These lines should be shot from each end, as shown in the graph. The arrivals on the Vibroseis record are, of course, timed at the correlation peaks. It is important that a bandwidth as wide as possible be used to improve the *definition* or *standout* of the refraction arrivals.

We look first at the theory for the horizontal, plane parallel layered case. It is a matter of simple geometry when the velocity increases with depth to see (Figure 5.18) that the time of refraction arrival at a geophone distance X from the source is given by

$$t = \frac{2h \cos i_1}{V_1} + \frac{2h_2 \cos i_2}{V_2} + \cdots + \frac{2h_{n-1} \cos i_{n-1}}{V_{n-1}} + \frac{X}{V_n}$$

When t is plotted against X a series of straight lines corresponds to the refracted

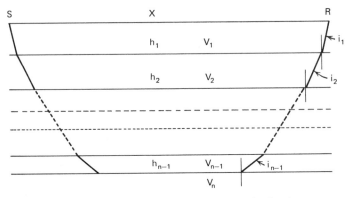

FIGURE 5.17 A diagrammatic model of an *n*-layered refraction system.

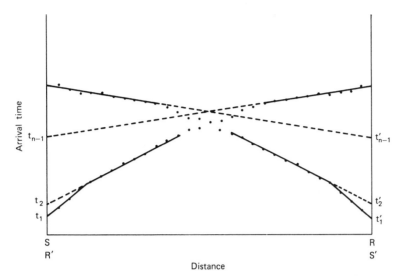

FIGURE 5.18 A typical time-distance graph for a reversed, horizontal, parallel-layered system. In the nonhorizontal system symmetry above the vertical center line would not exist.

arrivals from successive layers, these lines have slopes defined as (dX/dt) of $V_2, V_3, \ldots, V_n$ and intercept times of $t_1, t_2, t_3, \ldots, t_{n-1}$, where

$$t_1 = \frac{2h_1 \cos i_1}{V_1}; \qquad t_2 = \frac{2h_2 \cos i_2}{V_2} + t_1; \qquad t_3 = \frac{2h_3 \cos i_3}{V_3} + t_1 + t_2$$

and so on. Snell's law relates these angles:

$$\sin i_{j-1} = \frac{V_{j-1}}{V_j} \sin i_j; \qquad 2 \leq j < n$$

In practice, the main difficulty arises for one of two reasons:

1. It is difficult to distinguish unambiguously the various straight lines and to pair them up on the graphs obtained from surveys in opposite directions.
2. The velocity in the layers does not increase monotonically.

We can now examine the simplest case in which a survey is made in a single layer, the base of which dips in the direction of the survey. More complicated cases have been worked out and solutions are given in the literature.

Figure 5.19 is a composite diagram that shows a single source point S, used to survey along a down-dip direction SD or along an up-dip direction SC. A theoretical surface line SR is drawn to bisect the angle between SD and SC.

AB is the refractor surface, with velocity V_1 above and V_2 below.

For the single-layer case X_1 or X_2 is measured along the appropriate surface and is converted to a X_u or X_d along the theoretical horizontal surface. It can be seen from the geometry that

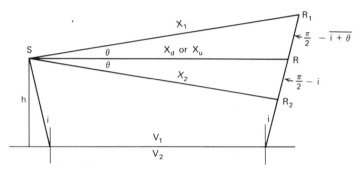

FIGURE 5.19 A diagram of a dipping, single-layer refracting system. The up-dip and down-dip have a single source point and the diagrams have been superimposed for easy comparison.

$$X_d = \frac{X_1 \cos \overline{i + \theta}}{\cos i} \; ; \qquad X_u = \frac{X_2 \cos \overline{i - \theta}}{\cos i} \tag{5.1}$$

For down-dip the total time is

$$t_d = \frac{2h \cos i}{V_1} + \frac{X_1 \sin \theta}{V_1 \cos i} + \frac{X_1 \cos \overline{i + \theta}}{V_2 \cos i} \tag{5.2}$$

and for up-dip

$$t_u = \frac{2h \cos i}{V_1} - \frac{X_2 \sin \theta}{V_1 \cos i} + \frac{X_2 \cos \overline{i - \theta}}{V_2 \cos i} \tag{5.3}$$

By differentiating (5.2) and (5.3) with respect to X we obtain *slownesses* (reciprocal apparent velocities):

$$\frac{dt_d}{dX} = \left(\frac{1}{V_{2d}}\right) = \frac{\sin \theta}{V_1 \cos i} + \frac{\cos \overline{i + \theta}}{V_2 \cos i} \tag{5.4}$$

$$\frac{dt_u}{dX} = \left(\frac{1}{V_{2u}}\right) = \frac{-\sin \theta}{V_1 \cos i} + \frac{\cos \overline{i - \theta}}{V_2 \cos i} \tag{5.5}$$

After a little manipulation

$$\left(\frac{1}{V_{2d}} + \frac{1}{V_{2u}}\right) = \frac{2 \cos \theta}{V_2} \tag{5.6}$$

$$\left(\frac{1}{V_{2d}} - \frac{1}{V_{2u}}\right) = \frac{2 \sin \theta \cos i}{V_1} = \frac{\sin \theta \cdot t_0}{h} \tag{5.7}$$

where t_0 is the (common) intercept time. Note also that $t_{u0} = t_{d0} = 2h \cos i/V_1$ (because the same shotpoint was used for each experiment, h is the perpendicular distance from S to the refractor).

Practically, for small angles of dip (θ), the slownesses may be averaged to get an approximate inverse velocity in the lower medium. The measure of h and V_1 is not so satisfactory, however. In fact, this is one of the difficulties in all refraction surveys.

The method is capable of being extended to more than one layer but is not discussed here.

5.8 SUMMARY AND CONCLUSIONS

Data-gathering methods for exploration seismology have changed from the early attempts to use isolated measurement points (common for geological well data) to continuous lines in which the measuring points are much closer together—thereby substituting decisions on correlations with several much smaller changes in record characteristics for a single, larger decision between points widely separated. Some reasons for this shift, which is undesirable from the exploration viewpoint, are the following:

1. The frequency bandwidth obtained with seismic data is insufficient to give a good chance of making undisputed correlations of reflections (and their associated geological boundaries).
2. A need existed to improve the signal/noise ratio on seismic recordings to prevent surface waves, multiple reflections, scattered interference, and random noise from complicating the process of picking corresponding reflection times. Much of this improvement has been achieved (after normal geometrical corrections) by stacking traces with a common reflection point but different source-to-receiver offsets.

An additional need for gathering three-dimensional information can be accomplished, of course, by running several closely spaced parallel CDP lines, but a swath method in which the source or receiver positions are common for several parallel lines of reflection points and are recorded simultaneously is preferred. Swath methods rely on the simultaneous recording of a large number of traces, and the methods of transmitting these data from the recording stations to a central recorder—or methods of recording several traces locally and interlacing the data later—are of great importance. It is likely that substantial advances in this art will soon be achieved.

In marine prospecting single-line CDP methods are easily carried out, but three-dimensional information is difficult to obtain. With improvement in navigation equipment it is now possible to position a boat accurately enough to allow 3-D marine surveys to be made by a sequence of closely spaced parallel lines.

REFERENCES

Beeston, H. E., and McEvilly, T. V. (1977), "Shear Wave Velocities from Down-Hole Measurements," in *Earthquake Engineering and Structural Dynamics*, Vol. 5, Wiley, Chichester, England, pp. 181–190.

Davis, J. L. (1975), U.S. Patent No. 3,890,593, June 17, 1975.

Ensing, L. (1983), "The Autobalancer—an Automatic System for the Reduction of Power-line Interference with Seismic Signals," *Geophysical Prospecting*, Vol. 31, pp. 591–607.

Hedberg, R. M. (1971), U.S. Patent No. 3,581,273, May 25, 1971.

Marr, J. D., and Zagst, E. F. (1967), "Exploration Horizons from New Seismic Concepts of CDP and Digital Processing," *Geophysics*, Vol. 32, No. 2, pp. 207–224.

Mayne, W. H. (1962), "Common Reflection Point Horizontal Stacking Techniques," *Geophysics*, Vol. 27, No. 6, Part II, pp. 927–938.

Mayne, W. H. (1967), "Practical Considerations in the Use of Common Reflection Point Techniques," *Geophysics*, Vol. 32, No. 2, pp. 225–229.

Michon, D. (1972), "Wide Line Profiling Offers Advantages," *Oil and Gas Journal*, Vol. 70, No. 48, pp. 117–120, November 27, 1972.

Smithers, M. A., and Pater, A. (1982), "Common Mode Interference Reduction in Seismic Recording," *Geophysics*, Vol. 47, pp. 1672–1680.

Waters, K. H. (1977), U.S. Patent No. 4033278, July 5, 1977.

Ziolkowski, A. (1979), "Seismic Profiling for Coal on Land," in *Developments in Geophysical Exploration Methods*, edited by A. A. Fitch, Applied Science Publishers, Ltd., Barking, England.

Ziolkowski, A., and Lerwill, W. E. (1979), "A Simple Approach to High Resolution Seismic Profiling for Coal," *Geophysical Prospecting*, Vol. 27, pp. 360–393.

SIX

Seismic Data Processing

6.1 INTRODUCTION

The data obtained in a typical seismic survey consist of measurements of the motion of the surface at a series of surface positions that follows the operation of a source of seismic waves. This motion may be particle displacement or velocity, which in its most general form is a vector function of time. In certain cases the scalar excess pressure can be measured as a function of time. An important property of sets of time functions is that they are recorded simultaneously or at least have an indication of operation of the source.

Usually the data most relevant to an exploration geophysicist, are the reflections from the geological strata. For this reason methods are needed for treating the raw data, which deemphasize all but the reflection information. As seismic reflection exploration methods developed, data processing became increasingly necessary. The advent of magnetic tape recording in the early 1950s began to demonstrate the possibilities of filtering, stacking, and rerecording. These methods dealt with continuous recordings of time functions (analog data).

The Geophysical Analysis Group (GAG) project, sponsored by several oil companies at M.I.T. (1953 onward), was the first major indication that digital computing with sampled values from continuous time recordings would eventually become a versatile means of seismic data processing. Many of the concepts of filtering with sampled data originated there (Treitel and Robinson, 1967), but a major advance in digital technology—originally called *convolution filtering equipment* (CFE) but later falling under the general heading of *array processors* (AP)—was necessary before a full range of data-processing techniques could be implemented economically.

It is probably not an exaggeration to say that the common depth point method, which involves correction for geometrical effects on offset traces and the addition of many corrected traces selected from a large number of those available to give a final CDP trace, owed its overwhelming success and acceptance to the speed and consistency afforded by the digital computer. In turn, it was the data-quality improvement provided by the CDP method that led initially to wholesale use of digital processing. Once computers were available and the speedup induced by array processors made other processes economical the digital revolution was on its way. Without taking anything away from the genius displayed by the inventors of the Vibroseis system, the full potential was fully evident only after digital computers and their accessories became fast enough (c. 1963) for correlation processes to be done digitally.

This chapter deals with the processes and methods of accomplishing them digitally that are now in use to improve seismic *reflection* data. In only a few examples of exceptional interest is reference made to analog methods.

6.2 SAMPLING (MULTIPLEX) PROCEDURES

Sampling of multitrace seismic records can be done directly in the field or by recording continuous time traces on analog magnetic tape and replaying these records later to provide the electrical signals. In either case the process normally involves taking a sample sequentially from a number n of traces and then returning to the first trace (well within the sampling interval) and recycling through the sampling process. This system provides a series of numbers in the sequence

$$\{a_{11}, a_{21}, \ldots, a_{N1}, a_{12}, a_{22}, \ldots, a_{N2}, \ldots, a_{NM}\}$$

where the first index is the trace number and the second index is the sample number (at a predetermined number of milliseconds apart) on a particular trace. The electronic consequences are easy to see (Figure 6.1). The allowed time for the digitization of N traces must be less than the sample interval (typically 0.002 or 0.004 s) which, in turn, is prescribed by the frequency spectrum to be retained. In the recording system itself an antialias filter is provided, which rejects very strongly (-50 to -60 dB) frequencies that with the selected digital interval may give rise to aliasing errors; for example, suppose a digital interval of 0.004 s is chosen. The Nyquist (or folding) frequency will be $1/0.004 \times 2$ or 125 Hz; that is, at 125 Hz there are two samples per cycle. Because this is a theoretical limit, it is usual to provide an antialiasing filter with a high reject capability above 62.5 Hz (the four-samples-per-cycle frequency). The exact filter cutoff points and the design of the filters are manufacturer's decisions.

Thus a decision to digitize at 0.004-s intervals makes the assumption that

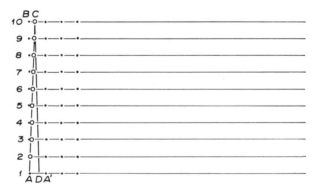

FIGURE 6.1   Deviation of digitizing from ideal sampling. Digital samples are required at the dots on the time traces 1 through 10. Actual samples are obtained at the positions of the open circles because of finite digitizing time (the path AC). Resetting the digitizer occurs along CD; DA' represents the waiting time until the next timing pulse initiates further digitizing.

energy with frequencies above (say) 62.5 Hz is mainly noise and is of no interest in the gathering of geological information. For digital recording in the field this important decision must be made before the survey is started. Usually the next step provided is 0.002 s, which raises the accepted frequency band to 125 Hz but also doubles the number of digital sample points for a given maximum recording time. In coal-exploitation seismic work sometimes frequencies as high as 500 Hz may be required, in which case the sampling interval must be 0.0005 s. The cost of digital processing is not always a linear function of the number of sample points, but the increased cost of processing double the number of points is never insignificant and is therefore an important decision.

It has been stated that the time for sampling N traces must be significantly less than the smallest digital interval available. For illustration purposes we suppose that the sampling time for 40 traces is 0.001 s or less. Then each trace must be digitized within 25 μs and the digitizer must step over to the next trace voltage to be sampled.

Details of the electronics are not within the scope of this book, but the general methods adopted are explained. Almost universally, digitization, recording, and subsequent digital computation are conducted in *binary* notation. This is a method of representing a number by using as a basic the number 2. As in decimal notation, in which we consider an integer (say) 461 as related to powers of 10,

$$4 \times 10^2 + 6 \times 10^1 + 1 \times 10^0$$

so can the same number be expressed as powers of 2

$$1 \times 2^8 + 1 \times 2^7 + 1 \times 2^6 + 0 \times 2^5 + 0 \times 2^4$$
$$256 \ + \ 128 \ + \ 64 \ + \ 0 \ + \ 0$$
$$+ 1 \times 2^3 + 1 \times 2^2 + 0 \times 2^1 + 1 \times 2^0$$
$$+ \ 8 \ + \ 4 \ + \ 0 \ + \ 1$$

or, in contracted notation,

$$111001101$$

where the power of 2, starting at 0 on the right, is indicated by the position of the digit in the number. A binary digit is called a bit. Sometimes eight bits together are called a byte.

These numbers can be continued as fractions because $\frac{1}{2}$ is represented as 2^{-1}, $\frac{1}{4}$ as 2^{-2}, and so on. Thus a binary fraction is written

$$110011 \quad (=0.796875)$$

In digital engineering practice the use of binary numbers is convenient because switching circuits can then be selected in which 1 (say) represents the circuit in one stable state and 0 represents the *only other* stable state. These *bistable* circuits are the basic components of all computer engineering. For most geophysicists the internal procedures are of little interest because they are concerned only with the visible recordings of electric signals generated in the reflection process. For them it is proper to point out that the computed results (still in binary form) are rendered tractable to plotters by the use of digital-to-analog converters.

After this brief diversion we return to the subject of sampling (or analog-to-digital) procedures.

If the total range of the voltage to be measured is considered 1 unit (perhaps 10 V) on the binary scale it will be 2^0 and 5 V will be $\frac{1}{2}$ unit $= 2^{-1}$. Alternatively, the smallest increment in voltage to be measured can be set equal to 2^{-N} units and voltages can be measured in terms of this increment. A voltage given as 1011 is 11 (decimal) increments.

The first method (see Figure 6.2*a*) consists in comparing a voltage provided against the required voltage. The comparison is a stair step which increases in equal increments according to the stepping of a binary counter. Thus the reading of the counter when the comparator indicates equality (or change in sign)

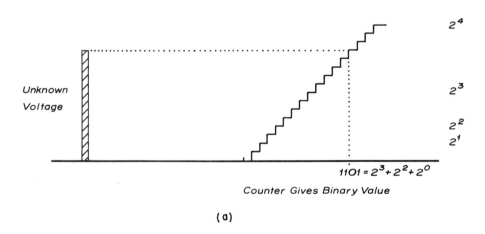

(a)

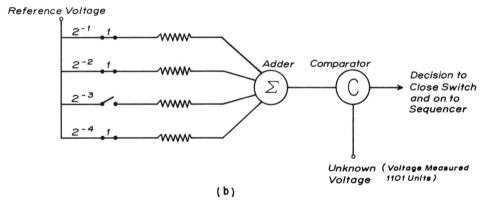

(b)

FIGURE 6.2 Digitization methods: (*a*) sequential counter method; (*b*) the faster sequential adder method.

provides a measure of the unknown voltage. This method is slow because it has to cover the entire range from zero to the unknown voltage each time. The sign of the voltage is an additional bit set separately. When a total of eight bits (one byte) is provided the range can be (almost) -2^8 to $+2^8$ or, in binary notation, (1)1111111 to (0)1111111; the digit in parentheses is the sign bit.

The second method (Figure 6.2b) is faster. The logic consists in dividing up the total voltage into ranges related by the binary scale and choosing a system in which these binary-related voltages can be added and compared with the unknown voltage. The most significant bit is tested first, and if it is less than the unknown voltage a switch is set to keep it permanently connected to the adder. The next most significant bit is added and the process, repeated. When all the bits have been added and tested the bit arrangement of the switches set is a measure of the unknown voltage. Although the method is sequential, it involves only N comparisons, compared with an average of 2^{N-1}.

Thus if N is 8 the ratio of times for same-speed electronic circuits is $8:128$ or $1:16$. If higher values of N are used the ratio increases.

For very high speeds parallel rather than sequential circuitry is available. In these circumstances digitizing speeds of 50 to 100 MHz can be realized.

The foregoing discussion applies only to seismic traces that have a known overall range for which the scale factor (or maximum range of digitization required) can be set. The precision of digitization, that is, the smallest incremental voltage change that can be detected, is set by the number of binary digits. The overall dynamic range between the largest number and the increment is also set by the number of bits; for example, a voltage range of ±255 to 1 implies a dynamic range of 48 dB in power. To deal with some of the larger ranges that may be required in seismic reflection work (e.g., when surface-wave and threshold-noise amplitudes must be correctly recorded in a linear manner) amplifiers with a binary ranging feature have been built. These wide-band amplifiers are arranged to permit the gain to be rapidly changed in steps of 2—as indicated schematically in Figure 6.3. By using these amplifiers it is arranged that the final output of the

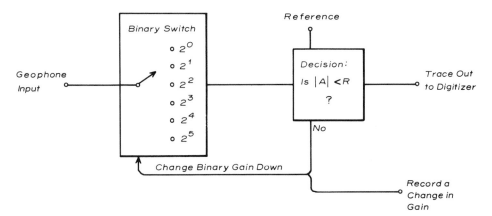

FIGURE 6.3 A schematic showing the simplistic action of the binary gain control. It is normally more important to charge the gain downward to prevent truncation of the trace amplitude than it is to change it upward. In an operational amplifier provision is made for both directions.

amplifier is a voltage that passes to the digitizer within a known range. The digitizer then produces a binary output to be recorded but, in addition, the scaling factor of the amplifier must be recorded. In logarithmic terms the digitizer output is the mantissa and the amplifier scale factor is the exponent. Because both are in binary notation, they can be adjoined in a (larger) binary work such as

$$\text{sign}$$

$$(1)\underbrace{0110}_{\text{exponent}},\underbrace{01111001}_{\text{mantissa}}$$

The exponent can be evaluated and used to shift the mantissa to the left. This number is equivalent to

$$(1)1111001000000 \qquad \text{(in binary notation)}$$

$$\text{sign}$$

or

$$-7744$$

The effective linear range of the amplifier (dynamic range) is therefore considerably increased, but the accuracy (in any binary range) is still the same.

It is obvious that these sampled values have to be stored in some manner. In the first instance [perhaps for one record, perhaps (in the field) for a plurality of records] they are held in the instrument's internal (or disk) memory, but at the first convenient time they are dumped onto the universal transfer memory afforded by magnetic tape.

The Society of Exploration Geophysicists established a committee in 1966 to examine the proliferating number of recording formats and to recommend some industry standards. In 1967 the subcommittees responsible for field tape and exchange tape formats produced their SEG-A, SEG-B formats for field tapes and SEG-X standard format for exchange tapes (Northwood et al., 1967). Since that time a number of instrumental advances have been made; for example, changes in nine-track tape speeds and field compositing needs. In 1972 a reconstituted committee formulated and published the SEG-C standard (Meiners et al., 1972).

The renamed digital recording commitee (now called technical standards committee) soon had to deal with a trend toward the use of minicomputers in the field and the necessity for a standard for demultiplexed field tapes. The result was the birth in 1975 of the SEG-Y format.

The latest efforts have been a family of formatting standards for tapes that can cope with the mushrooming use of more channels, finer sampling intervals, and data rates that are consistent with nine-channel tapes, all of which are contained in the new SEG-D formats.

It is by now obvious that this long history of valuable organization by the SEG and meritorious service by the committee members may not yet be over. The details of their work, however, can now be examined by reference to *Digital Tape Standards* published by the Society of Exploration Geophysicists (1980).

Careful perusal of these specifications reveals that each string of data is preceded by one or more *header* blocks. The first, the *required* general header, contains information from the field, trace number, gains, and filters. The extended headers provide additional areas for (1) equipment manufacturers to enter the necessary data (e.g., in the use of summing or compositing equipment in the field), (2) special user and survey material or a multiplicity of marine information.

6.3 DEMULTIPLEX PROCEDURES

The binary number sequence obtained in the digitization process, namely,

$$\{a_{11}, a_{21}, \ldots, a_{N1}, a_{12}, a_{22}, \ldots, a_{N2}, \ldots, a_{NM}\}$$

has multiplexed the information from N channels into a signal channel. For most digital processes it is necessary to demultiplex this single-channel sequence into a different single-channel sequence (e.g., on magnetic tape) in which the sequential values of a single trace appear in order, followed by the sequential values of the second trace, and so on, until all the N traces are recorded. The new sequence is

$$\{a_{11}, a_{12}, \ldots, a_{1M}, a_{21}, a_{22}, \ldots, a_{2M}, \ldots, a_{N1}, a_{N2}, \ldots, a_{NM}\}$$

This is a digital computer process in which numbers are read from a magnetic tape, reordered, and then recorded on a new digital tape. During the reordering process the format of the numbers may be changed; that is, numbers with binary gain may be changed to floating-point (computer-compatible) numbers or to straight binary integers or fractions. The main consequence of demultiplex programs is that the sampled values of each single trace are now in order, as are the sequential traces.

Demultiplexing is usually done off-line from the main computer—sometimes even in the field if a minicomputer or special equipment is provided. The tapes are used as input to the main computer, but this depends on the overall computer system design. In all later discussion of digital data in this book it is assumed that demultiplexing has been done and that trace-by-trace sequential sets of trace amplitudes are available. The headers are made large enough to allow new information on the trace treatment to be added as each process is completed. At the end of data processing the header will include a complete data-processing history of that trace. The format, however, may be different for each data-processing center; hence it is necessary to obtain a copy of the format when exchanged data are used.

6.4 LINE DESCRIPTION, VIBROSEIS® CORRELATION, AND REORDERING

Auxiliary information is provided in part by the land surveyor who lays out the seismic line or in marine surveys, by the electronic positioning equipment. On each trace there must be information that allows access to the elevation and position of the source and receiver for that trace. Information common to more than one trace can be kept in a central computer file for access during the entire data-processing run. The horizontal position of the source and receiver determines the offset distance and the coordinates of the reflecting point (for assumed zero-dip reflectors).

Part of the auxiliary information supplied from the field by Vibroseis crews is the group of parameters (beginning and end frequencies, type and length of tapers, and digital interval) that determines the control signal used by the vibrators. For this kind of work correlation is a necessary process after demultiplexing to provide the compressed records used in place of impulsive records for other type sources. The formula

$$F(t) = A(t) \sin 2\pi \left(at + \frac{bt^2}{2} \right) \qquad (0 < t < T)$$

generates the samples of the control signal of length T s once the form of $A(t)$ and $a, b, \Delta t, T$ is known

$$F(n) = \{ f_0, f_1, f_2, \ldots, f_i, \ldots, f_n \}; \qquad n \, \Delta t = T$$

where f_0 is the first value of the sweep used by the vibrator. It may correspond to the low-frequency or high-frequency end, depending on field practice. Then

$$\phi_{12}(i \, \Delta\tau) = \{ a_i f_0 + a_{i+1} f_1 + \cdots + a_{n+i} f_n \}; \qquad \Delta\tau = \Delta t$$

Note that the control signal is shorter than the field trace, which has the total time of reflections hoped for added on. If this is T_r and the control trace is T the field trace will have a length of

$$T_f = T_r + T$$

and will contain

$$\frac{T_f}{\Delta t} = q$$

samples. If i is incremented from 0 to $q - n$ this scalar product must be repeated for each delay time $i \, \Delta\tau$. In modern computers array processors do the correlation process in a reasonable time. As an example, a field trace 30 s long is correlated against a control signal 25 s long; the digital interval is 0.004 s for both and a cross-correlation function 5 s long is desired. The number of multiplications and additions is 25×250 for each point on the correlation function and $25 \times 250 \times 5 \times 250$ for the complete correlation function. Because each record contains, say, 48 traces, a record transformation from field type to correlation type will require $25 \times 250 \times 5 \times 250 \times 48 = 3.75 \times 10^8$ multiplications and additions. If they are done in a high-speed computer at 125 ns per multiply–add the result will take about 46 s to perform. Special purpose array processors do parallel operations, and even medium-speed computers with array processors can correlate one

48-trace record per minute. With the large amount of data, however, some attention must be paid to buffering input and output from the array processor; that is, these operations must be done in parallel with the actual computations, and the array processor must have fast access to the control signal and two field traces and must have room to deposit two correlated records. For signals of the size given above this amounts to 18,750 words of memory. Packing two values to a 32-bit computer word can usually be done to reduce this requirement to 9375 words.

After this correlation operation records are normal-length reflection records and for many programs are treated in the same way as impulsive records. They are, however, still in the order in which they were recorded; namely, a particular source point into many sequential receivers, followed by the next source point with receivers indexed upward by one. For other field geometries the sequence is different. For stacking operations, however, it is convenient to reorder the traces so that those that contribute to information at a given reflecting point occur in sequence and are followed by those at the next reflecting point, and so on. This reordering is strictly a sorting process handled by the computer and its peripheral gear under program control. The use of a few parameters enables the computer to do consistently a task that was formerly a hazardous tape-handling operation in the earlier analog tape days of CDP. In the full schedule of data processing reordering can be done at any convenient time before the removal of geometrical effects due to offset and stacking. This is the chief reason the traces carry header information with them. No matter what their number in a particular sequence, they can always be operated on by using the information they can supply.

6.5 CORRECTIONS FOR NEAR-SURFACE DELAYS AND GEOMETRICAL EFFECT OF OFFSET

These corrections are often referred to as static and dynamic because one is assumed to be constant for the entire trace, whereas the other, for offset, changes with record trace time. Although at first sight these are simple problems, the issue is often clouded by the fact that from an economic viewpoint no provision is made for special near-surface or velocity measurements. As shown later, this forces the system to determine velocities and near-surface corrections from the CDP data. This section deals first with methods of obtaining the velocity and the estimates of low-velocity-layer corrections and then shows how these values are used to correct the data in preparation for stacking.

Figure 6.4 shows the geometrical ray paths when a single constant velocity is assumed. From this we can see that the relationship

$$\frac{V^2 T_0^2}{4} = \frac{V^2 (T_0 + \Delta T)^2}{4} - \frac{X^2}{4}$$

holds and then reduces to

$$\Delta T = \frac{X^2}{V^2 (2T_0 + \Delta t)} \tag{6.1}$$

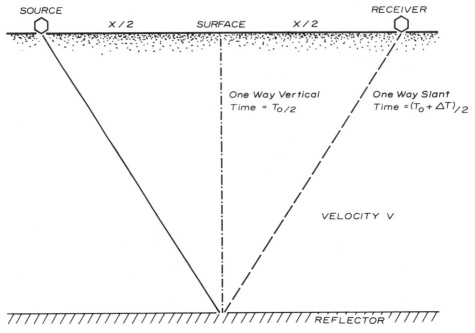

FIGURE 6.4 Geometrical ray paths for offset and perpendicular reflection paths.

For computer determination of the proper value of ΔT we can use a successive approximation:

$$\Delta T_i = \frac{X^2}{V^2(2T_0 + \Delta T_{i-1})} \qquad (6.2)$$

where the subscript denotes the order of the iteration; $\Delta T_0 = 0$ is the first assumption. The convergence is faster, however, if we write

$$\Delta T_i = \frac{2X^2}{V^2(4T_0 + \Delta T_{i-1} + \Delta T_{i-2})} \qquad (6.3)$$

because the values determined oscillate in a damped fashion about the true value.

It is usual, in applying these corrections, to prepare a table of ΔT versus T_0 for use in the computer instead of calculating the values needed each time. Figure 6.5 is a series of primary seismic profiles (sometimes called 100% records because there is no multiple fold stack) that differ from one another only in the constant velocity used to correct the NMO. For each reflection there is one velocity for which the best ΔT removal has been done. The reflection *pick* moves in a straight line across the record (the straight line may not be a constant time line because dip may be present). Thus in this simple way an average velocity to each reflector can be picked that will best align reflection events for later stacking. Because a change in delay times in the weathered layer may account for some curvature of the reflected event, this method has to be regarded as an approximation and *serves only for preliminary processing*. Several determinations are needed on each line to reach an acceptable velocity-time function for this preliminary processing.

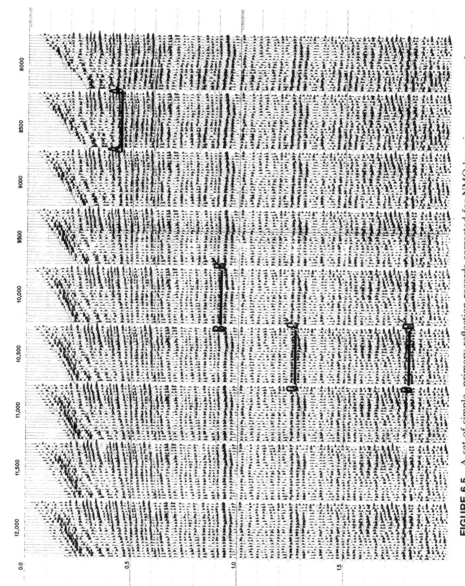

FIGURE 6.5 A set of simple, primary reflection records corrected for NMO by using a sequence of constant velocities.

213

Whenever possible a special experimental field method, known as an expanding reflection spread (Musgrave, 1962; Sattlegger, 1965), should be used to compensate for the effects of dip. It can be argued that the true velocity is not needed, but only a sufficient approximation to the true velocity that the data quality is not affected during the stacking operation to follow. Velocity variations in both horizontal and vertical directions are important, and the real need is to make a consistently good approximation to give these changes some credence.

Another approach to the measurement of velocity is to plot the square of the arrival time of a reflection against the square of the offset distance. Then, using the formula

$$\frac{V^2 T^2}{4} = Z^2 + \frac{X^2}{4} \tag{6.4}$$

the slope of the (near) straight line produces V^2. In Figure 6.6 the record is plotted to show that the distance down the record is proportional to T^2 and the distance across the traces is proportional to X^2. The best-fit straight lines can be drawn directly on this display.

Several more sophisticated computer programs have been written to compute (and display) the velocity as a function of time (Taner and Koehler, 1969). The early ones determined a velocity to a constant record time by considering a series of primary records and allowing a computer program to arrange for a comparison of the stacking effectiveness of various constant velocities for each record time. The decision criteria range from a simple measurement of the maximum sum taken across the traces within a given time interval to the more sophisticated which use the multiple-trace coherence coefficients discussed later. As computing speeds increase and unit costs of computer operations, drop, these velocity determinations are becoming more and more routine.

For some purposes connected with lithological implications of velocities there are two further needs. First, it is necessary to use a statistical velocity program

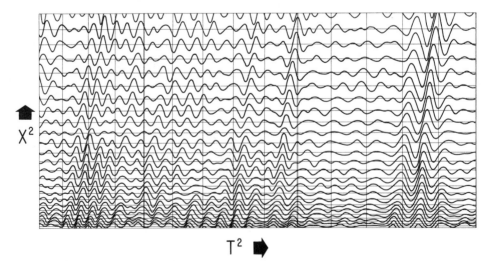

FIGURE 6.6 A seismic record is plotted to nonlinear scales. The offset X and amplitude A are plotted on a scale proportional to X^2. The arrival time is plotted on a scale proportional to T^2.

that measures velocities to a given reflection horizon (rather than a constant time) and, second, interval velocities computed from average velocities must be corrected for path curvature (deviation from the straight-path assumptions made previously). Dix (1955) showed that the true nth-layer interval velocity is related to two sequential average velocities by the formula

$$\frac{V_n^2 = \bar{V}_n^2 t_n - \bar{V}_{n-1}^2 t_{n-1}}{t_n - t_{n-1}} \tag{6.5}$$

where $\bar{V}_{n-1}$ and $\bar{V}_n$ = average velocities from the datum to reflectors above and below the layer t_{n-1} and t_n = the corresponding reflection times.

The reflection arrival times obey ideally the hyperbolic law

$$T^2 - \left(\frac{X}{2V}\right)^2 = T_0^2 \tag{6.6}$$

If, however, the reflecting horizon is itself curved this curvature will cause an incorrect hyperbola to be calculated as long as the shot point remains fixed or if the shot points differ but the receiver remains fixed. If the reflector is convex upward its curvature will be added to that of the wave front returned to the surface and a velocity that is too small will be calculated. The reverse is true for a surface that is concave upward. This fault was common to all early statistical methods and has been fully discussed by Dix (1955).

For CDP methods, however, when the receiver and source move equal distances in opposite directions on the surface and the normal to the reflecting surface is vertical at the CDP, this effect does not occur, and theoretically the correct velocity is calculated no matter what the curvature.

The effect of dip on statistically calculated velocities is not zero for CDP methods because the CDP migrates up the dip slope as the offset is increased. Levin (1971) showed that this effect is about 1% for dips of 8° but increases as the angle of the dip squared. The velocity obtained is equal to the true velocity divided by the cosine of the angle of dip. If dip and curvature are both present each will have an effect on a CDP-derived velocity that is similar to, but not so large as, the effect of curvature described earlier.

It is to be noted that any method of fitting a hyperbola by least squares to reflection times implies a method by which these times can be picked. Furthermore, the accuracy needed is such that the normal digital interval is too coarse a sampling interval and a method of interpolation is needed. The usual procedure is to pick the differential time between successive traces that corresponds to the maximum correlation coefficient between short *windows* of the seismic trace centered at the record time under consideration. Other criteria have been used and for details the reader is referred to articles listed in the references.

6.6 NMO CORRECTIONS AND THEIR EFFECT ON SEISMIC PRIMARY RECORDS

Figure 6.7 is a schematic set of impulses that might be received on a seismic record; the offsets of the traces vary from 0 to 10,000 ft and there is a constant velocity of 10,000 ft/s. The impulses denote the times of arrival only. The

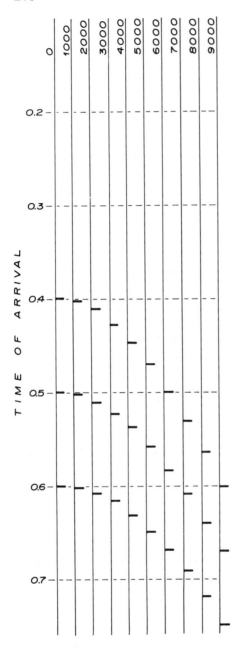

FIGURE 6.7 A schematic of an inpulse (nonreal) record show the comparative times of arrival: $X = 0 - (1000) - 9000$ ft and $V = 10,000$ ft/s.

individual reflection layer impulses form a hyperbola. These successive hyperbolas are always closer together on the far offset traces than on the near offsets. When impulses (infinite frequency response) are used the removal of the correct amount of NMO simply places all the impulses at a constant time for a given reflector regardless of offset distance.

When reflections have a finite bandwidth they overlap one another on the field record and all the traces are recorded with the same bandwidth. A serious wave shape and spectrum distortion then appears on the long-offset traces as NMO is removed. This is nonlinear distortion and filtering again after NMO removal does not give the result that would have been obtained if all traces had been recorded at wide bandwidth, the NMO removed, and followed by bandpass filtering. The latter procedure is to be preferred, but it means that a higher sampling rate on recording, followed by a resampling after NMO removal, is necessary. Another way of looking at this problem is to determine the NMO for a particular time. If the reflection event consists only of a spike the proper correction to render the reflection time independent of the offset distance may be applied. The true seismic event, however, in impulsive-type reflection work is a pulse possibly 0.1 s in length, with only its beginning occurring at the proper reflection time. The tail of the reflection event occurs at a later time, but its NMO is removed as though it *were* a later reflection. In fact, all velocities are incorrect because the main energy occurs with some delay after the true reflection time that applies to it.

It is possible to correct impulsive-type records by a filtering process known as *deconvolution*, so that, as in Vibroseis records, the main amplitude occurs at zero time. If this is done the main amplitude of a reflection will occur at the true reflection time. It is true that there are now positive and negative tails (precursors and after-runners), but they are minimal and should not cause the same problem as one-sided pulses in the same bandwidth. The reduction of impulsive-type records to zero-phase records is only as good as the information about the primary pulse shape. Deconvolution methods are considered later in this chapter.

The conversion of the offset trace to a pseudozero trace can be done in several different ways. One of the better ones is the following:

Each digitized amplitude A_i of the offset trace is mapped onto position $j + a$ of the pseudotrace, where a is a fraction. In fact, it maps between two digital positions. Its amplitude A_i is then proportioned between positions j and $j + 1$. Each position j on the pseudotrace receives contributions from more than one i position on the offset trace. It is therefore necessary to sum the contributions at the j positions before the trace is used. The pseudozero offset trace produced should be filtered with a zero-phase finite bandwidth filter. For some purposes this can be done after stacking, but the option should exist to filter here.

6.7 MUTING, TRACE ENERGY MEASUREMENT, AND STACKING

One further operation known as muting is common before the traces are ready to be stacked. It consists of zeroing out each trace to remove the high-energy, near-surface arrivals. Usually the trace is zeroed before a time given by the offset distance, divided by the muting velocity (sometimes a constant is added). Muting can be done at any convenient time.

It is also convenient to be able to measure the overall energy of a primary seismic trace. This is done by forming the sum of the squares of the amplitudes of the trace at each of the digitally sampled points:

$$E_j = \sum_{i=1}^{M} A_{ij}^2 \qquad (6.7)$$

where there are M values of the amplitude sampled on trace j.

Sometimes the sum of the absolute values is used instead of the square:

$$E'_j = \sum_{i=1}^{M} |A_{ij}| \qquad (6.8)$$

The traces that contribute to a single-stacked trace are usually scaled to a constant energy and the sample values at a constant time are averaged (where there are N traces to be added):

$$S_K(i) = \frac{1}{N} \sum_{j=1}^{N} \frac{C}{E_j} A_{ij} \qquad (6.9)$$

where C = a constant. Optionally, E_j can be replaced by E'_j.

The assumptions for treating the stacking procedure in this manner are that there will be a constant signal/noise ratio and that the individual traces will have different amplitudes because of other factors, such as the number of sources operative, the geophone placement, and gain setting. Of course, the far offset traces have traveled farther—more so at the beginning of the trace than at the end—and this should be compensated for. As shown later, when true amplitudes are discussed the overall assessment of various factors that affect amplitude is difficult.

In areas in which some traces may have considerable noise components simple scaling to constant energy allows the noisy traces too great an effect on the final trace, in which case each balanced trace can be weighted inversely on the assumption that most of the energy measured is noise energy. Then

$$S_K(i) = \frac{1}{N} \sum_{j=1}^{N} \frac{C'}{(E_j)^2} A_{ij} \qquad (6.10)$$

where C' = a constant. Optionally, E'_j, as defined before, can replace E_j. This method has been called diversity stack. When applied to CDP primary traces the reduction in the contribution of one or more traces because they are noisy may unbalance the system (as far as canceling unwanted events is concerned) and diversity stack must be used with caution, bearing in mind the unequal contribution of members of an array of traces.

Finally, the contributions of the individual traces to the final composite can be assessed on the basis of the coherence (similarity or semblance) between the trace under consideration and some standard. Because of the sensitivity of the coherence coefficient to shifts between events on the two traces, this criterion can be used only if the NMO and static shifts have been removed. The correlation coefficient is defined as

$$C_{12} = \frac{\sum_1^M a_{1j} a_{2j}}{\left(\sum_1^M a_{1j}^2 \sum_1^M a_{2j}^2 \right)^{1/2}} \qquad (6.11)$$

If the standard chosen is the average of N component traces the minimum value of C_{12} will be $1/N$ and its maximum value, unity. The stacking relation will then be

$$S_K(i) = \frac{1}{N} \sum_{j=1}^{N} \frac{C''}{C_{12}} A_{ij} \qquad (6.12)$$

where $C'' = $ a constant.

The range of determination of C_{12} need not be the full length of the seismic trace(s) but can be made over a window obtained by multiplying the seismic traces by a window function, the simplest of which is a boxcar:

$$\begin{aligned} W(t) &= 1 \qquad (t_1 < t < t_2) \\ &= 0 \qquad (t < t_1 \text{ or } t > t_2) \end{aligned} \qquad (6.13)$$

Other window functions and their advantages are discussed later.

6.8 EXPANSION OR GAIN CONTROL

There is evidently some need for a process that allows change in gain of seismic traces as a function of time. Spherical divergence of the seismic energy from the source gives rise to a change in amplitude with time that must be compensated for. The modern practice of using minimum offsets of several hundred to thousands of meters has reduced the range of data recorded, but compensation is still needed to provide traces in which relative amplitudes of reflected events are preserved. The method of gain compensation is simple. It consists of multiplying the amplitude of the trace, point by point, by the amplitude of the computed gain function:

$$A'(i\,\Delta t) = A(i\,\Delta t)G(i\,\Delta t) \qquad (6.14)$$

The simplest of the gain functions (apart from a constant multiplier) is a linear increase in gain with time:

$$G(i\,\Delta t) = i \qquad (6.15)$$

which should compensate for spherical divergence of the simplest type (homogeneous, constant-velocity medium without attenuation). However, true attenuation, loss of amplitude by passing through many thin layers (scattering), and instrumental effects of all kinds tend to make the loss of amplitude with time a more complicated function of time. Two additional approaches to compensation for these losses are given here in outline. Individual processing centers implement their own versions.

The first method is to determine the average amplitude over a series of windows whose shape, length, and location can be set by the geophysicist. In Figure 6.8 three different zones, A, B, and C, have been set up and the average amplitude for each zone is denoted by $|A_A|$, $|A_B|$, and $|A_C|$. The gain function is then set up to compensate for these average gains at the applicable times, T_A, T_B and T_C, and to grade between them; for example, in the range $T_A < T_i < T_B$ the gain at T_i is given by

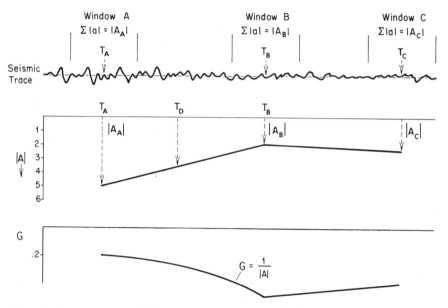

FIGURE 6.8 The average amplitude for three different zones is determined and the gain function is set up so that after application the average amplitude in the three zones will be equal.

$$G(i\,\Delta t) = g(T_A) + \frac{T_i - T_A}{T_B - T_A}\,[G(T_B) - G(T_A)] \tag{6.16}$$

Each interval is treated individually. If the gain function should be smoothly varying a cubic spline function can be fitted to the gain function values at each of the determination points.

The second method simply tries to fit the gain needs by simulating an automatic gain control in the computer (shown diagrammatically in Figure 6.9). A simple method starts as already shown, with the summation of the absolute values of the trace over a given interval of $2l$ values:

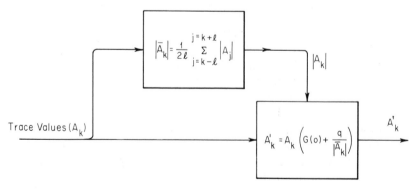

FIGURE 6.9 A simple form of automatic gain control. By varying the length $2l$ of trace used to determine the gain and its position on the trace different attack and release times can be obtained.

$$|\bar{A}_k| = \frac{1}{2l} \sum_{j=k-l}^{j=k+l} |A_j| \qquad (6.17)$$

Because of the use of the multiple points, the change in $|A_k|$ is smoothed. In this case some high or low values *still to come* have been used and a gain change can be anticipated; in fact, an attack time can be set.

The gain, if there is no more action than this, will be

$$G(k) = G(0) + \frac{2ql}{\sum_{j=k-l}^{j=k+l} |A_j|} \qquad (6.18)$$

where $q = $ a constant and $g(0) = $ initial gain.

As k advances a point is dropped off at the beginning and an additional point added at the end. A weighting system within the window can be used. More complex schemes will be evident to the reader but this one has the virtue of simplicity.

6.9   THE NEED FOR OR DESIRABILITY OF GAIN CHANGES

Before about 1971 the goal of geophysicists was to obtain seismic cross sections that were optimum from the point of view of continuity and signal/noise ratio. With the use of these sections continuous reflecting horizons could easily be followed and marked to permit structure and gross stratigraphic traps to be seen. Under these conditions traces were often expanded individually and then adjusted for constant power over the entire section.

The discovery of direct hydrocarbon indicators, particularly for use in gaseous hydrocarbon exploration, has placed much more emphasis on the preservation of true amplitude. Now, in this type of work amplitude variation with space and time is a useful parameter for making visible the large reflection coefficient changes that can occur when gas is present in the usually liquid-filled pores of the reservoir rock. The only gain changes that should be allowed are those (constant from trace to trace) that correct for gradual amplitude changes with time, such as spherical divergence, and those for which there is a physical basis, such as a change in the input power. Color Plate 1 shows how these amplitude changes can be translated into color changes that make them stand out strongly. It is not necessary to use color, but the standout is more visible than in monochromatic representations. The fallibility of these direct hydrocarbon indicators, or *bright spots*, is discussed later.

Coal exploitation seismic work relies strongly for interpretation on comparative amplitudes. The location of old mine boundaries, the recognition of coal-bed thickness and quality change, and some safety applications rely on the ability to display waveform amplitude in a quantitative manner. Changes in gain during processing should be strictly controlled and allowed to correct only earliest changes due to source strength change or to give a more usable average amplitude/time relationship by using an unchanging correction for spherical divergence.

6.10   FILTERING AND DECONVOLUTION IN GENERAL

It is important to stress the fact that at this point full seismic traces are available as digitized data, as are complete signal amplitudes over the time span. Processes that require future values of the trace can therefore be devised, whereas in a *real-time system* the filters and other operators can work only on data that have already been received. There are, of course, important differences. Correlation, for example, is possible only after a recording time at least equal to the duration of the input signal. Analog filters deal with electric signals as they arrive, and much of the power of the computer in processing seismic data derives from this property of *memory* by which the entire seismic signal is made available.

In Chapter 4 essentials of the description of a seismic trace in the form of a complex spectrum or as a sequence of amplitudes separated by a constant digital interval were given. We recapitulate some of the chief results.

First, in the frequency domain the seismic trace is described in terms of the amplitude and phase of a series of sinusoidal waves separated by a frequency of Δf, where $\Delta f = 1/T$; T is the length of the trace being considered. This description, if continued beyond time T, gives rise to a repeating signal. By *filtering* we mean changing in a systematic manner the amplitudes and phases of the constituent sinusoids. Thus we can describe a filtering operation as the complex multiplication of the spectrum of the trace by the spectrum of the filter:

$$T'(i\omega) = T(i\omega)F(i\omega) \tag{6.19}$$

where $(i\omega)$ means that the functions T', T, and F are complex.

Second, in the time domain the seismic trace is described in terms of real amplitudes at points separated by a constant time interval ΔT. At each one of these points we can visualize a delta function (Section 4.4) of the proper amplitude. Filtering is then done by systematic replacement of the delta function by a scaled pulse that has amplitudes at several digital sample times. This process is called convolution.

$$T'(t) = T(t) * F(t) \tag{6.20}$$

where $T'(t)$, $T(t)$, and $F(t)$ are now the time descriptions of the traces, or pulses, described in the frequency domain by $T'(i\omega)$, $T(i\omega)$, and $F(i\omega)$.

Because the seismic trace is normally given originally in the time domain, filtering by convolution can be done directly, and if T consists of n samples and F consists of m samples the amount of work done by the computer is $n \times m$ multiplications and additions. Filtering in the frequency domain, however, involves the following:

1. Fourier analysis, over the same time interval and using the same sample rate, of the original seismic trace and the convolving pulse.
2. Complex multiplications of corresponding frequency amplitudes.
3. A Fourier synthesis of the convolved trace.

A Fourier synthesis or analysis usually involves N^2 complex multiplications and the entire process should take about $3N^2 + N$ complex multiplications. In recent computer jargon the unit operation is usually designated a *flop*, a million operations *a megaflop*.

The discovery of the fast Fourier transform (FFT) (Brigham, 1974), based on an algorithm by Cooley and Tukey, has reduced the effort needed. By proper organization of the computer workload a single transform takes about $N \log_2 N$ multiplications and additions.

When N is large, as it is on seismic traces (up to 3000 points), the saving in time by using the FFT is considerable, particularly if N is equal to some integral power of two (say, 2^m). As an example compare the time taken for analysis of a 6-s trace digitized at 0.004-s intervals. There are 1500 sample points which would require about 2,250,000 multiplications and additions to perform the Fourier analysis in the old way. If, however, the trace *is increased* in length by the addition of zeros to $2024 = 2^{11}$ a FFT analysis will take about $11 \times 2024 = 22,264$ multiplications and additions or about one-hundredth of the time. Although filtering operations may still be done by convolution, filtering in the frequency domain is now a viable, possibly a preferred, alternative. If a sequence of operations is to be performed on a trace, all of which can be specified in the frequency domain, the switching *from* the time domain and *back* to the time domain need be done only once at the beginning and end of the operation sequence. In practice it is sometimes forgotten that the filtered trace will be longer than the original trace by the length of the filtering pulse. If insufficient

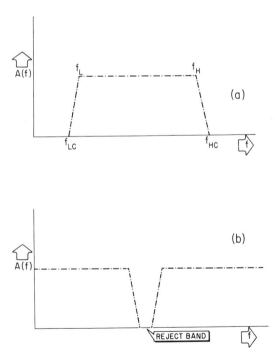

FIGURE 6.10 Amplitude spectra for a bandpass filter with a tapered cutoff (*a*) and a band reject filter with tapered cutoff (*b*).

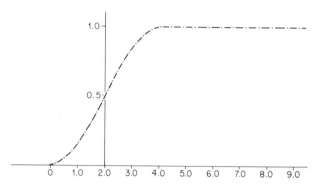

FIGURE 6.11 The form of the hanning cutoff for a spectrum, used to diminish the ringing that results from the application of a filter with a sharp cutoff. This form, or its mirror image, can be used instead of any discontinuity in the amplitude spectrum.

blank space has been left at the end of the original trace the filtered trace may have a tail—a part of which will *appear at the beginning* of the filtered trace. Doubling the length of the original trace (when using the frequency domain as an intermediate description) is often a small price to pay to prevent this problem from occurring. The circulating effect (aliasing) on data is particularly troublesome if the complex amplitudes in the frequency spectrum are generated from some formal relationship, to be followed by an FFT^{-1} (synthesis) to generate a time domain trace. Examples that have been encountered are synthetic record calculations and the generation of surface waves from their dispersive characteristics.

Filters commonly used in data processing of seismic traces are usually specified to have certain amplitude characteristics and zero phase in order not to change the time relationships of reflected events. Examples of these simple filters are shown in Figure 6.10. A sharp cutoff of any filter amplitude spectrum causes the time domain pulse to ring; that is, to contain an excessive amount of visible oscillation at the cutoff frequency. It is for this reason that cutoffs are tapered— usually by a straight line, but sometimes by a more sophisticated taper such as the hanning cutoff (Blackman and Tukey, 1958):

$$H = \frac{1}{2}\left(1 - \cos \pi \, \frac{f}{F}\right) \tag{6.21}$$

where f = range of frequencies over which the cutoff is desired (see Figure 6.11).

6.11   DETERMINISTIC DECONVOLUTION

Marine seismic exploration is now an important section of seismic work. It is unique in much of its equipment and many of its problems, but it also stands alone from a scientific point of view in that the input energy signals are generated in a homogeneous fluid medium. It is therefore possible to *measure the characteristics* of the downgoing signal.

Once this measurement has been made on equipment with the same receiver-amplifier characteristics as those of the standard marine recording system the pulse can be designated the starting point for *deterministic, spiking deconvolution*; that is, we would like to see what the seismic record would have been if the input had been as nearly as possible a delta function instead of a complex pulse. To accomplish this requires the replacement in the seismic trace (S) of each reflection event (P_i) by a zero-phase (symmetrical) wavelet with a flat amplitude spectrum between the frequency bandpass limits.

Two limitations to the calibration experiment must be made clear. First, it should be done in deep water to avoid near-source problems even though adjustments to the pulse shape can be calculated once the approximate amplitude spectrum of the downgoing pulse and the geometry of the calibration setup is known. Second, even in deep water proximity to the bottom sediments must be avoided. The measured pulse must not contain any energy from the sea-bottom reflection or subsequent multiple reflections. It *will* contain, and is meant to contain, near-source reflections from the water surface above the source system and fixed-geometry scattered waves from the boat.

It is realized that much of the production marine seismic work will not be done in such deep water. With the source characteristics removed (knowing the pulse from deep-water experimentation), however, the remaining reverberation problems of shallower water can be tackled in other ways.

The procedure for this deterministic deconvolution is described in the frequency domain and is illustrated by results of high-frequency SH wave propagation obtained on a granite block [Waters, Palmer, and Farrell (1978)]. Figure 6.12 is a portion of the waveform recorded directly beneath the source (on the far face of the rectangular granite slab). This waveform was used as the estimate of the

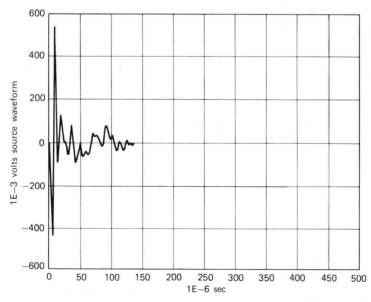

FIGURE 6.12 This waveform of a pulse propagated downward from the source was truncated at the point shown to avoid energy arriving from other than the straight downward path.

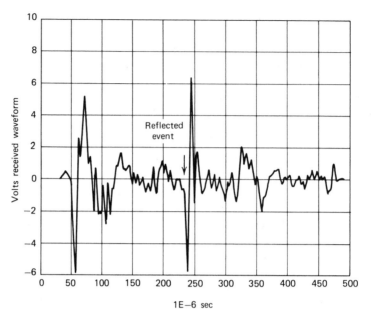

FIGURE 6.13 The original form of direct, reflected, and other waves arriving at an SH receiver from an SH piezoelectric source on a granite block.

source wavelet $f(t)$. Figure 6.13 shows the waveform recorded on the top surface of the granite slab for a particular source-receiver separation. This waveform was used to represent the received signal $s(t)$. It contains at least one reflection from the lower granite face. This procedure was the following:

1. The Fourier spectrum $\tilde{S}(\omega)$ of the received signal $s(t)$ was obtained.
2. A small amount of random noise was added to $f(t)$, thereby preventing a possible problem of dividing by zero when forming the complex quotient $\tilde{S}(\omega)/\tilde{F}(\omega)$.
3. The complex division $\tilde{S}(\omega)/\tilde{F}(\omega)$ was performed.
4. The quotient $\tilde{S}(\omega)/\tilde{F}(\omega)$ was edited to limit the spectrum to the frequency limits that exist in the source-generated wavelet. The editing used was simply a box-car window that spanned the frequencies from ω_1 to ω_2.
5. The inverse Fourier transform was applied to the quotient $\tilde{S}(\omega)/\tilde{F}(\omega)$ to obtain the deconvolved time-domain record shown in Figure 6.14.

The peaked event that occurred around 234 μs in the deconvolved record of Figure 6.14 corresponds to the reflected event marked on Figure 6.13. Furthermore, it is believed that this deconvolution now shows clearly the first multiple reflected event at 463 μs.

Although this deterministic deconvolution was done conveniently on a digital waveform processor in the frequency domain, time-domain processing could have been used equally well.

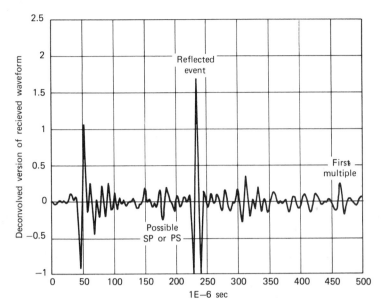

FIGURE 6.14 The reflection record of Figure 6.13, deconvolved by the direct pulse form of Figure 6.12. Note the ease with which the arrivals PS (or SP), SS, $(SS)^2$ can be picked. Some aliasing has occurred at the beginning of the record because of the limited capacity of the waveform processing system (512 points).

6.12 SPIKING AND GAPPED DECONVOLUTION OPERATORS

Another form of filter (or antifilter) is used to correct the trace for the effect of a filter already applied. This process is also known as deconvolution and takes several forms. We deal first with those that are applicable to impulsive-type sources. With this kind of source filtered pulses cannot realistically have any amplitude before zero time of the causal pulse. If the amplitude spectrum of this filter pulse is specified an infinite number of pulses can still be produced by specifying the phase spectrum in different ways. *One pulse, however, called a minimum phase pulse can be produced in which the phase is linked to the amplitude spectrum. In this pulse the energy occurs as close as possible to zero time while still satisfying the amplitude spectrum.* This is a most important concept, for, given any amplitude spectrum, it is possible by rather sophisticated mathematical methods to calculate the form of the minimum-phase pulse. From Bode's formula the phase of a minimum-phase pulse is given by

$$\phi(f_0) = \frac{1}{\pi} \int_{-\infty}^{\infty} \frac{\log A(f)}{f - f_0}\, df$$

where $\phi(f)$ and $A(f)$ are the phase and amplitude, respectively, of the pulse at frequency f.

This may be expressed as a convolution:

$$\phi(f) = -\frac{1}{\pi} \left[\log A(f)\right] * \frac{1}{f}$$

It is shown by these expressions that the phase at one particular frequency depends on the entire amplitude spectrum, although the amplitudes at frequencies remote from f_0 are relatively unimportant. Because it has been shown that the pulse resulting from a dynamite shot has a minimum-phase form (White and O'Brien, 1974), the procedure takes on real significance. Furthermore, the filtering of this pulse by a layered system is also minimum-phase. This is important because multiple reflections of the primary pulse in the near-surface layers have a significant effect on the shape of later reflections. All are part of the filtering effect we are seeking to eliminate.

The objective of deconvolution is to identify in the record traces themselves the filter that has been applied (as a result of passage through the earth, the recording system, and so forth) and then to select an inverse filter that will correct the effects of the filter(s). This means that the inverse filter must have the effects of bringing the phase back to zero and rendering the amplitude spectrum flat. The amplitude spectrum is rendered flat by multiplying all amplitudes by their inverses to give a constant resultant of unity. In so doing large multipliers are sometimes obtained for some frequencies. Because the amplitude spectrum is partly determined by noise, these high multipliers may not be correct and a noisy inverse may result. Consequently it is necessary to restrict the bandwidth of the inverse to the range in which the signal spectrum has appreciable amplitudes. The inverse is obtained in any convenient range but is then subjected to a flat-amplitude, zero-phase filter to eliminate noise outside the signal band.

Some problems are present, however. In the first place the spectrum of a seismic trace is not just the (smooth) spectrum of the initial pulse but contains ripples due to the interference effects of reflections that occur at different times. Any two positive reflections interfere constructively when their times of arrival are integral multiples of the period of a particular frequency. There is not just *one* resonant frequency; the spectrum also contains a ripple whose wave peaks are separated (in angular frequency):

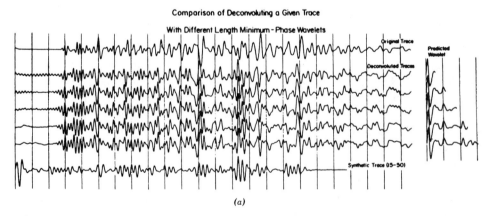

(a)

FIGURE 6.15   (a) An original record trace (top), followed by various deconvolved traces (using the predicted wavelet shown at their right). A synthetic record trace is shown below for comparison. (b) The effect of predicting minimum-phase wavelets of different lengths (left), the corresponding spectra (center), and the spectrum of the deconvolved trace (right). [Source: Ware (1964). Courtesy of CONOCO, Inc.]

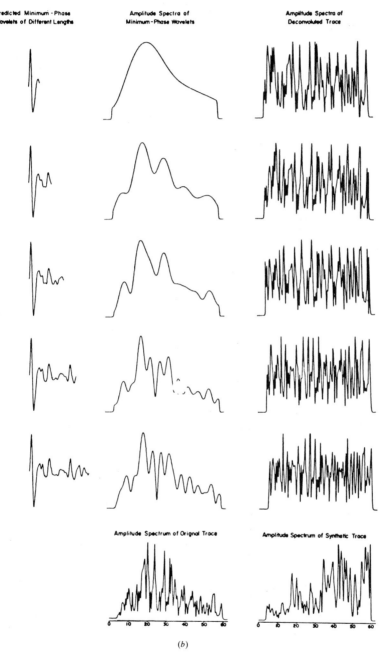

FIGURE 6.15 Continued.

$$\Delta\omega = \frac{2\pi}{T} \qquad\qquad (6.22)$$

where T = distance between the events. Note that events *far apart in time* give rise to *rapid ripples* on the spectrum and those *close together* show a *slow variation*. Thus, as illustrated in Figure 6.15 (Ware, 1964), it should be possible to smooth the amplitude spectrum of the trace and, to a first approximation, arrive at the amplitude spectrum of the convolving pulse.

This effect can be seen in the time domain if we perform an autocorrelation of the seismic trace, which is done by a convolution operation of the trace with the trace reversed in time. The autocorrelation shows a peak (obviously) at zero delay, followed by a decay and oscillation at nearby time delays—the shape of which is controlled by the average frequency content of the trace—followed by a series of peaks and troughs that correspond mainly to reflections correlating with one another. The near, truncated, shape of the autocorrelation waveform contains the information needed for estimating the amplitude spectrum of the initial pulse. If the assumption is then made that it is a minimum-phase pulse its phase can be determined from the shape of the amplitude spectrum and the problem of determining the initial pulse is formally solved.

The actual method of calculation has been described many times in the literature; for example, Burns (1968), Clark (1968), Ford and Hearne (1966), Foster, Sengbush, and Watson (1968), Robinson and Treitel (1967), Robinson (1967a), and Treitel and Robinson (1966, 1969). In matrix terms (see Appendix 6B), the process may be described as follows:

It is assumed that the signal $\mathbf{S}$ (usually a seismic trace) is made up of a series of pulses $\mathbf{K}$ of different amplitudes and different delay times. Noise is also contained in $\mathbf{S}$. It must also be assumed that the pulse $\mathbf{K}$ has approximately the same autocorrelation function for short delays as the signal $\mathbf{S}$.

The problem is to discover a minimum phase signal $\mathbf{K}$ with the correct amplitude spectrum. A further requirement is that we should be able to find an *inverse* $\mathbf{X}$ such that $\mathbf{KX}$ is as close as possible in a least squares sense to

$$\mathbf{B}^T = (1, 0, 0, 0, \ldots, 0)$$

Note that if this is done the combination $\mathbf{KX}$ will have a nearly flat amplitude spectrum and its phase spectrum will be near zero.

It can be shown that the inverse needed is

$$\mathbf{X} = (\mathbf{K}^T\mathbf{K})^{-1}\mathbf{K}^T\mathbf{B}$$

In this formula $\mathbf{K}^T\mathbf{K}$ is the autocorrelation matrix of $\mathbf{S}$, suitably edited so that it is very short and has a smooth amplitude spectrum. We do not know $\mathbf{K}$, but we do know that $\mathbf{K}^T\mathbf{B}$ must have the form of a constant times $\mathbf{B}^T$ or $(C, 0, 0, 0, \ldots, 0)$.

The autocorrelation function (edited), a Toeplitz form, is therefore inverted only as far as producing the first column and this is the required inverse $\mathbf{X}$. There are some potential stability problems but they can be avoided by multiplying the diagonal terms of $\mathbf{K}^T\mathbf{K}$ by a constant $1 + \epsilon$, where ϵ is a small constant of the order of 0.05. This action is equivalent to adding *white noise* to the trace $\mathbf{S}$ to ensure that there will be no frequencies for which the power spectrum is zero.

A word or two more may be needed about editing the autocorrelation function. One of the criteria is to make it short so that it is near enough to a δ-function to give it a smoothly varying spectrum. We cannot examine each of the autocorrelation functions produced because in a production operation it would take too long and it might be inconsistent, even with the most knowledgeable operator. Therefore various rules, depending on the design of the geophysical processing system of the computer, have been adopted, most of which have some weaknesses. One common rule has been to truncate the autocorrelation function at the second zero crossing away from the peak. Because of the sharp cutoff, however, this will *introduce* frequencies into the spectrum that were not present in the original trace. One alternative, to multiply the autocorrelation function by some waveform that decays fast (say, an *erf* function) has gained some acceptance. The truncation can then be performed consistently at a point at which the amplitude and its slope are small.

If a large ghost pulse occurs at some delay time τ and smaller ones are at 2τ, 3τ, and so on, this is indicative of multiple reflections on the original record and, because they are also minimum phase, could be removed by making the inverse pulse long enough (greater than 2τ). Yet this sometimes violates the original concept of smoothing the spectrum of the trace, and it is better to use a short deconvolution operator first, followed by another operator that has the characteristic of eliminating a single multiple-reflection path. These operators, called predictive or gapped, consist of a single spike at zero time, followed by a sequence of zeros, followed by a sequence of calculated values to the end of the operator. The gap and sequence of values thereafter are chosen in such a way that the near-in shape of the autocorrelation wavelet (hence the amplitude spectrum of the data) is preserved, but the ghosts or long-period reverberations that occur in the zone of the calculated values are minimized. As a sequence of sample values, the gapped operator looks like

$$1, 0, 0, 0, 0, \ldots, 0, 0, 0, a_1, a_2, a_3, \ldots, a_n$$

These operators are again derived from the autocorrelation pulse by a method described by Peacock and Treitel (1969) and by Kunetz and Fourman (1968). Reference should be made to these papers for further information.

Other routes may be taken to eliminate *some forms of multiple reflection*, one of which, designed specifically for the elimination of multiple reflections (reverberations) in water-covered areas, has proved to be a model on which even more complex treatment can be based. Figure 6.16 shows the paths taken by rays in a water-covered area. Multiple reflections occur at both source and receiver ends of this reflection trace.

It is seen that the downward traveling pulse from a *spike* source, referred to a point just below the bottom of the water can be written

$$S(t) = \delta(t)\{1 - R_0\Delta + R_0^2\Delta^2 - \cdots\}(1 - R_0)$$
$$= \delta(t)\{1 + R_0\Delta\}^{-1}(1 - R_0)$$

where R_0 = the reflection coefficient of the water bottom

Δ = the delay operator for a double travel path through the water

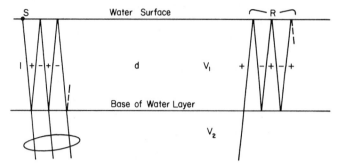

FIGURE 6.16 Multiple reflections within a water layer of depth d at both source and receiver. Each multiple is delayed by $2d/V$ and the sign is reversed.

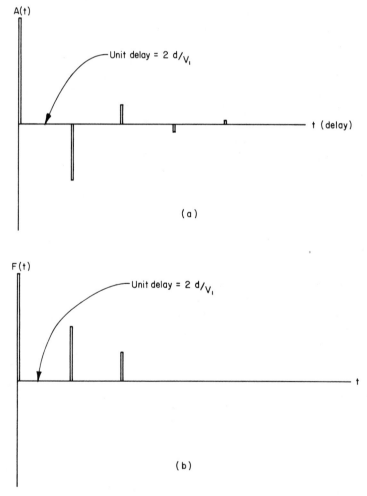

FIGURE 6.17 (*a*) The form of the transfer function of a water layer on the reflection response; (*b*) the form of the correcting filter. For both R_0 is assumed to be 0.25.

Just above the sea bottom near the receiver the transfer function is $\{1 + R_0\Delta\}^{-1}(1 + R_0)$ and a single spike put out by the source takes the form

$$R(t) = \delta(t)(1 - R_0^2)\{1 + R_0\Delta\}^{-2}$$

which is a series of spikes of alternating sign separated by delay Δ. This is illustrated for $R_0 = 0.25$ in Figure 6.17a.

Apart from the constant factor $(1 - R_0^2)$, it is possible to write, from inspection, the inverse filter

$$F(t) = (1 + R_0\Delta)^2 = (1 + 2R_0\Delta + R_0^2\Delta^2)$$

which is shown in Figure 6.17b.

It is obvious that this operator (once R_0 and Δ are measured) can be used to eliminate the reverberation due to the simple water layer assumed. The measurement of R_0 is not so easy as it might sound, nor are conditions such that the water depth at source and receiver(s) is always the same. More complicated models may be built, however, based on the same principles, and there is always the possibility of trial and error for the determination of R_0. The formation of a near-bottom layer containing gas is one reason for uncertainty and inconsistency in R_0.

6.13 DECONVOLUTION OF VIBROSEIS® RECORDS

As mentioned in Section 3.4, the Vibroseis® method of exploration differs from all others in that the signal introduced into the ground is a long train of quasi-sinusoidal waves of the general form

$$S(t) = A(t) \sin\left(at + \frac{b}{2}t^2\right) \qquad 0 < t < T \tag{6.23}$$

where $A(t)$ = a relatively slowly varying envelope function
 a and b = constants related to the initial frequency and the rate of change in frequency, respectively

Following the method of Section 4.8, we can expect, to a first approximation (neglecting attenuation), that if $I(t)$ is the impulse response of the earth the received reflection record will be given by

$$R(t) = I(t) * S(t) \tag{6.24}$$

In other words, each impulse in the impulse response is replaced by the swept frequency signal (or control signal). To generate a record $R_c(t)$ that can be interpreted in terms of geology it is necessary to change the control signal into a compressed form, the most convenient of which is the autocorrelation function that corresponds to the power spectrum of the control signal. This is done by correlating the field trace with the control signal. Correlation is nothing more than a convolution filter that uses a time-reversed or transposed control signal. Therefore

$$R_c(t) = I(t) * S(t) * S^T(t) \qquad (6.25)$$

and the last two filters can be combined to give

$$R_c(t) = I(t) * \phi_{ssT}(t) \qquad (6.26)$$

where ϕ_{ssT} is the autocorrelation function. At this point practice does not correspond to theory and a temporary diversion is necessary. In the foregoing discussion it was presupposed that if a vibrator is operated with a control signal $S(t)$ this is the form of the wave being produced in the earth. More specifically, we want to know the wave shape of the particle velocity produced in the earth because, with the geophones commonly used, it is a particle velocity that is being measured. When a vibrator is placed on the earth and the control signal $S(t)$ is fed into it the output signal is

$$S'(t) = B(t)S(t + \theta) \qquad (6.27)$$

where the phase angle θ is a function of rock type and frequency. Hence it varies with the position of the vibrator along the line and the time within the sweep; $B(t)$ is an amplitude gain function.

To introduce into the ground a phase-consistent signal under varying surface conditions a feedback loop (called a phase compensation loop) is normally used. In practice a signal from the baseplate (from an accelerometer whose output is integrated) and the control signal are fed into a phase comparator. As a rule, this circuit has an output unless the two signals are exactly 90° out of phase (a practical circuit consideration); then it acts on a phase shifter to control the input phase to the electrical servo input. Thus the baseplate velocity output to the earth is 90° out of phase with the control signal.

Although this 90° phase change is convenient at the vibrator because of instrument simplicity, in the final seismic traces it must be eliminated. This could be done directly in the correlation process but it is easier to proceed as follows:

1. The correlated trace with a 90° phase shift is obtained by correlating the control trace with the field trace:

$$R_c(t) = I(t) * S(t) * S(-t) \qquad (6.28)$$

2. A deconvolution (minimum-phase) pulse is obtained from the autocorrelation of the field trace. This pulse is Fourier-analyzed with a FFT and the amplitude spectrum is retained, whereas a constant $-90°$ phase shift is substituted for the phase angle that results from the analysis.
3. The deconvolution (constant 90° phase) pulse is resynthesized with a FFT.
4. This pulse is convolved against the Vibroseis correlated traces.

Vibroseis deconvolution was reexamined by Gurbuz (1972) and by Ristow and

Jurczyk (1975). Both authors have shown how to effect deconvolution entirely in the time domain after correlation with an antifilter:

$$\frac{1}{A_0(z)} = kS_0(z)S_0(z)c_0(z) \qquad (6.29)$$

where k = an unimportant constant

$C_0(z)$ = a spike deconvolution operator whose amplitude spectrum is that of the field trace

$S_0(z)$ = a minimum-phase pulse with the same amplitude spectrum as the *control* trace; it is applied twice

These operators and pulses have been stated as z transforms and are treated in an elementary way in Section 6.14. For the present it is understood that they are complex filters that can be specified in the time domain as convolution pulses or in the frequency domain as complex spectra.

The advantage of doing the deconvolution after correlation is that it is possible to make a deconvolution operator that varies with record time, thus compensating, to the extent possible with the noise present, for attenuation.

6.14 RECURSIVE FILTERS AND TIME-VARYING FILTERS

Because of the large number of multiplications and additions involved in computing the output of a digital filter, filtering can be time consuming and expensive in the overall processing of seismic recordings. Special convolution filtering equipment originally allowed these multiplications to be made in parallel and established the economic feasibility of digital processing. The newer array processors (AP), with much faster basic speeds, have now made these operations routine, even though they are attached as modules to basically slower machines.

As explained in Section 4.5, a sampled trace can be represented by a series of amplitudes, taken at some sampling interval Δt:

$$T \equiv \{a_0, a_1, a_2, a_3, \ldots, a_{n-1}, a_n\}$$

It can also be formally represented by a power series in the variable z (which represents a delay factor $\epsilon^{-i\omega\Delta t}$ that corresponds to the constant sampling rate Δt):

$$T(z) = a_0 + a_1 z + a_2 z^2 + \cdots + a_{n-1} z^{n-1} + a_n z^n$$

The polynomial $T(z)$ can then be expressed as the product of its factors

$$T(z) = a_0(z - r_1)(z - r_2) \cdots (z - r_i) \cdots (z - r_n)$$

where the roots $(r_1, r_2, \ldots, r_n)$ may be complex.

A formal test of the minimum phase system would show that if all of these roots were plotted in the complex plane they would lie *outside* the circle centered at the origin and having unit radius—the so-called unit circle. In other words, if

$$|r_i| > 1 \qquad \text{for all } i$$

As an example we apply this test to the pulse

$$\{64, 80, 40, 10, 1\}$$

$$64 + 80z + 40z^2 + 10z^3 + z^4$$

$$[z + (3 - i)][z + (3 + i)][z + (2 + 2i)][z + (2 - 2i)]$$

and we see that

$$r_1 = -(3 - i) \qquad |r_1| = \sqrt{3^2 + 1} = \sqrt{10}$$

$$r_2 = -(3 + i) \qquad |r_2| = \sqrt{3^2 + 1} = \sqrt{10}$$

$$r_3 = -(2 + 2i) \qquad |r_3| = \sqrt{2^2 + 2^2} = \sqrt{8}$$

$$r_4 = -(2 - 2i) \qquad |r_4| = /32^2 + 2^2 = \sqrt{8}$$

Hence the pulse $\{64, 80, 10, 1\}$ is minimum-phase. It can be noted further that it is the convolution of two pulses

$$\{8, 6, 1\} \quad \text{and} \quad \{8, 4, 1\}$$

Two different traces $T_1(z)$ and $T_2(z)$ can be convolved by multiplying their z transforms (as shown in Section 4.5) to give

$$T_3(z) = T_1(z) \times T_2(z)$$

and one trace, $T_3(z)$, can be deconvolved with $T_2(z)$ by dividing the z transforms

$$T_4(z) = \frac{T_3(z)}{T_2(z)} \qquad [= T_1(z)]$$

Now, by a suitable transformation of the desired convoluting pulse—as a rational fraction derived from the z transform—in most instances we can derive an equivalent computing algorithm that is significantly more efficient than digital convolution. Here we present the main features of the recursive filter, without going into much detail, because this method has an *economic* base rather than a scientific or geological base. It does nothing different from a normal convolution but enables it to be done quickly and economically. For more details the reader is referred to the excellent tutorial article by Shanks (1967).

In the z-transform domain a digital filter can be given as the ratio of two polynomials in z:

$$F(z) = \frac{A(z)}{B(z)} \tag{6.30}$$

$$= \frac{a_0 + a_1 z + a_2 z^2 + \cdots + a_n z^n}{b_0 + b_1 z + b_2 z^2 + \cdots + b_n z^n} \tag{6.31}$$

This cannot be applied directly as a convolution filter, but by the process of long division (Jury, 1964) it can be transformed into a single, infinite polynomial:

$$F(z) = f_0 + f_1 z + f_2 z^2 + \cdots$$

If the filter is stable the values of the coefficients will converge to zero and at some point the series can be truncated to give an approximate filter.

The rational fraction, as given in (6.30), can, however, be applied directly. We consider the simple filter

$$F(z) = \frac{a_0 + a_1 z}{1 + b_1 z + b_2 z^2} \tag{6.32}$$

If $X(z)$ is the input to the filter the output $Y(z)$ is

$$Y(z) = \frac{a_0 + a_1 z}{1 + b_1 z + b_2 z^2} X(z) \tag{6.33}$$

or

$$Y(z) = [(a_0 + a_1 z) X(z) - z Y(z)(b_1 + b_2 z) \tag{6.34}$$

This equation states that the output $Y(z)$ is equal to the input convolved with the discrete series (a_0, a_1), minus the *output* delayed by one sample interval convolved with the series (b_1, b_2). This is a feedback system, as shown in Figure 6.18; it can also be realized in a digital computer by using a suitable feedback algorithm.

There are numerous ways to compute the coefficients of rational filters designed for specific filtering operations. These are given in detail in Shanks (1967).

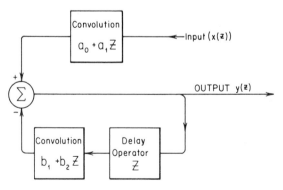

FIGURE 6.18 This feedback system is equivalent to the recursive filter $F(z) = (a_0 + a_1 z)/(1 + b_1 z + b_2 z^2)$.

From experience and from intuition it is known that recorded seismic traces have the appearance of generally decreasing in bandwidth (by losing high frequencies) as the record time increases. It can be said that the later reflection events are usually of lower frequency than the early ones. This remark has to be interpreted in the following sense.

If intervals of the seismic record (windows) are chosen to have the same characteristics but centered at gradually increasing record times and the seismic traces within these windows are Fourier-analyzed, there will be a gradual, but not necessarily constant, decrease in the high-frequency content of the amplitude spectra as the time of the window increases. Usually, in making these analyses, a time window is chosen that is long enough to provide the necessary frequency resolution; at the same time, however, overlap of the windows should be reduced to a minimum.

On the basis of this experience, it has proved desirable to design a filter system that changes its characteristics with time of application to optimize the signal/ noise characteristics of the seismic data. Although we are willing to optimize the signal power/noise power ratio, the foregoing discussion of deconvolution should have convinced us that we should not interfere with the phase characteristics of the record trace after deconvolution. Therefore any time-varying filter we design should be a zero-phase filter.

The design of time-varying filters has been accomplished in several ways, but for filtering seismic traces use has been made of the simple scheme shown in Figure 6.19 and described in the following.

This figure shows an unfiltered trace to be filtered in a time-varying manner; filter 1 applies to a time t_1, a mixture of F_1 and F_2, to t_2, a mixture of F_2 and F_3, to t_3, and so on.

Four filtered versions of the (relevant portions) signal are produced and then combined with the weights given on the lower graph. It is obvious that the convolution with F_1 need not extend beyond t_2 because its weight from that time on is zero. Similarly, the trace is filtered only with F_2 between t_1 and t_3, being given zero weight elsewhere.

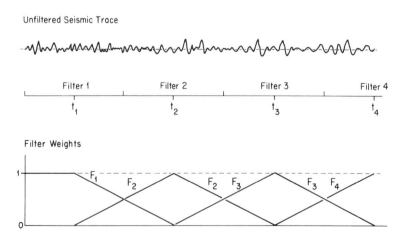

FIGURE 6.19   A simple method of achieving a time-varying filtered seismic trace.

F_1, F_2, F_3, and F_4 all consist of convolution operators that are zero-phase. Some care must be taken in scaling these operators because they may have different bandwidths, therefore different total energies, if the spectrum amplitude is kept the same. One method of approaching this problem is to compute the sum of the squares of the amplitude of the convolving pulses; for example,

$$\sum_{i=1}^{N} a_i^2 = A^2$$

$$\sum_{i=1}^{N} b_i^2 = B^2 \tag{6.35}$$

and then scale the amplitudes of the pulses by the square root of these energies. The new

$$a_i^1 = a_i \frac{C}{A}$$

$$b_i^1 = b_i \frac{C}{B} \tag{6.36}$$

where $C =$ a convenient constant.

6.15 HOMOMORPHIC DECONVOLUTION

The restriction of only being able to predict the pulse shape from the seismic trace when the pulse is minimum phase is an irksome one because there may be some filters we seek to neutralize that are not minimum-phase. The minimum-phase assumption has worked out reasonably well for impulsive-type sources. The material in this section is given mainly for the benefit of readers who may wonder if any thought has been given to removing the constraints associated with the minimum-phase assumption. Little routine use has been made of the method in exploration reflection seismology data processing.

The homomorphic method of deconvolution (Ulrych, 1971) has found some favor because it is supposedly free of the minimum-phase restriction. (Homomorphism is mapping from one domain to another.) In this particular case we start off with the standard frequency domain definition of a filtered trace (neglecting noise):

$$F(\omega) = I(\omega)P(\omega) \tag{6.37}$$

which states that the filtered trace spectrum (complex) is the spectrum of the impulse response $I(\omega)$ multiplied by the spectrum of the convolving pulse $P(\omega)$. We can transform this to another domain of the logarithm of amplitude by noting that

$$\log F(\omega) = \log I(\omega) + \log P(\omega) \tag{6.38}$$

It is to be noted that th pulse and impulse response log spectra are now *added* instead of being multiplied, as in (6.37).

It is presumed that it is easier to separate two factors that have been added together than two that have been multiplied. Common sense tells us that this is untrue unless some further criteria are involved. We give an example and then comment on the fallibility of the process.

Ulrych (1971) shows (Figure 6.20) a mixed-phase pulse (*a*), that is, a pulse whose phase spectrum cannot be deduced from a knowledge of the amplitude spectrum, convolved with a simple delay operator to give a pulse combined with a single echo (*b*). The process calls for digitization, that is, obtaining sample values at various multiples of the delay ΔT to find

$$(a_0, a_1, a_2, a_3, \ldots, a_n)$$

and forming the z transform by associating each amplitude with the appropriate power of z:

$$X(z) = a_0 + a_1 z + a_2 z^2 + a_3 z^3 + \cdots + a_n z^n$$

Form $\hat{X}(z) = \log X(z)$.

Now take the inverse z transform of $\hat{X}(z) \rightarrow \hat{x}(n)$. The latter is called the complex *cepstrum* from the work of Bogert, Healey, and Tukey (1963), who termed the power spectrum of the logarithm of the power spectrum the cepstrum by inverting the order of the first four letters. The word *complex* was added by Oppenheim, Schafer, and Stockham (1968) to emphasize the point that the cepstrum is formed by taking into account the amplitude and phase information contained in $\hat{x}(n)$.

In the specific case of the single pulse and echo shown in Figure 6.20, (*c*) shows

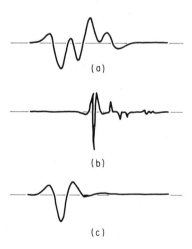

(a)

(b)

(c)

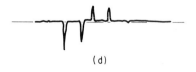

(d)

FIGURE 6.20 (*a*) Mixed-phase input; (*b*) complex cepstrum of (*a*); (*c*) low-pass output, the constituent pulse shape; (*d*) high-frequency output or impulse response. [*Source*: Ulrych (1971). Reprinted with permission from *Geophysics*.]

the complex cepstrum of (*b*). Note that the original pulse has been replaced by the large spike at zero time *n*, followed by smaller pulses of alternating sign. Ulrych considers these pulses to be well separated from the contribution of the main pulse; therefore the two effects can be separated.

One is struck by the sudden appearance in the complex cepstrum of high frequencies (of *n*) by means of which the times of occurrence are suddenly rendered discrete. The mathematics is too complex to analyze how these high frequencies occur, but we have endeavored to show that they are really manufactured in the process by the use of computer-generated examples or by the assumption that there is a specific origin of time and that all events must occur at a particular sample value.

First, we remark that if a seismic trace is a continuous function of time and the convolution of a continuous pulse and a continuous impulse response the amplitude spectrum of the trace will be the product of the two amplitude spectra. Even though the impulse response has high-frequency components they are much reduced in magnitude because the seismic pulse is band-limited in the low-frequency region. Of course, in theory the pulse may still have small high-frequency components (or they may be *manufactured* as a result of a digital analysis of finite accuracy data). It is axiomatic therefore that in the continuous case it is never possible to recover the impulse response at frequencies outside the effective bandwidth of the convolving pulse. The spectrum of the log spectrum is essentially a pseudotime trace, and the location of the times of the echoes, or impulses, cannot be determined closer than the pulse bandwidth allows. The best that can be done is to position an autocorrelation pulse of a flat-frequency pulse of the requisite bandwidth at the position of each echo, but this is precisely what is done in spike deconvolution working at its best.

In the digital examples the computer manufactures the output trace from the impulse response and from the pulse (utilizing finite error measurements of the pulse and giving it frequencies up to the Nyquist frequency associated with the sampling rate). It is then possible to turn around and, again using the small generated values of high frequency, reanalyze the entire system (which has a high bandwidth). This leads to the high-frequency output of the impulse response.

Second, we must examine the question of the exact location of the sampling points. It is possible to move the sampling-point zero (which is usually at a point of zero-amplitude output) by a small amount in the time domain. This causes all the samples to change in amplitude and, in turn, forces all the impulse response values to be located at different points in the time domain. This process can be imagined as being carried out at all possible points for the origin; therefore the impulse responses will now generate a continuous output. What we are saying is that in the real world there is uncertainty about the location of the impulse responses and uncertainty in all the phase information for all frequencies of importance. As far as the computer world is concerned, the information read in is exact and an exact answer is generated (according to the instructions given to the computer).

The introduction of noise into the system renders the high-frequency output even more unstable because the phase of small high-frequency components is changed considerably.

This method of homomorphic deconvolution remains interesting from the point

of view of the possibility of determining mixed-phase input pulses, but even this aspect is clouded by the admission that it is only in minimum-phase pulses that the separation of pulse and impulse response components is complete.

6.16 ZERO-PHASE NULL FILTERS

In Section 5.5 we considered a means of reducing power-line frequency pickup in the field. At that time the assertion was made that analogue null filters were undesirable because they have undesirable phase characteristics that tamper with the proper phase for *all* frequencies being recorded.

It is to be hoped that the alternative methods of reducing power-line interference will be effective, but as a last resort it is possible to produce digital filters that have no effect on the phase characteristics of the seismic trace. These filters are called zero-phase null filters and can be derived in a number of ways.

Figure 6.21 shows a characteristic passband—a constant amplitude spectrum from zero frequency to some upper frequency well below the Nyquist frequency for the sampling rate chosen—except for the null band tapered on both sides, and a slow ramp down to zero for the high frequency cutoff. The width of the null band depends on the constancy of the interfering frequency. The near exactitude of the 60 Hz power-line frequency in the United States contrasts with one of two Hz variations in the seismic frequencies induced in the earth by refrigerator motors (encountered frequently in surveys on city streets).

If the mathematical representation of this amplitude spectrum can be made relatively simple, then, for example,

$$F(f) = 1; \qquad 0 < f < f_1$$

$$F(f) = 1 - \left(\frac{f - f_1}{f_2 - f_1}\right); \qquad f_1 \le f \le f_2$$

$$F(f) = 0; \qquad f_2 < f < f_3$$

$$F(f) = \frac{f - f_3}{f_4 - f_3}; \qquad f_3 \le f \le f_4$$

$$F(f) = 1; \qquad f_4 < f < f_5$$

$$F(f) = 1 - \left(\frac{f - f_5}{f_6 - f_5}\right); \qquad f_5 \le f \le f_6$$

$$F(f) = 0; \qquad f_6 < f \le f_{NQ}$$

Then, having $\Delta f = 1/T$, the filter amplitudes for each frequency $(i\,\Delta f)$ within the spectrum can be calculated, entered into the FFT matrix with the phase $(\phi) = 0$ for all frequencies, and a synthesis performed to give the filter pulse in the time domain.

As mentioned previously, these filters may be very long and must be truncated at some point. They then have limited effectiveness.

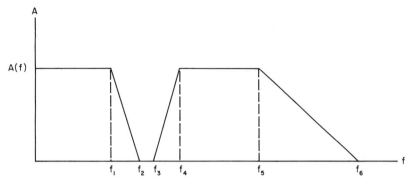

FIGURE 6.21 Schematic of the amplitude spectrum of a zero-phase null filter.

A second method is to calculate the minimum-phase null filter in the manner of Shanks (1967), thereby obtaining a recursive filter that can be used in a computer simulation of a feedback system already described (see Figure 6.18). By turning the trace around and reusing the recursive filter the phase introduced on the first pass is eliminated on the second *but the amplitude effect is squared.*

If the power-line frequency is not known some difficulties may obtain in the digital null filter design. Nyman and Gaither (1983) have introduced an adaptive rejection filter that (provided the power-line frequency is known approximately) works for all common mode frequencies—sequentially—and determines not only the amplitude and phase of the correction to be applied but the exact frequency as well. The assumption is made that the power-line contamination is uncorrelated with the seismic data. Therefore if the following definitions are adopted

$$r(t) = s(t) + A_0 \cos(\omega_0 t + \phi_0)$$

and

$$U = \int_0^T r(t) \cos \omega t \, dt; \qquad\qquad V = \int_0^T r(t) \sin \omega t \, dt$$

$$X = \int_0^T r(t)(\cos \omega t)\left(\sin \frac{2\pi t}{T}\right) dt; \qquad \epsilon \text{ is the deviation } \omega - \omega_0$$

$$Y = \int_0^T r(t)(\sin \omega t)\left(\sin \frac{2\pi t}{T}\right) dt$$

Then, to a first approximation, they have shown that

$$A_0 = 2T\sqrt{U^2 + V^2}$$

$$\phi_0 = \tan^{-1}\left[-\frac{(V + \pi Y)}{(U + \pi X)}\right]$$

$$\epsilon = \Delta \pi T^2 (X \sin \phi_0 + Y \cos \phi_0)$$

Finally, a correction to ω is made by ϵ and the process is repeated until $\omega \to \omega_0$ sufficiently closely.

6.17 TWO-DIMENSIONAL FILTERING OF
SEISMIC CROSS SECTIONS

The foregoing survey dealt with data-processing techniques that are applied, trace by trace, to the contituent traces of a seismic cross section. Apart from the stacking of 100% traces taken at different offsets, there was no assumption that any trace of a cross section had any relation to any other seismic trace on the cross section. Hope is always present, of course, that the combination of seismic traces, one example of which is given in Figure 6.22, will show only aligned seismic events that correspond to reflections from the geological interfaces below. In many cases this expectation is not realized and disturbing events cross over the cross section, which must be removed before the geological events can be seen clearly. This is particularly true on a 100% record, as shown in Figure 6.23, in which interference from air waves and near-surface waves often conceals the reflection information. They appear as lineups of pulses that travel across the

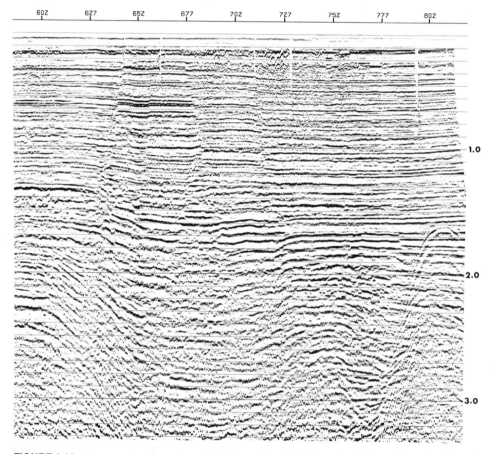

FIGURE 6.22 An example of a processed seismic section. The hope that this contains only reflected events is not realized in this case, as shown by the large diffracted event on the lower right. This event is probably scattered from an object in the water about 1.2 km (4000 ft) to the side of the line. [*Source*: Courtesy of Seiscom Delta, Inc. (1975).]

FIELD DATA FILTERED 1 - 5 CYCLES

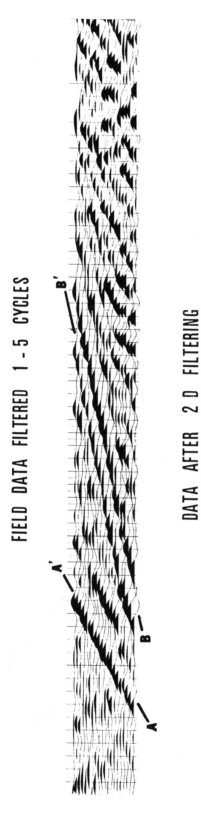

DATA AFTER 2 D FILTERING

FIGURE 6.23 The efficiency of two-dimensional filtering in removing a particular event (BB') due to interference. Other events are unchanged by this filtering action.

record with a slow apparent velocity. Two interferences are shown in Figure 6.23, one of which was to be removed. We anticipate the action of the two-dimensional filter to show the second record with the unwanted event removed. In full seismic sections unwanted events may originate because of scattering from linear weathering and elevation changes (e.g., scarps that almost parallel the line of section) or from linear outcrops, linear gas seeps, and other changes in conditions of the water layer in marine prospecting. There are enough causes; the trick is to find out whether there is false reflection information and to devise a system to remove it.

A restricted form of velocity filtering (pie slice) was originally offered by Geophysical Services, Inc., and has been described by Embree, Burg, and Backus (1963) and by Fail, Grau, and Layotte (1964). The method is as follows:

Just as it is possible to describe a one-dimensional seismic trace in the time domain by an equivalent amplitude and phase description in the frequency domain, it is also possible to extend this imagery to two (and more) dimensions. Thus a wave

$$F(x, t) = F(kx - \omega t)$$

which has a wavelength $2\pi/k$ in the x direction and a period $2\pi/\omega$ in the time domain, can be equally well represented as a function of k, the wave number, and ω, the angular frequency. Transformation between the two is done by a two-dimensional Fourier transform and, as in a single-dimensional function, the inverse Fourier transform also exists.

The transform is done first by making Fourier analyses of the traces in the time direction and then by analyzing these complex amplitudes in the x direction. We are concerned with a form of filtering that will reduce *amplitudes* of particular events as nearly as possible to zero, we are not concerned with the complex values but only with the amplitudes $A_i = \sqrt{a_i^2 + b_i^2}$, which are then displayed, much as the time traces are displayed in the form of a map in the ω-k plane (Figure 6.24).

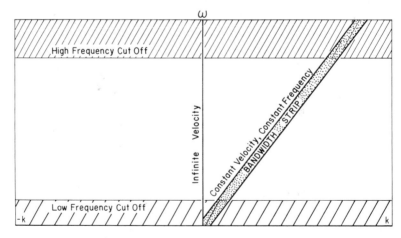

FIGURE 6.24 The ω-k plane, showing various regions that, if zeroed in amplitude, perform the designated filter operations. The boundaries need not be sharp and by this means ringing characteristics can be avoided.

On this map a line drawn from the origin in any direction represents all points for which ω/k, or the velocity, is a constant. The elimination of a sloping band in this ω-k plane therefore eliminates all energy that propagates across the record at a particular velocity.

In the pie-slice process the attitude adopted was that reflection events usually show up as having high apparent velocities and can therefore be left behind if a small wedge of energy on the ω-k plane is conserved (about the ω axis) and the rest of the ω-k plane is given zero amplitude (Figure 6.25). In the example shown the pie-slice method helps considerably to cut out low-velocity energy. Unless this is caused by steep reflections on the cross section the reflection data will be materially improved. If it is due to steeply dipping reflections the ω-k filter is poorly designed.

The pie-slice process has several characteristics:

1. It is a relatively simple filter and can be carried out in the time domain. Before the FFT this was a distinct economic advantage.
2. The boundaries of the triangular acceptance region are sharp (e.g., there is full acceptance inside and full rejection outside the acceptance region). This characteristic leads to a display of the energy surviving the filter on the time cross section in the form of crossing events with the velocity cutoff. These are caused by noise (e.g., energy outside the passband cut off with an infinitely fast rejection rate). (This is a characteristic of frequency filters which tend to ring at sharp cutoff points.)
3. Because of the symmetrical nature of the filter, it is inflexible as far as passing reflected events of other than zero dip.
4. At points on the cross section at which a reflection event suddenly terminates (the end of the section, some dead traces, or for any other reason) the energy from neighboring traces mixes in to give continuity to the interrupted event.

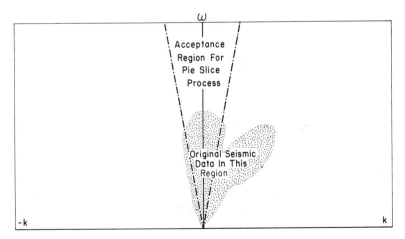

FIGURE 6.25 The acceptance in the ω-k plane of reflection data with a limited range of acceptable phase velocities ($-V_c < V < V_c$). The sharp velocity boundaries used in the original Pie-Slice® process produced a cross-hatched appearance on noisy sections.

Some of these characteristics are objectionable and have led to a more general, two-dimensional filter system; the FFT has made large transforms economical. The time domain cross section is, of course, a large rectangular matrix of points spaced at distance ΔX apart and Δt in time on each trace. This, after Fourier analysis, gives rise to a rectangular grid of points separated by $\Delta\omega$ and Δk. The same problems are experienced in sampling, for aliasing can occur in either direction. It is then necessary to adopt a program that provides not only for zeroing irregular regions in the ω-k plane but also a gradual cutoff—either a linear ramp or some more sophisticated scheme such as the hanning scheme. By this time the reader will have recognized the essential similarity in the treatment and characteristics of two-dimensional and one-dimensional transforms.

Finally, after filtering by multiplying by the transfer function of the two-dimensional filter pulse an inverse Fourier transform is performed to restore the data to the time domain.

It was by making use of this type of operation with a constant-velocity, constant-bandwidth filter that the results in Figure 6.23 were obtained. For low-velocity rejection it is most important that the sampling in the x direction be adequate.

Note that this method treats the entire section in the same manner. The energy for all sections of a particular ω-k is lumped together. If horizontally varying or time-varying frequency or velocity filters are required it will be necessary to remain in the X-t domain and to treat the section with rectangular convolving filters such that

$$A^1(m, n) = \sum_{k=-K}^{K} \sum_{l=-L}^{L} P_{k,l} F_{m-k,n-l}$$

where F is the cross section to be filtered. The method of two-dimensional filtering has been applied extensively to magnetic and gravity maps (Bhatta-charyya, 1972) but has not achieved its full potential in seismic work. There are several reasons why this is so, among them the fairly recent development of a two-dimensional FFT and the fact that seismic data are sometimes noisy by virtue of random uncorrected time errors due to weathering and other causes. If this potential is to be realized it must be done by using geological input as well as geophysical input to reach the goal.

6.18 USE OF THE TAU-p TRANSFORM FOR FILTERING SEISMIC RECORDS

The tau-p transform was introduced in Section 4.7 to convert seismic records, which include the long offset traces, into a domain that accepts reflections and refractions on an equal basis. It is not intended as a means of filtering seismic cross sections (after stacking and other data-processing operations) but rather as an intermediary in the cleaning up of *raw*, or 100%, records. There is an added advantage that refracted events can be represented in the same transform as reflections, thus adding to the knowledge available for interpretation if done at the multitrace, single-record level. Ordinarily, however, the tau-p transform must be regarded primarily as applicable to prestack filtering operations.

We see why this is so and how the transform works by examining a simple reflection/refraction situation.

Figure 6.26 illustrates the ray path of a seismic wave traveling through a horizontally layered acoustic medium and, in one of the layers, the details of a small element of the path. The velocity is V and the direction to the vertical is the incident angle i. Then the horizontal component of movement is

$$\Delta X = V \cdot \Delta T \cdot \sin i$$

the vertical component is

$$\Delta Z = V \cdot \Delta T \cdot \cos i \quad .$$

and the travel time Δt can be expressed as

$$\Delta t = p \, \Delta X + q \, \Delta Z$$

where $p = \sin i / V$ and $q = \cos i / V$; p and q can be thought of as horizontal and vertical components of the wave slowness

$$u = \frac{1}{V} = (p^2 + q^2)^{1/2}$$

We note that, due to Snell's Law, p has the same value for the ray in any of the layers, *even the terminal one*, in which head waves are formed.

For all layers, including reflections and refraction parts of the path,

$$T = pX + 2 \sum_{j=1}^{n} q_j Z_j \tag{6.39}$$

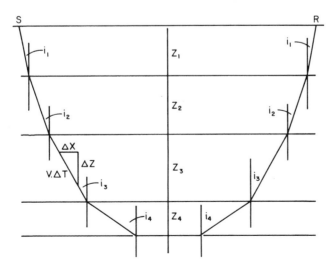

FIGURE 6.26 Typical ray path through a horizontally layered medium.

This equation describes a straight-line tangent to the travel-time curve at the point (T, X) with a slope p and intercept time τ so that

$$\tau = T - pX = 2 \sum_{j=1}^{n} q_j Z_j \tag{6.40}$$

Because

$$q_j = \left(\frac{1}{V_j^2} - p^2 \right)^{1/2} = (u_j^2 - p^2)^{1/2}$$

$$\Delta\tau_j = 2Z_j(u_j^2 - p^2)^{1/2} \tag{6.41}$$

If this equation is plotted in the τ-p plane we can see that the reflection/refraction energy of a seismic record (including long offset reception) will consist of a set of ellipses (Figure 6.27).

Although the analysis so far has been concerned with a single source and increasingly distant receivers, it should be obvious that common midpoint records can be dealt with in the same way if plane horizontal layers are being investigated. It is not so obvious, but equally true, that horizons with dips can be dealt with by using [Diebold and Stoffa (1981)] common midpoint results that have (Figure 6.28) the values of p and q differentiated according to up-dip (p_a and q_a) and down-dip (p_b and q_b). The thicknesses of the layers are measured at the surface midpoint. It can also be seen that $X_a = X_b = X/2$. Under these conditions

$$T_{CMP} = \frac{X(p_a + p_b)}{2} + \sum_{j=1}^{n} Z_j(q_{aj} + q_{bj})$$

and the observed travel-time slope dT/dX is the average of the slownesses in the

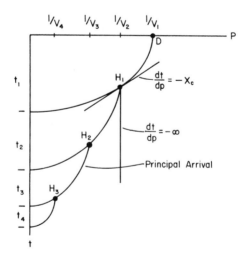

FIGURE 6.27 A tau-p diagram that corresponds to the ray paths in Figure 6.26; D is the location of energy from the direct wave and H_1, H_2, H_3 represent the refracted energy located at the point of critical reflections.

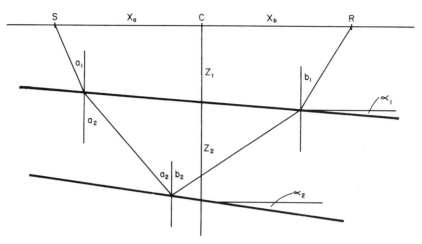

FIGURE 6.28 A reflection path in a medium with sloping boundaries. The angles are measured from the vertical; *a*'s are up-dip, *b*'s are down-dip.

up-dip and down-dip directions. This is more accurate than other techniques that average the apparent velocities.

It is, of course, quite common to segregate sets of traces such as these in the normal course of processing. Intercept times are functions of the vertical slowness multiplied by the layer thicknesses measured at the center. A schematic of the results in a dipping-layer situation on a τ-p plot is given in Figure 6.29.

Now, apart from the theoretical treatment of reflections and refractions in the (X, T) and the (τ, p) domains, we want to learn how to transform actual seismic data from one plane to the other. The operative transform, which is reversible, is the Radon transform [Radon (1917)] that relates the two sets of coordinates:

$$\tau = T - pX; \qquad T = \tau + pX$$

$$p = \frac{dT}{dX}; \qquad \frac{d\tau}{dX} = -p$$

(6.42)

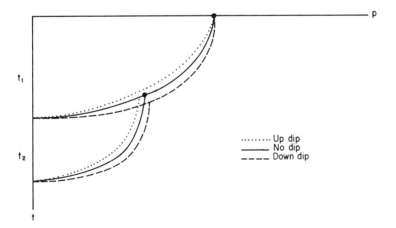

FIGURE 6.29 A tau-*p* plot of reflections that shows the influence of dip on the trajectories.

The inverse transformation (IRT) from a field in the (X, T) domain by the process of slant stack, a procedure defined as, "time shifting traces proportional to their distance from some reference point and then summing (stacking)," may be written formally as

$$\psi(\tau, p) = \int_{-\infty}^{\infty} \phi(\tau + pX, X) \, dX \qquad (6.43)$$

where ϕ is the observation wave field and ψ is the transformed field.

A forward Radon transform that produces a seismic profile from a slant stack is

$$\phi(T, X) = \int_{-\infty}^{\infty} \psi(T - pX, p) * \left[\frac{1}{2\pi} M_{\Omega}(T - pX) \right] dp \qquad (6.44)$$

where M_{Ω} is given [McMechan and Ottolini (1980)] as

$$\Omega \frac{\sin \Omega t}{\pi t} - 2 \frac{\sin^2 (\Omega t/2)}{\pi t^2} \qquad (6.45)$$

In the frequency domain the forward Radon transform is

$$\phi(T, X) = \frac{1}{4\pi^2} \int \int_{-\infty}^{\infty} F_{\psi}(\omega, p) \epsilon^{-i\omega(T - pX)} \cdot |\omega| \, d\omega \, dp \qquad (6.46)$$

and the inverse randon transform is obtained by

$$F_{\psi}(\omega, p) = F_{\phi}(\omega, -\omega p) \qquad (6.47)$$

In (6.47) the two-dimensional transform of the field data (ϕ) is evaluated along the slant line $-\omega p$, and this is the Fourier transform with respect to τ, $\psi(\tau, p)$.

Slant stacking involves multiple sweeps along lines in the data space by fixing p and sweeping over τ or fixing τ and sweeping over p. Either way, the sum of the amplitudes is plotted at the proper coordinate in the (τ, p) space.

Characteristics of Events in (τ, p) Space

For head waves and direct waves the travel velocity (V) is a constant over a large part of the path. In the (X, T)-domain they plot as straight lines, but because the slope is constant they plot (ideally) as *points* in the (τ, p)-domain.

For reflections inside the critical angle the amplitude plots at points on the ellipse that are not included in the overall envelope. This envelope is given the name *principal arrival*. For reflections beyond the critical angle the energy lies on the principal arrival.

An important characteristic in the τ-p-transformed data is that reflection events and others, travelling at wave speeds slower than the P-wave velocity, must lie *outside the principal event*; therefore an immediate discriminating factor is available. In the P-wave–SV-wave system normally used in exploration seismology the principal event divides the primary P events from the multiples and the SV events.

It is possible, naturally, to take extended offset seismic SH-wave records and, in theory, there should be no contamination with P waves; hence all energy *inside* the principal arrival should be due to subcritical SH arrivals.

Filtering in the τ-p Domain

Just as it is possible in the (ω, k) domain to alter in some justifiable manner the amplitudes of points on the plane, thereby filtering the data in a known way, it is also possible to make use of the characteristics of seismic events in the τ-p plane to provide justification for altering amplitudes thereon.

As an example we can compare actions to eliminate a refracted wave of known horizontal velocity V in each case.

For the (ω, k) representation a line with a slope $\omega/k = V_c$ will represent all arrivals with a single velocity. In this case in the (ω, k) domain the zeroing of a small band of wavelengths

$$\lambda_c - \Delta\lambda < \lambda < \lambda_c + \Delta\lambda$$

will eliminate the energy of the unwanted refraction or direct wave. As usual we note that the edges of this band should not be too sharp but tapered in an optimum manner to prevent the generation of false arrivals when this band passes through noisy data. Precautions are also taken to avoid aliasing in the ω and the k directions when we are using sampled data.

For the τ-p representation the refraction with velocity V_c will occupy a *point* (theoretically) on the τ-p plane, where the reflection ellipse for the lower medium becomes vertical (i.e., where it just reaches the upper ellipse). In a band-limited case, in which the arrival has some dimension in time (i.e., it is not a δ-function) the *point* is broadened to a zone on the τ-p plane. All points on the event must have the same velocity; hence it will have zero dimension in the p direction and the following characteristics for (a) Vibroseis arrival which should have the minimum time spread of any arrivals but will smear the energy on both sides of the expected τ_c in a symmetrical manner (see Figure 6.30*a*); (b) causal pulses which should smear the energy on the greater τ side of τ_c (Figure 6.30*b*).

In any case, for filtering purposes the line of the expected position on the τ-p plane must be multiplied by a *hole*, the bottom of which has zero amplitude with graduated sides up to unit acceptance over the remainder of the τ-p plane.

As a second example we can examine multiple reflections (which must arrive after the principal event) and shear waves which, on a τ-p plane ostensibly devoted to P events, must also arrive after the principal event. It would be well to note, however, that the principal event itself is smeared due to the limited bandwidth of the survey and the disposition of the smearing follows the rules enunciated in an earlier paragraph. After making allowances for the smearing the energy outside the principal event may be zero-ed, thereby removing from the P-wave record *some* noise-producing events.

The use of the τ-p domain is not confined to filtering. Other uses are explained in the appropriate locations.

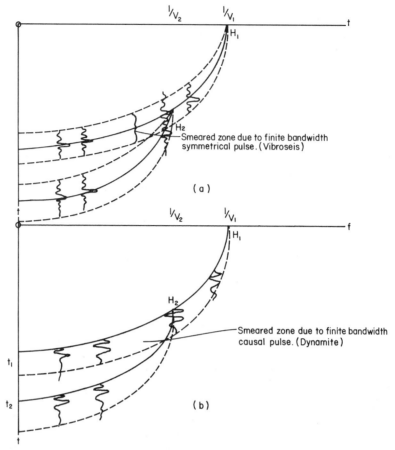

FIGURE 6.30 These tau-p diagrams show the smearing due to finite bandwidth in (a) Vibroseis results and (b) dynamite results.

6.19 SUMMARY AND CONCLUSIONS

During the last 20 years a revolution in the treatment of seismic data has been due largely to the consistency and flexibility of the digital computer. Data recorded in continuous and reproducible form on magnetic tape must now be represented by a series of discontinuous samples which are usually multiplexed because of the need for sampling across all input signals as they come in—almost simultaneously. High-speed analogue converters that have no difficulty in sampling up to 50 traces per millisecond are now available, and automatically ranging binary gain amplifiers have made recording in a floating-point system possible and have decreased the need for care in scaling.

For treatment of the data it is more convenient to have the numbers (amplitudes) arranged in trace sequential form. Then each trace can be treated on its own merits. For proper treatment auxiliary data from land surveying are incorporated in a line-description program that allows the calculation of trace offsets and primary near-surface corrections which are carried with the trace in a header—a few auxiliary words of computer information.

Primary near-surface corrections and a correction for the geometrical effects of offset are made for each trace; the relevant velocities are taken from the data. These are strictly first-estimate corrections, but they are sufficient to allow the proper data parts to be stacked together to provide a first look at the stacked CDP reflection cross section. Expansion of the amplitude as a function of time or a slow-working digital automatic gain control is sometimes incorporated to allow both early and late reflections to be seen.

Generalized filtering is much more flexible in digital computer programs, one of the most useful of which is called deconvolution—a program for undoing filtering that has already taken place. The input pulse, which may have been complicated, or the effect of shallow multiples can be compensated and effectively made into a zero-phase pulse, provided that the distortions in the first place were derived minimum-phase operations. It is the assumption here that for a given amplitude spectrum the energy in the pulse is optimally concentrated near the beginning of the pulse. It can be shown that multiple reflections within the shallow layers fall into this classification.

Deconvolution operators are obtained from primary data first by smoothing the amplitude spectrum (to eliminate the effects of reflections) and then by calculating the minimum-phase operator from this amplitude spectrum. The correction operator (the deconvolution operator) is obtained by spectrum inversion (within a given frequency range in which noise values are not enhanced) or by matrix inversion techniques. In the latter small amounts of white noise are added to improve the stability of the matrix inversion process. The deconvolution process for Vibroseis records must be treated in a special manner because the correlation step uses reflection pulses that are not minimum-phase in the restricted sense.

A form of deconvolution called homomorphic has been proposed to deal with situations in which the minimum-phase criterion is not met. Although it has been used with some success in other fields, there is little present (1985) application in reflection seismology.

Once seismic reflection results are presented in the form of cross sections several events can be seen which, from the pattern of arrival on successive traces, are judged not to be reflections. They can develop from instrumental problems (e.g., cable noise in marine systems) or from diffracted or scattered events. In some cases they can be removed by two-dimensional filters in which the basic assumption is that reflections are coherent, nearly constant time events. These filters work well, provided that the discrimination between other events and reflections map into two distinct regions in the frequency-wavelength plane and that they can be separated out by filters that have no sharp cutoffs.

The process of plane-wave decomposition of extended offset seismic records, that is, those that are long enough to include critical reflections and refractions, is the same as a mapping on another transform plane (the tau-p plane whose coordinates are intercept time and slope). This process can be accomplished by a weighted slant stack operation. In some problems filtering and other operations are simpler in the new domain.

Only simple, routine uses of digital computations have been discussed in this chapter, but these uses alone, taken with the reliability of a digital process once it is freed from procedural errors, have really constituted a revolution in the production of reliable seismic reflection data.

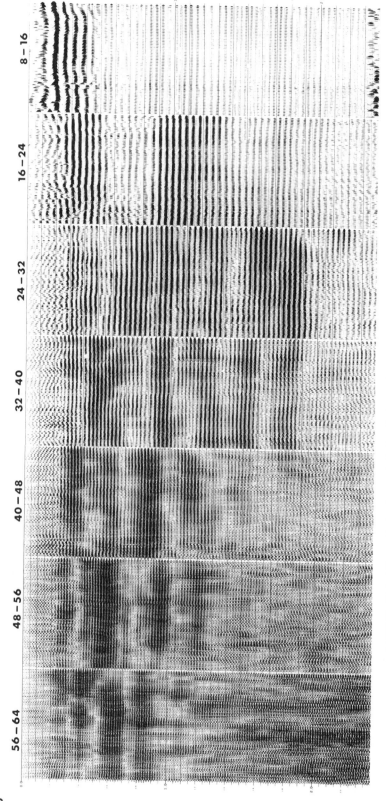

FIGURE 6A.1 The analysis of records in bands 8 Hz wide. In practice narrower bands, about 3 Hz wide, can be used. The 24 to 32 Hz and 32 to 40 Hz bands appear to have reflection energy down to 2.0 s, but both higher and lower frequency bands cut off at shorter times.

256

APPENDIX 6A: ADAPTIVE FILTERS

In the earlier days of data processing the final bandwidth of the processing filter (possibly varying as a function of depth) was determined largely by trial and error. The geophysicist in charge of the work would try different combinations of high- and low-frequency cutoffs until satisfied that an optimum signal/noise ratio had been obtained overall. All filters were, of course, zero-phase in order not to disturb the phase relationships in the seismic traces.

More definite information is obtainable by using narrow-band filters and subjective judgment to decide the total number of narrow bands to be kept at each depth. An example is given in Figure 6A.1.

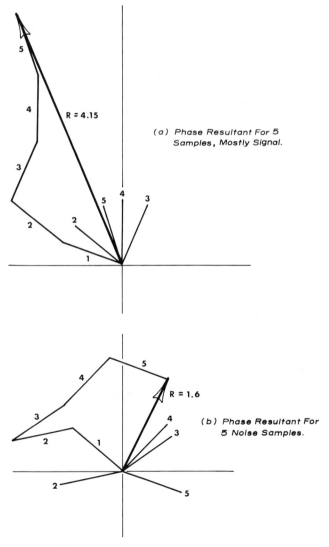

(a) Phase Resultant For 5 Samples, Mostly Signal.

R = 4.15

(b) Phase Resultant For 5 Noise Samples.

R = 1.6

FIGURE 6A.2 The results of adding five unit vectors derived from coherent traces, mostly signal (a), and from noisy traces, mostly noise (b). The vector R is the vector addition of the five unit vectors.

There has been a gradual trend toward more objective analyses of seismic traces to determine the bandwidth to be retained. This is useful because computer examination of the original data is involved and decisions regarding the final filters can be made within the processing stream and without the delays caused by trial and error or by the extra observational step involved.

One process has been described by Branisa (1975). Essentially, a window of a seismic trace is analyzed to obtain the amplitude and phase at each frequency. When two or more traces are examined the phase can be compared from trace to trace to establish the degree of similarity or phase coherence. Figure 6A.2 shows how unit-length vectors, plotted at the indicated phase angles, tend to add when the phases are nearly the same (mostly signal) and to give a small final summation vector when noise is being analyzed. The probability of a given amplitude of the resultant vector in random noise was analyzed by Rayleigh (1945) (Figure 6A.3), and Branisa (1945) found that experimental noise analysis follows the theory closely.

It is now possible to choose a phase coherence level believed to give the optimum compromise between high resolution (a wide-band, flat-amplitude spectrum) and a good signal/noise ratio. Finally, the flattened-amplitude spectrum of the original signals is adjusted (weighted) by the phase coherence to lower the amplitude of the more incoherent frequencies. In practice five or six consecutive (stacked) traces will determine the coherent spectrum at a series of points along a line of section. An average acceptable spectrum can then be computed for the entire line.

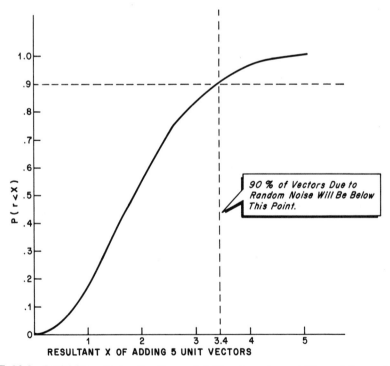

FIGURE 6A.3 Rayleigh's criterion for the probability of random addition of five unit vectors. Strictly, the theory is applicable only to N large. In practice it is applicable to five or more unit vectors.

Time-variant limits for adaptive filters evidently can be calculated with this procedure by using windows centered on a series of increasing reflection times. A compromise has to be made between having windows of sufficient length for resolution in frequency and rate of change in the filter with record time. In practice 1-s windows are adequate for most purposes.

APPENDIX 6B: MATRIX ALGEBRA OPERATIONS WITH EMPHASIS ON REFLECTION SEISMOLOGY APPLICATIONS

6B.1 INTRODUCTION

The following introduction to the definitions and rules of matrix algebra is necessarily condensed. It is provided for those readers whose formal training neglected this highly useful subsection of mathematics to give them an appreciation of its uses and to show why it has been adopted extensively in the geophysical literature of recent years. Its compact notation makes mathematical arguments much easier to interpret. Readers are urged to study this appendix before proceeding to later chapters. Although they will not be left out in the cold if they do not understand the matrix formulations, they will have a warmer appreciation of many different facets of their science if they do.

A rectangular array of numbers, which may be real or complex, is called a *matrix*. The following matrix has *elements* that could have been derived as the coefficients of a set of simultaneous equations and could equally well have been supplied by copying the *amplitude* of a set of seismic trace sample points (a_{ij}). The matrix is usually designated by

$$
\begin{pmatrix}
a_{11} & a_{12} & a_{13} & \cdot & \cdot & \cdot & a_{1n} \\
a_{21} & a_{22} & & & & & \cdot \\
a_{31} & & & & & & \cdot \\
a_{m1} & \cdot & & \cdot & \cdot & \cdot & \cdot & a_{mn}
\end{pmatrix}
$$

or a bold-faced letter such as **A**. The subscripts i, j are indices; i is the number of the row and j, the column (from top to bottom and from left to right).

Matrices may have only one row (with n columns) and is then called an n-dimensional row vector,

$$(a_{11}, a_{12}, a_{13}, \dots, a_{ij}, \dots, a_{1n})$$

or it may have as few as one column and m rows—and m-dimensional column vector,

$$
\begin{pmatrix}
a_{11} \\
a_{21} \\
\cdot \\
\cdot \\
\cdot \\
a_{m1}
\end{pmatrix}
$$

The operation of *transposition* is that in which the row vectors of **A** are made the column vectors of the *transpose* of **A** if

$$\mathbf{A} = (a_{jk})$$

Then

$$\mathbf{A}^T = (a_{kj})$$

For square matrices the formation of the transpose is equivalent to a rotation of the elements around the *principal diagonal*.

6B.2 SPECIAL TYPES OF MATRIX

A matrix of any size that contains all zeros as elements is called a *null matrix* and is designated **O**. A (necessarily square) matrix with all zeros except on the principal diagonal, in which the values are unity, is called the *identity* matrix and is written **I**. Other matrices with arbitrary numbers down the principal diagonal but with zeros elsewhere are *diagonal* matrices.

There is one other special type that is quite common in seismic work, one for which the largest terms of the matrix are concentrated in a band near the principal diagonal. It is symmetrical, with small values outside the band, and is called a *Toeplitz* matrix and occurs often in data processing. Figure 6B.1 is a symbolic illustration that shows the finite values *a* near the diagonal and zeros elsewhere. Its further properties are explained, as they are needed, in the text.

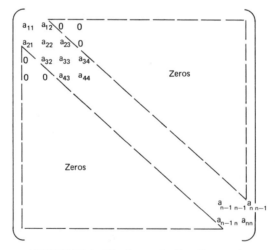

FIGURE 6B.1 The form of a Toeplitz matrix.

6B.3 MATRIX OPERATIONS

Matrices may be added (or subtracted) only if they have the same number of rows and columns, in which case

$$\mathbf{C}_{mn} = \mathbf{A}_{mn} \pm \mathbf{B}_{mn}$$

where the elements are constructed according to the rule

$$c_{ij} = a_{ij} \pm b_{ij}$$

Matrices may be multiplied only if two rules are obeyed:

1. Two matrices may be multiplied only if one has the same number of rows as the other has columns and then only if they are in the proper order; for example, given

$$\mathbf{A} = \mathbf{A}_{mn}$$

and

$$\mathbf{B} = \mathbf{B}_{nq}$$

then

$$\mathbf{AB} = \mathbf{C}_{mq}$$

We say that **B** is premultiplied by **A** or **A** is postmultiplied by **B**, in which case **BA** is not meaningful because rule 1 was not obeyed. However, when both **B** and **A** are square matrices of the same number of rows (they have the same order) both **BA** and **AB** will exist but they may not be identical.

2. The elements of the product matrix **C** are given by $c_{ik} = \Sigma_j (a_{ij} \times b_{jk})$ or $c_{ik} = \mathbf{a}_i \cdot \mathbf{b}_k$ where $\mathbf{a}_i$ is a row vector and $\mathbf{b}_k$ is a column vector:

$$
\mathbf{a}_i \rightarrow
\begin{pmatrix}
a_{11} & a_{12} \cdots a_{ij} & \cdots a_{in} \\
a_{21} & a_{22} \\
\cdot \\
a_{i1} & a_{i2} \quad a_{ij} \quad a_{in} \\
a_{m1} & \cdots \quad a_{im} \cdots a_{mn}
\end{pmatrix}
\begin{pmatrix}
b_{11} & b_{12} \cdots b_{ik} & \cdots b_{iq} \\
b_{21} & b_{22} \\
\cdot \\
& & b_{ik} \\
b_{n1} & \cdots & b_{nk} \cdots b_{nq}
\end{pmatrix}
$$

dimension $m \times n$ dimension $n \times q$, $\mathbf{b}_k$

6B.4 GEOPHYSICAL EXAMPLE 1

The construction, by matrix methods, of the autocorrelation function $A(\tau)$ may be illustrated as follows:

We construct a rectangular matrix $\mathbf{S}_{mn}$ by putting successive samples $S(j\Delta)$ of the seismic trace ($j = 1, m$) as the first column of the matrix **S**. The remaining $m - 1$ columns are made by shifting the same trace down by one sample position for each column. The resulting matrix is illustrated in Figure 6B.2. For clarity

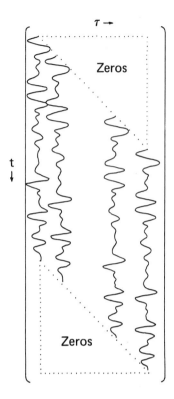

FIGURE 6B.2 A schematic of matrix (a_{mn}) which consists of seismic traces shifted by one sample value to the right as m increases.

most of the columns have been omitted and areas of zeros have been shown in outline only. To appeal to geophysicists a graphical seismic trace has been substituted for its sample values. The column number (n) is therefore a measure of the shift, measured in sample interval units (Δ). The number of columns needed (n) in the matrix is dependent on the number of samples needed in the autocorrelation function, whereas the number of samples in a column (the number of rows m) depends on the decision regarding the window of the seismic trace to be used.

The transpose of $\mathbf{S}$ ($\mathbf{S}^T$) has, of course, the same seismic traces plotted as rows. If the product

$$\underset{n \times m}{\mathbf{S}^T} \quad \underset{m \times n}{\mathbf{S}}$$

is formed, the resulting $n \times n$ square matrix will be symmetric about the main diagonal and will have the largest values along this diagonal. Generally, the amplitude of terms will decay rapidly with distance, symmetrically, from the diagonal. This is an autocovariance matrix.

The formation of the full $m \times m$ matrix is, of course, not needed in general because the values along one row or one column will give the full autocorrelation function, knowing it is symmetrical about the largest value.

6B.5 DETERMINANTS

The determinant is an array of numbers, real or imaginary, rather like a matrix except for the fact that it has *a value*. It is written as a square array bounded on the sides by vertical straight lines:

$$\det \mathbf{A} = \begin{vmatrix} a_{11} & a_{12} & a_{13} \cdots a_{1n} \\ a_{21} & a_{22} & a_{23} \cdots a_{2n} \\ \vdots \\ a_{n1} & a_{n2} & a_{n3} \cdots a_{nm} \end{vmatrix}$$

It derives from the solution of simultaneous equations in the evaluation of matrix inverses and in the determination of certain characteristic numbers (eigenvalues) that are defined later.

If one element of a determinant is chosen and its row and column are eliminated, the resulting determinant of order $n - 1$ is called the minor M_{ik} of the element a_{ik}.

It is then possible to evaluate the original determinant in terms of determinants of the order $n - 1$ by taking *any* row or column and writing

$$D = a_{i1}C_{i1} + a_{i2}C_{i2} + \cdots + a_{in}C_{in}$$

where C_{ik}, the *cofactor*, is obtained by taking the *minor* of the element and multiplying by $(-1)^{i+k}$; for example,

$$\det A = \begin{vmatrix} a_{11} & a_{12} & a_{13} \\ a_{21} & a_{22} & a_{23} \\ a_{31} & a_{32} & a_{33} \end{vmatrix}$$

may be expressed as

$$a_{11}(-1)^2 \begin{vmatrix} a_{22} & a_{23} \\ a_{31} & a_{33} \end{vmatrix} + a_{12}(-1)^{1+2} \begin{vmatrix} a_{21} & a_{23} \\ a_{31} & a_{33} \end{vmatrix} + a_{13}(-1)^{1+3} \begin{vmatrix} a_{21} & a_{22} \\ a_{31} & a_{33} \end{vmatrix}$$

and each of these lower order determinants may be resolved, for example, as

$$\begin{vmatrix} a_{22} & a_{23} \\ a_{31} & a_{33} \end{vmatrix} \begin{aligned} &= a_{22}\,a_{33} + (-1)^{1+2}a_{31}\,a_{23} \\ &= a_{22}\,a_{33} - a_{31}\,a_{23} \end{aligned}$$

6B.6 THE INVERSE OF A MATRIX

The operation *division* is not defined in matrix algebra. Rather, an *inverse* is defined for square matrices only, and an *equivalent* operation to division consists of multiplying one square matrix by the inverse of another of the same order.

The inverse $\mathbf{A}^{-1}$ of a square matrix exists if, and only if, the determinant of $\mathbf{A}$ is not equal to zero. We define $\mathbf{A}^{-1}$ as the matrix that, when multiplied by the original matrix $\mathbf{A}$, yields the identity matrix $\mathbf{I}$:

$$\mathbf{A}^{-1}\mathbf{A} = \mathbf{A}\mathbf{A}^{-1} = \mathbf{I}$$

The calculation of the inverse matrix $\mathbf{A}^{-1}$ is relatively simple in theory, but it is complex and requires computers for any order greater than about 2. The sequence of operations for its formation is as follows:

1. Find the elements (a_{ik}) and cofactors $A_{ik}(-1)^{i+k}$ of the original matrix. The cofactor must be evaluated to give a new element (b_{ik}).
2. Replace the original elements (a_{ik}) by (b_{ik}).
3. Transpose the resulting matrix and multiply by $1/\det \mathbf{A}$.

Two kinds of matrix are easy to invert: first, any nonsingular matrix of order 2.

$$\mathbf{A}^{-1} = \frac{1}{\det \mathbf{A}} \begin{pmatrix} a_{22} & -a_{12} \\ -a_{21} & a_{11} \end{pmatrix}$$

and, second, any diagonal matrix, if

$$\mathbf{A} = \begin{bmatrix} a_{11} & & \text{zeros} \\ & a_{22} & \\ \text{zeros} & & \ddots \\ & & & a_{nn} \end{bmatrix}$$

and

$$\mathbf{A}^{-1} = \begin{bmatrix} 1/a_{11} & & \text{zeros} \\ & 1/a_{22} & \\ \text{zeros} & & \ddots \\ & & & 1/a_{nn} \end{bmatrix}$$

Finally, the inverse of a matrix product is the product of the inverses, taken in reverse order; for example,

$$(\mathbf{AC})^{-1} = \mathbf{C}^{-1}\mathbf{A}^{-1}$$

$$(\mathbf{PQRS})^{-1} = \mathbf{S}^{-1}\mathbf{R}^{-1}\mathbf{Q}^{-1}\mathbf{P}^{-1}$$

The reader need not worry about the labor of inverting matrices or about the numerical evaluation of matrix operations because this is done by computers with matrix algebra programs supplied by most manufacturers of the hardware. It is the ideas and rules behind the operations that we are trying to convey, not skill in performing the numerical operations. For those interested Robinson (1967b) has included listings of subroutines for matrix operations in his book on multichannel filters.

6B.7 EIGENVALUES AND EIGENVECTORS

If we form a vector equation

$$\mathbf{AX} = \lambda \mathbf{X}$$

where λ is a number, we are, in effect saying that the result of premultiplying the vector $\mathbf{X}$ by the matrix $\mathbf{A}$ has no effect on the direction of the vector—only a change in the size. This equation is called the characteristic equation of $\mathbf{A}$, and any value λ for which the vector $\mathbf{X} \neq 0$ is called a characteristic value or eigenvalue of $\mathbf{A}$. The corresponding solutions $\mathbf{X}(\neq 0)$ are called characteristic vectors or eigenvectors of $\mathbf{A}$. Some problems of physics or engineering make use of this technique because, by its use, the natural modes of vibration and their amplitudes are found; for example, complex mechanical (or electrical) circuits with lumped elements (masses, springs, damping units or their equivalents, inductors, capacitors, and resistors) or mechanical lumped element systems (i.e., masses, springs, damping units, or their electrical equivalents) can be solved by the following:

1. Setting up a system of simultaneous differential equations.
2. These may be put in the form of a single vector equation (say) $\ddot{\mathbf{y}} = \mathbf{Ay}$.
3. Substituting $\mathbf{y} = \mathbf{X}e^{i\omega t}$.
4. Simplifying to form the characteristic equation $\mathbf{AX} = \lambda \mathbf{X}$, where $\lambda = \omega^2$.
5. Solving for the eigenvalues λ_i. These are simply the roots of the determinantal equation

$$\begin{vmatrix} a_{11} - \lambda & a_{12} & a_{13} \cdots a_{1n} \\ a_{21} & a_{22} - \lambda & a_{23} \cdots a_{2n} \\ \vdots & & \\ a_{m1} & \cdot & \cdot \quad \cdots a_{mn} - \lambda \end{vmatrix} = 0$$

6B.8 SOLUTION OF LINEAR EQUATIONS—GEOPHYSICAL APPLICATIONS

Consider this set of simultaneous equations:

$$a_{11}x_1 + a_{12}x_2 + a_{13}x_3 + \cdots + a_{1n}x_n = p1$$
$$a_{21}x_1 + a_{22}x_2 + a_{23}x_3 + \cdots + a_{2n}x_n = p2$$
$$\vdots$$
$$a_{m1}x_n + a_{m2}x_2 \quad \cdots \quad \cdots + a_{mn}x_n = p_m$$

It can be written in the form

$$\overset{\uparrow}{\underset{m}{\Big\updownarrow}} \begin{pmatrix} \xleftarrow{\ n\ } \\ \mathbf{A} \\ \\ \end{pmatrix} \begin{pmatrix} \overset{\uparrow}{\underset{n}{\Big\updownarrow}} \\ \mathbf{X} \\ \Big\downarrow \end{pmatrix} = \begin{pmatrix} \overset{\uparrow}{\underset{m}{\Big\updownarrow}} \\ \mathbf{p} \\ \Big\downarrow \end{pmatrix}$$

$$\mathbf{A} \qquad\qquad \mathbf{X} \quad = \quad \mathbf{p}$$

If $m = n$ we have the normal situation of n equations and n unknowns and the matrix $\mathbf{A}$ can have an inverse. Therefore we can solve this set of equations formally by the following process:

$$\mathbf{AX} = \mathbf{p}$$
$$\mathbf{A}^{-1}\mathbf{AX} = \mathbf{A}^{-1}\mathbf{p}$$
$$\mathbf{IX} = \mathbf{A}^{-1}\mathbf{p}$$
$$\mathbf{X} = \mathbf{A}^{-1}\mathbf{p}$$

The inverse of $\mathbf{A}$ and the postmultiplication by $\mathbf{p}$ are done in the computer.

A different situation sometimes occurs when, for example, the number of measurements made (the number of equations) exceeds the number of unknowns to be determined ($n < m$). In this overdetermined system we can again write

$$\mathbf{AX} = \mathbf{p}$$

but now $\mathbf{A}$ is not a square matrix and it has no inverse. To convert the system to one that can be handled we premultiply both sides by $\mathbf{A}^T$ to obtain

$$(\mathbf{A}^T\mathbf{A})\mathbf{X} = \mathbf{A}^T\mathbf{p}$$

Because $(\mathbf{A}^T\mathbf{A})$ is now a square matrix (the equivalent of the normal equations of statistics), an inverse may be found in which

$$(\mathbf{A}^T\mathbf{A})^{-1}(\mathbf{A}^T\mathbf{A}) = \mathbf{I}$$

and

$$(\mathbf{A}^T\mathbf{A})^{-1}(\mathbf{A}^T\mathbf{A})\mathbf{X} = (\mathbf{A}^T\mathbf{A})^{-1}\mathbf{A}^T\mathbf{p}$$

or

$$\mathbf{X} = (\mathbf{A}^T\mathbf{A})^{-1}\mathbf{A}^T\mathbf{p}$$

This condensed matrix treatment has performed all the operations necessary for us to obtain a best *least squares* estimate of the vector $\mathbf{X}$. Again the matrix multiplications and the inversion are standard matrix operations in the computer software package. We must enter the data and make sure that the operations are done in the order indicated.

This method is called obtaining a *generalized inverse* of the matrix **A**, and we shall use this technique to find the statistically best deconvolution operator for use in data processing.

More and more, matrix algebra is being used in data processing and other applications in seismology. The compact notation and a knowledge of the definitions and rules that govern the algebra will allow the nonspecialist to see that at least the derivations are plausible, although the mathematicians may wish to cavil over the seemingly callous way in which their techniques have been lifted into the applied sphere. All of this, of course, has been a relatively late development that, as far as seismology is concerned, could be used only after the technique of digital sampling raised the possibility of describing a sampled seismic trace as "an *n*-dimensional vector."

REFERENCES

Bhattacharyya, B. K. (1972), "Design of Spatial Filters and their Application to Aeromagnetic Data," *Geophysics*, Vol. 37, No. 1, pp. 68–91.

Blackman, R. B., and Tukey, J. W. (1958), *The Measurement of Power Spectra*, Dover, New York.

Bogert, B. P., Healey, M. J., and Tukey, J. W. (1963), "The quefrency analysis of time series for echoescepstrum, pseudo-covariance, cross cepstrum and saphe cracking": Proceedings of Symposium on Time Series Analysis. M. Rosenblatt, Ed., Wiley, New York.

Branisa, F. (1975), "An Automated Signal and Resolution Enhancement Filter," paper presented at the 45th Annual International Meeting. Society of Exploration Geophysicists, Denver, Colorado.

Brigham, E. O. (1974), *The Fast Fourier Transform*, Prentice-Hall, Englewood Cliffs, New Jersey. This book gives a readable account of the transform properties between the time and frequency domains, as well as a description of the FFT and algorithms for implementing it on a computer.

Brigham, E. O., and Morrow, R. E. (1967), "The Fast Fourier Transform," *IEEE Spectrum*, Vol. 4, pp. 63–70.

Burns, W. R. (1968), "A Statistically Optimized Deconvolution," *Geophysics*, Vol. 33, pp. 255–263.

Clark, G. K. C. (1968), "Time Varying Deconvolution Filters," *Geophysics*, Vol. 33, pp. 936–944.

Collins, R. E. (1968), *Mathematical Methods for Physicists and Engineers*, Reinhold, New York.

Cooley, J. W., and Tukey, J. W. (1965), "An Algorithm for Machine Calculation of Complex Fourier Series," *Mathematical Computation*, Vol. 19, pp. 297–301.

Diebold, S. B., and Stoffa, P. L. (1981), "The Travel-time Equation, Tau-*p* Mapping and Inversion of Common Midpoint Data," *Geophysics*, Vol. 46, pp. 238–254.

Dix, C. H. (1955), "Seismic Velocities from Surface Measurements," *Geophysics*, Vol. 20, No. 1, p. 73.

Embree, P., Burg, J. P., and Backus, M. M. (1963), "Wide Band Velocity Filtering—The Pie Slice Process," *Geophysics*, Vol. 28, No. 6, pp. 948–974.

Fail, J. P., Grau, G., and Layotte, P. C. (1964), "Amelioration du Rapport Signal—Bruit a l'Aide du Filtrage en Eventail—Etude d'un Cas Concret," *Geophysical Prospecing*. Vol. 12, No. 3, pp. 258–282.

Ford, W. T., and Hearne, J. H. (1966), "Least Squares Inverse Filtering," *Geophysics*, Vol. 31, pp. 917–930.

Foster, M. R., Sengbush, R. L., and Watson, R. J. (1968), "Use of Monte Carlo Techniques in Optimum Design of Deconvolution Process," *Geophysics*, Vol. 33, pp. 945–949.

Gurbuz, G. M. (1972), "Signal Enhancement of Vibratory Source Data in the Presence of Attenuation," *Geophysical Prospecting*, Vol. 20, pp. 421–438.

Jury, E. I. (1964), *Theory and Application of the Z Transform Method*, Wiley, New York.

Kunetz, G., and Fourman, J. M. (1968), "Efficient Deconvolution of Marine Seismic Records," *Geophysics*, Vol. 33, No. 3, pp. 412–423.

Levin, F. K. (1971), "Apparent Velocity for Dipping Interface Reflections," *Geophysics*, Vol. 36, No. 3, pp. 510–516.

McMechan, G. A., and Ottolini, R., (1980), "Direct Observation of a (p-τ) Curve in a Slant Stacked Wave Field," *Bulletin of the Seismological Society of America*, Vol. 70, pp. 775–789.

Meiners, E. P., Lenz, L. L., Dalby, A. E., and Hornsby, J. M. (1972), "Recommended Standard for Digital Tape Formats," *Geophysics*, Vol. 37, No. 1, pp. 45–58 (Digital Format C).

Musgrave, A. (1962), "Applications of the Expanding Reflection Spread," *Geophysics*, Vol. 27, No. 6, pp. 981–993.

Northwood, E. J., Weisinger, R. C., and Bradley, J. J. (1967), "Recommended Standards for Digital Tape Formats," *Geophysics*, Vol. 32, No. 6, pp. 1073–1084 (Formats A and B).

Nyman, D. C., and Gaiser, J. E. (1983), "Adaptive Rejection of High-Line Contamination," paper and extended abstract presented at S.E.G. Annual International Meeting, Las Vegas.

Oppenheim, A. V., Schafer, R. W., and Stockham, T. G. (1968), "Non Linear Filtering of Multiplied and Convolved Signals," *Proceedings of the IEEE*, Vol. 65, pp. 1264–1291.

Peacock, K. L., and Treitel, S. (1969), "Predictive Deconvolution: Theory and Practice," *Geophysics*, Vol. 34, pp. 155–169.

Rayleigh, Lord (1945), *Theory of Sound*, Vols. I and II (originally published in 1877), Vol. I, Dover, New York, p. 41.

Ristow, D., and Jorczyk, D. (1975), "Vibroseis Deconvolution," *Geophysical Prospecting*, Vol. 23, No. 2, pp. 363–379.

Robinson, E. A. (1967a), "Predictive Decomposition of Time Series and Application to Seismic Exploration," *Geophysics*, Vol. 32, pp. 418–484.

Robinson, E. A. (1967b), *Multi-Channel Time Series Analysis with Digital Computer Programs*, Holden-Day, San Francisco.

Robinson, E. A., and Treitel, S. (1967), "Principles of Digital Wiener Filtering," *Geophysical Prospecting*, Vol. 15, pp. 312–333.

Sattlegger, J. (1965), "A Method of Computing True Interval Velocities from Expanding Spread Data in the Case of Arbitrary Long Spreads and Arbitrarily Dipping Plane Interfaces," *Geophysical Prospecting*, Vol. 13, No. 2, pp. 306–328.

Shanks, J. L. (1967), "Recursion Filters for Digital Processing," *Geophysics*, Vol. 32, No. 1, pp. 33–51.

Taner, M. T., and Koehler, F. (1969), "Velocity Spectra—Digital Computer Derivation and Applications of Velocity Functions," *Geophysics*, Vol. 34, No. 6, pp. 859–881.

Treitel, S., and Robinson, E. A. (1966), "The Design of High Resolution Filters," *IEEE Transactions, Geoscience Electronics*, Vol. 4, pp. 25–38.

Treitel, S., and Robinson, E. A. (1967), "The M.I.T. GAG Reports," *Geophysics*, Vol. 32, No. 3. This entire volume is dedicated to selected topics from the GAG reports and gives a convenient summary of most of the work.

Treitel, S., and Robinson, E. A. (1969), "Optimum Digital Filters for Signal to Noise Ratio Enhancement," *Geophysical Prospecting*, Vol. 17, pp. 248–293.

Ulrych, T. J. (1971), "Applications of Homomorphic Deconvolution to Seismology," *Geophysics*, Vol. 36, No. 4, pp. 650–660.

Ware, J. A. (1964), personal communication.

Waters, K. H., Palmer, S. P., and Farrell, W. E. (1978), "Fracture Detection in Crystalline Rock Using Ultrasonic Shear Waves," L.B.L. Report 7051, N.T.I.S., U.S. Department of Commerce, Springfield, VA.

White, J. E., and O'Brien, P. N. S. (1974), "Estimation of the Primary Seismic Pulse," *Geophysical Prospecting*, Vol. 22, No. 4, pp. 627–651.

SEVEN

The Calculation and Measurement of Auxiliary Information

7.1 INTRODUCTION

In the preceding six chapters the principles of exploration seismology were explained and methods of data gathering and data processing, described. These principles and methods, when fully implemented, allow seismic data to be obtained and processed to yield an optimum cross section that is the basis for a first look at the stratigraphy below the line of section. In other words, it constitutes a picture, in a limited sense, of the true geological cross section.

On this cross section lines illustrative of structural geological features can be drawn and conclusions can be formed with respect to the trapping potential seen from this line of data. There is no information, however, on rock types, on the probability of higher than normal porosity, or on the probability of oil or gas filling the pores in the rocks at any point in the section.

Other parameters can be derived from these seismic data and some help to fill the gaps in our knowledge. In this chapter they are described as purely objective parameter measurements. Their (subjective) use to derive geological knowledge is deferred until later.

7.2 VELOCITY MEASUREMENT

In Section 6.5 the measurement of average velocity from the surface to the depth of a particular reflector was described. This velocity determination (stacking velocity) enabled a proper correction to be made for the normal (geometrical) increase in reflection time (NMO) due to offset of the receiver from the source. Many different offset traces, after NMO removal, are combined, to give a single CDP or stacked trace.

Although the seismic velocity of a compressional wave through a rock is not a definitive determinant of the rock type by itself, it is certainly an important clue and a partial classifier. For this reason much effort has been expended to find ways of obtaining accurate velocity values. Before we discuss these values in more detail it is necessary to review some of the properties of rocks, particularly as they affect the measurements made on velocities. A rock is not the homogeneous, isotropic medium of theory but an aggregate of mineral particles that have been compressed and cemented and subjected to heat and stress cycles that are not well known. As a consequence, the rock may (and almost always does) show different velocities for P waves, depending on their direction of travel, and waves can be attenuated because of nonlinear stress–strain cycles or because of scattering. In most rocks the dependence of velocity on frequency is small and negligible in the seismic frequency range. A considerable amount of work has been done to measure P- and S-wave velocities at ultrasonic frequencies in small samples taken from cores (Wyllie, Gregory, and Gardner, 1958; King, 1962; Banthia, King, and Fatt, 1965; Fatt, 1958, 1959). These experiments showed that the measured velocities, both P and S, are highly dependent on the external pressure P_e applied to the matrix of rock particles and to the internal pressure P_i of the fluid content. For sandstones it has been shown, experimentally and theoretically (Hicks and Berry, 1956; Brandt, 1955), that the velocity depends on a net effective pressure

$$P_n = P_e - nP_i \qquad (7.1)$$

where n is usually close to unity.

The most common way of measuring P-wave velocities, however, has been by ultrasonic logging in a hole. The method is simple in concept. Figure 7.1 is a schematic of a logging tool used in this process. The tool to which we refer consists of two sources, S_U and S_L (upper and lower), symmetrically disposed about a set of four receivers, R_1 through R_4. The sources are made of magnetostrictive or piezoelectric material that expands when a magnetic or electric field is suddenly applied. The acoustic pulse travels through the inner coupling fluid and a rubber seal into the borehole fluid, thence to the wall of the hole. There the energy is converted in part into a head wave and in part into an interface (Stoneley) wave; the remainder, of course, is transmitted into the earth. The part that is now a head wave travels along the rock near the borehole, is refracted back into the fluid, and picked up by a pair of receivers; S_U works with R_1 and R_3, S_L with R_2 and R_4, for the purposes of this discussion.

The time difference at the relevant two receivers is measured by triggering the electronic microsecond-counting circuit when the nearest receiver accepts a signal exceeding a preset gate level (Figure 7.2). When the second receiver picks up a signal that exceeds its preset level the counter is shut off.

The two sources are used alternately and the time differences are averaged. With this system P-wave velocities can be measured continually as the system is pulled out of the hole. The nominal frequency of the system is 20 kHz (pulse breadth) and the approximate tool distances involved are (a) source to nearest receiver, 1 m; (b) receiver to receiver, 0.3 m. The times involved are transmitted up-hole, where they are stored on magnetic tape and visually recorded on field logs. The usual log, in the recent past, has been a sonic slowness display, shown in microseconds per foot and accompanied on its left by a caliper log to permit the effects of rapid changes in hole diameter to be seen.

In a series of tick marks down the center the sonic log also shows a continuous integration of one-way wave travel time from the surface reference to the tool depth.

With this tool the up and down averaging automatically corrects for tilting the tool in the hole and, to some extent, washouts in the softer formations. The

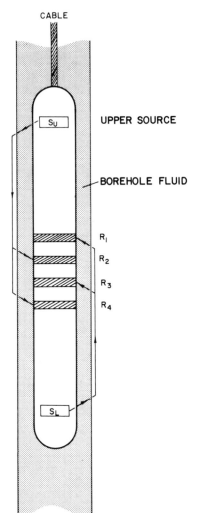

CABLE

S_U — UPPER SOURCE

BOREHOLE FLUID

R_1

R_2

R_3

R_4

S_L

FIGURE 7.1 Schematic of a modern logging tool. Refracted paths through the side wall of the hole from both upper and lower sources to central equispaced receivers are shown.

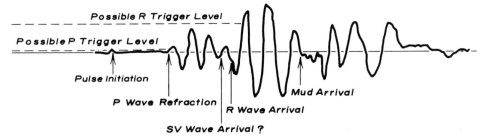

Possible R Trigger Level

Possible P Trigger Level

Pulse Initiation

P Wave Refraction

SV Wave Arrival ?

R Wave Arrival

Mud Arrival

FIGURE 7.2 Idealized trace of the acoustic wave train at the receiver in a borehole due to an initiated pressure impulse.

271

velocities appear to be reasonably representative of the vertical velocity of P waves in the corresponding formations. The exceptions are the following:

1. Invasion of the porous formation by drilling fluid is great or the formation such as montmorillinite-containing shale is altered, by the drilling fluid. The velocity recorded is not representative of the original formation. Hicks (1959) recommended tools with much larger spacing (source to receivers) and added source power to overcome this difficulty.
2. The hole diameter is large or irregular, in which case, there may be unequal distances in the borehole fluid for travel to the two relevant receivers. An alternative problem may be that the first (head wave) arrival is too weak to trigger the counters and it is operated only when a later, much larger, event occurs. In such cases *cycle skipping* causes incorrect velocity values to be recorded. The symmetrical tool just described has helped to reduce these and other problems.
3. The formation is lower than the P-wave velocity through the drilling fluid.

Innovations in tool design, the possibility of achieving high data-rate transmissions up-hole, and data-processing capabilities off-line have revolutionized the science of continuous velocity logging in holes [Aron, Murray, and Seeman (1978)]. In addition to plotting the full waveform over a relatively long time (5 ms) at each receiver, multiple receivers can be used to display the multiple traces and analyze them as arrays, as other forms of seismic record can (Figure 7.3). Here, three separate arrivals, with distinctive *step-outs* across the array, can be clearly seen.

An even later system design has been described by Morris, Little, and Letton (1984). The overall arrangement of the sonic array tool is shown in Figure 7.4*a* and more detail is given in Figure 7.4*b*.

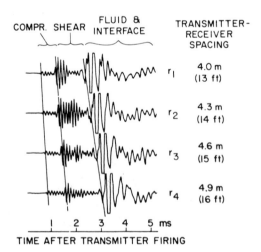

FIGURE 7.3   Full waveform record from modern logging tool. Types of wave, transmitter-receiver spacing, and a time scale are displayed. [*Source*: Aron, Murray, and Seeman (1978). Courtesy Schlumberger Inc.]

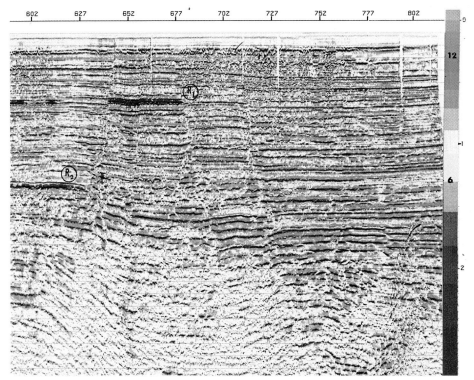

FIGURE 1.3 A color example of a finished reflection seismic cross section.

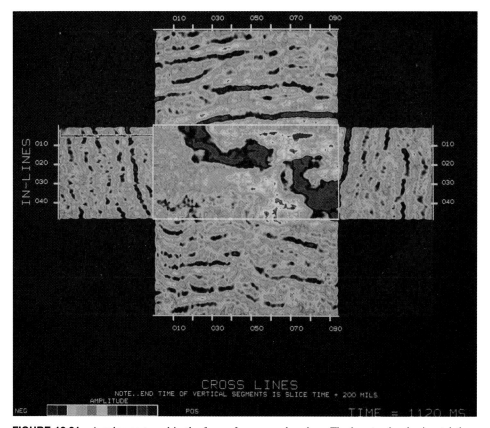

FIGURE 12.31 A color portrayal in the form of an opened-up box. The box top is a horizontal time slice at 1120 ms and the four hinged, vertical-section sides show 200 ms of added section.

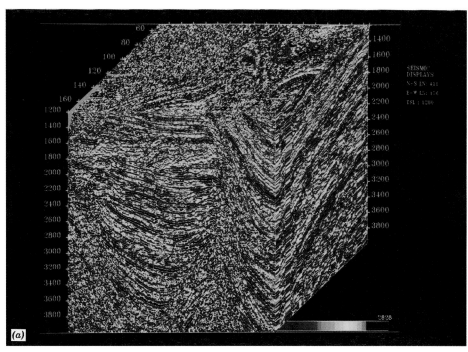

FIGURE 11.8 (*a*) The same data and presentation as Figure 11.7 on page 433, except that a color scale is used to represent amplitudes.

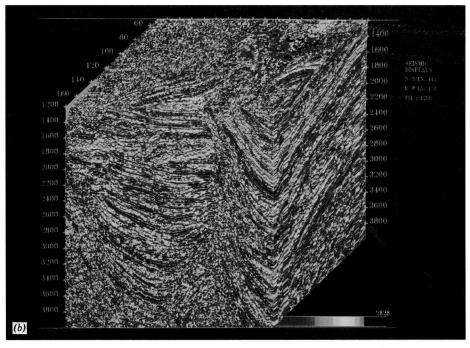

FIGURE 11.8 (*b*) The same data and presentation as Figure 11.7, except that a different color palette has been selected. Note the psychological importance of the subjective choice of the color palette, which can be made to suit individual preference.

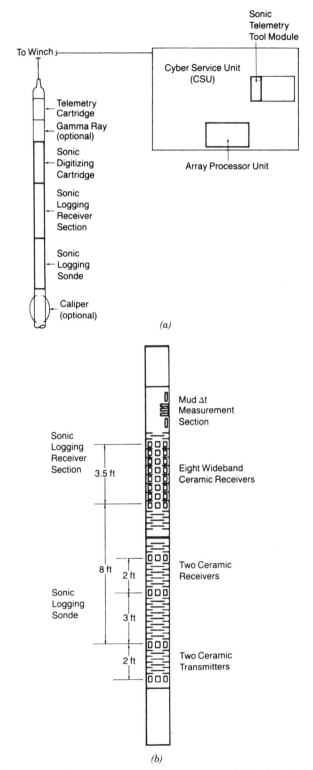

FIGURE 7.4 (*a*) Overall sonic array tool system arrangement. [After Morris, Little, and Letton (1984). *Source*: Courtesy Schlumberger Inc.] (*b*) Details of the sonic array tool. [*Source*: Morris, Little and Letton (1984). Courtesy Schlumberger, Inc.]

In addition to making provision for taking conventional short-spaced, borehole-compensated Δt logs (D D B H C) in open holes and others in cased holes, this tool also has a sonic logging receiver section that contains an array of eight wideband piezoelectric receivers at 8 to 11.5 ft (intervals of 6 in.) from the upper transmitter. The acoustic bandwidth of the system is 5 to 15 kHz.

These elements then allow an array similar to that in Figure 7.5 to be displayed. The on-site array processor computes the slowness of compressional, shear, and Stoneley waves to produce a display similar to that in Figure 7.6 if the shear and Stoneley modes are present.

These CVLs are available over some portion of the total drilled footage for thousands of wells and are cataloged by log service companies, along with electrical, neutron, gamma density, and other logs. In some cases, as shown in Figure 7.7, a check on total arrival time is made by placing a suitable receiver in the hole and arranging for shots to be fired in shallow holes near the well. The difference between shot and arrival times—the transit time of the wave—is corrected for elevation of the shot, offset distance and other relevant parameters and can then be compared with an integration of the elemental transit times measured by the CVL.

Dix (1939, 1945, 1946) has dealt in detail with the problems of velocity measurement by shooting in holes. The experimental method is not so foolproof as it sounds and care must be taken when it is used to calibrate a CVL.

The effects of water invasion, however, particularly in long shale sections, can be serious. Special long tools have been used to ensure that the first arrival corresponds to a wave that has traveled for at least part of its path in unaltered shale. This has resulted in agreement with the check-shot method and has also caused the synthetic seismograms to correspond better with the field records taken in the same location. Hicks (1959) showed that the invasion effect occurred

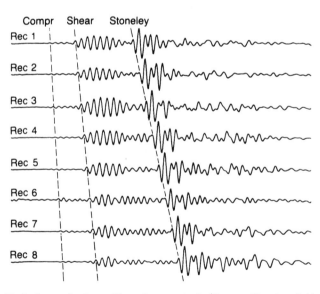

FIGURE 7.5   Typical record taken with sonic array tool. (*Source:* Courtesy Schlumberger, Inc.)

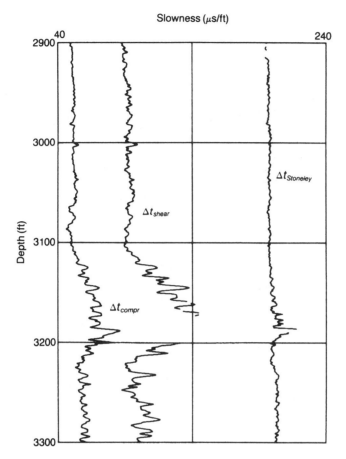

FIGURE 7.6 Typical log display derived from the results of the sonic array tool. Details are discussed in the text. (*Source*: Courtesy Schlumberger, Inc.)

between two- and six-hole diameters into the shale and that source-receiver spacings of 2.5 m upward are necessary to obtain the true shale velocity.

Check shots with vibrators have been used to calibrate or obtain P- and SH-wave velocities: the vertical and horizontal component geophone outputs, respectively, are correlated with the control signal of the appropriate vibrator. For shear wave measurements there appears to be no absolute necessity to clamp the down-hole tool (with two orthogonal horizontal geophones) rigidly to the wall, although it would undoubtedly be preferable. The hole itself is usually enough off-vertical that the sonde lies in contact with the hole and this contact is sufficient to give an adequate signal. Examples of such down-hole arrivals of SH waves have been given by Erickson, Miller, and Waters (1968).

The precision with which these times can be measured is still a matter for conjecture. An absolute travel time is dependent on the characteristics of the source, and, because shots are usually detonated below the weathered layer that allows more high frequencies to be generated and transmitted they are preferred. It is doubtful if accuracies greater than ±0.002 s can be expected. Somewhat

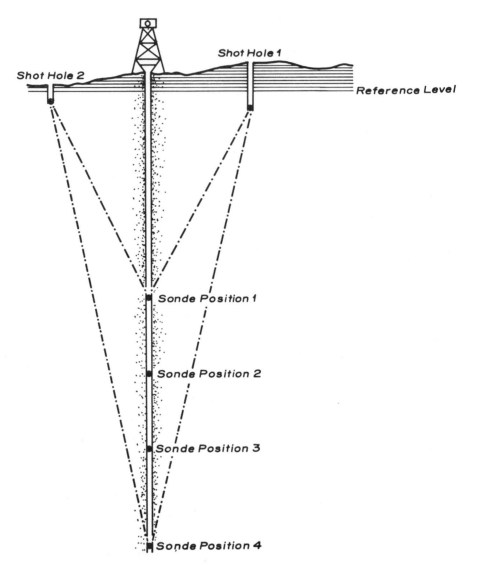

FIGURE 7.7 Schematic of well-shooting method. At least two shot holes are used on opposite sides of the well to check for well deviation from the vertical. Sonde positions are 100 to 160 m apart (300 to 500 ft). Vibrators may replace shots in holes.

better precision should be available for differential times if the same shot is used with two different receivers separated by several hundred meters (Kokesh, 1956) or if the source characteristics can be held consistent for different receiver locations in the hole. If a surface source does not move this can be accomplished with modern servocontrolled vibrators. The differential time accuracy should be obtainable to ±0.001 s. The limit of accurate interval velocities is therefore controlled by the time interval to be measured; for example, 400 ft of limestone may have a one-way transit time of 0.020 s and the precision obtained is, at best, 5%.

7.3   VERTICAL SEISMIC PROFILES

Investigations of the shape of the seismic pulse propagated deep inside the earth started with the early work of McDonal et al. (1958), and Levin and Lynn (1958). The latter studies can probably be regarded as a prototype of modern vertical seismic profiles (VSP) in that it established the method of clamping a geophone in a drilled hole at different levels, the interval being close enough to allow correlation of seismic events from level to level. The total profile of positions occupied had a depth that was an appreciable fraction of the total depth of the well. With this equipment and methodology Levin and Lynn were able to measure the initial arrival time, examine the initial pulse shape, detect reflections, trace them back to their sources, and differentiate them from some types of multiple reflection. Those successes constitute a great advance on the former sole use of the down-hole geophone—the determination of seismic travel times.

It is noticeable that this remarkable success did not spawn a host of reports of similar experimental investigations in the English technical literature. It is speculated that a limited research force was diverted to other, more pressing tasks apparently with a greater chance of immediate economic impact.

The Russian literature over the following 12 years is voluminous and has been summarized in a book by E. I. Gal'perin and made available in English under the auspices of the S.E.G. The translation was done by Dr. A. J. Hermont (1974) under the editorship of Dr. J. E. White.

With these few paragraphs as an introduction, it is possible now to pass directly to modern work on the VSP method, which has come into prominence as a commercially available contributory tool for seismic exploration only during the last 10 years. All of the earlier technological development of placing, clamping, and moving geophones in holes is, of course, available. The use of multiple geophone strings (i.e., vertical spreads by which the arrivals can be recorded simultaneously for a number of depths) does not yet seem to have become a reality, although the need for this development is obvious. It would then be possible to have consistency of source-pulse character and power for a number of depths as well as savings in cost and possibly time.

For wells situated offshore the source can be the same as that for conventional marine seismic surveying. It should also be noted that the pulse shape can be recorded by a special hydrophone in (deep) water.

For wells on land dynamite shots, vibrators, air guns, and other devices have been used. Because of their inherent safety and the fact that no holes or dynamite are required vibrators have a few advantages. Moreover, special vibrators for making shear wave VSPs are available. If air guns are used it is sometimes necessary to dig pits and fill them with water into which the air guns are lowered for operation.

In the discussion that follows we shall be talking about VSPs that are derived from a source on the surface close to the well and detected by geophones that clamp to the wall of the well, which is supposed to be vertical. Special considerations for offset sources or for deviated wells are dealt with later.

Figure 7.8 is (a) a schematic of the field arrangement and (b) a simplistic schematic of the record obtained from the VSP. The first arrivals give rise to a series of lines whose slopes are simply the interval velocities of the strata.

Reflections in each layer have paths that are mirror images about the vertical, and by following them downward to their intersections with the first arrivals the locations of the reflectors are found. The pulses that would be expected on an actual VSP have not been shown at each of the seismic traces—for clarity only. Provided that the arrivals are clear of interference, as they should be if they are first arrivals, amplitudes should be observed (after some corrections are applied) that obey the laws of reflection and transmission. The relevant corrections are the following:

1. Spherical divergence
2. Attenuation
3. Effectiveness of clamping to the wall
4. Source effectiveness

Thus, in theory at least, some estimates of velocity and density should be possible. Velocity and density logs are normally available for the same well; therefore the VSP's are not the primary source of that information. Some attempts have been made to derive attenuation estimates, however.

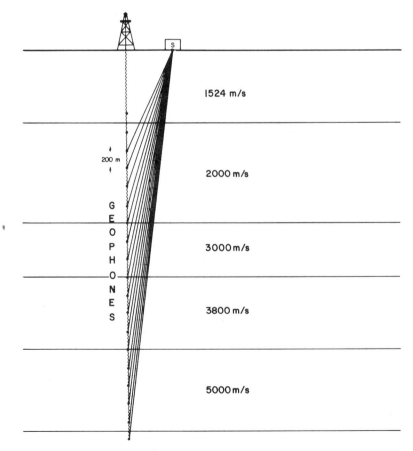

FIGURE 7.8 (*a*) General arrangements of equipment in a hole for VSP. Geophone positions are (at present) occupied sequentially. Offset of source from top of hole is minimal.

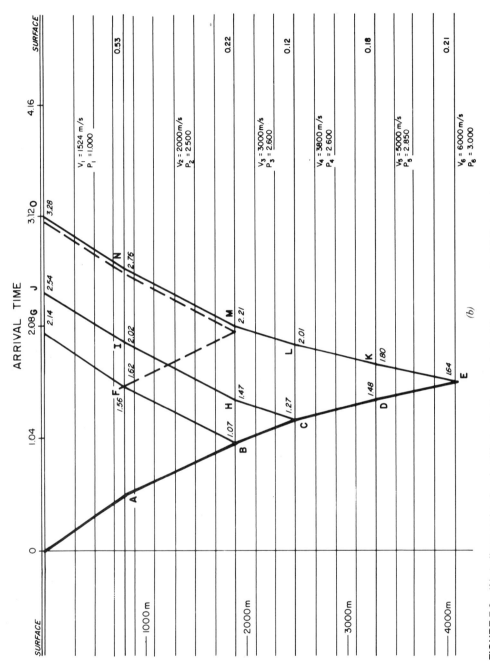

FIGURE 7.8 (*b*) A diagram showing simplified path diagrams (arrival time versus depth) that may be obtained from VSP. Note that downward, upward, and multiply-reflected waves are recorded.

Now it is possible to apply many of the data-processing steps already described in Chapter 6, steps that should be extremely helpful in obtaining the full potential of information from a VSP. The first is to shift the individual traces of the VSP by an amount that corresponds to the first arrival time. This is done in two distinctly separate operations, designed to accomplish different goals.

First, the shift is used in a negative sense (see Figure 7.9)—reducing arrival times—to align the first arrivals at a constant, near-zero time. Under this shift all downward propagating waves are aligned at a constant time and upward traveling waves have their step-out times increased. *Provided that the successive traces are close enough together that there is no aliasing* in the paths for reflections (remember that they now cross over the record at one-half the proper velocity), it is allowable to add all the traces together (weighted, if desired) to emphasize the downgoing pulse and to reduce the up-traveling waves. Note that the downward traveling parts of multiple reflections are emphasized but only in proportion to the distance traveled. Complexities (ghosts or multiple reflections in the near-surface and elsewhere) in the downgoing pulse may therefore be observed, and the possibility exists of using this estimate of pulse shape to deconvolve the VSP traces or (under certain restrictive conditions) nearby seismic reflection records.

Second, the shift may be used in a positive sense (see Figure 7.10)—increasing

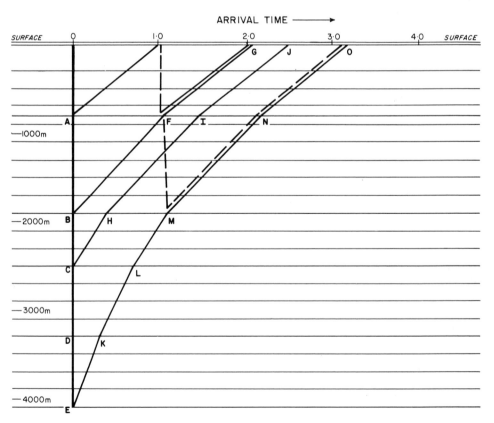

FIGURE 7.9 A diagram of arrivals for earth model in Figure 7.8*b*, but with all traces shifted *upward* to align the first arrival at zero time. A weighted horizontal sum of the traces will give an estimate of the down-traveling pulse.

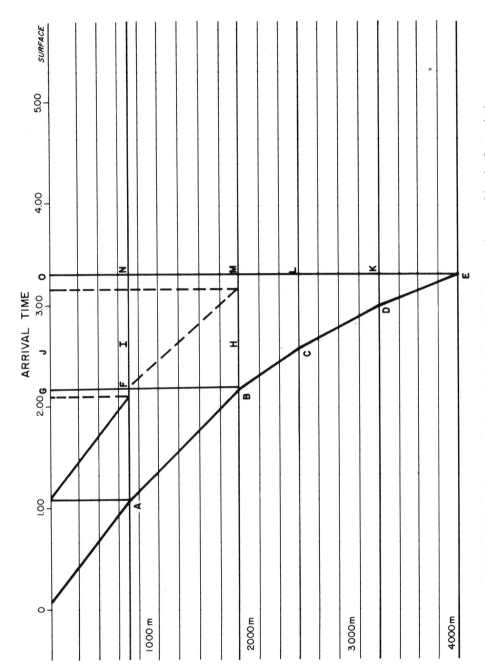

FIGURE 7.10 A diagram of arrivals obtained by shifting all traces *downward* by the first arrivals time. Reflections are now aligned across the record.

the arrival times—to align upgoing waves, the reflections, across the record. Again, with the proper field precautions against aliasing and by close geophone locations the resulting traces may be composited (or two-dimensionally filtered) to emphasize the reflections while optimizing the reduction of down-traveling waves.

To facilitate comparison with reflection records in the area one further VSP data-processing step is necessary to remove the phase of the downgoing pulse and to render the amplitude spectrum as nearly flat as possible. This implies deconvolution of the enhanced VSP reflection trace and can be done by using the downgoing pulse already obtained in deterministic deconvolution or by obtaining a spiking deconvolution filter from the VSP reflection trace itself. Because the latter method involves making an autocorrelation pulse from the trace, the AC can be examined for indications of multiple reflections. If they exist they may also be removed by using a gapped deconvolution operator after the spiking deconvolution has been done.

The final processed reflection trace can now be used for the identification of reflections with lithological changes and can, in turn, be transferred to seismic records.

P - Wave VSP

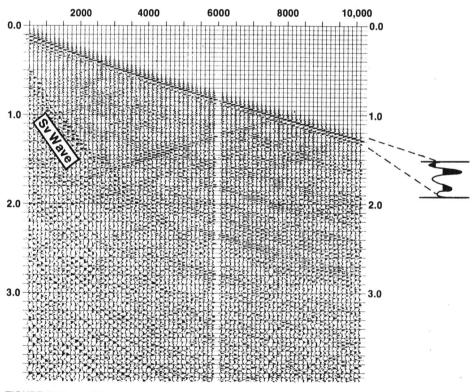

FIGURE 7.11 A field example of a P-wave VSP is shown. Correlation and gain adjustment is the only processing done. (*Source*: Courtesy of CONOCO, Inc.)

Finally, as far as zero-offset VSPs are concerned, this process can be carried out on land for P-wave or SH-wave sources. If the location is one in which there exists nearly horizontal plane-parallel stratification SV waves, by definition, are nonexistent. The combination of P and SH VSPs accomplishes (at the well location) the otherwise difficult task of matching P and S reflections. Once identified, the matching process can be carried to other parts of an area if the seismic records are clear in their continuity.

A field example of a P-wave VSP, after gain function has been applied, is shown in Figure 7.11 and, after shifting and deconvolution to show reflections, in Figure 7.12.

A similar example, but for shear waves, is shown for the field data with gain

P - Wave VSP

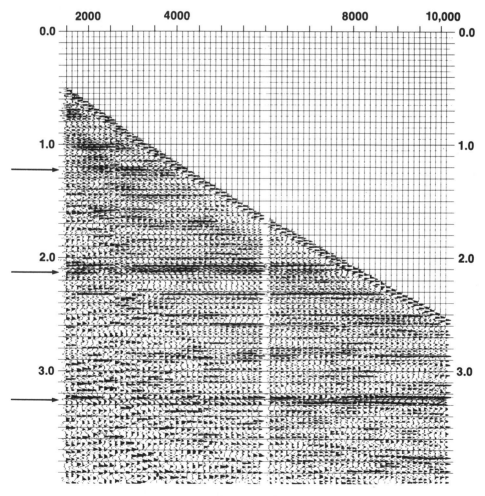

FIGURE 7.12 This section shows reflections, after deconvolution and upward shifting have been done on Figure 7.11. (*Source*: Courtesy of CONOCO, Inc.)

S-Wave VSP

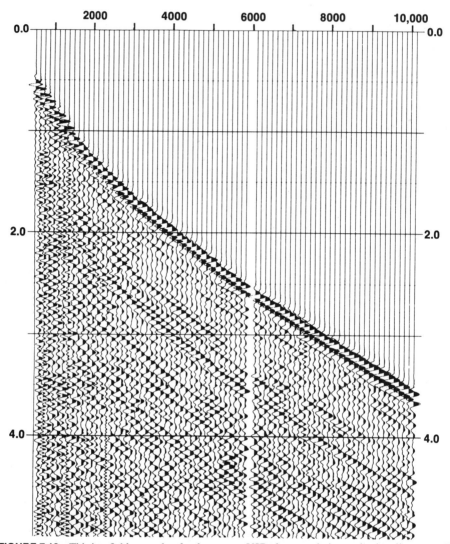

FIGURE 7.13   This is a field example of a shear wave VSP after correlation and gain adjustment. The output of two orthogonal horizontal geophones were combined at each level. (*Source*: Courtesy of CONOCO, Inc.)

function applied in Figure 7.13 and, after shifting, source signature deconvolution and removal of downgoing events, in Figure 7.14. For the shear wave records it was necessary to combine the outputs of two orthogonal horizontal geophones because the absolute orientation of the sonde was not known.

Although the depth scale is not shown in Figures 7.12 and 7.14, it is obvious that the depth of the reflections can be identified and the correspondence of

S-Wave VSP

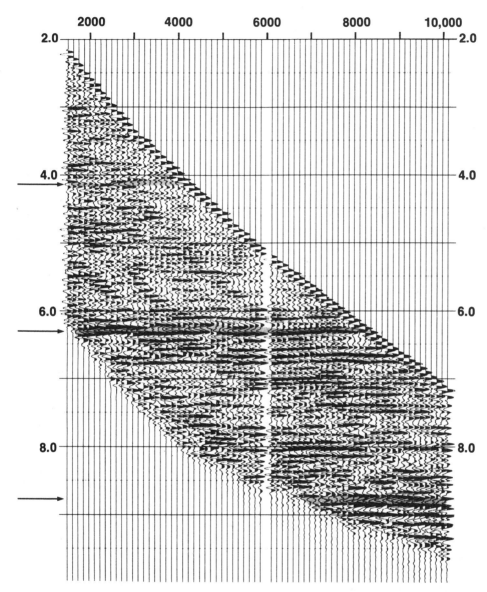

FIGURE 7.14 The result of shifting, source-signature deconvolution and the removal of downgoing events shows some shear reflections. (*Source*: Courtesy of CONOCO, Inc.)

certain P and SH reflections, established. Once this is done at one location in an area there is hope that this correspondence can be carried throughout most of the area concerned. In poor reflecting areas or those with a good deal of fracturing more VSP pairs may be needed to continue the correspondence to remote parts.

Many other examples of VSP's have been presented by Balch and Lee (1984); Balch et al. (1982); Cassell (1984); Fitch (1984); and Hardage (1983).

Offset VSPs

In the zero-offset VSPs described so far we have assumed that the reflection points are coincident with the points at which the well penetrates the strata. Apart from the possibility of locating reflectors below the well bottom, most of the claims for the VSP lie in their usefulness for calibrating seismic reflection data.

The possibility of gaining knowledge of the local structure around the well does exist, however, if the reflection points are made to shift away from the well by a shift in the shotpoints used. In a deviated well this is inherent in the geometry of the system. The shotpoint may usually be shifted to lie vertically above the individual geophone locations as required. The economics of the process, however, militates against wholesale use of this idea [as pointed out by Cassell (1984)]. Rig time costs for multiple offsets and individual receiver positions prevent this process from becoming a routine operation.

To present the data in a form that is reasonably familiar to all explorationists Wyatt and Wyatt (1981) and Wyatt (1981) have suggested a VSP–CDP transformation that images the VSP data into the equivalent of a CDP stacked section.

7.4 ESTIMATION OF SEISMIC VELOCITIES FROM REFLECTION RECORDS

It is against this background of direct measurement that we must judge the accuracy and value of indirect, reflection methods of velocity measurement. As described in Section 6.5, the average velocity $\bar{V}$ to a reflector should be determinable from the increase in reflection time as the offset distance (source to receiver) increases. The standard methods are an X^2-versus-T^2, from which $\bar{V}^2$ is obtained, or velocity obtained by an analysis of CDP data.

These analyses, which are often given proprietary names, consist of several discrete steps:

1. Selection of seismic traces, corrected as well as can be done for near-surface time differences, that apply to a single subsurface reflection point.
2. Application to each trace of a geometrical correction factor based on an assumed velocity.
3. Assessment of the correctness of this velocity and the output of a *goodness* factor for each time window of the record or, in some cases, for each window centered on known reflections.
4. A change in the selected velocity over a predetermined range.
5. An output in clearly visible form of the goodness factor versus the selected velocity for a series of window depths or for depths corresponding to particular reflections.
6. A subjective picking technique that selects the appropriate stacking velocity for each depth, record time, or reflection time.

In addition to these steps, it is often desirable and a programmed feature of the velocity determination to edit the results in different ways; for example, velocities determined at consecutive depths are compared to see if they represent values that will result in reasonable interval velocities. Other features, of greater complexity, have been deemed advisable in some programs.

Two examples (Figures 7.15 and 7.16a and b) have been given to illustrate the diversity in displays. The first is described by Taner and Koehler (1969) and the second is a proprietary program of CONOCO, Inc. called EVEL. In this section we examine factors that may limit the accuracy of these measurements.

The first factor is a lack of knowledge of the true path of the energy traveling from the source to the receiver. In Figure 7.17 a series of plane-parallel layers of alternating velocities is shown. This may be a series of sands and shales or an evaporite sequence commonly found in the Persian Gulf sedimentary section. The ray paths are refracted at the boundaries between the layers, and, as the offset is increased, the ratio of the paths in the higher velocity material to the path in the lower velocity material changes. Thus the composite material acts as though it were a single material with anisotropy. The average velocity of a composite ray at an angle to the vertical is higher than the velocity vertically through the layered system.

True anisotropy, that is, a velocity in the horizontal direction different from that in the vertical direction, is due to a preferred orientation of the matrix particles and has an effect that changes the pseudoanisotropic effect. In the particular case illustrated the shale anisotropy is higher than that of the sand, thus causing the shale paths to take a shorter time, the higher the angle of incidence. This then reduces the effect of the pseudoanisotropy because it occurs in the layer with the lower velocity. Cases can occur, however, as in the halite-anhydrite example in which the high anisotropic ratio applies to the material with the highest velocity. The pseudoanisotropic effect is then enhanced.

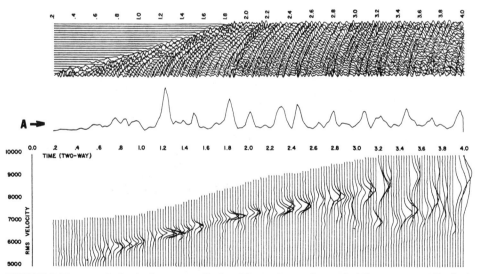

FIGURE 7.15 A typical velocity spectra display. The peaks on the vertical lines represent the best estimates of stacking velocity. Trace A shows the comparative energy of the reflections, hence gives an idea of the reliability of the velocity values. Criteria other than energy are sometimes used. [*Source*: Taner and Koehler (1969). Reprinted with permission from *Geophysics*.]

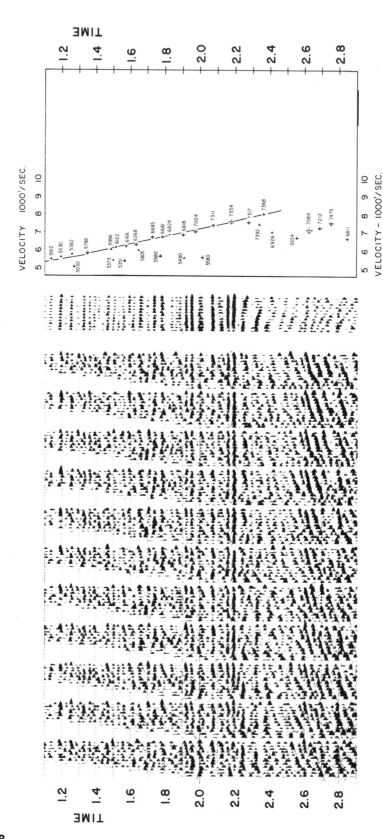

FIGURE 7.16 (a) The primary records (*left*), the stacked record, and the velocity associated with a number of events. (*Source:* Smith and Diltz. Courtesy of CONOCO, Inc.)

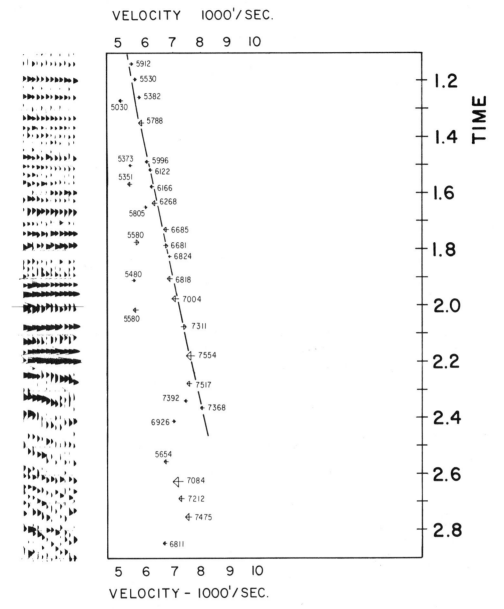

FIGURE 7.16 (b) Detail from Figure 7.16a. The velocity symbols point left or right, depending on whether the event used is a trough or a peak. The size of the symbol is related to the energy of the event. The number gives the associated *average* velocity in feet per second.

Wave propagation in an anisotropic medium has been the subject of several experimental and theoretical papers (Uhrig and van Melle, 1955; Postma, 1955; Van der Stoep, 1966, and others). The main effects are examined here.

The wave fronts in a homogeneous anisotropic medium are nearly elliptical (Postma, 1955), being characterized by the ratio of the major (horizontal or parallel to the bedding) axis to the minor axis, which is the ratio of the horizontal

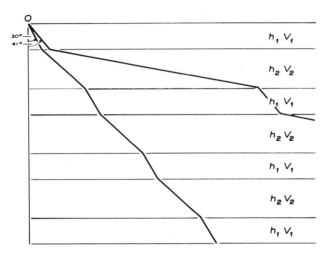

FIGURE 7.17   Ray paths in a layered medium show the pseudoanisotropy caused by differing P-wave velocities in the layers. The example is for $h_1/h_2 = 2/3$, $V_1 = 8000$ ft/s, $V_2 = 12,000$ ft/s. Average vertical velocity is 10,000 ft/s; average velocity at 30° incidence is 10,267.9 ft/s; average velocity at 41° incidence is 11,235 ft/s; critical angle is 41.81°.

to the vertical velocity in the medium—the anisotropic ratio. Figure 7.18 shows how Huyghens' principle leads to elliptic wave fronts of the same shape in which the ray path is not in general along the normal to the wave front.

The polar equation of an ellipse, with pole at the center and semiaxes a and b, is

$$r^2 = \frac{a^2 b^2}{a^2 \sin^2 \theta + b^2 \cos^2 \theta} \qquad (7.2)$$

and the velocity in a direction θ from the horizontal is

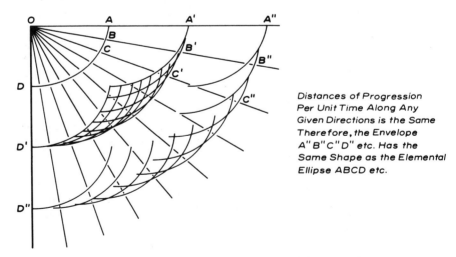

Distances of Progression
Per Unit Time Along Any
Given Directions is the Same
Therefore, the Envelope
A" B" C" D" etc. Has the
Same Shape as the Elemental
Ellipse ABCD etc.

FIGURE 7.18   Use of Huygen's principle to show that an elliptical wave front propagates as an elliptical wave front of the same major/minor axes ratio.

$$V_\theta = \sqrt{\frac{V_p^2 V_Q^2}{V_p^2 \sin^2 \theta + V_Q^2 \cos^2 \theta}} \qquad (7.3)$$

where V_θ = velocity at angle θ
$\quad V_p$ = velocity parallel to the bedding planes
$\quad V_Q$ = velocity perpendicular to the bedding planes

Note: This applies strictly to a homogeneous anisotropic material but will apply, in the large, to pseudoanisotropic material if the layers are thin compared with wavelength.

Another way to relate the time for a unidirectional inclined path to the vertical average velocity, anisotropic ratio, depth, and offset distance is (Uhrig and van Melle, 1955)

$$T^2 = \frac{X^2}{A^2 V^2} + \frac{Z^2}{V^2} \qquad (7.4)$$

For reflections from a constant depth reflector therefore we have

$$\frac{T_R^2}{4} = \frac{X_R^2}{4 A^2 V^2} + \frac{Z^2}{V^2} \qquad (7.5)$$

where T_R is the reflection time at offset X_R. A plot of T_R^2 against X_R^2 must have a slope of

$$\frac{1}{A^2 V^2}$$

and the anisotropic ratio A is not divorced from the vertical velocity. Only the *horizontal* velocity is determined from this procedure, which conforms to Postma's observation that the anisotropic ratio can never be obtained from surface work alone. Knowledge of the actual depth of the reflector is necessary as well. Levin (1979) examined in detail the propagation of seismic waves of different types (P, SV, SH) traveling in a medium layered on a much finer scale than the wavelength and found some deviations from the simple treatment just discussed. True, the seismic wave does act as if it were traveling in a single, transversely isotropic material. If the X^2–T^2 plot is a straight line the equation relating the horizontal and vertical velocities of travel is that of an ellipse. For SH waves the velocity obtained from the slope of the X^2–T^2 plot is the *horizontal* velocity, but for P waves and short spreads the slope of the X^2–T^2 curve describes the *vertical* velocity. For longer spreads the velocity can be anywhere between the vertical and the horizontal velocities. SV waves exhibit unusual behavior, and the data do not usually give rise to hyperbolic time–distance curves. Jolly (1956a and b) obtained results for a near-surface layered system of gypsum in weathered material. The X^2–T^2 plot for this system reveals an unusually high (2.0) anisotropic ratio, which can explain his experimental results. According to Levin, a water-bearing sand–gas-filled sand layered system has negligible anisotropy.

The effects of anisotropy are much larger for SH waves than for P waves. Therefore some care is necessary in using the velocities derived in this manner. The problem of relating SH-wave reflections to corresponding (?) P-wave reflections is more difficult than first supposed.

The second factor limiting velocity measurements from reflection data is the change in phase of a reflected beam, compared with an incident beam, as the angle of incidence is changed. This manifests itself in a change in wave pulse shape for different offsets and occurs at offsets that correspond to the critical angle when the P-refracted wave first occurs and later when the S-refracted wave occurs. These phase changes apply at a single interface, but there is a more common reason for a *gradual* phase change with offset in the case of layers. In reflections from thin layers it is the phase of the combined reflections from the upper and lower boundaries that is observed: strictly speaking, it is the combined phase of all multiple reflections within the layer.

Figure 7.19 shows how the path differences between the upper and lower reflections change. It is particularly noticeable at the longer offsets. In the case illustrated [layer thickness 30.5 m (100 ft) and velocities 2438.4 m/s (8000 ft/s) and 3657.6 m/s (12,000 ft/s)] there is a difference of phase at 50 Hz of 153.6° (0.0085 s), whereas at 25 Hz the difference is only one-half this phase angle, or 76.8°.

Of course, this effect is most noticeable in high-velocity layers in which the refracted ray is bent away from the normal and the distance in the high-speed layer increases rapidly near the critical angle.

A third factor is the dip of the reflecting horizon. A simple geometrical

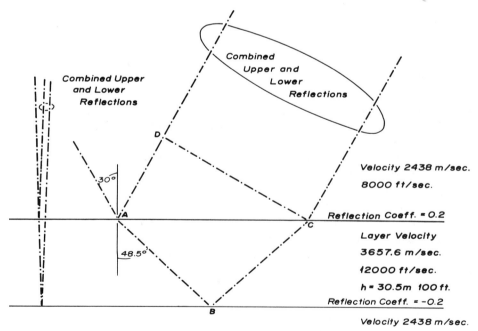

FIGURE 7.19 A change in phase difference between upper and lower reflections as the angle of incidence changes.

construction shows that the velocity determined by statistical methods based on CDP field work is always a true velocity divided by the cosine of the angle of dip. The following short table shows that this effect is usually important only at angles of 10° or above, but the velocity calculated is always greater than the true velocity if the medium can be regarded as homogeneous.

Angle (deg)	Ratio, V_{dip}/V_{flat}
0	1.0000
5	1.0038
10	1.0154
15	1.0353
20	1.0642
30	1.1547

Curvature by itself has no effect on the velocity retrieved because the CDP remains in the same location on the curved surface. It may be noted, however, that in dipping formations the common reflecting point migrates up-dip with increasing offset and the effects of curvature are then felt. It is difficult to give any general rules on the effect of curvature when dip is present.

7.5 EFFECT OF PORE FLUIDS ON ROCK VELOCITIES

Several theoretical studies have been made of the effects on elastic wave velocities of pore-filling fluids (Brandt, 1955; Gassman, 1951; Biot, 1955, 1956; Wyllie, Gregory, and Gardner, 1958; Domenico, 1974; Stolt, 1973, and others) and have been complemented by numerous ultrahigh-frequency measurements. Recent intense concentration on this subject has been brought about by the knowledge that gas-filled sands are often directly indicated on correctly processed seismic sections by large reflection amplitudes (and other seismically visible effects). These interpretive matters are dealt with later.

In this chapter only the simpler aspects of this subject are considered. The expressions for the velocity of sound waves in a two-phase medium are complicated and generally contain empirical constants that have to be set by comparison with measured values. The only equation that is comparatively simple, but the least effective [see comparison in Kuster and Toksoz (1974)], is the time-average equation of Wyllie, Gregory, and Gardner (1958) in which the random medium of matrix material and fluid-filled pore spaces is replaced by a series of plane-parallel layers of liquid and rock and the time through the rock is computed. Figure 7.20 shows how unrealistically this model portrays an assemblage of irregular rock particles with fluid in the interstices. The average velocity is then given as

$$\bar{V} = \frac{V_f V_m}{k V_m + (1-k) V_f} \tag{7.6}$$

where V_m = velocity of sound in the rock matrix
V_f = velocity of sound in the fluid
k = fraction of rock that is pore space

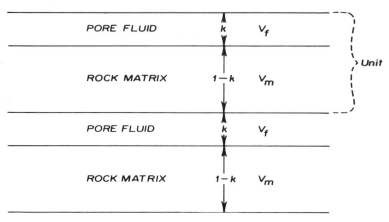

FIGURE 7.20 The system of layers corresponding to the time-average equation of Wyllie, Gregory, and Gardner (1968). The time through one unit of a repeating sequence is $T \simeq K/V_f + (1 - K)/V_m$.

The next equation to be considered (Brandt, 1955) was designed to give the velocity in a sandstone of randomly sized, roughly spherical rock grains filled with fluid. The sound velocity in this type of sandstone is

$$\bar{V} = \frac{F(P_r - CP_f)^{1/6}/K^{1/3}}{\sqrt{\phi}\ \sqrt{\rho_r(1 - \phi) + \rho_f\phi}} \frac{(1 + 17.5B^{3/2}K/\sqrt{P_r - CP_f})^{5/6}}{(1 + 26.3B^{3/2}K/\sqrt{P_r - CP_f})^{1/2}} \qquad (7.7)$$

where P_r and P_f = rock and fluid pressures (lb/in.^{2})

         ρ_r and ρ_f = rock and fluid densities (g/cm^3)

         ϕ = porosity of the rock

         B = bulk modulus (reciprocal of compressibility) of the fluid (lb/in.^{2})

         C = fraction of the pore pressure *felt* by the rock, normally $C \approx 0.9$

         K = a constant determined by the mechanical properties of the rock, $K \sim 10^{-7}$

         F = a constant depending on Poisson's ratio of the rock

For consolidated sandstone Brandt took F to be 5.75. To allow a description of a shallow, unconsolidated sand F can be $\leqslant 5.75$.

This equation shows reasonable agreement between laboratory measurements and calculations for porosities higher than 10% (Hicks and Berry, 1956), but it must be remembered that three empirical constants can be set. It does, however, predict an increase in velocity with pressure and a decrease as porosity or fluid compressibility are increased. Sands containing highly compressible fluids, such as oil and gas, have lower seismic velocities than water-bearing sands.

If a mixture of hydercarbons and water is permitted in the pores the Brandt velocity equation can be modified. The bulk density ρ_B is expressible as

$$\rho_B = (1 - \phi)\rho_r + \phi(1 - \mu_w)\rho_h + \mu_w\rho_w \qquad (7.8)$$

where ϕ = fractional porosity
$\quad\mu_w$ = fractional water content by volume
$\quad\rho_r$ = specific gravity of the matrix
$\quad\rho_h$ = specific gravity of the hydrocarbons
$\quad\rho_w$ = specific gravity of water

The bulk modulus of the two fluids coexisting in the pore space is

$$B_f = \frac{1}{[(1 - \mu_w)/B_h] + (\mu_w/B_w)} \tag{7.9}$$

where B_h = bulk modulus of the hydrocarbons (psi)
$\quad B_w$ = bulk modulus of water (psi)

and the other parameters are as defined in (7.7). Then

$$\bar{V} = \frac{F(P_r - CP_f)^{1/6}/K^{1/3}}{\sqrt{\rho_B \phi}} \; \frac{(1 + 17.5 B_f^{3/2} K/\sqrt{P_r - CP_f})^{1/6}}{(1 + 26.3 B_f^{3/2} K\sqrt{P_r - CP_f})^{1/2}} \tag{7.10}$$

It can be seen intuitively that for a two-phase system the compressibility increases significantly when even a small amount of gas is present. This conjecture has been verified by Domenico (1974) and has important consequences in the interpretation of reflections of high amplitude from *gas* sands.

Geertsma and Smit (1961) developed from Biot's comprehensive theory (1956) an equation for the velocity of sound of low frequencies through a fluid-filled porous medium. These frequencies are such that the wavelengths of the sound are much greater than the pore dimensions. This longitudinal velocity is given as

$$V = \frac{1}{\rho_b^{1/2}} \left[\left(\frac{\beta}{C_s} + \frac{4}{3} G_b \right) + \frac{(1 - \beta)^2}{(1 - \phi - \beta)C_s + \phi C_f} \right]^{1/2} \tag{7.11}$$

where C_s = compressibility of the matrix material
$\quad C_b$ = compressibility of the empty reservoir bulk material
$\quad\beta = C_s/C_b$
$\quad G_b$ = shear modulus of the reservoir bulk material
$\quad C_f$ = compressibility of the fluid
$\quad\phi$ = porosity

This equation was also derived earlier by Gassman (1951).

Curves calculated by Domenico (1974) have been reproduced as Figure 7.21, in which the sharp change in velocity as small amounts of gas are admitted to the system is clearly shown.

In general, the Biot–Geertsma–Gassman equation gives good results in comparison with experiment, as indeed it should, because it is based on the solid matrix and fluid-filled pores that constitute two interacting thermodynamic systems.

Other velocity equations proposed by Stolt (1973) and Kuster and Töksöz

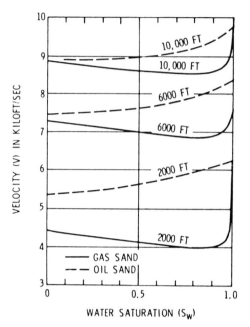

FIGURE 7.21 Longitudinal velocity V as a function of water saturation S_w for gas and oil sands at depths of 2000, 6000, and 10,000 ft. [*Source*: Domenico. Courtesy of World Petroleum Congresses (1975).]

(1974) are based on scattering theory and in general have the same velocity properties as the Gassman–Biot–Geertsma equation.

It may be interesting to compare the results for shear waves (based on the Kuster–Töksöz scattering model) with those for P waves. This theory shows that for a shear wave traveling in rock that contains a two-phase pore-filling fluid the velocity is

$$\beta = \sqrt{\frac{\mu^*}{\rho^*}} \qquad (7.12)$$

where β = shear wave velocity
    μ^* = effective shear modulus
    ρ^* = effective density

For spherical fluid-filled pores

$$\rho^* = \rho(1 - c) + \rho'c \qquad (7.13)$$

$$\mu^* = \frac{\mu(1 - c)}{1 + c\left(\dfrac{6K + 12\mu}{9K + 8\mu}\right)} \qquad (7.14)$$

where ρ = density of the rock matrix
  μ = shear modulus
  K = bulk modulus of the rock matrix
  ρ' = fluid specific gravity
  c = porosity

For any reasonable rock the shear modulus equation reduces nearly to $\mu^* = \mu[(1 - c)/(1 + c)]$, which relates the effective shear modulus to the shear modulus of the rock matrix and porosity. This is in accord with common sense because the fluid should contribute nothing to the shear stiffness of the rock—it is simply weakened by the holes.

Thus the shear wave velocity can be expressed as a function of porosity:

1.

$$\beta(c) \cong \frac{\beta(0)}{\sqrt{(1 + c)\left(1 + \frac{\rho'}{\rho}\frac{c}{1 - c}\right)}} \tag{7.15}$$

2. For a water-filled rock this equation can be written

$$\beta(c) \cong \frac{\beta(0)}{\sqrt{(1 + c)\left(1 + \frac{4c}{1 - c}\right)}} \tag{7.16}$$

3. For a partly gas-filled rock,

$$\beta(c) = \frac{\beta(0)}{\sqrt{(1 + c)\left(1 + 4S_w\frac{c}{1 - c}\right)}} \tag{7.17}$$

where S_w is the fractional water saturation. Thus, although there is some change in shear wave velocity with gas saturation, the effect is chiefly that of the density change and does not change to anything like the extent that the P-wave velocity does; for example, in a rock with 20% porosity the value of $\beta(c)/\beta(0)$ changes only from 0.908341 to 0.870784 when the water saturation changes from 10 to 99%.

7.6   ATTENUATION OF SEISMIC WAVES IN ROCKS

Observations of seismic data in some oil and gas areas (particularly offshore Gulf of Mexcio) have shown that some correlation exists between the presence of gas (and perhaps oil) and anomalous absorption of seismic waves. Mateker et al. (1970) have suggested that attenuation measurements can be made to estimate the sand/shale ratio within a given part of a geological section.

In Section 2.9 a calculation was made which suggested that the classic attenuation from solid friction is not a large factor in homogeneous rocks, even in those of the lower velocity rocks of the Tertiary. It is difficult to imagine that the solid friction effects, small as they are over a few hundred feet, be measurable to an accuracy that would allow differentiation between sands and shales.

From the standpoint of a general relationship Figure 7.22, which shows measured values of $1/Q$ plotted against velocity (in feet per second), it can be seen that, roughly, a line of slope -2 can be drawn through the data, the consequence of which is that the relation between $1/Q$ and V is

$$\frac{1}{Q} = \frac{C}{V^2} \tag{7.18}$$

where $C \approx 10^6$. It is interesting to speculate that, if this relation were true for shear waves as well as P waves, we would expect that the attenuation for shear waves would be four times that of P waves (of the same frequency) because shear wave velocities in consolidated sediments are about one-half the P-wave velocities. On a wavelength basis S waves of the same wavelength should be attenuated about twice as much as P waves.

If it is assumed that there is no dissipation in a material during a purely compressional cycle it can be shown (Anderson and Archambeau, 1964; Kennett, 1975) that the specific Q's for compressional and shear waves are related by

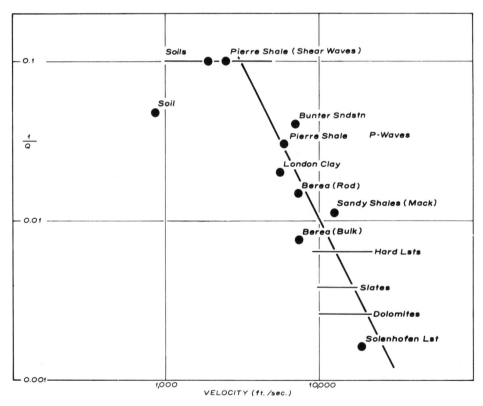

FIGURE 7.22 A log-log plot of $1/Q$ as a function of velocity for several rocks. The slope of the best fit line is roughly -2.

$$\frac{Q_\beta}{Q_\alpha} = \frac{4}{3}\left(\frac{\beta^2}{\alpha^2}\right)$$

where α = compressional wave velocity

β = shear wave velocity

Although this relation was derived with competent rocks in mind, as, for example, in earthquake seismology, it appears that it is reasonably accurate for the measurements in Pierre shale (McDonal et al., 1958) of attenuation for compression and shear waves.

Another form of energy loss due to scattering, which may be the major cause of attenuation, results from the loss of amplitude of a seismic wave passing in both downward and upward directions through a seismic interface. O'Doherty and Anstey (1971) have suggested that this is indeed the major cause of observed attenuation on seismic records. Schoenberger and Levin (1974) experimented with synthetic seismograms and confirmed the importance of these transmission losses, but they estimated that the attenuation due to layering accounts for approximately one-third to one-half the total frequency-dependent attenuation derived from field seismograms at well locations.

If the primary pulse of unit amplitude is reflected by the nth interface of a series with reflection coefficient r_i its amplitude is given by

$$A_n = r_n \prod_1^{n-1} (1 - r_i^2) \tag{7.19}$$

It is the continued product part of this equation that is of interest. First, we establish some orders of magnitude of the effect. Table 7.1 was calculated on the basis that all r_i are of the same magnitude. Because this is known to be untrue and is a nonlinear term in reflection coefficients, we would expect the effect calculated to be less severe than is actually the case; that is, the substitution of an average reflection coefficient gives a conservative answer. The number of layers has been varied to show the cumulative effect as the number of layers increases.

TABLE 7.1

AMPLITUDE OF A UNIT PRIMARY PULSE TRANSMITTED THROUGH N INTERFACES IN BOTH DIRECTIONS VERSUS THE REFLECTION COEFFICIENT

N	Reflection Coefficient, r					
	0.01	0.1	0.15	0.20	0.25	0.30
10	0.9990	0.9046	0.7964	0.665	0.524	0.389
	−0.0087 dB	−0.87 dB	−1.977 dB	−3.55 dB	−5.61 dB	−8.19 dB
20	0.998	0.8179	0.634	0.442	0.275	0.1517
	−0.0173 dB	1.746 dB	−3.953 dB	−7.09 dB	−11.21 dB	−16.38 dB
50	0.995	0.605	0.321	0.1299	0.0397	0.00895
	−0.043 dB	−4.36 dB	−9.883 dB	−17.73 dB	−28.03 dB	−40.96 dB
100	0.990	0.366	0.1027	0.0169	0.00157	0.00008
	−0.087 dB	−8.73 dB	−19.77 dB	−35.46 dB	−56.06 dB	−81.92 dB
200	0.980	0.133	0.0106	0.000285	0.0000025	6×10^{-9}
	−0.174 dB	−17.46 dB	−39.53 dB	−70.91 dB	−112.11 dB	−163.83 dB

The drop in power (as well as the amplitude ratio) is given in decibels for those who feel more at home with that logarithmic measure. It is noticeable that the decibel loss of amplitude is proportional to the *number* of interfaces and that even for reflection coefficients of 0.1 (acoustic impedance changes by 5%) the loss is appreciable.

O'Doherty and Anstey have discussed the situation in a readable, semiquantitative manner and have pointed out that this loss of amplitude by the primary event is counterbalanced (to some extent) by the fact that the peg-leg multiples carry the energy for deep reflections. Nevertheless, it is usually the case that the number of peg-leg multiples to be used, which contain most of the energy, is limited to about five and that *all* these paths have to traverse the layers above if they are to have anything to do with the reflections from the layer under consideration.

From the point of view of attenuation, therefore, there may be sequences of layers (alternating sedimentation) that attenuate strongly and others (transitional sedimentation) that attenuate little. This is very likely the type of attenuation measured in a given geological section. We can see that it is not the sand/shale ratio that is measured but the number of depositional cycles.

It is not safe to say that the reflection coefficients have to be small in sand-shale sequences because we have now seen how the presence of a little gas can give rise to large velocity and density changes. Of course, we do not expect that these gassy formations will necessarily be plane-parallel layers, but they will have high *scattering* coefficients and will divert much of the energy out of the signal beam.

It is easy, however, to observe the effect of attenuation in seismic record sections that have gas indications (bright spots) on them. The energy carried by reflections beneath the bright spots is sharply reduced. White (1975) showed that in an unconsolidated sand with partial gas saturation there is a 20% increase in compressional wave velocity from 1 to 100 Hz and an attenuation of 27 dB/1000 ft at 31 Hz and 82 dB/1000 ft at 123 Hz. Fluid flow waves are shown to be responsible for the dispersion and attenuation at low frequencies. Although this appears to be an important loss mechanism for heterogeneous porous rocks, it can be shown that (say) a 50-ft-thick sand, which contributes 100 ft of travel path for reflections from below, would reduce the amplitude of these reflections by only 2.7 dB. This is not sufficient to explain the observations. Two reflection boundaries, each having a reflection coefficient of 0.4, would provide a further loss of another 3 dB. Therefore the occurrence of small layers within the sand still seems to be plausible. More discussion is given later.

It is noted here, however, that the shear wave velocity and attenuation are not affected. Because the reflection coefficients for shear waves are not so large as those for P waves, the attenuation due to layering is also reduced.

A classic case of cyclic sedimentation occurs under Sabkha conditions (periodically flooded and evaporated tidal lagoons) in the Persian Gulf area and evidence from velocity logs taken in the vicinity shows that extensive sequences occur in the Upper Fars formation, where 200 or more interfaces with reflection coefficients of 0.3 (between anhydrite and halite layers) can exist. Here is a built-in mechanism for the known difficulty of obtaining reflections from below this level. The multiple reflections that exist because of this layer and the reflection above simply compound their dips to produce deep multiple reflections that apparently have a structure different from that of the shallow events—but it is not the correct structure for formations at this depth.

McDonal et al. (1958) carefully measured the attenuation constant in Pierre shale (a near-homogeneous shale) for both P and S waves. No allowance was made for the possibility of boundaries (or scattering of any kind). It is now commonly accepted that their work forms a basis for considering that rocks have internal friction that gives rise to an attenuation proportional to the first power of the frequency rather than considering them viscoelastic solids that cause an attenuation proportional to the second power of the frequency. Their attenuation constants for Pierre shale are given in terms of α (decibels per 1000 ft) times the frequency. This is not the α defined in Section 2.9 and it may be advisable to clear up this lack of consistency. The α used in this book is defined as the symbol in the equation

$$A = A_0 e^{-\alpha x} \sin 2\pi f \left(t - \frac{x}{c} \right) \tag{7.20}$$

which is for a plane wave of frequency f traveling at a velocity c. If we take the ratio A/A_0 and its logarithm we will have

$$\log_{10} \frac{A(f)}{A_0(f)} = \log_{10} e^{-\alpha x} = \frac{-\alpha x}{2.303}$$

The power loss in decibels, however, is $20 \log_{10}[A(f)/A_0(f)]$; hence $-8.684\alpha x$. The attenuation per foot is -8.684α dB/ft and per 1000 ft (McDonal's α), $8684\alpha = \alpha_M$.

Now we can take McDonal's value α_M for shear waves ($\alpha_{MS} = 1.05f$ dB/1000 ft) and P waves ($\alpha_{MP} = 0.12f$ dB/1000 ft) and compute equivalent Q values.

For shear waves

$$Q_s = \frac{8684\pi}{C_s} \left(\frac{f}{\alpha_{MS}} \right) = 9.99 ; \qquad C_s = 2600 \text{ ft/s}$$

For P waves

$$Q_P = \frac{8684\pi}{C_p} \left(\frac{f}{\alpha_{MP}} \right) = 32.02 ; \qquad C_p = 7060 \text{ ft/s}$$

The discrepancy between this value and that given in Table 2.2 apparently arises in averaging the values in Bradley and Fort's original tables.

Perhaps the chief value of studying McDonal et al. (1958) is to see how meticulously attenuation studies must be done to give the values some meaning. It emphasizes the difficulty encountered in making *differential* attenuation measurements with normal reflection seismograms.

7.7 CALCULATION OF SIMILARITY, CORRELATION, OR SEMBLANCE COEFFICIENTS

There is often a need for some objective measure of the optimum line-up of reflections on a seismic record. In this section we examine a few of those available.

One example of their use was described in Section 7.4 in an estimation of velocities from operations on the seismic records themselves. An attempt is made, by removing NMO from the traces under a range of different constant velocities, to cause the reflections to arrive at the same time on all traces. This then points to the proper velocity for a particular velocity. The objective measure of optimum line-up is used.

These is also a need, over a given time range, to examine and measure the degree of similarity of two seismic traces. If the source and receiver characteristics remained constant and the geological section did not change as the measuring points moved over the surface of the earth it would be expected that the seismic reflection traces would also be constant. That is to say, there would be at the maximum a constant scale factor difference between corresponding values on successive seismic traces. We know, however, that this does not happen in practice and that each trace is different from the preceding or following one. The degree of coherence, or the correlation coefficient, is defined as being related to the cross-correlation coefficient between equivalent time windows of the traces:

$$C_{xy} = \frac{\phi_{xy}}{\sqrt{\phi_{xx}\phi_{yy}}} \tag{7.21}$$

where $\phi_{xy} = \Sigma_{i=1} a_i b_i$ and the traces are

$$x(i) = a_1, a_2, a_3, \ldots, a_i, \ldots, a_N \tag{7.22}$$

$$y(i) = b_1, b_2, b_3, \ldots, b_i, \ldots, b_N \tag{7.23}$$

The terms

$$\phi_{xx}^{1/2} = \sqrt{a_1^2 + a_2^2 + a_3^2 + \cdots + a_i^2 \cdots + a_N^2} \tag{7.24}$$

$$\phi_{yy}^{1/2} = \sqrt{b_1^2 + b_2^2 + b_3^2 + \cdots + b_i^2 \cdots + b_N^2} \tag{7.25}$$

are normalizing factors to facilitate a comparison of the two traces without regard to the amplitude at which they are displayed. Normally, the samples are taken in the windows of the original traces, which start and stop at the same absolute (corrected) time; that is, no delay is allowed between the traces. (This procedure is contrasted with that used in the Vibroseis® system, in which the correlation value, unnormalized, is played out for two time series as the delay is progressively increased.)

We plan to use a coefficient of this type to quantify, and possibly display prominently, portions of the seismic section in which the geology is changing, as opposed to portions in which the layer thickness and velocities and constituent rocks are constant. This, however, necessitates an allowance for inadvertent shifts between traces. Insensitivity to straight shifts between consecutive traces is made by allowing the delay to change over a small range and then choosing the maximum value of the correlation coefficient (Figure 7.23).

$$\gamma_{xy} = (C_{xy})_{\text{max}} = \max \frac{\phi_{xy}(\tau)}{\phi_{xx}^{1/2} \phi_{yy}^{1/2}} \qquad -\tau_0 < \tau < \tau_0 \qquad (7.26)$$

Naturally it is desirable to minimize the effect of the window used, but it is also required that the changing of end values be prevented from influencing the correlation coefficient unduly. For this reason a window shape that gives small comparative weight to the end values of the window is often used. Foster and Guinzy (1967) contains a table of windows and their properties.

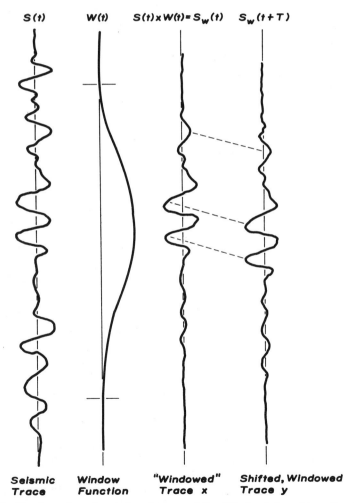

$S(t)$ $W(t)$ $S(t) \times W(t) = S_w(t)$ $S_w(t + T)$

Seismic   Window   "Windowed"   Shifted, Windowed
Trace   Function   Trace x   Trace y

FIGURE 7.23 The formation of a windowed trace and the shifted, windowed trace for which γ_{xy} is a maximum when the cross correlation has a lag of τ.

It should be noted that the *coherence* is equivalent to the correlation coefficient but is a function of frequency. It is therefore difficult to deal with for different time delays. In spite of the prevalence of this term to denote the relations between seismic traces, it is really applicable only as a phase relationship between one or more constant-frequency wave trains.

Semblance is another measure of multichannel correlation in the time domain. It is used specifically to measure the degree to which multiple traces are aligned when they are stacked after NMO removal. It is the normalized output/input energy ratio that results from stacking several channels and is again normally used as a measure of similarity of time of a limited number of phases of seismic events.

If $f(i, j)$ is the jth sample of the ith trace the semblance coefficient S_c is

$$S_c = \frac{1}{M} \frac{\sum [\sum_{i=1}^{M} f(i, j)]^2}{\sum_{j=K-1/2N}^{K+1/2N} \sum_{i=1}^{M} (f_{(i, j)})^2} \qquad (7.27)$$

where M channels are summed. The coefficient is evaluated over a window of width N, centered at point k on the original time-sampling scheme. It is equivalent to the zero-lag value of the autocorrelation of the sum of the traces divided by the mean of the zero-lag values of the autocorrelations of the component traces (Neidell and Taner, 1971). As for the correlation coefficient, this formula must be modified to allow delays between traces.

Our purpose here is to describe the application of correlation coefficients (or other measures of similarity) to finding areas of the seismic cross section in which there is a lack of similarity and to do so on a quantitative basis. We are seeking geological answers and have to modify our mathematical methods to correspond to geological reality; for example, even if we allow shifts between traces to compensate for the fact that there may be uncorrected shifts originating in the incorrect assessments of the near-surface corrections, the only time we will obtain perfect correlation (maximum correlation coefficient $\gamma_{xy} = 1$) is when all the seismic events suggest that the geological section consists of parallel layers of constant-velocity sediments. Note here that we have said *suggest* rather than *prove* because the seismic spectrum is not wide enough to *see* all of the detail in the earth's lithological variation. This question of resolution of detail by seismic methods is discussed at greater length in Chapter 8.

The systematic computation of γ_{xy} between pairs of traces for windows of constant length and shape but different centers and for an entire cross section allows the development of a matrix of values that can be contoured (see Figure 7.24). On this figure lines have also been drawn to indicate the time differences between the traces that give the maximum value γ_{xy} at the centers of the windows. It can be seen that this system can be a basis for the automatic picking of seismic cross sections. Merdler and Paulson (1968) developed a system that incorporated several additional rules to deal with problems of ambiguity.

One other problem remains, nevertheless. The trapping of hydrocarbon accumulations occurs almost exclusively on the sides of geological basins, where experience tells us that the sedimentary layers are likely to be increasing in thickness toward the center of the basin. The correlation coefficient is sensitive to the increase in time between individual reflection events and may therefore vitiate its usefulness in discovering other types of horizontal lithological gradients more

TRACE NUMBER ⟶

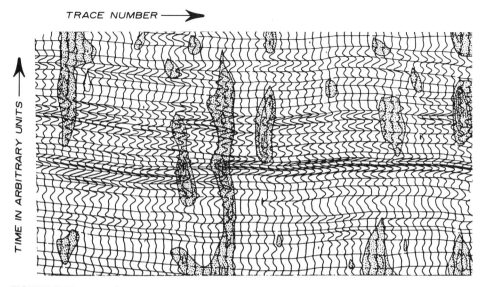

TIME IN ARBITRARY UNITS ⟶

FIGURE 7.24 A portion of a record section that shows similarity coefficient contours superimposed on the section. The nearly horizontal lines show the time relation between traces for maximum similarity. [*Source*: Courtesy of the World Petroleum Congresses (1975).]

favorable to oil trapping and accumulation. This problem can be circumvented by a transformation that can be explained by reference to Figure 7.25a and b.

A sequence of impulses a_i and random delay times τ_i gives rise to an amplitude spectrum whose shape is determined by the amplitudes and time separations of the impulses. The amplitude spectrum contains events e_j that occur at frequencies f_j. If the original sequence is expanded uniformly in time by the factor k, such that all preceding delay times τ_i become $k\tau_i$, it will be obvious that all the events e_j on the spectrum occur at frequencies given by f_j/k.

If the original and expanded trace spectra are plotted on graphs with the logarithm of frequency as the horizontal axis

$$\log \frac{f_j}{k} = \log f_j - \log k$$

and the events e_j on the expanded spectrum are shifted to the left of corresponding events on the original spectrum. It is now possible to treat the spectra with the maximum correlation coefficient concept, except that $\log f$ replaces time as the independent variable. This revised concept of similarity under expansion gives rise to the similarity coefficient—which is equal to unity when all events are stretched out in a uniform manner. The amount of expansion can be estimated.

It is obvious that if the spectra were taken just as they come out of the Fourier analysis program considerable influence would be applied by low and high cutoffs. It is therefore necessary to remove the average amplitude of the spectra and to taper the ends of the spectra before the similarity coefficient is determined. To obtain a fine enough division of the spectrum the windowed portion of the trace may have to be extended by the addition of supplementary zeros.

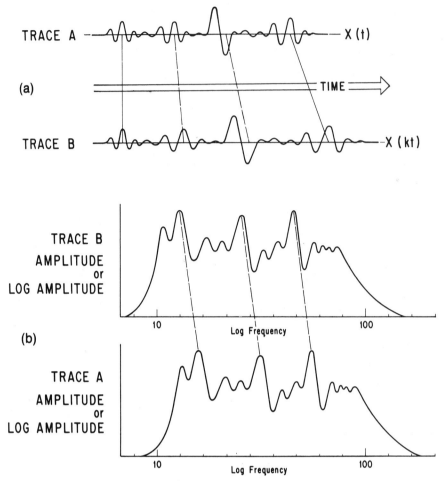

FIGURE 7.25 (*a*) Trace *B* is similar to trace *A*, except that the time scale of all events has been multiplied by the factor *K*. (*b*) The shift in the frequency spectrum due to convergence of the reflecting horizons. [*Source*: Courtesy of the World Petroleum Congresses (1975).]

7.8 SUMMARY AND CONCLUSIONS

It is evident, from measurements made in boreholes, that seismic velocity is a valuable clue to the identification of rock-type, P-wave velocities, and approximate S-wave velocities can be measured over small intervals of a few feet in a continuous manner. These velocities, however, are subject to errors when the rocks are altered by the drilling fluid, either chemically or by invasion of the pores. In addition, the velocities are often incorrect when the hole diameter changes rapidly or when the actual rock velocity is near that of the drilling fluid. Sometimes this can be discovered by the use of check shots—but this procedure can give rise to considerable error. The invasion or alteration-of-rock problem can be overcome by the use of long-interval logging devices, which, from the strictly seismic point of view, are to be preferred.

It has recently become worthwhile to record seismic events by placing receivers at close intervals down-hole (sequentially) with the source at the surface near the well. By taking a record of arrivals for several seconds not only are the first breaks recorded but the after-events, both up- and down-traveling waves, as well. Appropriate data processing of these VSPs can separate the down-traveling from the up-going waves. The down-traveling waves constitute the down-going pulse and can be used for deconvolution of the entire data set, whereas the deconvolved up-going waves can be used for reflection identification and timing. If both P and SV VSPs are taken automatic matching is achieved and there is a chance to establish the V_s/V_p ratios for a number of intervals of interest in this area.

The use of CDP reflection data allows the computation of velocities and average velocities computed from the surface to the depth of each reflector that stands out above the background noise. It has been shown that this stacking velocity is in reality the average *horizontal* velocity in an anisotropic rock sequence. The anisotropic factor cannot be separated from the velocity by the use of surface reflection data only. The anisotropic factor therefore cannot be used as a rock-type indicator. These statements concerning anisotropy apply, of course, only to surface data and especially to P-wave data. Anisotropic factors can be derived from VSPs for P and SH, taken in combination with P and SH reflection data. As well as being a rock type discriminant, anisotropic information potentially indicates the presence of aligned fractures. More information is given in the chapter on shear waves.

The velocities associated with porous rocks depend to a significant extent on the fluids filling the pores. With fluids as diverse as water and gaseous hydrocarbons the velocity varies over a wide range. This is the basis for the bright-spot change in reflection amplitude, which, although there are changes in amplitude when oil replaces water, is chiefly associated with gas in the reservoir. These changes in reservoir properties are explained quite well by theoretical treatments like the Gassmann–Biot–Geertsma equation for P waves and the Kuster–Töksöz treatment for shear waves. Much care must be taken in assigning quantities of gas to a bright spot reservoir because of the theoretical, and experimentally verified, fact that small quantities of gas are effective in changing the elastic properties of a reservoir.

As expected, the change in fluid in a reservoir rock has only a marginal effect on the shear wave velocity—a distinction that may have considerable utility in the future.

The attenuation of waves propagated through consolidated rocks is usually not large enough to prevent exploration in the widest sense. Theory suggests that it is much greater in rocks that contain gas, but, with the thickness of gas layers usually encountered, the effect is much smaller than the experimentally observed change in the amplitudes of deeper reflections. An additional cause of change in amplitude is the transmission losses that occur at each boundary between rocks. In some cases, notably in sequential gas rock-shale and anhydrite-halite layering, it may be large enough that penetration through the system is impossible by known methods.

Attenuation variations in sand-shale sequences are insufficient to be accurately measured and used as a sand/shale ratio indicator. The largest attenuation effect is likely to be the transmission loss mechanism, and this is responsive to the number of interfaces or the thinness of the alternating beds.

Similarity between successive seismic traces, or multiple sets of traces, can be calculated in several ways. With the use of proper windowing techniques the similarity coefficients become relatively free from end effects. As a function of horizontal position along the profile and the time or depth on the cross section the similarity coefficient can be contoured and used as an indicator of rapid changes (departures from the normal) of the sedimentary pattern—locations favorable to the occurrence of stratigraphic traps.

APPENDIX 7A: MAXIMUM ENTROPY SPECTRAL ANALYSIS

We have remarked several times that the process of taking the spectrum of a portion of a seismic trace (or of the whole trace, for that matter) involves splitting out (or windowing) sample values of the portion being investigated. If the values are just separated out it is as though the trace were being multiplied by a rectangular function

$$W(t) = 1 \qquad (\tau_1 \leq t \leq \tau_2)$$
$$= 0 \qquad \text{(elsewhere)}$$
(7A.1)

The isolated portion of the trace may be too short to give the needed resolution in the frequency spectrum; for example, a 0.5-s portion of a seismic trace allows amplitudes to be computed only at frequency differences of 2 Hz. In this case the custom is to extend the windowed portion of the trace to make it as long as necessary for the required resolution by adding additional values, all equal to zero, inside the needed interval T.

If we let $F(t)$ be the function to be examined in the interval between τ_1 and τ_2 this portion is obtained by multiplying $F(t)$ by $W(t)$:

$$F_W(t) = F(t)W(t)$$
(7A.2)

and when we obtain the spectrum it is that of $F_W(t)$. The effect of the window can be determined from the general rule that multiplication in the time domain is equivalent to convolution in the frequency domain; therefore

$$F_W(\omega) = F(\omega) * W(\omega)$$

The spectrum $F(\omega)$ has been convolved with the spectrum of the window $W(\omega)$.

It is possible to choose a different window $W(t)$ so that when this convolution is done a minimum effect is produced on the desired spectrum. In any event, there will always be some smearing of the spectrum by the convolution procedure.

Burg (1967, 1968, 1972, 1975) suggested an alternative procedure. The conventional methods already described sometimes generate power spectra, some values of which are negative, or the spectral estimates do not agree with the known autocorrelation function. Because these conditions must not occur, the new approach was adopted.

From the given limited portion of the time series to be analyzed an autocorrelation function can be obtained:

$$A(t) = F(t) * F(-t)$$

This autocorrelation function, of course, has a limited number of values and an infinite number of spectra agree with them—but produce additional time-lagged values outside the known range. Maximum entropy spectral analysis (MESA) is based on the choice of all possible spectra that maximize the unpredictability of a time function that agrees with the known values of the autocorrelation function. This assumption corresponds to the idea of maximum entropy used in the physical sciences.

The characteristics of maximum entropy spectral analysis are the following:

1. The spectral estimates are nonnegative functions of frequency.
2. The resolution of the MESA is greater than that obtained conventionally.
3. MESA can be calculated from $F(t)$ values in little more time than is required for conventional spectral estimates.

Andersen (1974) has described a fast, simple procedure for MESA estimation for a set of data $\{x_1, x_2, x_3, \ldots, x_N\}$ with equal spacing Δt. It is done by recursive methods that make the computations much easier. His paper contains the operative formulas and a flow diagram of the recursive procedure.

As far as is known, few tests have been taken of MESA in practical seismic reflection work. Andersen, however, reports that the procedure is superior to the FFT because of its greater resolution for a given series of data. An example is the analysis of 128 points of a sinusoid mixed with 10% white noise. After five iterations the MESA, in agreement with approximate formulas derived by Lacoss (1971), was better than that obtained by FFT analysis.

REFERENCES

Anderson, D. J., and Archambeau, C. B. (1964), "The Anelasticity of the Earth," *Journal of Geophysical Research*, Vol. 69, pp. 2071–2084.

Andersen, N. (1974), "On the Calculation of Filter Coefficients for Maximum Entropy Spectra Analysis," *Geophysics*, Vol. 39, No. 1, pp. 69–72.

Aron, J., Murray, J., and Seeman, B. (1978), "Formation Compressional and Shear Interval, Transit, Time Logging by Means of Long Spacings and Digital Techniques," paper presented at the 53rd Annual Fall Technical Conference, Society of Petroleum Engineers of A.I.M.E. SPE 7446.

Balch, A. H., and Lee, M. W. (1984), *Vertical Seismic Profiling Technique, Applications and Case Histories*, International Human Resources Development Corporation, Boston.

Balch, A. H., Lee, M. W., Miller, J. J., and Taylor, R. T. (1982), "The Use of Vertical Seismic Profiles in Seismic Investigations of the Earth," *Geophysics*, Vol. 47, pp. 906–918.

Banthia, B. S., King, M. S., and Fatt, I. (1965), "Ultrasonic Shear-Wave Velocities in Rocks Subjected to Simulated Overburden Pressure and Internal Pore Pressure," *Geophysics*, Vol. 30, No. 1, pp. 117–121.

Biot, M. A. (1955), "Theory of Elasticity and Consolidation for a Porous Anisotropic Solid," *Journal of Applied Physics*, Vol. 26, pp. 182–185.

Biot, M. A. (1956), "Theory of Propagation of Elastic Waves in a Fluid-Saturated Porous Solid. 1. Low Frequency Range. 2. Higher Frequency Range," *Journal of the Acoustical Society of America*, Vol. 28, pp. 168–191.

Biot, M. A., and Willis, D. G. (1957), "The Elastic Coefficients of the Theory of Consolidation," *Journal of Applied Mechanics*, Vol. 24, pp. 594–601.

Brandt, H. (1955), "A Study of the Speed of Sound in Porous Granular Materials," *Journal of Applied Mechanics*, Vol. 22, pp. 479–486.

Burg, J. P. (1967), "Maximum Entropy Spectral Annalysis," paper presented at the 37th Annual International Meeting of the Society of Exploration Geophysicists, Oklahoma City, October 31, 1967.

Burg, J. P. (1968), "A New Analysis Technique for Time Series Data," paper presented at the NATO Advanced Study Institute on Signal Processing, Enschede, Netherlands.

Burg, J. P. (1972), "The Relationship between Maximum Entropy Spectra and Maximum Likelihood Spectra," *Geophysics*, Vol. 37, pp. 375–376.

Burg, J. P. (1975), "Maximum Entropy Spectra Analysis," Ph.D. Thesis, Stanford University.

Cassell, B. (1984), "Vertical Seismic Profiles—an Introduction," *First Break*, Vol. 2, No. 11, pp. 9–19.

Dix, C. H. (1939), "Interpretation of Well Shot Data I," *Geophysics*, Vol. 4, No. 1, pp. 24–32.

Dix, C. H. (1945), "Interpretation of Well Shot Data II," *Geophysics*, Vol. 10, No. 2, pp. 160–170.

Dix, C. H. (1946), "Interpretation of Well Shot Data III," *Geophysics*, Vol. 11, No. 4, pp. 457–461.

Domenico, S. N. (1974), "Effect of Water Saturation on Seismic Reflectivity of Sand Reservoirs Encased in Shale," *Geophysics*, Vol. 39, No. 6, pp. 759–769.

Erickson, E. L., Miller, D. E., and Waters, K. H. (1968), "Shear Wave Recording Using Continuous Signal Methods—Part II," *Geophysics*, Vol. 33, No. 2, pp. 240–254.

Fatt, I. (1958), "Compressibility of Sandstones at Low to Moderate Pressures," *Bulletin of the AAPG*, Vol. 42, pp. 1924–1957.

Fatt, I. (1959), "The Biot–Willis Clastic Coefficients for a Sandstone," *Journal of Applied Mechanics*, Vol. 26, pp. 296–297.

Fitch, A. A. (1984), "Interpretation of Vertical Seismic Profiles," *First Break*, Vol. 2, No. 6, pp. 19–23.

Foster, M. R., and Guinzy, N. J. (1967), "The Coefficient of Coherence—Its Estimation and Use in Geophysical Data Processing," *Geophysics*, Vol. 32, No. 4, pp. 602–616.

Gal'perin, E. I. (1974), *Vertical Seismic Profiles*, translated by A. J. Hermont, ed. J. E. White, published by S.E.G. (Tulsa, OK).

Gardner, G. H. F., Wyllie, M. R. J., and Droschak, D. M. (1965), "Hysteresis in the Velocity Pressure Characteristics of Rocks," *Geophysics*, Vol. 30, No. 1, pp. 111–116.

Gassman, F. (1951), "Elastic Waves through a Packing of Spheres," *Geophysics*, Vol. 16, pp. 673–685.

Geertsma, J., and Smit, D. C. (1961), "Some Aspects of Elastic Wave Propagation in Fluid Saturated Porous Solids," *Geophysics*, Vol. 26, No. 2, pp. 169–181.

Hardage, B. A. (1983), *Vertical Seismic Profiling*, Geophysical Press, London.

Hicks, W. G. (1959), "Lateral Velocity Variations Near Boreholes," *Geophysics*, Vol. 24, No. 3, pp. 451–464.

Hicks, W. G., and Berry, J. E. (1956), "Application of Continuous Velocity Logs to Determination of Fluid Saturation of Reservoir Rocks," *Geophysics*, Vol. 21, pp. 739–754.

Jolly, R. N. (1956a), "Investigation of Shear Waves," *Geophysics*, Vol. 21, pp. 905–938.

Jolly, R. N. (1956b), "Seismic Wave Surfaces in Stratified Earth Materials," M.S. Thesis, University of Tulsa, Tulsa, OK.

Kennett, B. L. N. (1975), "The Effects of Attenuation on Seismograms," *Bulletin of the Seismological Society of America*, Vol. 65, No. 6, pp. 1643–1652.

King, M. S. (1962), "Ultrasonic Shear Wave Velocities in Rocks Subjected to Simulated Overburden Pressure," *Geophysics*, Vol. 27, No. 5, pp. 590–598.

Kokesh, F. P. (1956), "The Long Interval Method of Measuring Seismic Velocity," *Geophysics*, Vol. 21, No. 3, pp. 724–738.

Kuster, G. T., and Töksöz, M. N. (1974), "Velocity and Attenuation of Seismic Waves in Two-Phase Media, Part I: Theoretical Formulations," *Geophysics*, Vol. 39, No. 5, 587–606.

Kuster, G. T., and Töksöz, M. N. (1974), "Velocity and Attenuation of Seismic Waves in Two-Phase Media: Part II: Experimental Results," *Geophysics*, Vol. 39, No. 5, pp. 607–618.

Lacoss, R. T. (1971), "Data Adaptive Spectra Analysis Methods," *Geophysics*, Vol. 36, pp. 661–675.

Levin, F. K. (1979), "Seismic Velocities in Transversely Isotropic Media," *Geophysics*, Vol. 44, pp. 918–936.

Levin, F. K., and Lynn, R. D. (1958), "Deep Hole Geophone Studies, " *Geophysics*, Vol. 23, No. 4, pp. 639–664.

Mann, R. L., and Fatt, I. (1960), "Effect of Pore Fluids on the Elastic Properties of Sandstone," *Geophysics*, Vol. 25, pp. 433–443.

McDonal, F. J., Angona, F. A., Mills, R. L., Sengbush, R. L., van Nostrand, R. G., and White, J. E. (1958), "Attenuation of Shear and Compressional Waves in Pierre Shale," *Geophysics*, Vol. 23, No. 3, pp. 421–439.

Merdler, S. C., and Paulson, K. V. (1968), "Automatic Seismic Reflection Picking," *Geophysics*, Vol. 33, No. 3, pp. 431–440.

Morris, C. F., Little, T. M., and Letton, W. III (1984), "A New Sonic Array Tool for Full Waveform Logging," paper SPE 13285 presented at the 59th Annual Technical Conference of the Society of Petroleum Engineers of A.I.M.E., Houston, Texas.

Neidell, N. S., and Taner, M. T. (1971), "Semblance and Other Coherence Measures for Multichannel Data," *Geophysics*, Vol. 36, No. 3, pp. 482–497.

O'Doherty, R. F., and Anstey, N. A. (1971), "Reflections on Amplitudes," *Geophysical Prospecting*, Vol. 19, pp. 430–458.

Postma, G. W. (1955), "Wave Propagation in a Stratified Medium," *Geophysics*, Vol. 20, No. 4, pp. 780–806.

Schoenberger, M., and Levin, F. (1974), "Apparent Attenuation Due to Intrabed Multiples," *Geophysics*, Vol. 39, No. 3, pp. 278–291.

Stolt, R. H. (1973), personal communication.

Taner, M. H., and Koehhler, F. (1969), "Velocity Spectra—Digital Computer Derivation and Applications of Velocity Functions," *Geophysics*, Vol. 34, No. 6, pp. 859–881.

Uhrig, L. F., and van Melle, F. A. (1955), "Velocity Anisotropy in Stratified Media," *Geophysics*, Vol. 20, No. 4, pp. 774–779.

Ulrych, T. J. (1972), "Maximum Entropy Power Spectrum of Long Period Geomagnetic Reversals," *Nature*, Vol. 235, pp. 218–219.

Ulrych, T. J. (1972), "Maximum Entropy Power Spectrum of Truncated Sinusoids," *Journal of Geophysical Research*, Vol. 77, pp. 1396–1400.

Ulrych, T. J., and Bishop, T. M. (1975), "Maximum Entropy Spectral Analysis and Autoregressive Decomposition," *Review of Geophysics and Space Physics*, Vol. 13, No. 1, pp. 183–200.

Van der Stoep, D. M. (1966), "Velocity Anisotropy Measurements in Wells," *Geophysics*, Vol. 31, No. 5, pp. 909–916.

White, J. E. (1975), "Computed Seismic Speeds and Attenuation in Rocks with Partial Gas Saturation," *Geophysics*, Vol. 40, No. 2, pp. 224–232.

Wyatt, K. D. (1981), "Synthetic Vertical Seismic Profile," *Geophysics*, Vol. 46, pp. 880–891.

Wyatt, K. D., and Wyatt, S. B. (1981), "The Determination of Subsurface Structural Information Using the Vertical Seismic Profile," paper presented at the 51st Annual International Meeting of the S.E.G., Los Angeles.

Wyllie, M. R. J., Gregory, A. R., and Gardner, G. H. F. (1958), "An Experimental Investigation of Factors Affecting Elastic Wave Velocities in Porous Media, *Geophysics*, Vol. 23, pp. 459–493.

EIGHT

Resolution and Diffractions

8.1 INTRODUCTION

In optics an instrument is always qualified by a number that describes to the initiated the limit of fineness of detail that can be seen when the instrument is used. In a telescope this is the angular separation between two points of light that can just be resolved, that is, distinguished as separate. In light and electron microscopes the qualification is similar but sometimes translated into a linear measure; for example, it may be possible to resolve two points (say) 0.001 mm apart. Films used in cameras have a limit of resolution of perhaps 200 lines per millimeter. That is to say, if black lines on the film were closer together than that the area would appear to be a uniform gray rather than a number of separated lines.

It must be remembered that in optics we are dealing with visible light in wavelengths of about 4000 to 7000 Å (4 to 7×10^{-7} m), usually very small compared with the detail of the objects being investigated, and the eye is sensitive only to variations in light intensity, not to variations in phase. Now it is known that even with the most perfect optical system, free of all forms of aberration, the image of a point source is not a point but illumination over a central finite area, followed in a radial direction by successive dark and light rings. These are the famous Fraunhofer diffraction patterns (Born and Wolf, 1959, p. 391), the intensity of which is given as a function of radial distance from the center in Figure 8.1a. Although this distribution is not quite a $[(\sin Kx)/Kx]^2$ distribution, it approaches it and, in fact, a small rectangular aperture does give rise to such a distribution in both directions parallel to the sides of the rectangle.

In Figure 8.1b two Fraunhofer diffraction curves have been added together with a separation that allows the peak of one curve to lie at the first minimum of the other—this separation occurs at an angle equal to $0.61(\lambda/a)$, where λ is the wavelength of the light and a is the diameter of the telescope or microscope aperture.

There is no need for further detail here because we have already noted that the eye is sensitive to intensity (amplitude); the radiation used is monochromatic and there is no possibility of waveform variation. Although optical theory suggests a resolution of two neighboring point sources, which is dependent on the wavelength of the light, there are few other analogies we can use.

312

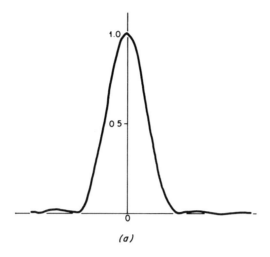

(a)

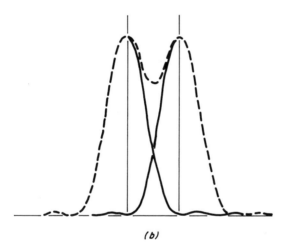

(b)

FIGURE 8.1   (*a*) Fraunhofer diffraction at a circular aperture; the function is $y = [2J_1(x)/x]^2$. (*b*) The sum of two Fraunhofer diffraction curves at a separation of $0.61\lambda/a$ (Rayleigh's criterion.)

8.2 SEISMIC RESOLUTION

The question can be asked: What do we wish to distinguish from what? In a certain (physical) sense we do not know which part of a seismic cross section or which part of a geological section we wish to examine in detail. Until a specific objective has been stated from other evidence it can be argued that in most cases it is the greater part of the geological section that may need to be investigated. The aim of all seismic prospecting is to learn as much as possible about a geological section—the manner in which the rocks are laid down, folded, fractured, and faulted, their mineral constitution, and the amount and type of fluid contained in the pores.

The most direct form of seismic measurement uses the following logs obtained in holes drilled through formations:

1. P-wave velocity logs.
2. S-wave velocity logs.
3. Density logs.

A perfectly logged single hole provides information applicable to a small volume around the hole, and many holes are necessary to yield some of the geological information sought. A seismograph would approach perfection if it could, in a vertical sense, measure the same quantities as the three logs listed in the same degree of detail.

One answer to the general question of resolution is that we want to maximize the detail with which the vertical variation of the two seismic velocities and the density are obtained. In other words, we would like to be able to determine, as closely as we can with well logs, the depths at which the lithology and connate fluids change.

In another sense, however, the seismic method has been used to acquire information on the lateral variation in some or all of the quantities listed earlier. This has become a cheaper alternative to the drilling of many wells. We must therefore expect to be able to detect horizontal changes in the same elementary parameters—not only to detect them but to place them correctly in space. We can ask: How closely can this be done? And this is another form of resolution about which information must be forthcoming.

Finally, with only three parameters, the bulk and shear moduli of elasticity and the density, that can affect the seismic waves physically, even in the perfect case, how much knowledge of the economic factors associated with hydrocarbon production can be obtained? And what is the precision to be expected?

These are complex questions that have no easy answers.

8.3 VERTICAL SEISMIC RESOLUTION

Earlier, in Section 4.8, the characteristics of reflections, as related to the velocity and density logs, were given. It was shown in (4.19) that a relation exists between laminar velocity changes in the earth and reflection coefficients:

$$\delta R = \tfrac{1}{2}\delta(\ln V)$$

and this can be extended, if the density also varies, to

$$\delta R = \tfrac{1}{2}\delta(\ln Z) \tag{8.1}$$

where Z is the acoustic impedance of the rock—the product of the proper velocity and the density—which is a function of the depth h or the two-way reflection time t. It is possible to define a piecewise continuous function called the reflectivity:

$$r(t) = \lim_{\delta t \to 0} \frac{\delta R}{\delta t} = \frac{1}{2} \frac{d}{dt} (\ln Z) \qquad (8.2)$$

It may be advisable to point out that the reflection coefficients we have been dealing with have been obtained at constant sampling rates. They really represent the product of the reflectivity and the sampling rate, although this has not been stated explicitly.

Going back now to (8.2), we note that it is not a linear function of the acoustic impedance (a small change in impedance when the average impedance is low has a higher reflectivity than the same change when the average impedance is high). The reflectivity, leaving out some complications for the moment, is the quantity that gives rise to the seismic reflection record, but it is not so easy to interpret geologically as the actual rock property—the impedance would be. Thus we transform (8.2) by two steps:

$$2 \int_{t_0}^{t} r(t) \, dt = \int_{Z(0)}^{Z(t)} \frac{d}{dt} (\ln Z) \, dt = \ln \frac{Z(t)}{Z(0)}$$

and

$$\frac{Z(t)}{Z(0)} = \exp \left(2 \int_{t_0}^{t} r(t) \, dt \right) \qquad (8.3)$$

This gives a relationship that allows calculation of $Z(t)$ as a function of t and an assumed, or known, impedance at the beginning t_0 of the log. It does, however, assume a knowledge of reflection coefficients. There are some difficulties.

1. In Section 4.8 it was shown that it is the impulse response (including all multiples and transmission losses) that is actually responsible for the seismic reflection trace.
2. This impulse response is wide-band in frequency and, to obtain the seismic record, it must be filtered with the effective bandwidth pulse received by the seismic system.
3. It is assumed that it is possible to produce a seismic trace as though it had been generated by plane waves (i.e., the spherical divergence has been removed exactly).
4. It is assumed that the *exact* values of the reflection coefficients are known. The process is nonlinear and responds in a different manner to large reflections than to small ones. Thus a knowledge of the reflection coefficient scaled by some unknown constant does not allow the exact acoustic impedance log to be determined.

The effect of all these restrictions is to limit the fidelity with which the acoustic impedance log can be displayed. It is a seismic approximate impedance log (SAIL).

These restrictions are now examined in more detail. It is interesting, since we are concerned with questions involving thin layers, to examine the case of an isolated thin layer, that is, thin compared with the wavelengths of the seismic pulse. The impulse response, as shown in Figure 8.2, consists of a series of rapidly diminishing pulses, equally spaced in time by the two-way transit time for the layer $2d/V$, where d is the thickness and V is the relevant velocity. The signs of the separate pulses are the following:

1. A first impulse from the upper surface, followed by a second of the same sign, and then later ones of alternate sign (for a layer of intermediate acoustic impedance between two extremes).

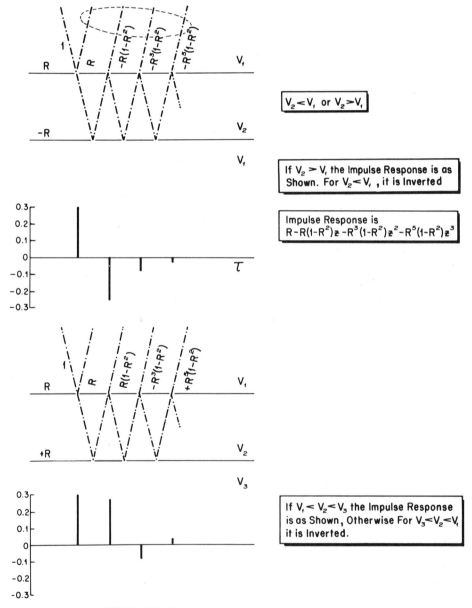

$\boxed{V_2 < V_1 \text{ or } V_2 > V_1}$

If $V_2 > V_1$ the Impulse Response is as Shown. For $V_2 < V_1$, it is Inverted

Impulse Response is
$R - R(1-R^2)z - R^3(1-R^2)z^2 - R^5(1-R^2)z^3$

If $V_1 < V_2 < V_3$ the Impulse Response is as Shown, Otherwise For $V_3 < V_2 < V_1$ it is Inverted.

FIGURE 8.2 The impulse response of isolated thin beds.

2. A first impulse from the upper surface, followed by decreasing pulses all of the opposite sign (for a layer whose impedance is less than, or greater than, the impedance on either side).

The sequence of impulses decays rapidly and in practice is important only for very large velocity and density contrasts which occur, for example, in thin coal beds in a siltstone matrix or in a gas sand in a shale–water sand environment. The values plotted in this and subsequent diagrams are approximately 0.3—which is not a common reflection coefficient in the earth.

Next, we have to deal with the situation in which each pulse in the impulse response has been filtered by the combination equipment-earth filter. It is assumed that the Vibroseis system has been used or that the equipment phase has been removed by an inverse filter such as deconvolution or a special filter designed for the purpose. In either case each constituent pulse is a symmetrical (zero-phase) pulse. For the examples given it is the autocorrelation pulse which corresponds to a particular amplitude spectrum.

We look first at the effect of integrating the wide-band reflection coefficient. A perfect integrator can be regarded as a process or circuit by which a delta function is transformed into a positive step function. This is easily understood; if the delta function occurs at a time τ integration up to that time yields zero by the definition of the delta function. At time τ the integral yields a unit value, which is retained for all times thereafter. Mathematically,

$$\int_{-\infty}^{T} \int_{-\infty}^{\infty} \delta(t - \tau)\, dt\, dt = 0 \qquad (T < \tau)$$

$$= 1 \qquad \text{(otherwise)}$$

If this integrator is applied to a sinusoidal wave train

$$\int_{0}^{t} \sin \omega t\, dt = \frac{1}{\omega} \sin\left(\omega t - \frac{\pi}{2}\right)$$

which shows that the integrated output has an amplitude that decreases inversely as the frequency and is retarded in phase by $\pi/2$ radians. Thus, for constant-amplitude input, doubling the frequency (an increase of one octave) reduces the amplitude of the integrated output to one-half.

Figure 8.3a shows the effect of approximating the delta function by a narrow rectangular pulse to obtain an approximate step function—with a steep ramp replacing the step because of the finite width of digital sampling. The amplitude is in arbitrary units because, not knowing the size of the reflection coefficient, we cannot evaluate the exponential function. In Figure 8.3b the integral of a nominal 5- to 100-Hz, flat-spectrum, autocorrelation pulse is shown. Actually, the spectrum had a taper effective over the frequency ranges 3 to 7 Hz and 98 to 102 Hz. Tapering a spectrum of constant bandwidth at the two ends has the following effects:

1. The high-frequency oscillations tend to die out more rapidly with time, the more slowly the tapering is done at the high-frequency end.

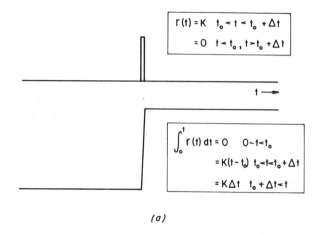

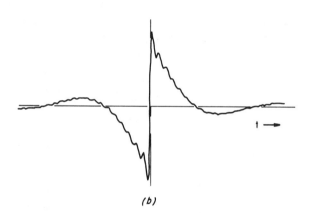

FIGURE 8.3 (*a*) Integration of an approximation to a delta function. (*b*) Integration of a band-limited autocorrelation function. The bandwidth is nominally 5 to 100 Hz and the ends are tapered linearly from 3 to 7 Hz and from 98 to 102 Hz.

2. The low-frequency oscillations on either side tend to die out more rapidly with time, the more slowly the tapering is done at the low-frequency end.
3. The central breadth of the autocorrelation function tends to increase, the more tapering is done.

Various forms of taper have been used, but in practice the linear taper is almost always selected.

Figure 8.3 shows that, although the integration of a reflection impulse (whose spectrum contains frequencies from zero to some high value connected with the thickness of the pulse) gives an acoustic impedance change over a small interval of time and the change remains after the impulse has passed, it is obvious that the integration of the band-limited zero-phase pulse results in a fast change in apparent impedance (controlled by the high-frequency end of the spectrum) preceded, and followed, by slow decays controlled by the low-frequency end of the spectrum. We are thus forcibly made aware of the need—in order to achieve our goal of acoustic impedance fidelity—of a wide band of frequencies. Not only are the high frequencies important, but also the lows. This statement is reinforced later by several examples.

We now go one stage beyond in the process of deriving the practical SAIL trace from the reflection trace because we know that the absolute values of the reflection coefficients are not likely to be known (exceptions are sometimes possible in certain geological situations in which isolated interfaces between thick constant-velocity rocks exist).

Instead of the true impedance $Z(t)$, we are now dealing with an approximate impedance function $\tilde{Z}(t)$, and the reflection coefficient trace must be replaced by a special processed record trace $S(t)$. Because the absolute scale of $S(t)$ is not known, the 2 in front of the integral sign must be replaced by a more general constant A. Thus we have

$$\tilde{Z}(t) = Z(0) \exp \left(A \int_{t_0}^{t} S(t) \, dt \right) \qquad (8.4)$$

Provided that $A \int_{t_0}^{t} S(t) \, dt$ is always small, which, in practice appears to be nearly true, the exponential term on the right can be expanded; only the constant and first-order terms are retained:

$$\tilde{Z}(t) = Z(0) \left\{ 1 + A \int_{t_0}^{t} S(t) \, dt + \cdots \right\}$$

or $\qquad\qquad\qquad\qquad\qquad\qquad\qquad\qquad\qquad\qquad\qquad\qquad (8.5)$

$$\frac{\tilde{Z}(t) - Z(0)}{Z(0)} = A \int_{t_0}^{t} S(t) \, dt + \cdots$$

Thus the fractional change in impedance (indicated by the SAIL trace) is simply the seismic trace integrated and displayed on an arbitrary scale.

In the remainder of this section the question of resolution is examined from the point of view of the accuracy with which the variations in the actual acoustic impedance can be forecast (from the surface) by the integration of properly processed seismic traces.

Returning to the isolated thin-layer case, we note, as a suitable starting point, that each of the impulses in the impulse response has to be convolved with an integrated symmetrical autocorrelation pulse. In the examples that follow two pulses with the same low-frequency spectrum (7 Hz, tapered by 4 Hz, centered on 7 Hz) are used, but to show the effect of the higher frequencies in one case the upper frequency is 43 Hz and in the other 90 Hz, with a 20 Hz taper centered on the nominal frequency. These pulses, their amplitude spectra, and the integrated pulses resulting from them are shown in Figure 8.4. Note that in the case of the 7-43I pulse the rise time is near 0.020 s, whereas for the 7-90I pulse the rise time is approximately 0.012 s.

As far as the spectra are concerned, we know that integration results in a 6-dB decrease in power (half-amplitude) for every doubling of the frequency (an increase of 1 octave). One way of integrating the pulse is to make this change in the amplitude spectrum, change the phase by 90°, and Fourier-synthesize the pulse.

Figures 8.5 and 8.6 show the results of a model computer program which computes the reflection response from a generalized layered system. Models corresponding to 10, 20, 40, 60, 80, and 200 ft are computed, the two-way times

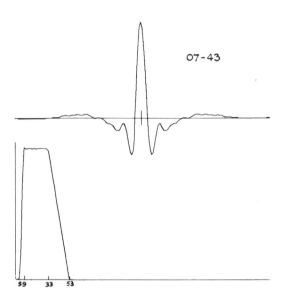

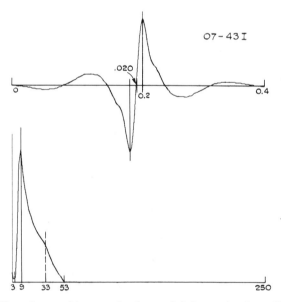

FIGURE 8.4 Zero-phase and integrated pulses and their associated amplitude spectra.

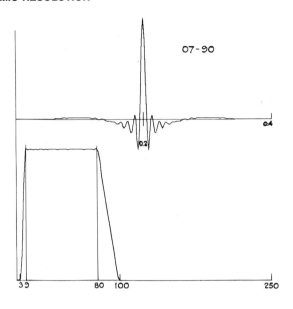

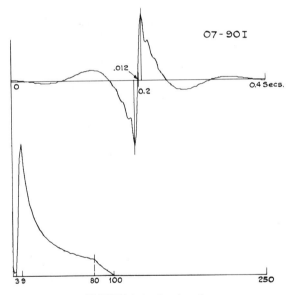

FIGURE 8.4 Continued.

being 0.002, 0.004, 0.006, 0.008, and 0.020 s, respectively. The velocities outside and inside the layer were 5000 ft/s (1524 m/s) and 10,000 ft/s (3048 m/s), respectively; the result is a reflection coefficient at the layer boundary equal to 0.33.

In these plots the velocity log is displayed to show the different layer thicknesses and the velocities just mentioned. The digitization of velocity as a function of time has caused some problems in describing properly the system of thin beds of *constant* velocity. These strata should have been shown as rectangular

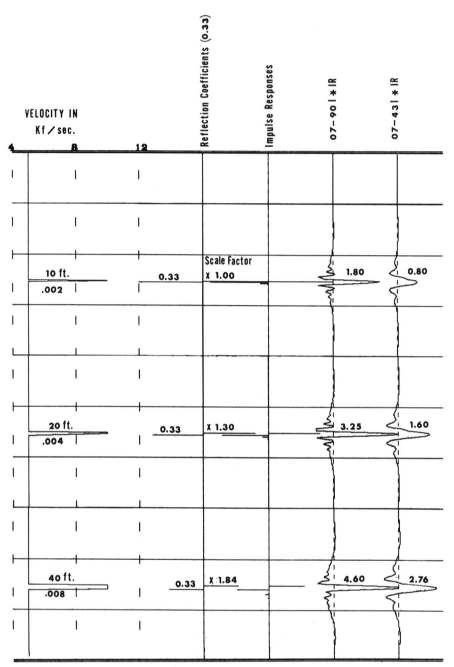

FIGURE 8.5 Responses of thin beds (10 to 40 ft) to two different integrated pulses.

pulses of constant velocity change—with no tapering to the right. The apparent loss of resolution is even more apparent in Figures 8.10 to 8.12. It does not affect the results given in the text. The first column then shows reflection coefficients, positive for down-going waves to the right. In this display all vertical logs, except the velocity log, are given their time amplitude in relation to one another, although for each layer the maximum excursion is limited to 1 in. Thus we can use

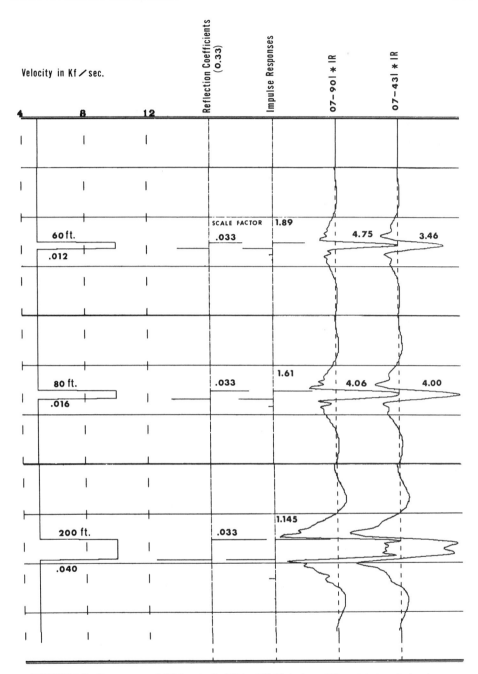

FIGURE 8.6 Responses of thicker beds (60 to 200 ft) to two different integrated pulses.

the first reflection coefficient trace for scaling purposes. Some features to be noted are the following:

1. All beds up to the 60-ft (18.3 m) bed are thin as far as both of the pulses are concerned. The amplitudes continue to grow with bed thickness (although not linearly).

2. For all these thin beds the higher frequency pulse output is larger than the lower frequency output, but the shape does not change. Here we have the first difference between the seismic and the light. For thin beds amplitude is a measure, in some sense, of the reflecting layer thickness. Resolution, in the light case, does not change with intensity.

3. We do not see any appreciable separation based on the Rayleigh criterion between the two sides of the SAIL trace until the layer thickness is greater than 80 ft (24.4 m) but visible at 200 ft (61 m).

4. Between 60 and 80 ft the higher frequency pulse output decreases in size, whereas the lower frequency pulse output increases.

5. The output pulses from the thin layers are close to the original autocorrelation pulses before integration. This is because a positive pulse, followed closely by an equal and opposite pulse, effectively differentiates the input pulse. Hence the integration has been countered by the differentiation performed by the impulse response.

6. Both pulses resolve the separate edges of the 200-ft layer to produce an almost square-topped SAIL trace, although it is evident that the layer indication starts from a base much lower than the reference line. The lack of low frequencies and zero frequency makes their effect felt.

7. The impulse response shows that the predicted multiple reflections within the layer cause a slight tail, but their effect is barely noticeable.

In general, a single frequency is reflected from the lower side of a thin layer with a phase change of 180°. Therefore to emerge in phase with the reflection from the upper surface a phase change of an additional 180° must occur because of the travel path. This happens when the layer has a thickness of one-quarter wavelength for the frequency and velocities in question. We would expect the maximum amplitude reflection to occur when the layer is one-quarter of the wavelength of the average frequency in a pulse. For the 7-90I pulse the average frequency must be near 48 Hz, and this, in a medium of 10,000 ft/s (3048 m/s) velocity, has a wavelength of 204 ft (62.2 m); the optimum reflection should be obtained with a layer about 51 ft (15.5 m) thick. This corresponds closely to the example given, even though this too simple analysis has been given for a single arithmetic average frequency.

An output almost one-half the maximum, however, is given by a layer only one-sixth of the optimum thickness, and it is evident that smaller reflection indications can be picked when the signal/noise ratio is good; *but are these small events real?*

One other example has been presented (Figure 8.7) in which three different thickness layers are included at different depths in the section. There is no significance to be attached to the actual bed thicknesses chosen nor to their relative depths. Although the outputs from the three layers appear to be similar to those computed earlier (in value, so to speak), the impulse response trace can be used as a clue to see that, as the waves pass through each layer, the down-going energy pulse becomes more and more complex and results in changed reflection outputs from the lowest layer and a series of low-frequency events of greater than expected amplitude trailing the last layer output. These can easily be

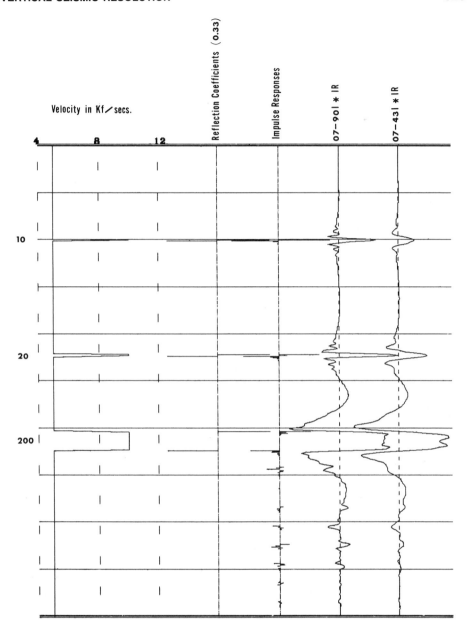

FIGURE 8.7 The effect of multiple reflections on the response of a succession of thin and thicker beds for two different integrated pulses.

interpreted as thin layers (of alternating velocity contrasts) in their own right. The presence of only a few thin, high-contrast layers can make the SAIL indications ambiguous in meaning.

Discussions of resolution are usually based on the seismic pulse itself, and some additional distinctions are made with respect to *resolution* and *definition*. The former is the ability of an object to give a high-frequency output, which may, however, be ringy if it is derived from a spectrum that is too narrow (the light case can be regarded as an example in which only the envelope of a narrow-band frequency wave is effective for resolution (see Figure 8.8); the latter is a measure of discreteness of the pulse itself in which the waveform is useful in defining two closely spaced objects. The discussions run somewhat parallel, but the direct explanation in terms of the integrated waveform not only involves both definitions implicitly but also served the direct interpretation of the seismic data problem more effectively.

So far these discussions of resolution have been given in the vacuum of a noise-free environment. It is obvious that changes in amplitude of reflections from thin beds—or indications of layer velocity changes—are interpretable only when random changes of the same nature due to geological or other noise are inappreciable. In Section 8.4 the influence of random noise is discussed.

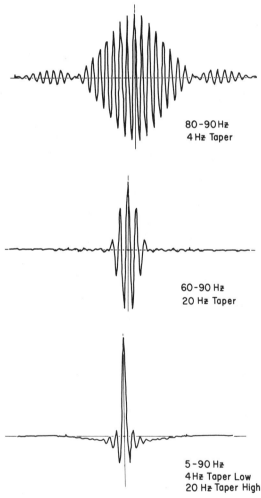

80-90 Hz
4 Hz Taper

60-90 Hz
20 Hz Taper

5-90 Hz
4 Hz Taper Low
20 Hz Taper High

FIGURE 8.8 The increase in definition as the bandwidth increases. All pulses have the same upper-frequency limit.

8.4   VERTICAL RESOLUTION AND RANDOM NOISE

Any seismic trace taken in the real world consists of a mixture of signal and other events that consist of all other energy unrelated to the problem under discussion. On a single seismic trace there is nothing to indicate which is which, unless some of the noise lies outside the frequency band associated with the reflections. The latter may be falsely generated impulses in the computer processing, but they are usually tracked down vigorously and eliminated. For a single source set of traces, or other genetically related system, identification of false events can often be inferred from their rate of change in arrival time at the trace group—their apparent velocity along the surface. If these false events have no relation to the reflection process under study they can at least be partly eliminated by velocity f-k filtering or, in specific instances, by a method to be described.

In the case of random noise, however, the shape, the time of occurrence on any one trace, and the pattern of arrival times are all unpredictable. For most of the remainder of this section the seismic trace is assumed to consist of the time reflection signal and additive noise:

$$T(\omega) = S(\omega) + N(\omega) \qquad (8.6)$$

The coefficient of coherence (Foster and Guinzy, 1967) is given by

$$\gamma_{xy}(\omega) = \frac{|f_{xy}(\omega)|}{\sqrt{f_{xx}(\omega)f_{yy}(\omega)}} \qquad (8.7)$$

where $x(t)$ and $y(t)$ are stationary time series with power spectra $f_{xx}(\omega)$ and $f_{yy}(\omega)$, respectively, and cross spectrum, $f_{xy}(\omega)$. It must be emphasized that *estimates* of coherence are made and great care is needed. The coefficient of coherence is related to the signal/noise ratio by the expression

$$\frac{S(\omega)}{N(\omega)} = \frac{\gamma_{xy}(\omega)}{1 - \gamma_{xy}(\omega)} \qquad (8.8)$$

Thus, if the coefficient of coherence is estimated, the average noise power can be estimated in terms of the average signal. This provides a tool that we may need in predicting whether an event seen on a trace is related to geology or whether it is noise. Discussion of the evaluation of this probability and the proper techniques for estimating the coherence are outside the scope of this chapter and should be sought in Foster and Guinzy (1967).

One problem, which often occurs in all discussions of seismic data, is the proper handling of near-surface corrections. This problem is aggravated if higher than normal frequencies are used because the accuracy of time is related to the pulse duration. Coherency measures in the frequency domain appear to have been unaffected by small time origin inconsistencies, but unfortunately the estimation of the coherence coefficient requires the establishment of a lag window within which the spectrum is estimated and the phases of the various frequencies are related to the time origin of the relevant window.

Corrected seismic traces to be composited should have the same geology-related reflections at the same time. They can be added together after being scaled in amplitude by some preselected weighting function. At most, the composited trace will have a signal/noise ratio that has been improved by the square root of the number of traces added together. Generally, the weighting constants are equal to unity, but in some cases of high noise level the traces can be weighted inversely as their power. When the noise level is that high, however, it appears to be inadvisable to pursue the objective of high resolution; that is, the attainment of high fidelity of the predicted acoustic impedance compared with that attainable by logging.

If a well is available in an area and seismic reflection work is performed in its neighborhood the coefficient of coherence can be obtained between the synthetic trace that results from the well log information and the field trace(s) recorded in the vicinity. Because it is evidence of the existence on the field traces of information at a particular depth that is required, the synthetic record made should start from reflection coefficients rather than from the impulse response. The reflection coefficients can be filtered to correspond in power spectrum to the field trace, but Foster and Guinzy (1967) have warned against the dangers with some forms of filter. The minimum requirement for this analysis is sufficient correspondence between the synthetic and field traces to permit a lag window with consistent starting times to be used. In some cases, in which formation alteration may be a problem, particular attention may have to be paid to obtaining velocity information by the use of a long-offset velocity tool in the hole and to correcting the density measurements for any formation alteration.

8.5 VERTICAL RESOLUTION AND DEPTH OF PENETRATION

At some depths the random noise level eliminates the chance of obtaining adequate fidelity of the seismic acoustic impedance log compared with the real log. The level of random noise is controlled by such factors as wind noise acting directly on the geophones or indirectly by shaking the ground, scattering horizontally traveling waves by random inhomogeneities, and so on. The transmission characteristics of the water layer, for example, are excellent, and irregularities of bottom, scatterers (fish, gas bubbles, or man-made artifacts) combine to give incoherent scattered energy that persists up to high frequencies.

It has been shown, however, that penetration of the earth directly by high frequencies and the return of high-frequency energy from the target layers is limited by the following:

1. Attenuation by solid friction.
2. Loss of energy or change in frequency by reverberations and transmission losses due to many layer boundaries.

To a first approximation, the loss of amplitude follows an exponential law:

$$A(z) = A_0 \epsilon^{-\alpha z}$$

where α = the attenuation constant that combines both effects and can be written
as $\pi f / Qc$
f = frequency
$1/Q$ = a specific dissipation constant (effective for both effects)
c = velocity

The loss in decibels in traveling a distance d through the earth is

$$\text{decibel loss} = 8.686 \, \frac{\pi f d}{Qc}$$

the graph of which is a straight line. Note that the loss is proportional to d and f. Figure 8.9 illustrates diagrammatically how both factors change the spectrum and the high-frequency effective output drops below the noise level as the frequency or depth is increased.

As an example, suppose that we wish to propagate 200 Hz to the layers just below Pierre shale, which has a thickness of 5000 ft, a velocity of 7800 ft/s, and an effective Q of 20. Suppose also that an effective signal/noise ratio of 10 has been established for a frequency of 30 Hz; we would like to know how much higher the signal input must be to achieve the same signal/noise ratio at 200 Hz.

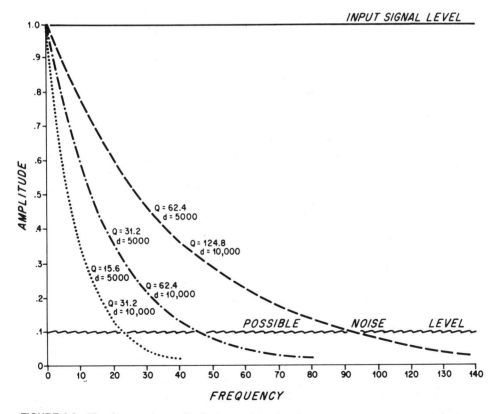

FIGURE 8.9 The decrease in amplitude due to Q or d variation. An increase in Q by a factor of 2 gives the same drop in attenuation as a decrease in distance d traveled by a factor of 2.

The decibel loss at 30 Hz from some reference level is

$$\frac{8.686\,\pi \times 30 \times 10,000}{20 \times 7800} = 52.48 \text{ dB}$$

The decibel loss at 200 Hz from the same reference level is

$$\frac{8.686\,\pi \times 200 \times 10,000}{20 \times 7800} = 349.84 \text{ dB}$$

A required gain in signal of 297.36 dB, or about 10^{15} in amplitude, would be necessary to accomplish our objective—an impossibility.

Because this section is not meant to be negative toward the accomplishment of high-frequency penetration, it is pointed out that in a section that has an effective Q of 100 the difference in input for 100 and 30 Hz is about 24.5 dB, an increase in amplitude of only 16.8, an achievable result.

Experiments with shear waves have illustrated how drastic this exponential decay can be, and caution is required to ensure that an objective is within the realm of possibility before attempting costly experimentation. At best, the results may be attainable only in some lithological sections.

8.6 VERTICAL RESOLUTION AND THE CHARACTERISTICS OF NEAR-SURFACE VELOCITY LAYERING

It has been stated that near-surface layering, because of the reverberations produced and their effect on the character of the down-going pulse, can cause loss of high frequencies. This loss is further accentuated by transmission of the reflection pulse upward.

This situation is illustrated by three model studies (Figures 8.10 through 8.12), in which the layering near the surface was chosen to be alternating in characteristics of velocity and the layer thicknesses were chosen randomly between 0 and 100 ft by a suitable computer program. The only difference in input to these three examples is the addition of 10 layers at a time—from 10 to 20 to 30. Reflection coefficients and impulse responses were computed by using a zero reflection coefficient at the surface to avoid extra complications. It can be seen that the additional layers tend to make the impulse response for the deepest boundary deteriorate as far as high-frequency character is concerned. The impulse response is chiefly a single spike for the 10 layers, although followed by some small consecutive positive values, but the single spike continually diminishes until at 30 layers the succession of small consecutive positive spikes becomes dominant.

This impulse response change is reflected in the synthetic reflection records. Although for the 10-layer case the 07-90I filter gives a sharper indication of velocity increase at a time of 0.53 s, by the time the 30-layer case is considered there is no more resolution than with the 07-43I filter. Although this may seem extreme, it must be remembered that the usual shallow section has many more layers (to represent it properly) than the 30 chosen here, although the reflection coefficients may not always be as large.

The synthetic technique of inserting a large, isolated velocity boundary below part of a log is a useful one for studying attenuation due to just reverberations.

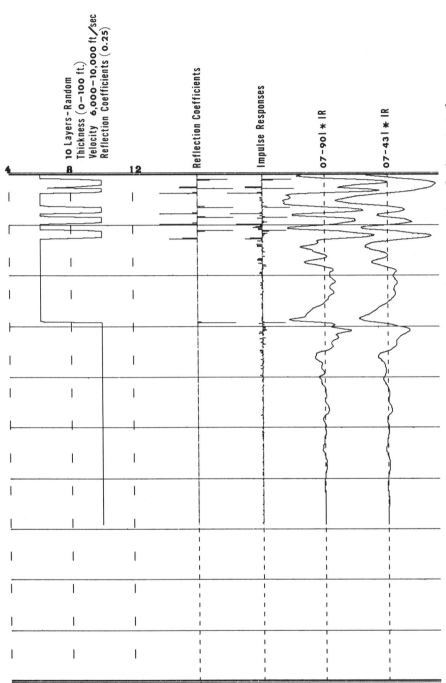

FIGURE 8.10 The effect of 10 thin shallow beds of high velocity contrast on the reflection output of a single deeper velocity increase.

331

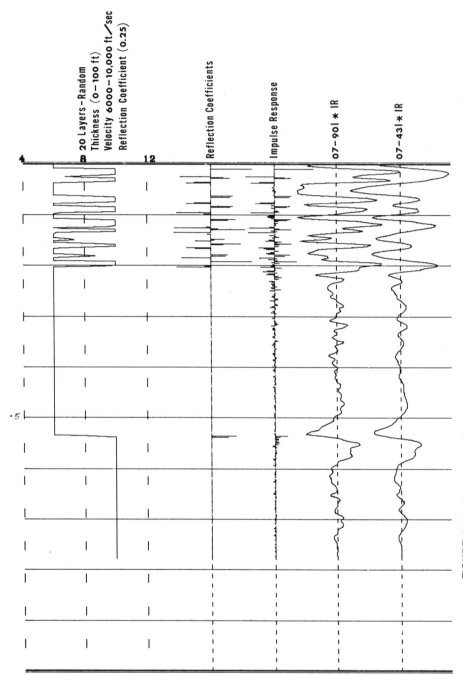

FIGURE 8.11 The effect of 20 thin shallow beds of high velocity contrast on the reflection output of a single deeper velocity increase.

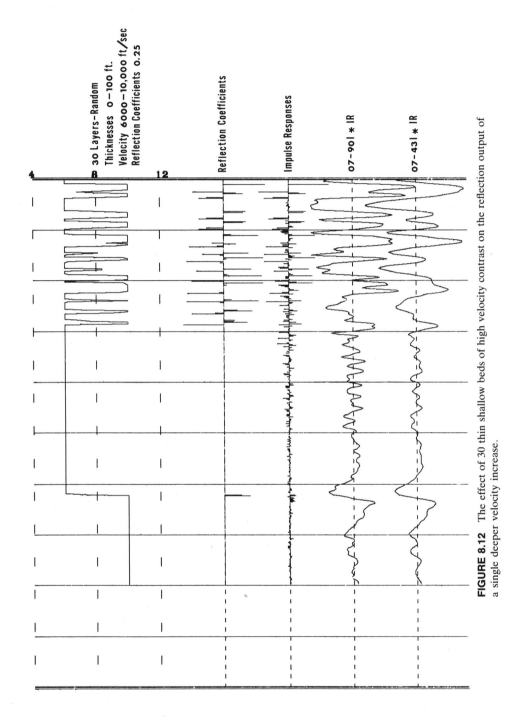

FIGURE 8.12 The effect of 30 thin shallow beds of high velocity contrast on the reflection output of a single deeper velocity increase.

333

8.7 SCATTERING AND DIFFRACTIONS FROM REFLECTING DISCONTINUITIES

The task of detecting lateral changes in reflection qualities and placing them in correct relation to one another is another form of resolution; for example, for an unspecified reason one portion of a reflector may possess a sharp boundary at which the reflection coefficient changes suddenly. If this is a closed boundary or if there is more than one boundary it may be necessary to determine how large the enclosed area must be before it can be distinguished or whether a boundary of small area can be distinguished from one with a larger boundary. It is a more specific question than the general one of placing all reflectors in their proper places. It is called migration and is dealt with in a later chapter.

We shall see that a discontinuity in the reflection capability of an interface does not mean that there is a discontinuity in the seismic events returning energy to the receivers. In fact, this discontinuity in energy returns does not occur. The general property of returning energy, or diverting energy from the incident energy beam, is called scattering. An account of scattering an acoustic wave by a small obstacle in a homogeneous medium has been given by Officer (1958). The small obstacle gives rise to two different types of scattering. The first, called a simple source, is due to compression and expansion of the obstacle and this emits radiation uniformly in all directions. The second form of source is double and is due to the density difference between the scatterer and the material in which it is embedded. This scatterer is then accelerated along a direction perpendicular to the incident wave front and gives rise to directional scattering.

The amplitude of the secondary wave is proportional to the incident field, the difference in bulk moduli, and in densities, the volume of the scatterer, and the *square* of the frequency of the incident wave. This frequency dependence is often referred to as Rayleigh scattering. The secondary wave is spherically divergent away from the scatterer. Its amplitude falls off inversely with distance. For the double source there is an angular dependence of $\cos \theta$, where θ is the angle between the incident beam and the scattered direction.

For more complex problems of scattering resort often has to be made to numerical calculation. The principles that control these calculations are now given. If a plane wave impinges on a rigid obstacle each point of the surface, by Huyghens' principle, will act as an emitter of seismic waves. As shown in Figure 8.13a, energy is received, even though the receiving and source points are not in a relation that is proper for a reflection. This scattered energy, which is organized by the geometry of the scatterer, is called a diffracted wave. It appears that some of the energy comes from the edge itself. A small element of length δl is irradiated, and it is noted that pairs of point sources on either side of the center have equal travel times from the source to the receiver—as long as they are in position for a reflection. If, however, the directions are such $(S'OR')$ that Snell's law is not obeyed there is a time difference between rays irradiated and emitting at one end and at the other end (see Figure 8.13b). The rate of change in the time delay (along the direction of the obstacle) is then $(\sin \phi - \sin \theta)/c$. If the obstacle has a more complex form the problem becomes one of dividing up into a series of small elements whose individual contributions at the receiver are known and then to sum up all the contributions, taking into account the phase due to variation in distance.

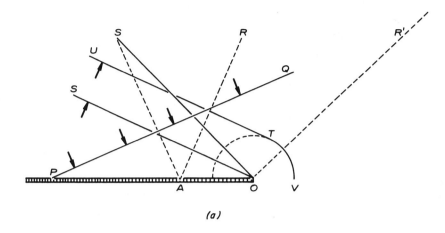

(a)

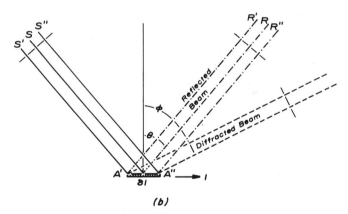

(b)

FIGURE 8.13 (*a*) The incident wavefront *PQ* on the reflector *AO* is reflected as a plane wave taking positions SO and, later, UT, but a circular portion of the wavefront TV has a center at *O*, the termination of the reflector. Energy reaching *R* appears to come from *O*, but in fact comes from elemental contributions from all positions on the reflector. (*b*) Illumination of a small strip (δ*l*) and reflected and diffracted beams. The delay per unit length at diffractor in the *l* direction is $dt/dl = (\sin \phi - \sin \theta)/c$.

One way of looking at this problem is that it is analogous to a two-dimensional pattern of radiators, each of which is illuminated by a different amount (both in amplitude and time).

The summing procedure often cannot be done formally but has to be numerical, and several artifices have been introduced by various workers in this field to aid in this task. One, used by Hilterman (1970) and Trorey (1970), is to make the source and the receiver occupy the same point. This has the advantage of simplifying the calculations, but at the expense of solving unreal problems unless the source receiver distance is small, which is seldom the case in CDP reflection profiling. Thus their results, although probably correct, are not accurate for the general CDP case. Mitzner (1967) has devised more accurate methods, but they are expensive in computer time.

We can, with advantage, however, examine some of Trorey's examples because they provide general guidelines and some evidence needed later when we finally come to lateral resolution.

The first result is that the sign of a diffracted wave must change as it moves from one side of an edge to the other; for example, Figure 8.14 shows a strip with diffractions at each boundary. Each branch (right and left) from each edge follows the same geometrical curve in time and distance, but on the right-hand edge the diffraction is positive to the right and negative to the left. The opposite occurs on the left-hand edge. The reflection itself is positive. It is noted that the diffraction causes the composite reflection and diffraction (they occur at the same time) to have only 50% amplitude right over the edge. An illustration of diffraction from a single edge, in the form of a seismic cross section, is given in Figure 8.15. The drop to 50% amplitude over the edge is clearly shown. Second, the rate of decay of amplitude with the distance of the source and the receiver from the edge is greater than could be accounted for by inverse spreading.

Finally, the shape of the diffraction waveform differs more and more from that of the reflection waveform as the source or receiver point moves horizontally away from the surface projection of the discontinuity. It should be noted that Trorey's solution is for the velocity potential (a scalar value) and that to evaluate the particle velocity measured at the surface the derivative in the vertical direction must be taken. It is evident that the vertical component of velocity diminishes at least as fast as the sine of the angle the diffracted ray makes with the horizontal and must therefore diminish even faster than in the solution given by Trorey.

We give a further illustration of how the shape of the pulse is altered when a diffraction from a half-plane is investigated. Some simplifying assumptions are made:

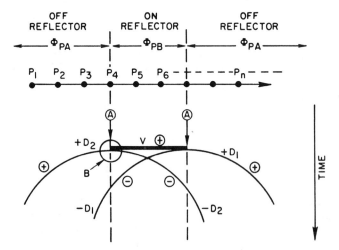

FIGURE 8.14 A diffraction must undergo a 180° phase change on each side of a diffracting edge: P_1 and P_2 are successive source or receiver positions, D_1 and D_2 are diffractions. The circled signs show the phases of the various parts of the response. [*Source*: Trorey (1970). Reprinted with permission from *Geophysics*.]

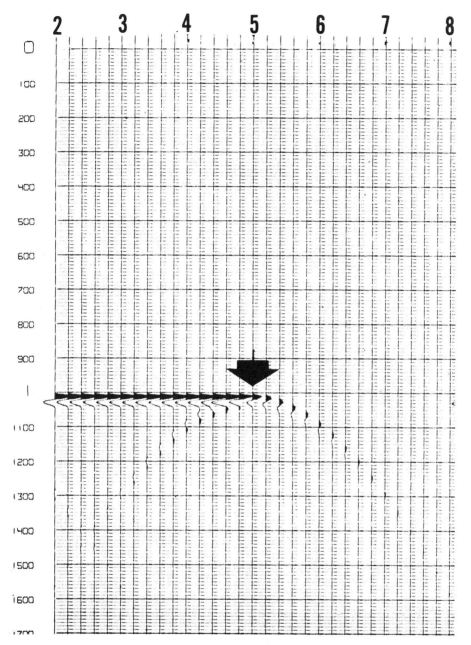

FIGURE 8.15 The appearance on a seismic record of the diffractions from a single edge. Because correction for spherical divergence has already been applied, the amplitude dies off more rapidly than $1/t$. [After Trorey (1970). Reprinted with permission from *Geophysics*.]

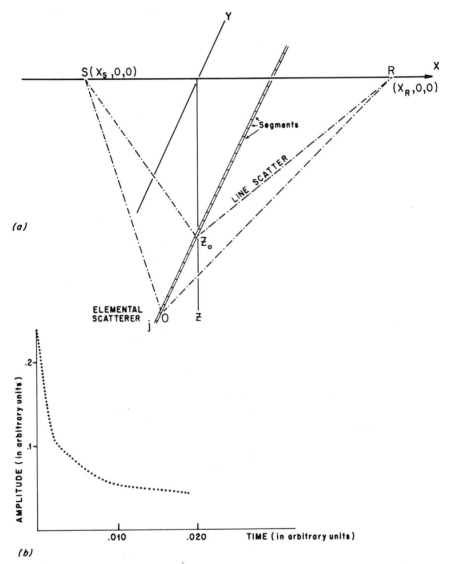

FIGURE 8.16 (*a*) The geometry associated with scattering from a line perpendicular to the *X-Z* plane. The line is divided into segments whose time SOR obeys the equation $T_j = T_0 + j \cdot \Delta t$. (*b*) Output pulse at *R* for specific source and receiver locations.

1. The plane is a series of strips, as illustrated in Figure 8.16*a*, each of which can be split up into elements with arrival times that differ by one sample time.
2. The strips are chosen such that their arrivals differ from the nearer strip by one sample time.
3. All elements contribute inversely as the travel distance and inversely as the distance from the source but are otherwise unidirectional in strength. (This assumption causes the pulse shape to fall off in amplitude more quickly than the rate calculated.)

4. All elements emit a delta function of velocity and this must be multiplied by the sine of the angle of emergence to obtain the vertical amplitude of particle velocity.

The mathematical manipulation is too cumbersome to be included here, but Figure 8.16*b* shows the shape of the output pulse at *R* for a specific source-receiver pair. It is assumed here that most of the energy for the diffracted pulse comes from a region on the diffracting plane near the edge. This pulse shape for a delta function input can be approximated by a spectrum in which the amplitude falls off with frequency at a 3-dB/octave rate and the phase changes by $\pi/4$. It could be called the result of a *half-integration*.

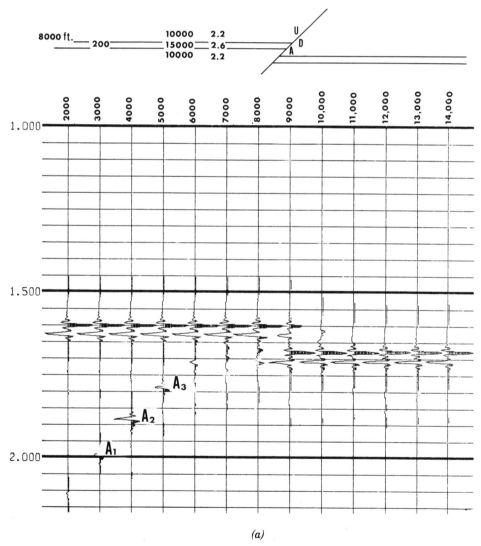

(a)

FIGURE 8.17 (*a*) Reflections and diffractions from a reverse fault, by computer, two-dimensional model. The large diffraction amplitudes A_1, A_2, and A_3 are probably due to the inclined fault face *A*.

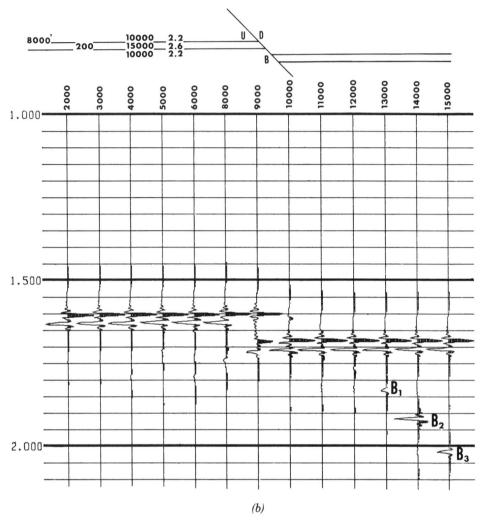

(b)

FIGURE 8.17 (*b*) Reflections and diffractions from a normal fault. The large amplitudes B_1, B_2, and B_3 are probably due to the inclined face B.

A reflection from the main part of the plane reflector results from convolution of the incident pulse with the delta function assumed in the calculation to be the source function; in other words, a reproduction of the incident pulse. The diffraction pulse then corresponds to a convolution of the original input pulse with the wave shape shown in Figure 8.16*b*. Because it is comparatively richer in low frequencies, the diffracted wave gradually becomes lower and lower in frequency as it is found at greater and greater distances from the reflector edge.

In the earth real faults displace the ends of the formations, in general both vertically and horizontally. With normal faults the horizontal motion is such that a vertical line through the formation at the fault location encounters a thinner part of that formation, or none at all; for reverse faults the opposite is true and duplication of the section along a vertical line may occur (see Figure 8.17). The effect of the diffractions is to continue the apparent reflected energy laterally past the formation ends. Hilterman's examples show this very well. There is, of course, a modeling problem here in that the near-vertical wall is sometimes modeled as just being cut in the same rigid material used for the reflector. In the calculations, as in the analog model, the elemental areas of the wall also radiate. In practice the fault is often displaced enough that different formations be laterally adjacent to one another. The fault zone is often fractured and becomes an anomalous zone in porosity (density) and elastic constants.

8.8 LATERAL RESOLUTION

There is an ever-increasing demand for additional fine detail from the seismic reflection method. Whereas formerly it was used as a reconnaissance tool and an adjunct to geology in specifying wildcat well locations, it has now become a prime tool for providing evidence of exploitation (i.e., development) wells. The exact location of faults and the estimation of their hade and throw are now requested. Moreover, the seismic reflection method is finding increasing use in other technologies; for example, there are requirements for the location of old, water-filled mines and for the location of fire fronts in the new technology of underground gasification of coal—where seams are too thin for economical mining. Although the latter is an esoteric venture, it can provide an example of the new lateral resolution requirement. A coal seam is set on fire along the line of a hole directionally drilled to traverse the seam horizontally. Oxygen or air is pumped in through the hole, and the combustion products are obtained by percolation through cracks in the coal seam to neighboring, parallel holes, where they are pumped away for use. Figure 8.18a shows the situation diagrammatically. Now, to avoid drilling additional monitoring holes from the surface to the coal seam a requirement is made that the position of the fire front be monitored (making use of the additional reflectivity of the seam after burning) by seismic stations permanently mounted on the surface. For ease of consideration surface topography is, at first, disregarded here.

It can be seen that to a first approximation the burned coal area is a narrow strip, more or less parallel to the original oxygen intake hole. The questions to be asked are the following:

1. Where at any time is the boundary of the burned-out zone? Within what accuracy can it be specified?
2. What is the minimum boundary change that can be detected? Can deviations from the thin strip be readily located to prevent (or have knowledge of) fire breakthrough into the collector holes?
3. What are the seismic characteristics needed to make this system feasible in terms of delineation accuracy, seismic frequency bandwidth, and seismic surface arrays?

Evidently the knowledge required is of the seismic field at the surface for a given source, a given reflector distribution within the coal seam, and the rate of change in the seismic field when the burned area changes in size.

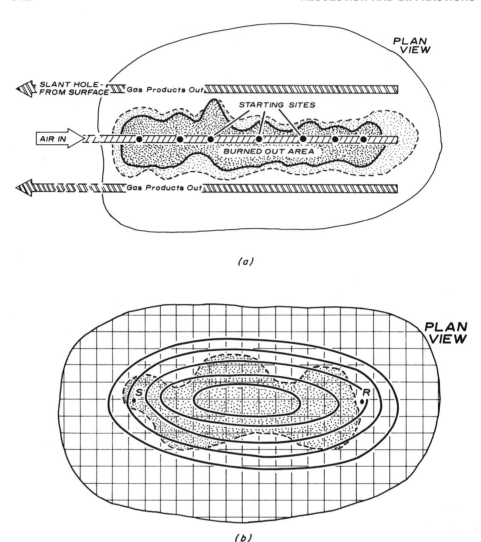

FIGURE 8.18 (*a*) A burned-out area (dark boundary) and the new boundary after additional burning. This moving boundary is determined continually. (*b*) The contoured subsurface area shows zones of equal *S-R* time. The hypothetical burned zone is superimposed. Evaluation of the response is done at the grid corners.

The steps involved are, essentially, these:

1. For a particular source-receiver combination zones of equal time are drawn on the subsurface. In a separated source and receiver and a constant-velocity overburden they are ellipses.
2. The points of a grid that fall within each contour zone are evaluated for response at a receiver by use of the general diffraction formulas given for acoustic media by Mitzner (1967).

3. Step 2 is repeated for each contour zone and the results are added—with the appropriate sample time delay. In this way the time response at a particular receiver is obtained.

4. Steps 1 through 3 are repeated for each source-receiver combination required.

5. The impulse response at each receiver is convolved with the source pulse equivalent. (This may be an autocorrelation pulse in the case of Vibroseis.)

6. The entire process is redone with a new, or revised, burned-out zone to find the sensitivity or resolution. Note that this process is additive and, for a given source-receiver combination, earlier results can be added to, to avoid redundancy in computer calculations.

As a preliminary to these expensive investigations, however, it is possible to examine the situation from the point of view of time and amplitude of events received from an infinite strip on a line traversing the strip at right angles. Trorey's example in Figure 8.14 sets the stage. For any general trace near the middle′ of the cross section there are three events: the reflection and two diffractions of opposite sign and different amplitudes and times of arrival. The lateral resolution of strip width then turns into a problem of vertical resolution of the pulses coming from three different processes. From this point of view the amplitude spectrum bandwidth and the phase spectrum are important. The depth is also important because it controls the curvature of the diffracted events, hence the differences in arrival times of the events on any given seismic trace. Details of these events near the strip are shown in Figure 8.19 for a particular strip width and depth. The time scale is controlled by the seismic velocity above the strip. This type of analysis is comparatively rapid and inexpensive but provides no answer to more refined questions that may be asked, such as those dealing with local fast-burning locations.

The three different seismic impulse responses that could arise from this model are shown in Figure 8.20. Note that at Section 1, compared with Sections 2 and 3, the arrival at the reflection time is the mean value. The later diffraction arrivals are plotted approximately to true amplitude, and it is evident that for low-frequency pulses their effect is algebraically added to the reflection, whereas under sufficiently wide-band conditions they can be separated. At section distances up to 200 ft from the edge of the strip the positive events require time separation of peaks only 0.01 apart. Because this corresponds to the first zero of a $(\sin 2\pi ft)/2\pi ft$ pulse, the upper frequency of the pulse has to be 100 Hz.

Now this strip is quite wide in comparison with its depth. The amount of energy contributed by the strip is controlled by the proportion of the energy given out by the source and intercepted by the strip (this is true of all obstacles) and the amount of total energy emitted by the strip that is intercepted by the receiver. Geometrically, the output is dependent on the solid angle subtended by the obstacle at both source and receiver. Thus, as the strip diminishes in width, the ratio of anomalous energy to normal reflected energy received diminishes rapidly (as the width of the infinite strip squared or the area of an obstacle squared in the general case). This is consistent with the Rayleigh scattering principles.

The resolution needed to separate all the diffracted events from the reflection also increases as the square of the ratio $Z/(A_1 - A_2)$.

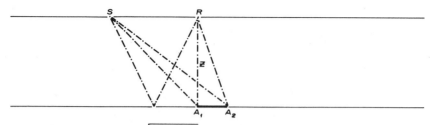

Time for reflection is $2\sqrt{\left(\frac{R-S}{2}\right)^2 + Z^2}/C$

Time for diffraction from A_1 is $\sqrt{(A_1-S)^2 + (R-A_1)^2}/C$

Time for diffraction from A_2 is $\sqrt{(A_2-S)^2 + (R-A_2)^2}/C$

For the example $S-R = 6$ units, $Z = 6$ units, $A_1-A_2 = 2$ units

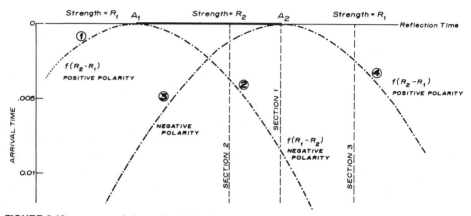

FIGURE 8.19 events and times of arrival of reflections and diffractions from an infinite strip. Times are in seconds for 1 unit = 100 ft and velocity $C = 10,000$ ft/s.

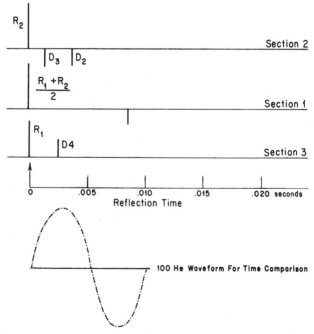

FIGURE 8.20 The events in approximate amplitude and relative time for the example given in Figure 8.19.

It is hoped that the application of principles in this example will allow other cases of particular interest to the reader to be worked out.

Experimental evidence has been obtained over a shallow coal seam (200 m deep), which suggests that faulting (Figure 8.21) can be located to an accuracy of a few tens of meters and that a mined area can be distinguished from an unmined area within 20 m from amplitude alone (Figure 8.22). The individual mined-out passages cannot, however, be differentiated from the coal pillars left to support

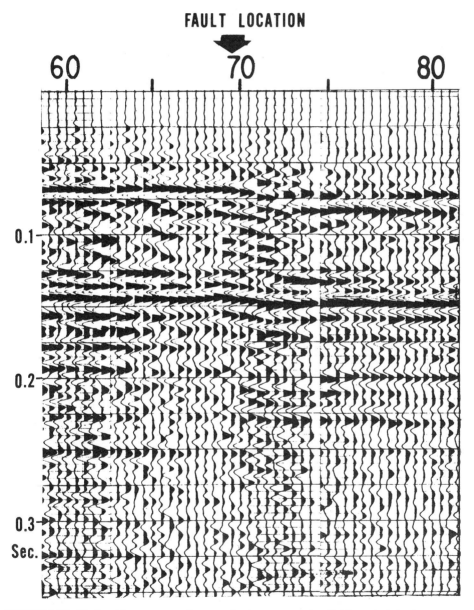

FIGURE 8.21 Fault evidence on a coal-seam reflection cross section (0.15 s arrival time). The throw is known to be about 8 ft and the time shift of 0.0035 s appears to be due to velocity change and depth change.

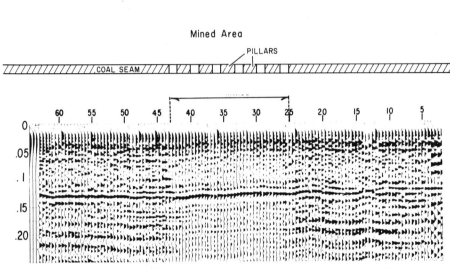

FIGURE 8.22 A high-frequency (40 to 170 Hz) seismic survey over a recently mined area of a coal mine. The mine depth is 550 ft, the coal-seam thickness is 7 ft, surface shotpoints are 40 ft apart, and the mined area is air-filled. The section was processed for equal total power in all traces; hence the large reflection over the mined area gives rise to a diminished background. (*Source*: Courtesy CONOCO, Inc.)

the roof, even though these pillars have dimensions of 20 to 30 m. The experimental parameters are the following:

Trace spacing (subsurface 6.5 m)
Frequency spectrum, 40 to 170 Hz
Fivefold stack
Source pattern, 15 positions over 20 m
Receiver pattern, 20 geophones over 20 m

Diffraction problems, except for those involving the simplest geometries, are difficult enough in the acoustic case without bringing in the added complications of an elastic medium in which two types of wave are possible and in which transfer of energy between the two types occurs at boundaries. Knopoff (1959) has dealt with incident plane P or SH waves impinging on a spherical obstacle of a different material. For problems like the one discussed, which, essentially, involve plane layering with discontinuities, the use of SH waves eliminates any concern connected with energy conversion and can be treated by the existing scalar wave-diffraction theory.

8.9 SUMMARY AND CONCLUSIONS

In seismic reflection work the distances encountered are always of the same order of magnitude as the wavelength. The whole composite waveform in closely spaced events is observed, and, because we are dealing with a linear system, these composite waveforms are simply additions of the individual event waveforms. Although in the optical case resolution is controlled by the envelope of the intensity curve, in seismic reflection work we have to deal with the composite waveform itself.

Resolution has been dealt with in a form that requires the recovery of an approximate acoustic log from the seismic trace, and it is the (local) faithfulness with which the actual acoustic impedance log can be recovered that constitutes vertical resolution. We need to be able, a priori, to examine a seismic cross section and to determine from it locations at which oil or gas accumulation is most probable.

The difficulties, some of which may be overcome in time, in achieving this fidelity in reproducing the acoustic impedance log are the following:

1. The presence of multiple reflections.
2. The difficulty of recovering true amplitude because of attenuation and reverberation in thin-layered contrasting sections.
3. The transfer function of the near-surface regions on the wave shapes (at both source and receiver).
4. The need for extreme bandwidth. This is coupled with the difficulty of introducing low frequencies and the attenuation with depth of high frequencies.
5. The presence of random noise and interference.

Scattering and diffractions from discontinuities constitute the means by which the discontinuities can be located. The acoustic diffraction theory is reasonably well established but complex, and only very simple, not very meaningful, examples of seismic work have been formally solved. Computer methods can be used to solve specific lateral resolution problems, however. For investigation of the size or shape of small scatterers the time resolution needed (and the high frequency of a very wide bandwidth signal) increases as the square of the ratio of depth to a scatterer dimension. The energy return also diminishes as the square of the scatterer area (fourth power of a scatterer dimension) if shape is maintained. In many cases this may have to be observed in the presence of a strong reflection. Otherwise the signal/noise ratio for the whole required spectrum is important.

Field examples of perturbations in the characteristics of a coal seam have been shown. At a depth of 200 m the edge of a mined-out area can be established to about 20 m, but individual pillars and gallaries of these dimensions in the mined-out area cannot be distinguished at an input frequency of 40 to 170 Hz.

Although the acoustic diffraction theory is in reasonably good, but complex, condition, few attempts have been made to deal with elastic wave diffraction except under simple, nonrealistic conditions. The use of shear waves, which can be propagated to the distances necessary, should make comparison of acoustic diffraction theory with experiment justifiable in cases that involve discontinuities in near-planar reflectors.

REFERENCES

Born, M., and Wolf, E. (1959), *Principles of Optics*, Pergamon, New York.

Foster, M. R., and Guinzy, N. J. (1967), "The Coefficient of Coherence: Its Estimation and Use in Geophysical Data Processing," *Geophysics*, Vol. 32, No. 4, pp. 602–616.

Hilterman, F. J. (1970), "Three Dimensional Seismic Modeling," *Geophysics*, Vol. 35, No. 6, pp. 1020–1037.

Knopoff, L. (1959a), "Scattering of Compressional Waves by Spherical Obstacles," *Geophysics*, Vol. 24, No. 1, pp. 30–39.

Knopoff, L. (1959b), "Scattering of Shear Waves by Spherical Obstacles," *Geophysics*, Vol. 24, No. 2, pp. 209–219.

Mitzner, K. M. (1967), "Numerical Solution for Transient Scattering from a Hard Surface of Arbitrary Shape," *Journal of the Acoustical Society of America*, Vol. 42, pp. 391–397.

Officer, C. B. (1958), *Introduction to the Theory of Sound Transmission*, McGraw-Hill, New York.

Trorey, A. W. (1970), "A Simple Theory for Seismic Diffractions," *Geophysics*, Vol. 35, No. 5, pp. 762–784.

NINE

Migration—The Correct Placement of Reflectors and Diffractors

9.1 INTRODUCTION

The fiction of plane, near-horizontal layering has served a useful purpose, but we must now return to reality and take nature as it is found—for this is the condition under which the real seismic reflection method works. Three important elements of structural geology have been ignored:

1. The layers may not be horizontal.
2. Discontinuities exist that give rise to diffracted events.
3. The curvatures of the formations are not negligible.

Nevertheless, all processing described to this point has assumed that these effects are absent and that the seismic reflection record has a one-to-one correspondence with a depth display. It is the purpose of this chapter to trace the development of methods of placing the various events on a seismic time section in their proper place on a seismic depth section. Little time is spent on the older methods in which the reflections were picked and dip bars—lines on a new cross section—were computed and plotted. Only passing reference is made to the aplanatic method, although it is really the progenitor of more modern procedures. Some attention is paid to unsolved problems because, with the lithology becoming (at least) two-dimensional, the velocities to each layer are no longer constant. This is a problem, even now.

The advent of new methods was triggered by the arrival of the fast digital computer, laser plotters, and similar devices that can plot two-dimensional digital data of great informational content in a reasonable amount of time.

9.2 EARLY MIGRATION METHODS

Figure 9.1a shows a hypothetical set of seismic traces in which, after correction for near-surface delays and NMO, the reflections increase constantly in time from one trace to its successor. Because of the assumption that NMO has been removed from these seismic traces, it can be taken that they represent traces picked up at the point at which the energy was generated. Under the assumption

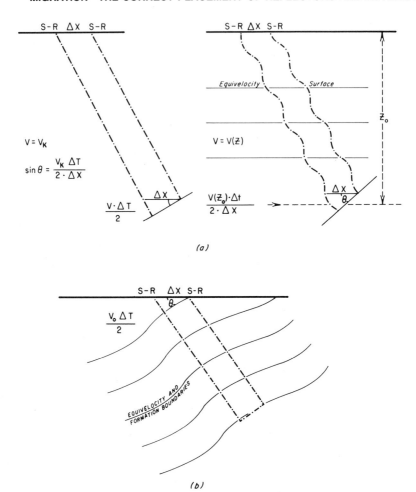

FIGURE 9.1 (*a*) Examples of angle of dip determination when the equal-velocity surfaces are horizontal or if a constant-velocity medium exists. (*b*) Equal velocity and reflector surfaces parallel.

of constant velocity before the reflector is reached, the wave fronts are circles and the rays are straight lines perpendicular to the reflector. Even under a lesser restriction of a velocity that is a function only of depth, the reflected and incident rays must coincide and be perpendicular to the reflector if the source-receiver coincidence is retained. It is then easy to see that the differential time between traces for these assumptions is given by

$$\Delta t = \frac{2 \sin \theta \, \Delta x}{V}$$

or   (9.1)

$$\sin \theta = \frac{V \, \Delta t}{2 \Delta x}$$

The velocity V used here must be the interval velocity at the depth z of the reflector. By Snell's law the angle of emergence is $\sin \theta_0 = (V_0/V) \sin \theta$. By making

V a function of z only the assumption is that the formation type does not control the velocity at all and that it is strictly controlled by depth of burial. This, of course, is nonsense. It would be unusual geology if all the formations down to a particular level were horizontal and had the same velocity for a given depth; this was then followed by a gross divergence from these assumptions.

Another possible assumption is that all the formation velocities are constant and independent of depth (Figure 9.1b), in which case the dip of a a particular reflector would be given by

$$\sin \theta = \frac{V_0 \, \Delta t}{2 \Delta x} \qquad (9.2)$$

where V_0 = near-surface velocity. This is equally ridiculous. The truth evidently lies somewhere in between (Waters, 1941; Faust, 1951). The equal-velocity surfaces are a function of x and z and tend to follow the structure to some extent but are not parallel to it.

Thus we are in the unenviable position that we know that the dipping reflections shown on the time section do not come directly from below the shot point but have to be moved in a direction up-dip. At the same time, we do not know what velocity function should be used to calculate this movement. Because the average velocity is usually intermediate between the surface and interval velocities, most early forms of migration used the average velocity and many newer forms still do. It is possible to grade velocities between CDP or well velocity determinations. Some care is necessary, however, because (unlike the removal of NMO) a change in migration velocity can give rise to structural change and errors of velocity correspond to errors of structure.

It is wise to keep these uncertainties in mind in the remainder of the discussions in this chapter. Methods have changed but the problems have not.

After choosing a velocity depth function (or selected one from a graded set related to velocity determinations) the first method was to pick the records, compute the angle of dip, swing the center of the reflection sequence up-dip by the calculated amount, and draw a dip bar of the appropriate length on the cross section (see Figure 9.2). This was repeated for all reflections that could be picked, most of the time grading the reflections for continuity and indicating this by the

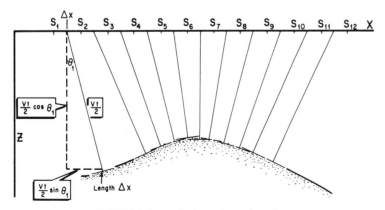

FIGURE 9.2 A dip bar migrated section.

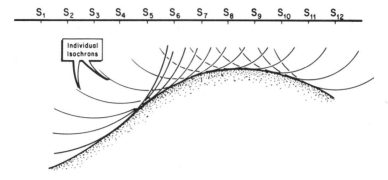

FIGURE 9.3 Development of an aplanatic surface—thus the reflecting surface—from individual isochron surfaces.

quality of the line drawn on the cross section. If the cross section was not to be made with equal-distance scales on the X and Z axes the proper modifications can be made in the angle of dip.

The second method, sometimes adopted, was called the aplanatic surface method. This, again, was a manual construction. If from any surface pair of source-receiver reflection times a set of constant time lines is drawn (Figure 9.3), these will represent all possible positions of a reflecting surface that will receive energy from the source and return it to the receiver at a particular time. The form of these aplanatic surfaces must be consistent with the velocity function assumed to apply; for example, if the source and the receiver are coincident, either a linear increase in velocity with depth or a constant velocity will give rise to a series of concentric spheres (which are cut by the plane of observation to give circles). The height of the center above the source depends on the rate of increase in velocity with depth. If the source and the receiver are separated the isochron (equal-time) lines will be ellipses. When a reflection can be identified and correlated from trace to trace it is possible to draw these isochrons, which correspond to the reflection, at each pair of source-receiver points. In Figure 9.3 they are shown as circles with centers on the reference line at the source-receiver position. The aplanatic surface is the common tangent to all the isochron lines, as shown, and this is the reflecting surface. This construction, of course, makes use of the stationary time principle of Fermat—in this case, minimum time paths. As a manual construction this process is time-consuming but, except where there are strongly concave reflecting surfaces, produces a continuous reflecting curved surface. It is now chiefly of interest as the progenitor of the wave-front migration method.

9.3 MORE DETAILED VELOCITY CONSIDERATIONS

In an outstanding monograph published by the S.E.G. Hubral and Krey (1980) considered the kinematic aspects (i.e., travel times and velocities) of seismic waves that travel through the earth. Although limited to discussions of layers of discrete thicknesses, this work covers in detail most of the questions that are basic to processes of common datum point stacking and migration of nonhorizontal dips into their proper places.

The full considerations are too long and too complex for inclusion here and interested readers are referred for all the details to the original monograph. Here we state only those facts, derived from their mathematical analysis, that appear to be preeminent in the migration process.

First, for one plane horizontal layer the relation between arrival time and offset distance can be written

$$T^2(x) = T^2(0) + \frac{x^2}{V_1^2} \tag{9.3}$$

For the multilayer horizontal case

$$T^2(x) = T^2(0) + C_1 x^2 + C_2 x^4 + C_3 x^6 \cdots \tag{9.4}$$

where C_1, C_2, C_3 are time-weighted, average-velocity quantities along the *vertical* path that results for $x = 0$. In particular

$$C_1 = \frac{1}{V_{(2)}}$$

and

$$V_{(2)} = \left[\frac{2}{T_0} \sum_{i=1}^{N} V_i^2 \, \Delta t_i \right] = V_{\text{RMS}}^2 \tag{9.5}$$

Thus, *to a first approximation only*, the slope of the x^2-T^2 curve gives the square of the root-mean-square velocity. It is seldom necessary to use added terms, but some are given by Hubral and Krey (1980) and Marschall (1975).

It has been shown that for sloping layers a similar expansion can be made in terms of a time-weighted average, not on a vertical path but on a path that is *normal to the relevant reflector*. This path has its origin on the common source-receiver (SR) position and ends at the proper reflector. For all possible positions of SR on the surface there is a *parallel normal ray to that same reflector*. The path above the reflecting horizon is in accord with Snell's laws.

Both the normal ray and another called *the image ray* are shown in Figure 9.4. Now we can see that the conditions are reversed. The image ray meets the surface perpendicularly, can be traced downward through the (supposed known) formation boundaries by Snell's law, and arrives at a point scatterer on the reflector. Reversing the argument but staying with the image ray, we may say that the image ray arrives at the surface at the vertex of the hyperbola (2-D) which describes the scattered event from the scatterer S. In three dimensions, of course, the arrival times at the surface from S form a hyperboloid of revolution, provided only that the aperture (the angular spread of rays starting out from S) is small.

Going back to the CDP case temporarily, we can show that

$$V_{\text{NMO}}^2 = \frac{2V_1 R_0}{t_0 \cos^2 \beta_0} \tag{9.6}$$

where R_0 is the radius of curvature of the wave front from a sloping reflector

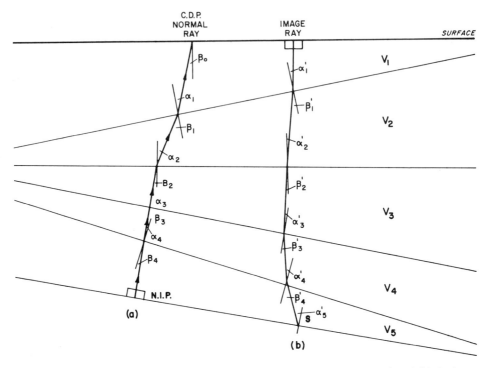

FIGURE 9.4   The construction of (a) the CDP normal ray (normal to the reflector) and (b) the image ray (normal to the surface). NIP is the normal incidence point.

whose normal emerges at an angle of β_0 to the vertical. If this is now generalized to N sloping layers it can be shown that

$$R_0 = \frac{1}{V_1} \sum_{j=1}^{N} V_j^2 \, \Delta t_j \prod_{k=1}^{j-1} \frac{\cos^2 \alpha_k}{\cos^2 \beta_k} \tag{9.7}$$

where Δt_j is the one-way time in the jth layer *along the normal ray*, α_j is the angle of incidence of the ray downward onto the interface, β_j is the corresponding angle of refraction, and V_j is the velocity in the jth layer.

Substituting (9.7) in (9.6),

$$V_{\text{NMO}}^2 = \frac{2}{t_0 \cos^2 \beta_0} \sum_{j=1}^{N} V_j^2 \, \Delta t_j \prod_{k=1}^{j-1} \frac{\cos^2 \alpha_k}{\cos^2 \beta_k} \tag{9.8}$$

we see first that for *plane horizontal layers* this reduces to the square of the RMS velocity.

Second, the deviation from the RMS velocity increases with depth because of the continued product of the squares of $\cos \alpha_k / \cos \beta_k$ down to the layer concerned.

Third, that this formula requires a knowledge of α_k and β_k, which are not easily determined accurately from the seismic data. For smallish angles, however, the ratio $(\cos^2 \alpha_k)/(\cos^2 \beta_k)$ is not sensitive to an exact knowledge of α and β.

Now, returning to the principal subject here—that of time migration, particularly by the Kirchhoff integral method—the sum of the effects of individual scatterers along the hyperboloid of the arrival time—we find that it is necessary to determine the best form of that hyperboloid or, the equivalent, the best migration (*MIG*) velocity.

The image ray now takes the place of the normal ray and we have the formula

$$V^2_{MIG,N} = \frac{1}{T_N} \sum_{j=1}^{N} \prod_{k=0}^{j-1} \frac{\cos^2 \alpha_k}{\cos^2 \beta_k} V_j^2 t_j \tag{9.9}$$

where $T_N = \Sigma_{j=1}^{N} t_j$ is the total travel time *along the image ray* from the diffractor to the surface

t_j = travel time for the jth layer along the image ray path

α_k = angle of incidence of the down-going wave at the kth interface

β_k = angle of refraction on the image ray downward from the kth interface

If now we set

$$\sigma_j = \prod_{k=0}^{j-1} \frac{\cos \alpha_k}{\cos \beta_k} \cdot V_j \tag{9.10}$$

then $V_{MIG,N}$ takes a form similar to that for the R.M.S. velocity

$$V_{MIG,N} = \left[\frac{\Sigma_1^N \sigma_j^2 t_j}{\Sigma_1^N t_j} \right]^{1/2} \tag{9.11}$$

Although theoretically this will give the best migration velocity to use for constant velocity dipping layers, we still cannot deal with lateral velocity variations within each of the constituent layers, dipping or not. The broad considerations of the effect of lithology and compression by burial with which we started Section 9.2 are still valid.

9.4 THE WAVE-FRONT MIGRATION METHOD

A seismic trace is digitized so that every sample time is represented by an amplitude of the measured parameter (usually particle velocity or excess pressure). Now, if we had the time we could draw isochron lines on the cross section for any one of these sample points and for all the seismic traces and then draw the common tangent lines. The following method, which requires a digital computer, results in a migrated seismic cross section.

In Figure 9.5 a single isochron line crosses over a rectangular grid formed by vertical lines midway between the individual traces and by the points midway between the sample points. Other definitions could have been used. As indicated, the cells through which each isochron passes are assigned (on some weighting basis) the amplitude of the seismic trace sample. For more isochrons the amplitudes within the cells are added algebraically.

Additional criteria may be added; for example, not all parts of the isochron line may carry equal weight. It may be decided that dips greater than N_0 ms per

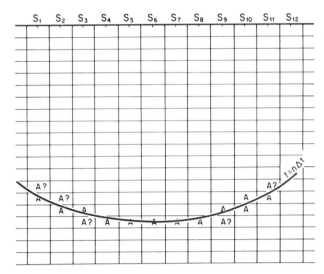

FIGURE 9.5 One isochron line passes through a gridded system. A weighted portion, dependent on the vertical angle, of the amplitude of the digital record sample is assigned to each cell through which the isochron passes. For more than one isochron passing through a cell (for another time sample or source position) the amplitudes assigned are added.

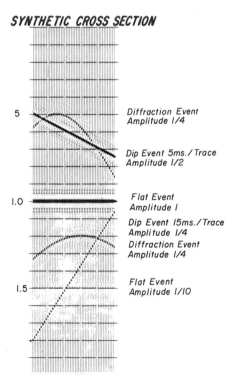

FIGURE 9.6 A synthetic cross section shows various types of seismic arrival.

trace should not be admitted, in which case, the isochron line is discontinued when its slope reaches that dip limit. There is another probability of a dip occurring proportional to cos θ (or some more complex function of vertical angle) and of weighting the amplitude to be assigned to the cells accordingly.

Finally, after all the isochron sample lines have been used in this manner the accumulated amplitudes in the vertical columns of cells are played out as new seismic traces in which the events are migrated. Essentially, this has been done by the aplanatic surface method. The justification for the name *wave front method* lies in the fact that the aplanatic surfaces and the wave fronts (for half the reflection times) are coincident if the sources and receivers are coincident, and this is virtually true if all traces have been properly corrected for NMO.

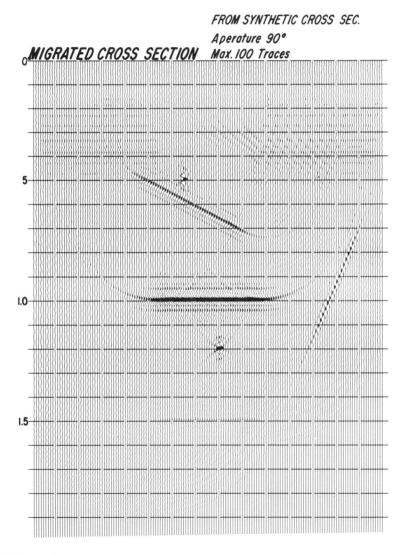

FIGURE 9.7 A migrated cross section by wave-front method from data given in Figure 9.6. (*Source*: B.J. Heath. Courtesy of CONOCO, Inc.)

Figures 9.6 through 9.8 describe some theoretical possibilities, that have the following characteristics:

1. Constant velocity.
2. Events, such as dipping reflectors and diffractions, have been included in the unmigrated data in Figure 9.6.
3. All events have been migrated by using the full half-circle isochrons in Figure 9.7. Note that the hyperbolic diffraction events have been condensed to a minimum area consistent with the band-pass characteristics of the seismic pulse. Amplitudes (except for the diffractions) have been roughly conserved.
4. All events, except the one with the steepest dip, have been migrated in Figure 9.8, in which the aperture for dip migration is limited to ±30°.

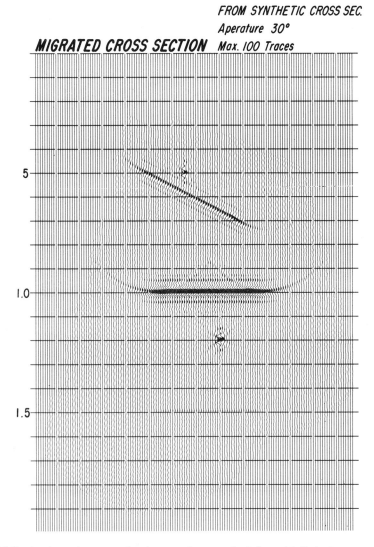

FIGURE 9.8   A migrated cross section by wave-front method (but with limited aperture). (*Source*: B.J. Heath. Courtesy of CONOCO, Inc.)

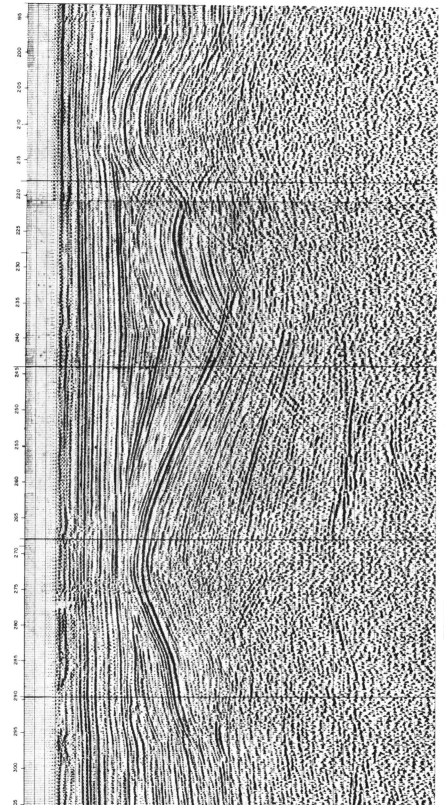

FIGURE 9.9 An original, nonmigrated marine seismic cross section. (*Source:* Courtesy CONOCO, Inc.)

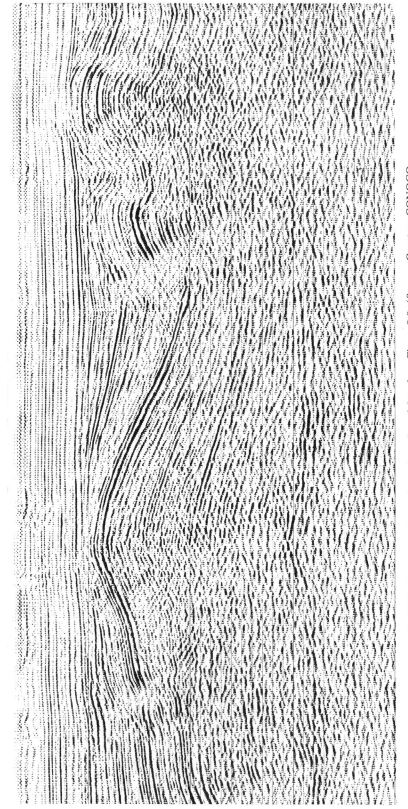

FIGURE 9.10 A wave-front migrated version of data from Figure 9.9. (*Source:* Courtesy CONOCO, Inc.)

Note the characteristic, when migrating a finite length of reflector, of adding an inverse diffraction at the ends of the reflector; this characteristic is especially noticeable in practical examples when seismic noise is present (Figures 9.9 and 9.10). These examples are provided by courtesy of CONOCO, Inc.

Some modifications in the form of editing the original sections before or during the migration process are obviously necessary, particularly when the signal/noise ratio is not large. This editing consists of applying an algorithm that allows only sequences of dips that can be followed over several seismic traces to be picked. Thus large noise amplitudes are not migrated in the form of curved events of large amplitude. The criteria used as a basis for editing are often controversial—as are all attempts to substitute machine picking of events for picking by human interpreters (Paulson and Merdler, 1968)—but even a few simple rules, which can generally be agreed on, help considerably to reduce the overall confusion.

Synthetic results show that, apart from the inverse diffractions, the wave-front method performs well as long as the correct velocity information is used. Because energy from the postulated reflectors and diffractors is spread out all over the cross sectional plane, where it tends to cancel if the number of reflectors is large and randomly spaced, the noise background increases. For this reason the migration method should be attempted only with seismic cross sections that have a high signal/noise ratio. It has been suggested that migration, by any automatic method that produces a migrated seismic section retaining the seismic waveforms on the traces, can be used on the primary records that make up the CDP records. As a result of the multifold stacking now adopted these primary *migrated* cross sections can be added to reinforce the true data and to reduce the noise level. It is not obvious that this will be the result achieved. The 100% records are much noisier than the CDP record (on the basis of random noise and N-fold stacking, they should have a signal/noise ratio of $S_F\sqrt{N}$ if S_F is the final signal/noise ratio achieved after CDP stacking). The stacking of the cross sections may achieve a $\sqrt{N}$ improvement over the 100% cross sections and get back to the situation achieved by migrating the CDP traces. Examples tried have not resulted in any appreciable improvement and the cost is much higher.

Wave-front-migrated sections can be made if the velocity function varies smoothly and laterally; that is, each trace is treated by using an interpolated velocity function. The shape of the isochron lines under these conditions is not symmetrical, however, with respect to some vertical line, and caution is needed in regard to aperture angle used and the rate of lateral change in velocity.

9.5 MIGRATION BY DIFFRACTION STACKING (KIRCHHOFF SUM METHOD)

In what appears to be an almost inverse method diffraction stacking relies on the Kirchhoff integral method for solution of the wave equation. In essence, each portion of the subsurface is regarded as a scatterer of acoustic waves (the simplification of the elastic wave problem is necessary). In Chapter 8 we saw that scattering gives rise to a hyperbolalike curve when plotted on a T-versus-X graph, the exact shape depending on the variation in acoustic velocity with depth. It is therefore reasonable to try to perform the migration by summing up the energy

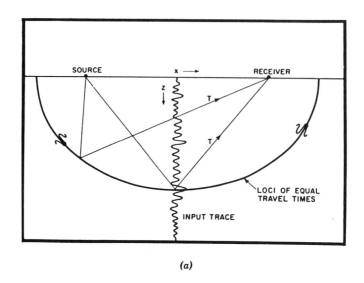

(a)

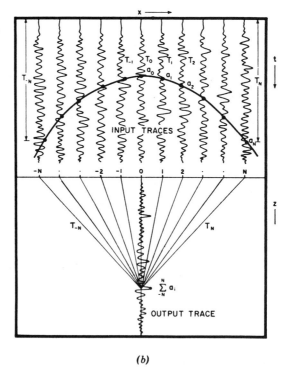

(b)

FIGURE 9.11 Illustrating the difference in philosophy between the wave-front method (*a*) and diffraction curve method (*b*). In (*a*) a single trace amplitude is distributed over the proper wave front. In (*b*) many traces are sampled at times appropriate for a single scatterer and added to give a single trace amplitude. [*Source*: Schneider (1971). Reprinted with permission from *Geophysics*.]

ON SHORE SEISMIC DATA

Unmigrated Time Section

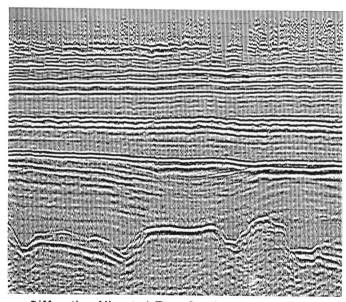

Diffraction Migrated Time Section

FIGURE 9.12 Original unmigrated, on-shore, seismic data and a diffraction-migrated time section. (*Source*: Courtesy of Prakla-Seismos, GmbH.)

that appears on the seismic cross section along these diffraction curves. This summing produces a high-amplitude event at the (T, X) position of the scatterer. On the basis of Huyghens' point of view that a reflector can be treated as a linearly arranged set of scatterers, each of which gives its output independently of the others, we see that the procedure of summing along the appropriate diffraction curve should work as well for reflectors.

The two different processes can be compared on the diagrams (Figure 9.11) due to Schneider (1971). The proper diffraction curves must be calculated for each sample time in the vertical ray path and the proper samples to select at offset traces must be determined. For a generalized velocity function this can be done by the method outlined in Appendix 9A. Examples of diffraction curves for a simple velocity depth function are given in Figure 9A.1.

When velocity functions are measured at different points along the line of section the velocities can be interpolated and the diffraction curves, calculated for migration use at each source point. In the process of stacking by this method it is evident that for events received on traces laterally removed from the scatterer the energy is reduced because of the spherical divergence of the wave front (in true reflection). For real scatterers that are lines the appropriate divergence rate is more nearly cylindrical, although the initial divergence from the source point to the line scatterer is spherical. It is therefore preferable to prepare true-amplitude cross sections before migration, and, in fact, it is realistic to weight the amplitude of each trace according to a scheme that gives preference to traces near the vertex of the curve. Weighting inversely as the arrival time may be suitable but, as in wave-front migration, the aperture must be limited to avoid an increase in noise in a low-signal area.

Figure 9.12 is an excellent example of a time section and a migrated time section for some onshore data (courtesy of Prakla-Seismos, Gmbh).

9.6 FINITE DIFFERENCE METHODS BASED ON THE ACOUSTIC WAVE EQUATION

In Chapter 8, when considering scattering and diffraction, we saw that complex effects are often introduced into the time cross section because of the various paths in the plane of the depth section that can arrive on a particular trace; for example, a syncline that has too great a curvature gives rise to three events, all from the same reflector and, in general, at different times. This situation always occurs when the radius of curvature of the (concave) structure is less than the radius of curvature of the wave impinging on it—that is, the depth.

The situation would be much better if the surface of observation were down in the earth far enough for the radius of curvature of the subsurface to be less than the depth (for a particular reflector). The best possible situation (from an academic standpoint) would be when the plane of observation is *coincident* with the reflector or scatterer in question. The distance of travel would then be small enough that there would be no migration involved. It is the purpose of this section to show how this can be done by a method related to the wave equation. Most of this work was done by Jon Claerbout and his associates at Stanford University, whose aim was to produce practical methods of data processing of use to exploration seismology. Because the formal solution of the acoustic wave equation in inhomogeneous media is generally an impossibility, resort has been made to numerical methods, in particular, that are related to methods of finite differences; that is, the solution is a finite array of values in which the rows represent the x position of the seismic traces and the columns are the record time t just as if the traces represented depth and had been converted to time by using the known velocity function. The starting point can be a CDP seismic cross section, an array of the same kind with columns at every observation point, and event amplitudes as a function of observation time. The observation time, of course, consists of the distance traveled by the wave, divided by an appropriate velocity over the path taken, which need not be, and in general is not, vertical.

The phrase *downward continuation* has been used for many years in gravity and magnetics interpretation to denote a method by which the observed fields at the surface can be manipulated to generate the field as it would appear if the observation were at a lower level, thus giving more detail and resolution. It is a method, however, dealing with time-invariant potential fields and is much simpler to visualize than the present method. It amounts to a boost of high-frequency components of the signal.

We can imagine the process in a different way that is fundamental to the downward-continuation process for acoustic waves. Consider that each point on a subsurface inhomogeneity is a source, in the manner of Huyghens, and let these sources radiate upward but with only half the velocity normally associated with the geological section. Then, apart from considerations of the directionality of these secondary sources, the ensuing signals received at the surface are close to those that would have been generated by a simultaneous set of sources at the observation points $(S/R)_1$, $(S/R)_2$, and so on. Thus only the up-going wave of the solution to the wave equation needs to be dealt with—using one-half the normal velocities.

Now if it were possible for the observer to travel upward with the wave an even greater simplification would result, for then the wave field would change only slowly with time. This can be done in the mathematical formulation of the basis for migration by changing the coordinate axes to reference axes that move upward with the wave field.

If we take two columns of amplitudes it is possible by using the known physical laws embodied in the wave equation to predict from any three elements (two in one column and another in the second column) the size of the fourth adjacent element in the second column. To do this the wave equation has to be put in a suitable form—a finite difference form that relates the amplitudes of nearby elements to those already known. Although we have discussed this as though we were going to do it one cell at a time, it has been found possible for grosser approximations (e.g., all dips limited to 15°) to proceed by larger steps. The step consists of convolving the existing data with a calculated pulse or transfer function.

In the first step transformation from record time t to migrated time τ is done down to a depth Z_1; this is followed by a similar step down to Z_2 and so on. These steps may be equal or unequal, but the first is usual for reasons of economy of computer effort and simplicity. The transfer function appropriate to the transformation can then be standardized for a given depth. Because this is a finite difference approximation to the wave equation, several errors will be made that depend on the size of the steps, the range of dips of the reflectors, the variability of the velocity function, and so on. These errors have been investigated by Claerbout (1971), Alford, Kelly, and Boore (1974), and Loewenthal, Roberson, and Sherwood (1976).

After the application of each transfer function the section is migrated fully down to $\Sigma\, Z_i$ (the sum of the depths of the range of the transfer functions) but only partly migrated (transformed from t to τ) below this depth.

Claerbout and Johnson have given the (phase) transfer function across a layer of thickness Δz as

$$\phi(Z) = \frac{1 - [aT(1 + Z)/(1 - Z)]}{1 + [aT(1 + Z)/(1 - Z)]} \tag{9.12}$$

where $a = (C\,\Delta t\,\Delta Z/8(\Delta X)^2)$
    $Z =$ transform operator $\exp(-2\pi i f\,\Delta t)$

Because only shifting of seismic energy is involved in the migration process, the operator has only the *phase* of the signal and the amplitude is left alone; $T/(\Delta X)^2$ represents some finite difference approximation to the differential operator $-(\partial^2/\partial x^2)$. This may be the simple double difference operator D that corresponds to the three-point convolver $(-1, 2, -1)$ or it may be a better approximation such as (Loewenthal et al., 1976):

$$T = \frac{D}{1 - D/6} = D + \frac{D^2}{6} + \frac{D^3}{36} + \cdots \tag{9.13}$$

A full justification of the transfer function (9.12) is not within the scope of this chapter. Reference may be made to the original works of Claerbout and Doherty for more details.

Although the finite difference method is in use by several companies, its efficacy compared with the diffraction stacking method is open to question. It is probably a question (as, in fact, are many decisions in exploration seismology) of the combination of data from a particular area, the original signal/noise ratio, the personal preference of the interpreter, and the economics of the process. Examples given (courtesy of Western Geophysical Company) of some offshore California seismic data (Figure 9.13) show little difference except the noise produced by the Kirchhoff sum method above the water bottom.

Nevertheless, research in migration methods has continued to develop new and more satisfactory methods. Stolt (1978), Hood (1978), and Bolondi et al. (1978) have shown how many of the problems of dispersion and different approximations (to the wave equation) to deal with different degrees of dip may be avoided by transforming the entire problem into the two-dimensional, frequency-wave-number domain. This transformation allows a great increase in the speed of processing because two successive fast Fourier transforms (from time to frequency, then from distance to wave number) quickly bring the initial field data into a form in which the migration procedure is nothing but a shift of frequency plus a change of scale. The inverse fast Fourier transforms then bring the data back to the migrated cross section. An outline of the entire procedure is given in Appendix 9B.

CALIFORNIA OFFSHORE DATA

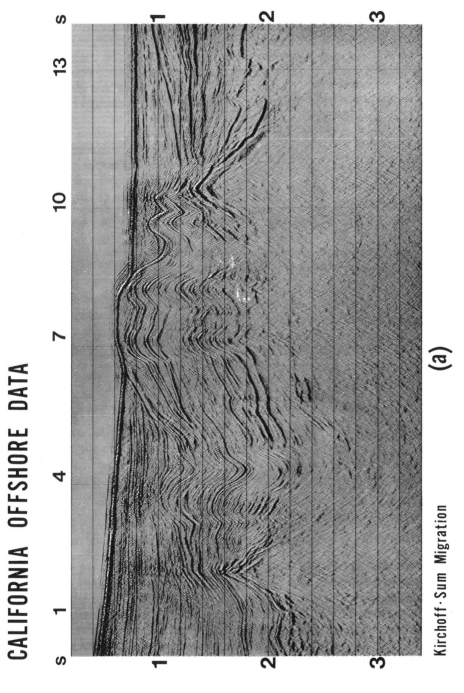

Kirchoff·Sum Migration

(a)

FIGURE 9.13 (a) Shows a Kirchhoff-Sum migration of some California offshore data.

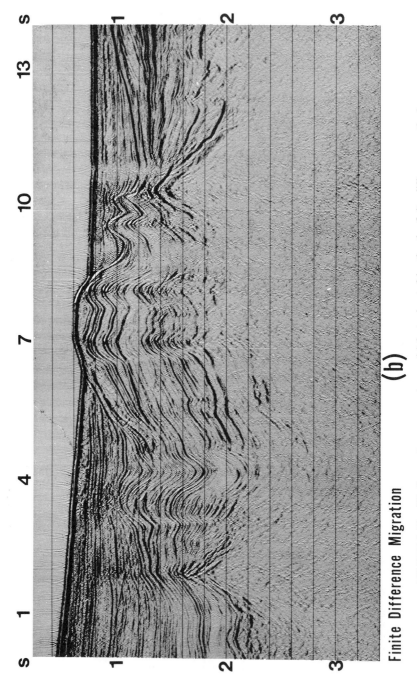

Finite Difference Migration

FIGURE 9.13 (*b*) Shows the same section as (*a*) but migrated by the the finite difference method. The results of methods (*a*) and (*b*) are similar. All dips are less than 25°. (*Source:* © 1975, Western Geophysical Company of America.)

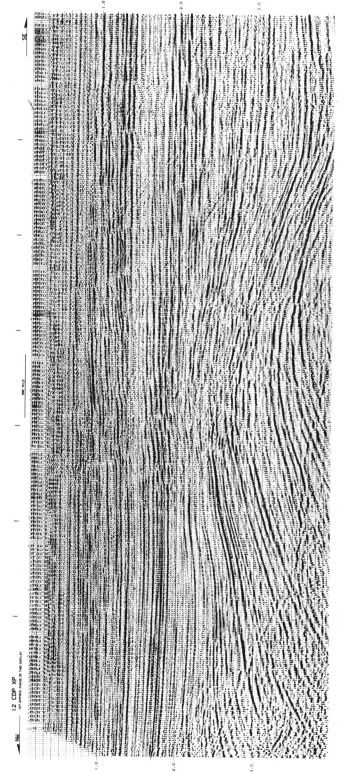

FIGURE 9.14 Original stacked offshore Texas seismic line. (*Source:* Courtesy of Geophysical Service, Inc.)

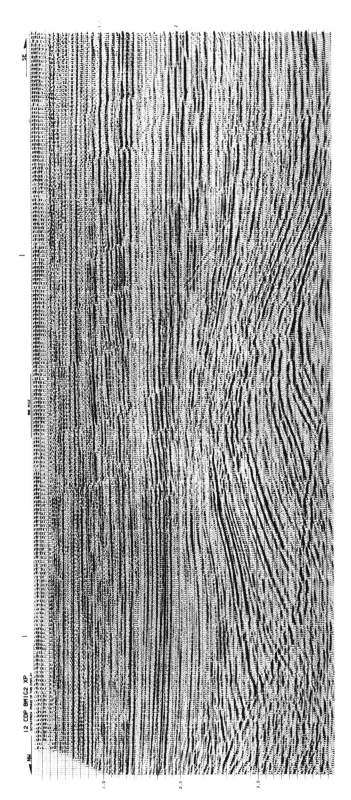

FIGURE 9.15 A cross section migrated by the *f-k* method from data shown in Figure 9.14. (*Source:* Courtesy of CONOCO, Inc.)

It is even possible to migrate the original unstacked data; the process of stacking is an integral part of the procedure and the final product is a migrated stacked section.

As an example of an $f - k$ migration Figures 9.14 and 9.15 show, respectively, a stacked conventional section and a stacked migrated section made by the frequency-wave-number method.

Finally, it is almost trivial to extend this method to three dimensions, because this involves only a three-dimensional FFT, shifting and scaling in the frequency domain, and a three-dimensional inverse transform.

Dip Move-out Processes

Most migration processes are applied to stacked data on the assumption that the result of stacking offset traces from which NMO has been properly removed is equivalent to the use of zero-offset traces. The assumption is sufficiently valid for areas in which the dip is small.

Levin has shown that the RMS velocity (V_{RMS}) in a dipping situation is related to the *true* velocity (V) by the equation

$$V_{RMS} = \frac{V}{\cos \theta} \tag{9.14}$$

where θ is the component of dip along the traverse on which V_{RMS} is determined.

If it is required that an intermediate stacked section be obtained before a migrated section a number of processes have been described that will make the correction for the dip. Among them are Satlegger (1975), Dohr and Stiller (1975) and Yilmaz and Claerbout (1980).

A new process of dip move out by Fourier transform has been published by Hale (1984). If $p(t, y, h)$ denotes the seismic trace initially recorded as a function of true time (t), midpoint coordinate (y), and half-offset (h) the zero-offset time t_0 can be written

$$t_0 = \frac{2}{V} (y - y_i) \sin \theta \tag{9.15}$$

and the true time

$$t = \left(t_0^2 + \frac{4h^2 \cos^2 \theta}{V^2} \right)^{1/2} \tag{9.16}$$

which is the move-out equation corrected for dip. The normal move-out time (t_n), not corrected for dip, is

$$t_n = \left(t_0^2 - \frac{4h^2 \sin^2 \theta}{V^2} \right)^{1/2} \tag{9.17}$$

For any particular angle of dip (θ) it is evident that

$$\frac{\Delta t_0}{\Delta y} = \frac{2 \sin \theta}{V} = \frac{k}{\omega_0} \tag{9.18}$$

By arguments that are too extensive to be treated fully here Hale (1984) found that the entire process of dip move-out correction can be condensed into a step-by-step process, starting from the non-NMO corrected data $p_n(t_n, y, h)$. First, a Fourier analysis (on y) is performed to change the data to the k domain. This is followed by a numerical integration with respect to the time variable (t_n) for all ω_0 and k which involves the calculation of a quantity A (a function of dip, time, and offset):

$$A = \left(1 + \frac{k^2 h^2}{\omega_0^2 t_n^2}\right)^{1/2} \tag{9.19}$$

The final step is to take the frequency-wave-number data which have been corrected for dip and use a 2-D inverse FFT to transform back to the zero-offset time, space domain

$$p_0(t_0, y, h) = \frac{1}{4\pi^2} \int d\omega_0 \, e^{-i\omega_0 t_0} \int dk \, e^{iky} \cdot P_0(\omega_0, k, h) \tag{9.20}$$

It should be noted that $P_0(\omega_0, k, h)$ need be computed only for $k/\omega_0 \leqslant 2/V$ to avoid imaginary angles.

The data should be presorted into sets of constant offset (h) for ease of one of the transforms. Finally, at the end of the dip move-out correction procedure the data can be reshuffled into CMP sets ready for stacking.

The reader who wishes to apply this method practically can find all the necessary details in Hale (1984) in which excellent *with* and *without* dip move-out examples are shown. There is little doubt that the method is both theoretically sound and a practical success, but the fact remains that this prestack process is computer-intensive and quite costly. Therefore careful justification is needed.

9.7 DEPTH MIGRATION

The *time* migration methods we have described work reasonably well, provided that there is little lateral change in velocity. Small lateral velocity gradients can be accommodated in the time migration methods by changing the migration velocity slowly as progress is made along the traverse. We have already noted that there is a theory available to allow change of velocity with depth by an additional change of variables (in the *f-k* method).

For fast lateral velocity changes, however, proceeding in this manner lead to considerable structural distortion. To prevent this distortion a method of *depth* migration was introduced by Schultz and Sherwood (1978) and Larner et al. (1981).

The method derives from Hubral's observation that the apex of the diffraction hyperbola occurs at the position (on the recording surface or equivalent) of a ray from the diffracting point that reaches the surface vertically. Figure 9.16 makes this clear.

Arguing in the opposite direction, the diffractor (P) must be on a ray that departs from the minimum time-point on the surface, moves vertically downward, and is refracted according to Snell's law at each of the boundaries. This is the

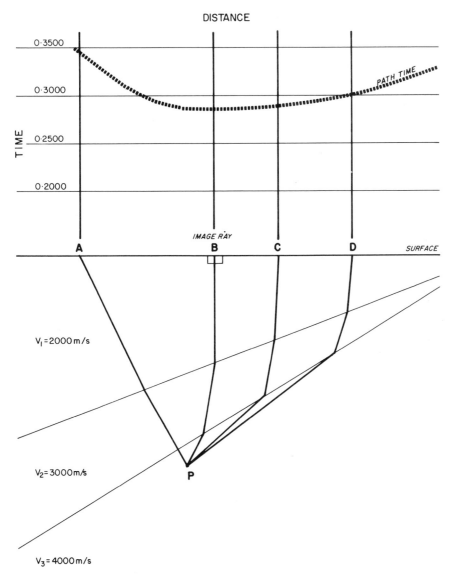

FIGURE 9.16 An illustration of the minimum time property of the image ray from a scatterer P in a medium with dipping interfaces.

image ray described earlier. The steps in making the depth migration are the following:

1. Make a conventional wave-equation migration in time. This will collapse diffractions and move events to the image location at least approximately. For this step it is claimed that ordinary velocity estimates are sufficient.
2. Draw an interval velocity model on this migrated section. It is recommended that the velocities be chosen from a geologically plausible model and that the velocities be extrapolated from well log data or velocity analyses in a nearby uncomplicated part of the basin.

3. An inverse-ray tracing procedure is used in which the image ray starts vertically downward. The depth model is computed by observed (migrated) travel times and interval velocities. The spatial dip of the interface is computed from a time gradient observed for the horizon. Snell's law then governs the refraction of the ray into the next layer, the interval velocity of which is known. This process continues for all required horizons. On a depth section plot the lateral deflections of the image rays from the vertical give an estimate of events observed on the conventional migrated section.

4. Finally, as indicated in Figure 9.17, the amplitudes may be plotted along the computed image rays.

It is quite evident that this process relies heavily on intervention from the geophysicist, which is quite different from the automated time migrations described earlier. It could therefore be claimed quite legitimately to be one phase of semiautomated interpretation. As such, the description of the process should possibly have been left until later. The purpose of the discussion here is to indicate that a means has been found to tidy up *some* loose ends that are left after all the *standard* migration processes have been fully and accurately used.

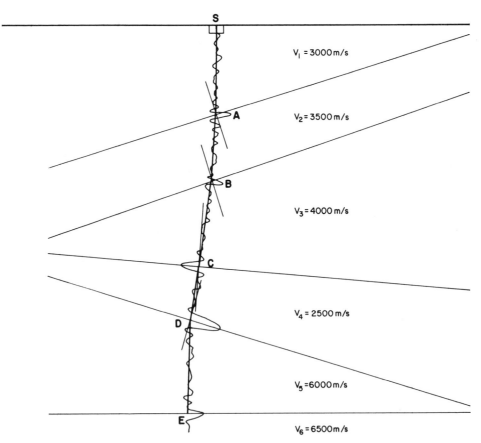

FIGURE 9.17 Plotting a migrated trace along the image ray in a medium with dipping interfaces.

It has now been found possible to do wave migration in depth. This involves changing the partial differential equations that are the basis for finite difference or *f-k* migration. The theory is difficult and is not pursued here. Those readers who are interested will find the details, as well as some examples, in Hatton et al. (1981).

9.8 THREE-DIMENSIONAL MIGRATION AND ACOUSTIC HOLOGRAPHY

All the methods discussed so far have been two-dimensional. They can, of course, be adapted to three-dimensional migration except for two problems:

1. The data are not usually available on a continuous basis for giving the extra experimental information necessary.
2. The extension of these two-dimensional programs to three-dimensional would increase the computer time by (at least) an order of magnitude and may increase the storage necessary to uneconomical limits.

Thus, although we may wish for three-dimensional migration, it is unlikely to be feasible except in special cases. It is, desirable, however, to know how its lack can affect the appearance of migrated data in a two-dimensional presentation. We deal with this subject first before touching on the subject of acoustic holography, which, for special purposes, may have some use in defining three-dimensional objects with two spatial dimensions and one time dimension.

French (1974) has given some (model) examples that show the influence of out-of-plane structures on the seismic results obtained by a single line profile. Two-dimensional migration is not the answer to all these problems. Essentially, the difficulty lies in the fact that the three-dimensional reflecting surface is, in a sense, concave laterally; that is to say, legitimate reflections are possible from points on the subsurface that are laterally offset from the vertical plane through the line of source and receiver points. If the data are available (e.g., see the three-dimensional swath method described in Section 5.5) the common tangent to ellipsoids or spheres takes the place of that to ellipses and circles or the common surface generated by a set of near hyperboloids of revolution (about the vertical axis) takes the place of the common plane generated by a set of near hyperbolas. The principles are not different—only the difficulty of organization of the data and the time required to place the contributions of each (x, y, t) cell into the appropriate (x, y, z) cell. Three-dimensional programs are now in use by major oil and geophysical contracting companies.

The term *holography* originated with Gabor in 1948. It was proposed as a two-step method of optical imagery (see Born and Wolf, 1959, p. 452; Gabor, 1948, 1949, 1951) in which the first step is to illuminate the object and a recording plate with coherent light (or electron waves) of a single frequency. The object scatters the electromagnetic energy falling on it and the photographic plate records the scattered energy as it interferes with the primary energy. This diffraction pattern is called a hologram, the process, holography. If the plate, suitably processed, is illuminated by the background wave alone the wave transmitted through the plate contains information about the original object, and with the proper optical system an image that appears to be three-dimensional is seen.

First, we note the requirement of this original holographic method that the illumination be single-frequency and coherent and that the *phase* be preserved in the form of an interference map in a particular plane. The object itself is extremely large compared with the wavelength of the illumination. It was evident in our discussion of resolution that the problems of resolution for light and seismic waves are quite different. In the seismic case the wavelength and the objects to be measured may easily be the same size; in addition, it is possible to measure the phase of the seismic signal directly. This is done by recording the source and geophone output signals on the same recording medium and, by comparison, the phase difference can be recovered from the recorded data. This is different from the optical case because the photographic plate records only amplitudes (or intensities).

Acoustic holography refers to a method of using high-frequency acoustic waves (usually for medical purposes) for investigating abnormalities in an otherwise homogeneous medium. If the medium has an acoustic velocity like that of water (1500 m/s) the wavelengths can be subcentimeter, therefore usually smaller than the object sought.

A further difference, when seismic methods are considered, lies in the type of object. In the optical and acoustic cases these are three-dimensional objects, whereas in the seismic case the holographic method has to deal with smooth sheets of material that are almost two-dimensional and, in terms of the wavelengths used (of the order of 150 m), mirrorlike on their surfaces. Under these conditions an image of the source is seen in the mirror—distorted, of course, by the curvature of the mirror. To see such an image is not desired. The requisite output of any system is to *see* the structure of the rock layers.

Finally, the seismic case is characterized by a velocity function in the earth that in comparison with other holographic media is highly variable. In near-surface layers, for example, the velocity often varies by factors of 4, and irregular thicknesses give rise to phase noise that may be many times that being sought to elucidate the structure of subsurface objects. The layered structure and the number of thin layers make the reverberation problem much more difficult.

Having cited some of the difficulties, we shall nevertheless see how a holographic method can be used and how it compares with existing methods. In Figure 9.18a the general layout of the system is shown. A source S of monochromatic elastic (P) waves is assumed to be present, without generating surface waves or shear waves. Over a large area, compared with the wavelength of the seismic waves, a set of receivers (R) is set out. Their output is recorded, along with the initial signal generated by S. With a vibrator these recording characteristics are easy to obtain, but providing a P-wave-only generator may be extremely difficult (see notes on the output from a surface source in Section 3.5). A source array and receiver arrays may be needed before the signals received are essentially P wave.

The surface array of geophones receives part of the output from a scatterer O on the object, and for a constant-velocity medium with no weathered layer (Figure 9.18b) the output constitutes a circular pattern of constant-phase lines that are unique to the scatterer O (since the center is at O', vertically over O, and

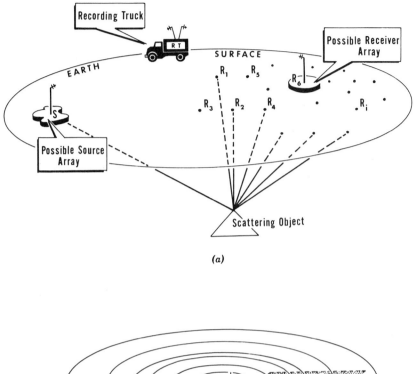

(a)

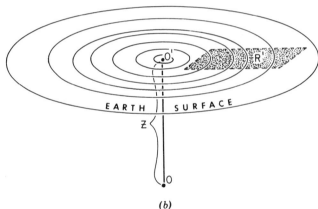

(b)

FIGURE 9.18 (*a*) A possible field setup for a holographic method, using seismic waves. Note that this is a monochromatic, two-dimensional surface version of standard VIBROSEIS practice. (*b*) A series of constant-phase lines on the surface due to a scatterer *O* below the surface. Note that theoretically the information within a limited portion of the surface *R* should be sufficient to establish the location of *O*.

the rate of change in phase with distance from O' is unique for the depth Z of the scatterer). Thus, in theory at least, a correlation of the output pattern received by the set of receivers with a set of theoretical patterns (similar to the Fresnel zone plate in optics) establishes the position of O in the earth. A set of correlations made for three different levels by Chapman (1967, personal communication) reproduced in Figure 9.19 (courtesy of CONOCO, Inc.). It is shown that the overall response at a point in the earth reaches a peak and diminishes, the peak of the envelope being at the proper depth. At three different levels, the proper one

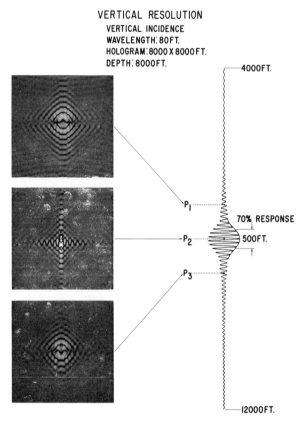

VERTICAL RESOLUTION
VERTICAL INCIDENCE
WAVELENGTH: 80 FT.
HOLOGRAM: 8000 X 8000 FT.
DEPTH: 8000 FT.

4000 FT.

P_1

70% RESPONSE

P_2 — 500 FT.

P_3

12000 FT.

FIGURE 9.19 Vertical and horizontal resolution achieved with a single-frequency (100-Hz) signal at three different levels (*left*). On the right is the continuous output down the vertical axis. (*Source*: Chapman. Courtesy of CONOCO, Inc.)

and ±10%, the response over a horizontal plane has been computed to show the lateral resolution. The example is somewhat extreme for a field case because it corresponds to a 100-Hz wave investigated without aliasing over an area equal to the depth squared. For proper detection without aliasing 400×400 receivers would be necessary. In principle, the more of the amplitude and phase pattern intercepted by the set R of receivers, the more exactly Z can be determined. It is possible, however, to derive an approximate value of Z from an array R' that only partly intercepts the output of O. Let us examine the case R', which is a line of receivers along a radius from O'. The phase and amplitude can be measured and, provided that the proper phase line can be established from one receiver to the next, the diffraction method or the wave-front method of migration can be used, assuming different values of Z (t_0, time vertically to the surface) until the proper value of t_0 is found when all the indicators cross at a point. In the Vibroseis method a continual change in frequency with respect to time allows this method to become more positive. After correlation with the sweep signal, which is recorded on the same recording medium, the events accepted by the geophones R becomes pulselike, their times are uniquely available, and the process becomes the standard migration process. Deviations in time of arrival due to the presence of the weathered layer give, in both holographic and direct migrated Vibroseis methods, uncertainty with regard to the position of the scatterer O.

Now, going back to Figure 9.18a, if there is a multitude of scatterers like O all of them will be found because they have different centers O' and different diffraction patterns. In the monochromatic case the resolution is limited in a vertical and lateral sense by the size of the array (aperture) and by the wavelength of the illuminating signal.

Chapman (1967) also investigated the depth resolution that can be obtained by using a series of frequencies (50 to 100 Hz), and this is illustrated in Figure 9.20. The ghosts on the trace labeled "Sum, $\lambda = 80'$ to $180'''$" are caused by the limited(?) aperture, but it is shown that a much greater definition of the proper depth is obtained. This definition would be still further improved by widening the swept frequency band on the low side—as is the normal practice in Vibroseis work. The ghosts may be eliminated, or at least reduced, by tapering the array.

It is obvious that with an array of the magnitude described the noise level is considerably reduced and the effects of phase noise caused by variable near-surface residual errors tend to cancel. however, they reduce the amount of usable high-frequency component.

Other methods advocated for making digital reconstruction of the object follow the lines of the optical theory or are called the method of conjugate phase. The results achieved differ little from those of the correlation procedure, except that they appear to be consistent with the use of absolute amplitudes. The envelope of

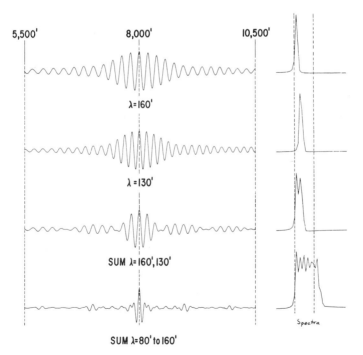

FIGURE 9.20 The increase in definition obtained at the 8000-ft level in Figure 9.19 by summing different wavelengths (80 to 160 ft) that correspond to frequencies of 50 to 100 Hz. The corresponding spectra are shown on the right. (*Source:* Chapman. Courtesy of CONOCO, Inc.)

definition is, however, little changed. All methods make use of the Huyghens–Kirchhoff diffraction theory to work backward from the diffracted field on a given surface to the causative scatterer within the medium.

We have seen that seismic holographic methods, when extended to finite bandwidth signals, achieve the same results by the same diffraction methods as those of normal migration. Any apparent advantages accrue from increasing the size of the reception array and making it two-dimensional. These factors can, of course, be enhanced in normal seismic methods, and in the seismic case holography is another name for reflection seismology with adequate migration.

9.9 SUMMARY AND CONCLUSIONS

It is a mistake to regard a seismic reflection cross section as a one-to-one display of the depths to the reflecting horizons. This is obvious in some cases that display steeply dipping events because continuous reflection (diffraction) events cross over one another—an impossibility for solid rock layers.

Subjective methods that consist of reflection picking and individual dip migrations according to a simple velocity function have now been replaced by automatic, digital computer-performed procedures that do not need the picking of correlated events from trace to trace. In many cases, nevertheless, this reflection picking helps to achieve cross-section simplification, but its efficacy is questionable unless all the criteria of computer picking are agreed on and their effects fully understood.

The computer methods available are of three general types. First, the wavefront migration method takes the energy at a particular time on a given input trace and distributes it in space (two dimensions) according to the constant-time curves that apply between the source and receiver. These are near circles for coincident source and receiver (simulated by stacked traces) or near ellipses for offset sources and receivers. The entire process consists of summing all energy contributions moved into elemental sections of space (usually rectangular elements with the dimensions of sample time rate and trace distance separation).

Second, it is possible to look at the process as one in which the proper elemental energies of a series of traces are summed to give the energy associated with a scatterer in the migrated section space. This addition has to be done with selection times that correspond to the proper diffraction curves. This Kirchhoff sum or diffraction scattering method is applied to every time element on every trace; the result is the migrated section. It makes use of the Kirchhoff-Huyghens diffraction hypothesis that every element of a reflector acts as a separate scatterer.

Jon Claerbout and his associates have evolved a method that relies on the use of a difference equation (based on the acoustic wave equation) to continue the surface-observed field data downward. Various approximations are possible, depending on the amount of dip to be dealt with. It involves a mapping from reflection time domain to migrated time domain.

The final methods, to date, are based on a transformation from the time–distance domain to the frequency-wave-number domain. This is followed by modification of the kernel by a frequency shift downward and a change of scale. An inverse two-dimensional transform brings the data back to the time–distance domain. In this f-k domain treatment the extension to three dimensions is easily accomplished, and it is even possible to migrate the raw reflection data before *stack*, finishing with a stacked, migrated section. Lack of dispersion and relatively low cost of processing have now made this approach popular.

It must be emphasized, however, that seismic records always contain energy that is not generated or scattered from the vertical plane of the seismic cross section. In marine shooting this is probably more evident because the water layer is an excellent transmission medium and the seismic cables used have no means of discriminating between events that come from the side and those that come from below. The presence of the water surface, which introduces an image of the scattering point within the water layer, does, however, cause the overall response of the cable to be different for near-horizontal events compared with near-vertical events. The specific difference is hard to define because it is related to the size of the scatterer, its depth, the roughness of the sea surface, the turbidity or aeration of the water, and many other factors. It is the purpose of this note to warn the interpreter of the need for alertness when unusual events are present on the seismic cross section. Not all should be migrated, and if a specific cause can be ascertained the adulterating noise should be removed before migration.

In marine exploration during the last few years attempts have been made to minimize to some extent those diffracted events in the water layer. The source now consists of several units, spread out laterally by the use of paravanes, otter boards, or other devices. The overall lateral coverage is usually limited to 150 to 200 m and the number of units is small, usually four or less. Under these conditions the cancellation of events from laterally disposed diffractors is determined by the array directivity, the frequency spectrum of the event, and the location of the diffractor in relation to this source array.

Seismic holography has been brought up as a possible, distinct method of producing three-dimensional images of the subsurface. On careful examination, it is found to rely on the formation of a scattering function distribution in the earth by operating on the diffraction pattern measured at the surface. Because the phase differences in seismic work can be measured directly, there is no need for the artifice of mixing in the original signal and the production of a photographic plate, followed by illumination by a coherent light source—as in optics. Rather the amplitudes and phases of the samples, spread over a wide area, can be used to reconstruct the image digitally by applying the principles of the Huyghens-Kirchhoff diffraction theory. The definition of the subsurface scatterer is inadequate for even the highest single frequency used, and resort must be made to a series (or a continuum) of frequencies. This brings the method back to standard seismology by Vibroseis methods if the results are treated by migration subsequent to recording.

APPENDIX 9A: CALCULATION OF DIFFRACTION CURVES FOR A GIVEN VELOCITY DEPTH FUNCTION

We assume that the interval velocity is known as a function of depth. A special case considered later is $V(z) = V_0(1 + kz)$.

The time of travel from S to D to R (R and D coincident) is (see Figure 9A.1)

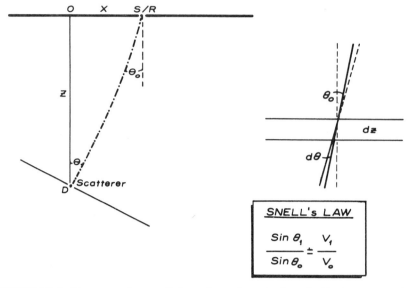

FIGURE 9A.1 The assumed ray path from the scatterer to the coincident source receiver.

$$T = \int_0^z \frac{ds}{V(z)} \quad \text{where} \quad ds = \frac{dz}{\cos \theta} \quad \text{and} \quad dx = dz \tan \theta$$

$$= \int_0^z \frac{dz}{V(z) \cos \theta} \tag{9A.1}$$

Similarly,

$$X = \int_0^z dz \tan \theta \tag{9A.2}$$

These two parametric equations, together with Snell's law,

$$\frac{\sin \theta(z)}{V(z)} = \frac{\sin \theta_0}{V_0} = p \tag{9A.3}$$

determine the T, X curve.

Since $\cos \theta = (1 - \sin^2 \theta)^{1/2} = (1 - p^2 V^2)^{1/2}$

$$V = V(z)$$

$$T = \int_0^z \frac{dz}{V(1 - p^2 V^2)^{1/2}} \tag{9A.4}$$

$$X = \int_0^z \frac{pV \, dz}{(1 - p^2 V^2)^{1/2}}$$

Usually these equations are solved by calculating (in the digital computer) the values of T and X for various values of $\theta_0 = n\,\Delta\theta_0$ and then interpolating to give values of T for equal increments of X.

For the relatively simple case of

$$V_z = V_0(1 + kz) \tag{9A.5}$$

it can be shown that

$$T = \frac{1}{V_0 K} \ln\left(\frac{V_z(1 + \cos\theta_0)}{V_0(1 + \cos\theta_z)}\right)$$

$$\tag{9A.6}$$

$$X = \frac{1}{K \sin\theta_0}\,(\cos\theta_0 - \cos\theta_z)$$

where V_z is the velocity that corresponds to the particular depth chosen and

$$\cos\theta_z = \sqrt{1 - \sin^2\theta_z} = \sqrt{1 - \left(\frac{V_z}{V_0}\right)^2 \sin^2\theta_0} \tag{9A.7}$$

By taking $\theta_0 = n\,\Delta\theta$ the values of T_n and X_n can be determined and plotted

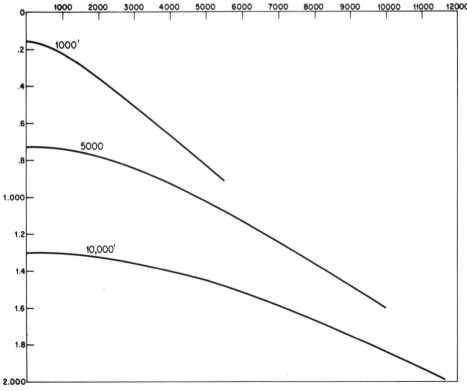

FIGURE 9A.2 Diffraction curves for particular vertical depths for the interval velocity function $V_z = 6000(1 + 0.0006z)$.

against one another. The curves shown in Figure 9A.2 are calculated for the velocity function

$$V_z = 6000(1 + 0.00006z)$$

and are for vertical depths of 1000, 5000, and 10,000, respectively. These values can be in feet or meters.

APPENDIX 9B: FUNDAMENTALS OF MIGRATION USING THE ACOUSTIC WAVE EQUATION

Although it will be necessary for the reader to go back to original papers to get all the details for implementing a seismic migration program, it is believed to be worthwhile to attempt to provide a synopsis of the essential physical and mathematical concepts of those methods that rely on the use of the acoustic wave equation. Emphasis should be placed on the word *acoustic*, for there are still no methods that attempt to take into account the much more complex problems associated with an *elastic* medium. We suppose that the medium through which the waves will travel is simply endowed with a single, slowly varying acoustic velocity (in most of the work it is assumed to be a constant). An exception is made, however, in a few localities that have a sufficient change of velocity or density, or both, to give them the property of scattering the waves. If there were a sufficient number of these scatterers and they were properly arranged, the formation of a reflector would be possible. It can be assumed that the overall velocity is $C(x, z)$ and the reflectivity, $R(x, z)$ (assumed omnidirectional) in terms of the original two-dimensional Cartesian coordinates. Multiple reflections are conveniently ignored (because $R \ll 1$).

Looking now at the initial data, we know that it has been obtained at a series of (equally spaced) measurement points along the surface, where the source and receiver of a typical seismic trace can occupy two of those points. The set of data we have to work with consists of a large number of these offset seismograms. Although the positions along the surface are discrete points, the seismic signals themselves are continuous functions of time. It is commonplace today, however, to have them in the form of amplitudes taken at a set of time sample points.

In the earth itself the acoustic waves have measurable parameters; for example, excess pressure (p), displacement u, v, or particle velocity $\dot{u}$, $\dot{v}$ or convenient mathematical parameters such as potential (ϕ) related to the physical ones. It is known that the variation in time and space must be described by the wave equation

$$\phi_{xx} + \phi_{yy} - \frac{1}{c^2} \phi_{tt} = 0 \tag{9B.1}$$

Here subscripts are used to denote partial derivatives. This equation may be regarded as a description of the relation between rates of change of ϕ with respect to space and time—in any neighborhood. The function of the wave equation in migration may, perhaps, be described best by saying that it is used under conditions in which ϕ is slowly varying (in both space and time) to *predict* from a set of known values what the values are in neighboring locations. Thus we shall take a set of parameter values obtained on (and corrected to) a horizontal surface and then project them downward.

Figure 9B.1 illustrates this action. The coordinates of the source and receiver points are (x_s, z_s) and (x_r, z_r) with measurements being taken at $(x_s, 0)$ and $(x_r, 0)$. Generally the fields measured for any locations of source and receiver will be represented by

$$\psi(x_s, z_s, x_r, z_r, t)$$

where t is the two-way time of reflection. We may now try to predict (Claerbout, 1971; Claerbout and Doherty, 1972) what ψ would be for sources and receivers situated on horizontal planes down in the earth by using the prediction properties of the wave equation. Eventually, by extension of the *downward continuation*, the point is reached at which

$$x_s = x_r = x; \quad z_s = z_r = z$$

$$t = 0$$

and the output $\psi(x, z, x, z, 0) = R(x, z)$.

If we take the case of stacked sections a first approximation would be to let the source and receiver positions be the same point. *Midpoint* and *relative offset* coordinates can be used, as follows:

$$\left. \begin{array}{ll} X = (x_s + x_r)/2 & Z = (z_s + z_r)/2 \\ x = (x_r - x_s)/2 & z = (z_r - z_s)/2 \end{array} \right\} \tag{9B.2}$$

as shown in Figure 9B.1.

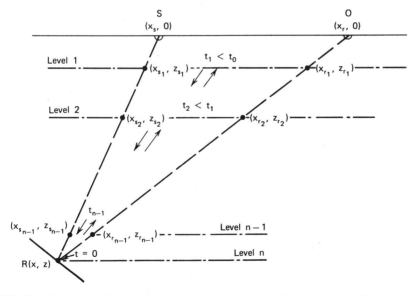

FIGURE 9B.1 Illustrating the stepwise process of downward continuation of an offset seismic trace. The source and observation point coordinates approach one another at successive depths, whereas the reflection time t diminishes to zero when the reflecting point R is reached.

If we represent the field data by $\psi(x_s, z_s, z_r, t)$ in the new coordinates it is $\psi(X, x, Z, z, t)$ and the stacked section is $\psi(X, 0, 0, 0, t)$, whereas the migrated section is $\psi(X, 0, Z, 0, 0)$. Under certain limited conditions the wave differential equation becomes

$$\psi_{XX} + \psi_{ZZ} - \frac{4}{c^2(X, Z)} \psi_{tt} = 0 \qquad (9B.3)$$

which differs from the original equation only in the 4 in the numerator of the last term on the left-hand side. We can correct this difference by redefining t as the one-way travel time.

The original Claerbout scheme uses a difference equation equivalent [to use the sampled values of $\psi(X, t)$], which gradually allows movement of the midpoint coordinates into the earth. The waves traveling in the earth vary rapidly in amplitude with depth, and to make the scheme practical new coordinates, *which move with the wave*, have to be defined. The variation with the new depth coordinate is now much slower and relatively large steps downward can be taken. The new coordinates are

$$D = ct/2 + Z; \qquad d = Z \qquad (9B.4)$$

The differential equations becomes:

$$\phi_{XX} + \phi_{dd} + 2\phi_{dD} = 0 \qquad (9B.5)$$

The stacked section (source and receiver at the same point) is

$$\phi(X, 0, D) = \phi(X, 0, ct/2) = \psi(X, 0, 0, 0, t)$$

and the migrated section,

$$\phi(X, D, D) = \phi(X, Z, Z) = \psi(X, 0, Z, 0, 0)$$

These are bases for Claerbout's original 15° approximation, and better approximations to the differential equation (9B.5) gave subsequent higher approximations.

Substitution of a difference equation for the differential equation needs a scheme of sampling:

$$D_j = j \cdot \Delta D; \qquad d_k = k \cdot \Delta d$$

Using the compact notation

$$\phi(X)_j^k = \phi(X, d_k, D_j)$$

two possible forms of difference equation are

$$\phi_j^{k+1} = -\phi_{j+1}^k + \left(\frac{1 + T}{1 - T}\right)(\phi_{j+1}^{k+1} + \phi_j^k) \qquad (9B.6a)$$

and, even less accurately,

$$\phi_j^{k+1} = -\phi_{j+1}^k + (1 + 2T)(\phi_{j+1}^{k+1} + \phi_j^k) \qquad \text{(9B.6b)}$$

where

$$T = \frac{\Delta D \, \Delta d}{8} \frac{d^2}{dX^2} \qquad \text{(an operator in } X) \qquad \text{(9B.6c)}$$

It is more economical to use (9B.6b) than (9B.6a) because the latter involves the inversion of the operator $(1 - T)$. From Figure 9B.2 it is obvious that by pursuing this scheme and letting j vary a new set of values at $k + 1$ can be produced. Thus j will vary rapidly, whereas k will jump slowly.

For more accurate approximations to the differential equation (9B.5) and discussions of accurate and dispersion the reader is referred to the original papers of J. Claerbout and his collaborators and to the convenient summaries in R. H. Stolt (1978).

We now break away from the time–distance domain treatment and describe briefly the frequency–wave-number method originated by Stolt (1978). P. Hood (1978) and Bolondi et al. (1978) have described similar procedures independently. Convolution in the time domain is equivalent to multiplication in the frequency domain—and this general principle carries over to two or more dimensions. It therefore seems reasonable that a simplified treatment of migration might evolve by changing the original field data over to the frequency–wave-number representation by using the FFT, operating on this data in a manner prescribed by the wave equation and finally performing a two-dimensional inverse transform to arrive at the migrated section.

In essence, the surface results are $\phi(X, 0, D)$ and

$$A(p, \omega) = \frac{1}{2\pi} \int dX \int dD \, e^{i(pX - 2D/c)} \phi(X, 0, D) \qquad \text{(9B.7)}$$

$$\phi(X, 0, D) = \frac{1}{2\pi} \int dp \int d\omega \, e^{-i(pX - 2\omega D/c)} A(p, \omega) \qquad \text{(9B.8)}$$

the standard two-dimensional Fourier transform pair.

It can be shown that for upcoming waves at positive depth d (9B.8) can be generalized to

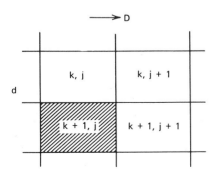

FIGURE 9B.2 The value in the shaded location is found by the difference equation selected from three adjoining (unshaded) locations. The areas actually represent sample points of the (d, D) array.

$$\phi(X, d, D) = \frac{1}{2\pi} \int dp \int d\omega \; e^{-i(pX+qd-2\omega D/c)} A(p, \omega) \qquad (9\text{B}.9)$$

where (to satisfy the wave equation)

$$q = \frac{2\omega}{c} - \sqrt{\frac{4\omega^2}{c^2} - p^2}$$

Because the migrated section must be $\phi(X, D, D)$, it has the form

$$\phi(X, D, D) = \frac{1}{2\pi} \int dp \int d\omega \; A(p, \omega) e^{-i\left(pX - \sqrt{\left(\frac{4\omega^2}{c^2} - p^2\right)}D\right)}$$

To put this is the standard form of a Fourier integral involves one more substitution:

$$k = \sqrt{\frac{4\omega^2}{c^2} - p^2}$$

Then

$$\phi(X, D, D) = \frac{1}{2\pi} \int dp \int dk \; B(p, k) e^{-i(pX-kD)} \qquad (9\text{B}.10)$$

where

$$B(p, k) = \frac{1}{\sqrt{1 + p^2/k^2}} \; A\left(p, \frac{kc}{2}\sqrt{1 + p^2/k^2}\right) \qquad (9\text{B}.11)$$

Stolt notes that this transformation amounts to shifting data from a frequency ω to a lower frequency

$$\omega' = \sqrt{\omega^2 - p^2 c^2/4}$$

which amounts to a move-out correction and also a change of scale

$$\frac{k}{\sqrt{k^2 + p^2}} = \frac{\omega'}{\omega}$$

Interpolation in the frequency domain can be dangerous, but it is usually helped by a fine sampling induced by adding many zeros to the original time data. Fortunately the use of FFTs does not involve a severe penalty for adding zeros up to the next (or even the next but one) power of 2.

It is now reasonably obvious that in the frequency-wave-number domain three-dimensional migration can be done as a reasonably simple extension of the two-dimensional method.

Migration can be accomplished before stack and can be used to effect move-out correction, stack, and migration of data at the same time. Moreover, although the preceding scheme is dependent on a constant velocity, transformations of coordinates that render a variable velocity system workable can be made. Details are contained in R. H. Stolt's publication.

REFERENCES

Alford, R. M., Kelly, K. R., and Boore, D. M. (1974), "Accuracy of Finite Difference Modeling of the Acoustic Wave Equation," *Geophysics*, Vol. 39, No. 6, pp. 834–842.

Bolondi, G., Rocca, F., and Savelli, S. (1978), "A Frequency Domain Approach to Two-Dimensional Migration," *Geophysical Prospecting*, Vol. 26, No. 4, pp. 750–772.

Born, M., and Wolf, E. (1959), *Principles of Optics*, Pergamon, New York.

Claerbout, J. F. (1971), "Toward a Unified Theory of Reflector Mapping," *Geophysics*, Vol. 36, No. 3, pp. 467–481.

Claerbout, J. F., and Doherty, S. M. (1972), "Downward Continuation of Moveout Corrected Seismograms," *Geophysics*, Vol. 37, No. 5, pp. 741–768.

Dohr, G. P., and Stiller, P. K. (1975), "Migration Velocity Determination, Part II: Applications," *Geophysics*, Vol. 40, pp. 6–16.

Faust, L. Y. (1951), "Seismic Velocity as a Function of Depth and Geologic Time," *Geophysics*, Vol. 16, No. 2, pp. 192–206.

French, W. S. (1974), "Two Dimensional and Three Dimensional Migration of Model-Experiment Reflection Profiles," *Geophysics*, Vol. 39, No. 3, pp. 265–277.

Gabor, D. (1948), "A New Microscopic Principle," *Nature*, Vol. 161, p. 777.

Gabor, D. (1949), "Microscopy by Reconstructed Wavefronts," *Proceedings of the Royal Society*, *A*, Vol. 197, p. 454.

Gabor, D. (1951), "Microscopy by Reconstructed Wavefronts: II," *Proceedings of the Physical Society*, *B* Vol. 64, p. 449.

Hale, D. (1984), "Dip-Moveout by Fourier Transform," *Geophysics*, Vol. 49, No. 6, pp. 741–757.

Hatton, L., Larner, K. L., and Gibson, B.S. (1981), "Migration of Seismic Data from Inhomogeneous Media," *Geophysics*, Vol. 46, No. 5, pp. 0751–0767.

Hood, P. (1978), "Finite Difference and Wave Number Migration," *Geophysical Prospecting*, Vol. 26, No. 4, pp. 773–789.

Hubral, P., and Krey, T. (1980), *Interval Velocities from Seismic Reflection Measurements*, Society of Exploration Geophysicists, Tulsa, Oklahoma.

Judson, D. R., Schultz, P. S., and Sherwood, J. W. C. (1978), "Equalizing the Stacking Velocities of Dipping Events via Devilish," paper presented at the 48th Annual International S.E.G. Meeting, San Francisco.

Larner K. L., Hatton, L., Gibson, B. S., and Hsu, I. C. (1981), "Depth Migration of Imaged Time Sections," *Geophysics*, Vol. 46, No. 5, pp. 734–750.

Laski, J. D. (1970), "Simultaneous Estimation of Parameters of Reflection Events (Depth, Dip, Velocity) and Relative Static Corrections, *Geophysical Prospecting*, Vol. 18, No. 2, pp. 269–276.

Loewenthal, D., Lu, L., Roberson, R., and Sherwood, J. W. C. (1976), "Wave Equation Applied to Migration," *Geophysical Prospecting*, Vol. 24, No. 2, pp. 380–399.

Marschall R. (1975), "Einige Probleme der Benutzung grosser Schuss-Geophon-Abstande und deren Anwendung auf Unterschiessungen," Ph.D. thesis, Leoben.

Paulson, K. V., and Merdler, S. C. (1968), "Automatic Seismic Reflection Picking," *Geophysics*, Vol. 33, No. 3, pp. 431–440.

Sattlegger J. W. (1975), "Migration Velocity Determination, Part 1: Philosophy," *Geophysics*, Vol. 40, pp. 1–5.

Schneider, W. A. (1971), "Developments in Seismic Data Processing and Analysis," *Geophysics*, Vol. 36, No. 6, 1043–1073.

Schultz, P., and Sherwood, J. W. C. (1978), "Depth Migration Before Stack," paper presented at the 48th International S.E.G. Meeting, San Francisco.

Stolt, R. H. (1978), "Migration by Fourier Transform," *Geophysics*, Vol. 43, No. 1, pp. 23–48.

Waters, K. H. (1941), "A Numerical Method of Computing Dip Data Using Well Velocity Information," *Geophysics*, Vol. 6, No. 1, pp. 64–73.

Yilmaz, O., and Claerbout, J. F. (1980), "Prestack Partial Migration," *Geophysics*, Vol. 45, No. 12, pp. 1753–1779.

TEN

Near-Surface Corrections

10.1 INTRODUCTION

The concept of a *weathered zone* whose characteristics are strikingly different from the *consolidated* rocks farther from the surface is based on observations made in holes, in rail and road cuts, and in excavations for massive buildings or engineering projects. It was observed that the soil and near-surface rocks had been affected by the elements (rain, frost, ice, elevated temperatures, and wind), and modifications made in the cementation of the rocks caused a loss of strength and change of density. In some areas secondary effects such as variation in depth to the water table had induced further complications—at times leaching more or less soluble minerals, at others the deposition from the groundwater of cementing materials. Nearly always these changes, which had occurred over long time periods, were superimposed on local, sporadic deposition of lake, river, or near-shore sediments.

When the concept was borrowed for use in geophysics it was highly simplified and, as generally used, the weathered zone consisted of a single, low velocity zone that had a marked, sharp interface with the lower, unweathered material. Small wonder, then, that areas are found in which the near-surface variations are much more complicated than can be accounted for by a simple, uniform velocity layer of variable thickness.

There is no question that for *compressional wave* seismic exploration some of the variations of the weathered layer have been smoothed out by the presence of a water table. A water table that is very shallow fills in the rock pores with a relatively high, compressional velocity material and allows little change of velocity with depth. A seasonal water table, on the other hand, leaves behind it, as it drops, a zone in which water can rise by capillary action wherever the cracks and pore spaces are small but connected and will not rise where the pores are large or large fractures are present. Thus there can be fast lateral changes of velocity (and density) within this seasonally affected zone.

In some areas the rapidly varying sedimentation pattern of geologic time has been changed by weathering to allow thin, high-velocity stringers to exist. It is well known that these stringers can give rise to *refracted* waves but in a manner sensitive to frequency and hazardous to the results of refraction methods of weathering control.

All of this background makes it evident that the problem of the weathered layer, or *near-surface zone* as it is sometimes called, is a challenge in every area.

It runs the gamut from being a problem easily overcome to one that in a few areas has not yet been conquered.

At this relatively late stage in the description of the seismic reflection method we have spent little time on the effects of surface-elevation variations or near-surface changes that affect the reflection results. In Chapters 5 and 6 just enough was said to indicate that proper corrections are highly desirable, but the actual procedures, except in one simple case, were bypassed. The need, strangely enough, was not felt strongly during the early times when use was made of dynamite. For one thing, it was considered necessary to make corrections only at locations in which holes existed and for which the time-delay data forthcoming was believed reliable. For another, the visual impact of the display of traces close together had not yet received the attention that it has today. Judgment of *roughness* or *time noisiness* of the cross sections had not been developed.

When it is possible the best correction results should be obtained from assessments of the weathered zone delays by their effects on the reflection results themselves. The ultimate success must be judged (subjectively) by the positive effect on reflection quality (continuity) and geologic reasonableness. When these intrinsic measurements are insufficient there is a case for *special* methods [Handley (1954); Dobrin (1942); Dusha (1963); Patterson (1964)]. In this chapter we consider most of the *standard* methods that have been tried for compressional wave reflection work. Consideration of the special effects and problems that exist when shear waves are used is deferred until a later chapter.

10.2 SIMPLE METHODS USING SHOT HOLES

In the case of explosives, when only structure was considered, the methods and results were usually straightforward. A hole was drilled through the weathered layer and the depth of shot, elevation of the top of the hole, and the up-hole time for the seismic wave were recorded. In many areas it was standard practice to drill at least one deep hole and, by a series of shots at different depths, measure the vertical velocity through the subweathering material. This is important information, not usually available when surface sources are used. It allowed a time correction to be made to a reference surface. The concepts are simple and the corrections are (see Figure 10.1) as follows: T_r is the uncorrected reflection time for a shot at A_1 received at E_2 and $T_r^{(c)}$ is the corrected time for the same shot and receiver but given with reference to the reference plane E_s by

$$T_r^{(c)} = T_r - \frac{(E_1 - d_{s_1} - E_s + E_2 - d_{s_2} - E_s)}{V_0} - t_{uh_2}$$

The same reference plane is used throughout the area and is usually taken as near as possible to the average shot elevation to avoid the slant path problems for long-offset traces. It should be obvious that the exact elevation chosen for the reference plane is of little consequence, provided that the correctional velocity is constant.

Ziolkowski and Lerwill (1979) emphasized the need for the reference surface to be as close as possible to the average bottom-hole elevation. This implies a lack

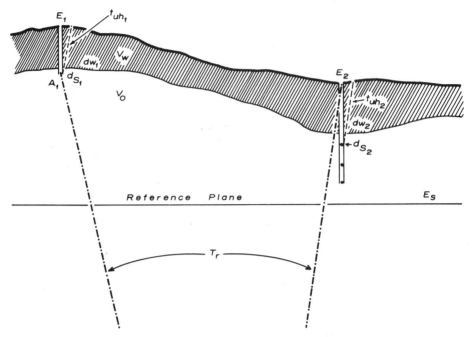

FIGURE 10.1   Corrections when shot holes are used.

of confidence in the correctional velocity used or the knowledge that in nature it is bound to be variable. By minimizing the depth differences to the reference plane it is true that the effect of a wrong choice of velocity will be reduced, but, as the calculation of a simple example will show, the *real* problem of velocity variation has not been solved and will be only when the reference surface is at a level at which the velocity is constant and the transmission times above it are *measured*.

Near-surface corrections are naturally more important in seismic surveys for coal because of the higher frequencies used in seeking better resolution.

There are therefore velocity problems that cannot be treated by these shallow holes; for example, near the Fort Peck Reservoir in North Dakota the U.S. Corp of Engineers drilled holes in the flood plain of the Missouri River and found recent, unconsolidated sands and gravels as deep as 600 ft (183 m). Other discrepancies appeared in areas of West Texas and New Mexico, where mesas caused by uneroded areas of resistant rock several hundreds of feet above the surrounding plains influenced the velocity in the subsurface. This effect, due to the static loading of the surface by the excess weight of the mesa, gave rise to a distribution of velocity abnormalities that existed not only below the mesas but for thousands of feet to the side and in decreasing percentage with depth.

This problem usually shows up when the time differences between deep and shallow reflections are mapped and contoured. The resulting contours partly correlate with the surface elevation.

A correlation coefficient can be calculated between the variations in the surface and the time variations in the subsurface reflections:

$$k = \frac{\Sigma_i^N (E_i - \bar{E})(t_i - \bar{t})}{[\Sigma (E_i - \bar{E})^2 \, \Sigma (t_i - \bar{t})^2]^{1/2}} \tag{10.1}$$

If we adopt a geological postulate that the surface topography bears no relation to the subsurface structural variations corrections based on the correlation coefficient and the surface variations can be applied to the reflection times. This method, of course, is subject to the danger that the postulate may be wrong (in some cases surface lows correspond to subsurface anticlines) and must be used with care.

The assumption of a reference plane is obviously incorrect in country with large-scale elevation changes because the assumed constant-elevation correction velocity V_0 is most likely affected by loading effects and changes in lithology. For time cross sections a horizontal reference plane is an obvious necessity; however, when reflection times were to be corrected to depth by some velocity function and the horizon levels referred to a sea-level datum a floating time reference was sometimes used. Often the level of this floating surface was a constant plus an estimated fraction of the change in shot elevation:

$$(E_{R_1} - E_{R_2}) = k(E_{S_1} - E_{S_2}) \tag{10.2}$$

where E_{R_1} and E_{R_2} = elevations of the shots at positions 1 and 2
E_{S_1} and E_{S_2} = elevations of the shots at positions 1 and 2
k = constant $0 < k < 1$

This device, however, begged the question because it was then necessary to apply a (constant?) velocity down to the reference surface and then convert the corrected reflection times to depth, assuming that the entire velocity log (possibly determined from a well survey) moved up and down with the reference surface. It has been remarked that the intermediate traces at the various receivers between the widely spaced shot points served only to correlate, to follow the desired reflection from shot point to shot point. There was no redundancy in these data.

As soon as the CDP method came on the scene there was redundancy and further reason for the correct estimation of static corrections. The data to be stacked to give a better signal/noise ratio and some multiple-reflection cancellation came from different shot points and receivers and had to be properly corrected before signal/noise ratio improvement could be realized or proper velocities, calculated. In the following sections various methods of obtaining near-surface corrections from redundant CDP reflection data are examined. They have their dangers, too, and it is not surprising even to the initiated that full success has not yet been achieved. These CDP methods can, of course, be used with shots in holes as sources. It is assumed for simplicity, however, that all sources and receivers are on the surface. If further information, such as up-hole times, is available, so much the better. These data can be used as a further control.

10.3 SOURCES AND RECEIVERS ON SURFACE (CDP)

Although on the basis of surface elevations and an estimate of the *near-surface* subweathering velocity it is possible to make corrections for elevation change— and this is supposedly done in the remainder of this section—there is still a necessity for making an estimate of the effect of the weathered layer on the reflection record. This effect is most often regarded as a time-delay constant for all frequencies in the reflection passband. In a later section we take up the question whether it is possible to treat a frequency dependence in the weathering effect.

The first, and still common, practice was to estimate the weathering time by the *ABC* refraction method described in Section 5.7. This procedure, of course, adds to the cost of the survey and does not permit any determination of the possibility of change in velocity of the weathered layer. In practice the velocity in the weathered layer is not constant, for it is very much a function of water saturation of loose, unconsolidated soils and rocks. There is economic pressure, always, to determine the residual static *corrections* (i.e., the surface-consistent extra time delays after the elevation corrections have been made as well as possible) from the redundant data collected during practice of the CDP method. In this section the principles involved in determining these residual static corrections (by three separate methods) are reviewed.

A typical set of seismic traces is shown in Figure 10.2 for a sixfold stack system. Note that the traces are numbered with the shot point and the receiver numbers separated by a slash. When arrayed in this manner we can classify the traces by the way in which they are aligned; for example, all traces with a common reflection (depth or basement) point are aligned vertically, and the columns under SP 16, which consist of the traces 14/18, 13/19, 12/20, 11/21, 10/22, and 9/23, all have a basement point under SP 16.

Similarly, the constant receiver traces are lined up downward to the left, constant shot point traces, downward to the right, and finally constant offset traces occupy horizontal rows such as 1/10, 2/11, 3/12, 4/13. Each method to be described makes use of one or more of these arrangements and all have their weaknesses.

To find comparative times it is usual to cross correlate selected windows of the traces concerned to determine at what (algebraic) time difference the best match is obtained. This consists of plotting the cross-correlation function

$$\phi_{fg}(\tau) = \int_{t_1+\tau}^{t_2+\tau} f(t)g(t+\tau)\, dt \tag{10.3}$$

for a range of delay times (say, -20 to $+20$ ms) and over the range of the window t_1 to t_2, moved by the corresponding shift τ in one trace. It is important to keep the number of samples in the window constant and to minimize the effects of the end samples. This is usually done over a window as long as possible and the window itself is tapered. It is presumed that the following precautions have been taken:

1. The window is taken over prominent reflections.
2. The window is not allowed to include long intervals of noise. It may be better under some circumstances to use two or more windows and to combine or average the time delays found.
3. The normal move-out has been removed as closely as possible (this applies to all but one of the methods to be discussed). Correlation is sensitive to stretching or shrinking of a trace as we noted when considering the Vibroseis method.

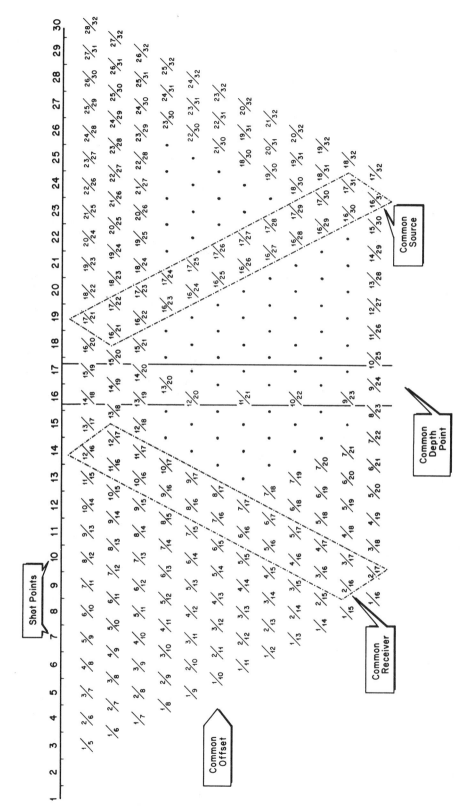

FIGURE 10.2 The characteristics of depth points, offsets, sources, and receivers in a specific, sixfold CDP system.

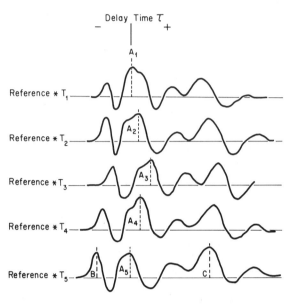

FIGURE 10.3 Typical cross-correlation functions. The picked peaks at A_1, A_2, A_3, A_4, and A_5 have varying delay times; B and C represent some higher peaks that can be machine-picked if the data are noisy or if the criteria are not strict enough. The asterisks indicate cross correlation.

It is usually necessary to resample the individual traces at a finer sampling rate in order that the time delay for maximum cross correlation can be obtained accurately; that is, with better precision than the original sample rate of the trace. Linear, quadratic, or cubic interpolation may be done, but it is best (see Appendix 10A) to use the $(\sin x)/x$ method.

The outcome of the cross correlation is a curve (Figure 10.3) with its peak (hopefully this is unambiguous) shifted from zero delay; this delay can be determined. These are then the time differences that can be entered into an array similar to the one already shown in which every trace has its place.

Note that these determinations of delay time can be subjected to a statistical analysis to give the mean μ and the variance σ^2 [e.g., see Wine (1964)]. The standard deviation σ is the distance from the mean (for a normal distribution of observations) within which approximately two-thirds of the observations fall. Ninety-five percent of the observations in a normal distribution fall inside the range $\mu - 2\sigma$ to $\mu + 2\sigma$, and it is a commonly accepted practice to throw away, as suspicious, any observations that fall outside this range.

Method 1. Common Reflection Point

Hileman, Embree, and Pflueger (1968) appear to have been the first to publish this method. The τ discovered from the cross correlation of two traces is made up of several individual contributions:

$$\tau = \tau_d + \tau_{WS} + \tau_{WR} + \tau_{NMO} + \tau_N \tag{10.4}$$

where τ_d = dip time difference

$\quad \tau_{WS}$ = weathering at shot point

$\quad \tau_{WR}$ = weathering at receiver

$\quad \tau_{NMO}$ = time difference due to incorrect NMO calculation

$\quad \tau_N$ = noise

If a method that involves the CDP is chosen the term due to dip should disappear.

First, however, we have to select a reference trace that preferably should be a near offset trace in which the NMO is smallest. It is probably the best single one to use if the signal/noise ratio is high. Other reference traces have been used, such as the average of all nonshifted traces in the group. If, however, the static shifts between the traces are large compared with a period of any constituent frequency there is the danger that the character of the reference trace may be so altered that the cross-correlation function between the reference and any other trace will have an altered shape and give an incorrect time shift. Remember that this function has the spectrum common to both traces compared—and the difference of their phases.

Rice and Burnett (1978) have suggested that the reference trace should be calculated from the entire group to be compared (or stacked) by a method that combines the matrix and factor analysis methods. In essence, this method makes use of the fact that any one of a group of signals can be considered as a linear combination of characteristic functions (or eigenvectors). These eigenvectors are obtained from the group of traces by matrix methods and correspond to eigenvalues (characteristic values) that indicate the importance of each one in making up the entire group.

The primary eigenvector, at this stage, is nearly the same as a straight average of all traces, but signal/noise estimates are introduced and a process known as *rotation* is invoked (corresponding to a time shift between components) to produce an even better *average*. The signal/noise ratios limit the number of (decreasingly important) eigenvectors considered.

Each of these secondary eigenvectors is tested to determine whether it is a time-shifted version of the primary; if so, it is aligned and added. Otherwise it is discarded. The final result of this process is the reference trace.

Now that the reference trace has been selected or calculated all individual traces for a common depth point are cross correlated against it and the time shift (τ) of the peak of the cross-correlation function is obtained. If the reference trace is, in fact, one of the individual traces then N values of τ, including a zero τ for the reference trace correlated against itself, are obtained. If, however, the reference trace is one calculated from the group N generally nonzero τ values are obtained and adjusted, if necessary, to have a zero mean.

By this procedure the entire table (like Figure 10.2) can be filled out. The set of τ values along the *common receiver* line, if averaged, results in a delay at this receiver alone (it is assumed that the shot delays average to zero), and the set along the *common shot* line, if averaged, gives rise to a value that can by similar reasoning be the delay (or residual correction) for this shotpoint.

If the values that lie along the line of the CDP are examined, the statics may vary in a random manner about a near-zero mean or they may vary systematically with offset (either an increase or a decrease). This is a sign that the NMO has been incorrectly removed.

Gross errors in the estimation of shot point or receiver static can be perceived and removed. These errors usually develop through the possibility of two possible maxima (peaks of almost the same height in the cross-correlation function) and one or both must be rejected.

Systematic changes due to incorrect NMO are removed by low-order polynomial fitting and retaining the residuals.

Finally, the statics applied to the individual traces making up the stacking set are the sum of the source point and receiver point statics. *Surface consistency* is thus achieved, which means that a particular point has an applicable source or receiver static and is independent of the offset or particular combination of source and receiver used.

It is intuitively obvious that because the sampling is good within a spread length but the combined shot point-receiver redundancy is poor outside the spread length, this method measures high-frequency statics well (i.e., statics that vary rapidly along the surface) but long-period statics poorly. This, in fact, is what is found and because there is a limit to the spread length that can be used statics that affect overall structure are poorly estimated and false structure can be inserted. Remember that the averages are taken over the spread length for both source point and receiver values. Thus, although CDP traces are used to avoid errors due to dip, errors can be made that affect the dip.

Method 2. Equivalence of Source and Receiver Statics

Another variant of the statistical automated statics system is one in which advantage is taken of the common shot point or common receivers. The table in Figure 10.2 can be filled in by correlating each trace of a common shot point record with a reference trace. The same remarks made earlier apply to the choice of the reference trace and in general the same cross-correlation procedure is adopted. The advantage here may be in the common spectrum given off by the source-earth combination; consistency is difficult to achieve in practice, even with the Vibroseis system. Thus the table is filled in and the same precautions can be taken as before:

1. The receiver τ's must form a zero-mean set.
2. Values that deviate more than two standard deviations from the mean may be (at least) questioned at this stage.

The same procedure is, of course, available when the traces are sorted into sets with different shot points but a common receiver. Again, the justification may be that the near-surface conditions at the receiver are common (i.e., the effect of near-surface reverberation is the same). These values then allow the table to be filled out again. Note that these values fill the table along the lines down to the right (common source) or down to the left (common receiver). Effects of NMO removal error are still present. Effects of dip are present in each method, except that a different dip occurs for each determination and there is some hope that τ_{dip} will cancel. Note that there are $2N - 1$ determinations of the common source and

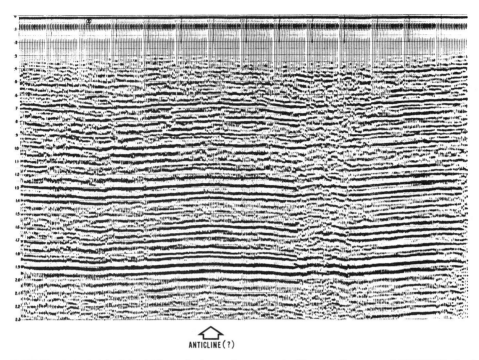

ANTICLINE (?)

FIGURE 10.4 Original fivefold, stacked test data area *A*. (*Source*: Courtesy of CONOCO, Inc.)

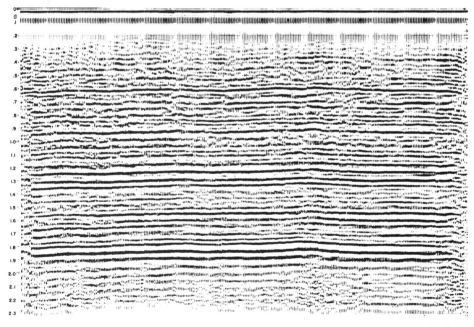

FIGURE 10.5 Test data in Figure 10.4 treated with statistical correction program 2 described in the text. Note the change in the reflection structural pattern and the deterioration near the ends at which the full stack has not been obtained. (*Source*: Courtesy of CONOCO, Inc.)

$2N - 1$ values of the weathering difference between a given pair of surface points. If an average is taken (on the assumption that *time differences* are independent of whether the surface point is a source or receiver) surface-consistent values of the weathering can be obtained. This method provides more statistics (both sources and receivers are treated as equivalent and dubious values can be checked) and the statistical error can be calculated more accurately.

The difficulties of remnant τ due to incorrect NMO and dip have not been solved, however, except that we can expect high-frequency statics to be well determined and those of long surface period, which are the dangerous ones, undetermined. Figures 10.4 and 10.5 show fivefold field data in the original stacked version and after application of method 2 statics before stacking. Also well illustrated in these two figures is the ambiguity of structure and the dilemma of the interpreter. Is the anticline shown by the arrow in Figure 10.4 present or not? The answer is not in either of these sections.

Method 3. Constant Offset Static Corrections

Returning now to (10.4),

$$\tau = \tau_d + \tau_{WS} + \tau_{WR} + \tau_{NMO} + \tau_N$$

we can reduce a different term to zero if we take only those traces that have a constant offset distance to correlate with one another. Then, as long as the average velocity to the reflection stays constant, $\tau_{NMO} = 0$. In Figure 10.2 these traces occur along the horizontal lines. Reflection seismic traces from adjacent sources with the same offset are cross-correlated:

$$\tau_{i, i+1} = \tau_d + \tau_{WS, i} - \tau_{WS, i+1} + \tau_{WR, i+P} - \tau_{WR, i+P+1} + \tau_{N, i} - \tau_{N, i+1} \quad (10.5)$$

where P is the offset. All offsets can, of course, be utilized. The dip still stays in the time difference determined, a characteristic of all methods, even though CDP method 1 *apparently* avoids this complication. It has to be dealt with in an interpretive way by assuming, for example, that a mean of a sufficient number of determinations averages out the dip for the short-period variations. In practice it is usual to fit a low-order polynomial

$$\Delta t = a_0 + a_1 x + a_2 x^2 + a_3 x^3 + \cdots \quad (10.6)$$

to the τ's computed and then to use only the residuals (values determined less the polynomial calculated values) as the proper weathering corrections. As before we assume that τ values due to noise cancel. Finally, we state the hypothesis that the source point statics are equivalent to the receiver statics (there is surface consistency, whether the point is a source or a receiver) and we obtain

$$\tau_{i, i+1} = \tau_d + \tau_{W, i} - \tau_{W, i+1} + \tau_{W(i+P)} - \tau_{Wi+P+1} + \tau_{N, i} - \tau_{N, i+1} \quad (10.7)$$

This is more of a mixture of terms than we have encountered before because different receivers and different sources are involved in each determination of $\tau_{i, i+1}$. Because there are $2N$ equations and $2N + P_0$ weathering values to be

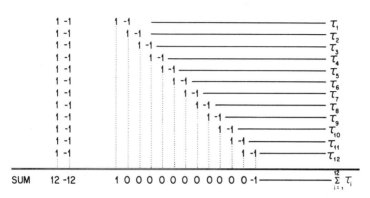

FIGURE 10.6 Data from the cross correlation of adjacent pairs of common-offset traces (common shot points 5 and 6) arranged to show how the sum lends weight to the shot-point.

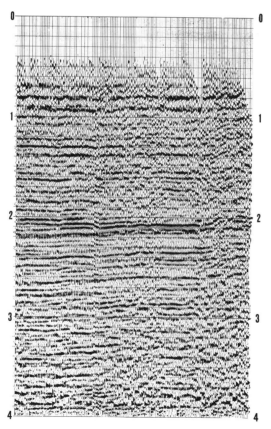

FIGURE 10.7 Original tenfold-stacked data, area B. (*Source*: Courtesy of CONOCO, Inc.)

determined, the values appear to be underdetermined in any set. This is by no means the case, however, because the redundancy caused by the shift of only a partial spread at a time makes the chief values overdetermined and the usual procedures apply. Of course, at the beginning and end of lines some points have very little or no redundancy, and as in other weathering measurements the determinations are made so that one value may arbitrarily be given some constant value. Figure 10.6 shows that the set of $2N$ equations, if added, exhibits nearly $2N$ times the difference between $\tau_{W,i}$ and $\tau_{W,i+1}$, two other weathering values entering in (the nearest offset and the farthest offset) with a weight of only 1. It is usually sufficient to adopt this simple approach, but a correction may be made later after the first approximation has been established to all of the weathering corrections. The higher the value of N, the less this second approximation is needed.

The calculation of the residuals is then made by a least-squares best fit to a low-order polynomial (or any other suitable long-period function of offset distance). The standard (pseudoinverse matrix) method is suitable.

Uncorrected and corrected data, together with a graph of the calculated and residual values applied to the data, are given, respectively, in Figures 10.7 through 10.9. Even the residuals are large, ranging from 17 to $-36\,\text{ms}$, but these

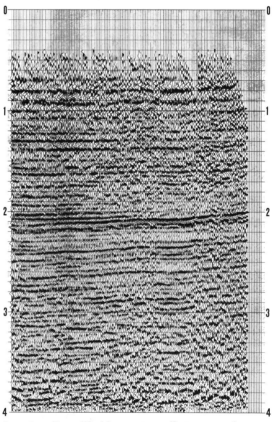

FIGURE 10.8 Data from Area B modified by constant-offset, near-surface corrections and linear dip left in. (*Source*: Ware. Courtesy of CONOCO, Inc.)

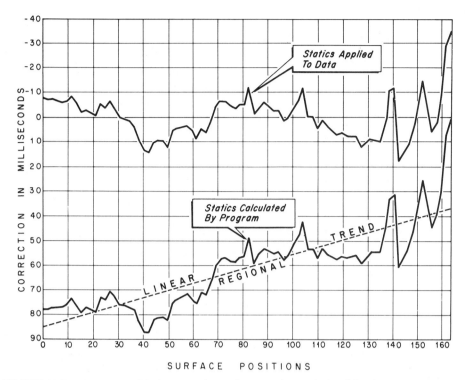

FIGURE 10.9 The relationship between the static corrections calculated by the program and the values applied to the data to leave in linear dip.

areas are not uncommon and the data illustrate the need for exceptionally good near-surface corrections to improve data quality. The structure is, of course, a function of the polynomial fitted. It should be pointed out, however, that the statics calculated here do include elevation corrections, a procedure that is contrary to the recommendation made earlier. This was necessary because of changing subweathering velocity.

In some areas of immature erosion in which valleys collect debris from ongoing erosion of the topographically high areas, these near surface delays may be incorrectly estimated. The corrections of the long offset seismic traces will be different from those of the short offset traces. (In the latter both source and receiver may be located in the same valley.) Under these conditions it is instructive, and often constructive, to determine the near-surface corrections separately from the long and near-offset traces. The proper set of corrections is used for each reflection trace set before stacking.

10.4 SOURCES AND RECEIVERS ON SURFACE—ITERATIVE AND ADVANCED STATISTICAL METHODS

It is evidently necessary—particularly when high-frequency records are processed—to obtain the near-surface correction times as accurately as possible. An error of (say) 0.005 s causes the 100-Hz component of one trace to be out of phase with another with which it is being compared or stacked. The chief purpose of great accuracy in near-surface corrections is simply to preserve the high-frequency components of the reflections when the CDP components are added together. These records are a better reflection of the geological boundary space relationships but naturally give the changes in the individual boundary positions more clearly. Thus, although the methods outlined so far may be adequate for most areas and for the routine requirements of a present-day reflection survey, there remain some locations and some final requirements that demand better corrections.

Inaccuracies may be present in one or other of the methods cited as a result of the following:

1. Residual NMO
2. Dip between basement points (particularly of the slowly varying type)
3. Incorrect assessment of time difference between traces (cycle skipping or a change in character of the central peak)
4. Errors due to noise
5. Different estimates for different frequency bands

Once an estimate of the static shifts has been made it can be incorporated in the original data and a new set of velocity determinations entered along the line. On the basis of these new velocities (which will be more accurate if the original traces are better aligned after preliminary static correction), the NMOs can be improved and the process of static determination again performed. This is an iterative procedure but there is no guarantee that it will be convergent. Care is needed in the comparison of successive iterations. Saghy and Zelei (1975) have given a method that includes repeated averaging according to depth, source, and receiver points and have found the effectiveness of the original Hileman, Embree, and Pflueger (1968) method considerably improved. It appears to be possible to examine the variance of the time delays when the velocity determination is made, to terminate the iterations at the point of minimum variance, and to give some indication of the expected standard deviation of the weathering determinations.

Wiggins, Larner, and Wisecup (1976) have set up the problem as a set of linear equations in which the near-surface time T_{ij} is written as the sum of four terms:

$$S_i + R_j + G_k + M_k X_{ij}^2 = T_{ij} \qquad (10.8)$$

The indices i and j refer to the position of receiver and source and $k = i + j - 1$ is the CDP index position,

where S_j = the travel time from shot point j to reference plane
$\quad R_i$ = the travel time from receiver i to reference plane
$\quad G_k$ = the normal incidence two-way time to a subsurface position at the kth CDP position (in other words, the geology or structural term)
$\quad M_k$ = the time-averaged residual, normal move-out coefficient
$\quad X_{ij}$ = the distance between the ith shot point and the jth receiver

Later Larner, Gibson, Chambers, and Wiggins (1979) added a term to account

for a change in reflection time due to cross dip, thereby generalizing the situation
to accommodate crooked lines:

$$S_i + R_j + G_k + M_k X_{ij}^2 + D_k Y_{ij} = T_{ij} \qquad (10.9)$$

It has been found desirable to abandon the determination of T_{ij} from a long
window cross correlation between component traces in favor of two or more
individual short window cross correlations, centered on what appear to be
potential good reflection zones, and this upgrades the residual-static correction
determinations.

Normally, in earlier determinations of residual statics by the methods of
Section 10.3 there were more equations than unknowns, and the standard method
of determining the least-squares, best-fit values to these equations is by solving
the matrix equation

$$\mathbf{Ap} = \mathbf{t} \qquad (10.10)$$

where $\mathbf{t}$ = reflection times observed
  $\mathbf{p}$ = the unknown parameters
  $\mathbf{A}$ = a coefficient matrix

The solution (see Appendix 6B) is known to be

$$\mathbf{p} = (\mathbf{A}^T\mathbf{A})^{-1} \mathbf{A}^T\mathbf{t} \qquad (10.11)$$

which involves the inverse $(\mathbf{A}^T\mathbf{A})^{-1}$. There is no problem, provided that there are
more independent equations (observations) than unknowns (p_i), but in the new
set of linear equations there are more unknowns than equations and a different
technique has to be used. These observations and unknowns are fully explored in
the papers cited and only the physical aspects of the solution need be of concern
here. Part of the trouble lies in the ends of the lines where incomplete stacking
must result.

These problems are best handled by advanced spectral decomposition
methods, which lie outside the scope of this book. A few remarks can be made,
however, which may give the flavor of the method. First, the data and the
variables are spectrally decomposed to produce a set of relationships between
statics and other parameters that are variable along the surface. Each characteristic function found is independent of any other characteristic function and each can
be handled independently. In fact, these functions are much like sinusoidal
variations along the surface and can be thought of as having low to high
wavelengths. If a gross estimate of the uncertainty of the observations can be
given a property of the decomposition is that the accuracy of the statics solutions
can be assessed *as a function of the wavelength*.

The solutions are found by an iterative method (of matrix inversion) but now
the number of iterations is known in advance. As expected, errors in the shorter
wavelength, more rapidly varying statics converge rapidly to zero, whereas those
in the longer wavelengths converge slowly or not at all. This is consistent with
earlier observations that the use of redundant data provides little control over

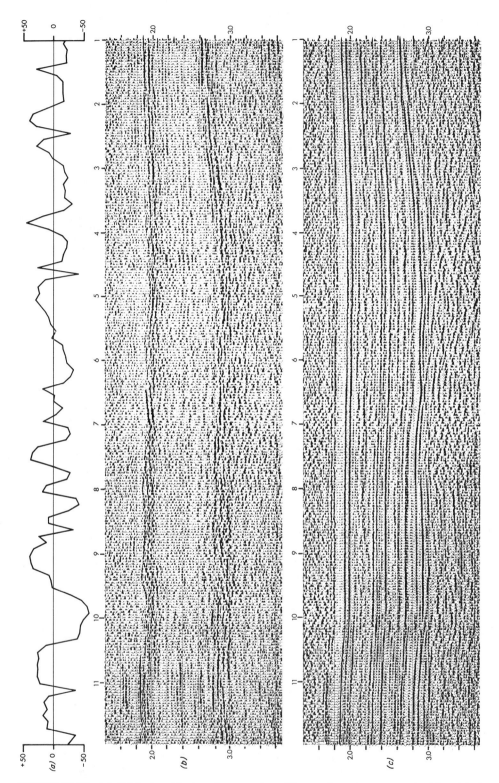

FIGURE 10.10 (*a*) Original residual static correction; (*b*) record section corrected with original residual statics; (*c*) same record corrected with MISER static correction (*Source*: © Western Geophysical Company of America.)

406

gradual statics changes of the order of the length of the spread (Booker et al., 1976).

This method (known as MISER, a Western Geophysical proprietary program) gives excellent results like those seen in Figure 10.10, in which the total receiver static corrections, the field static record section (elevation corrections only), and the MISER-conditioned section are shown.

10.5 WEATHERING DETERMINATIONS BY SURFACE WAVE MEASUREMENTS

One universal problem that plagues the solutions obtained from CDP data is the inability, except by the introduction of extraneous criteria, to separate out the structure of the near-surface corrections from the structure of the geological formations. Because the latter is of vital importance—a prime part of the oil-finding information being sought—there is still a need for a method independent of the reflection data to be used as a reference to check the CDP-derived data at intervals. If only the weathered layer is thought to be the problem then the ABC refraction method has already been described.

A second method, which may be useful in some areas, involves the pseudo-Rayleigh waves generated in the weathered layer by a surface source. A single source and a short spread of single vertical geophones is all that is needed, as shown in Figure 10.11. The waves are usually very strong and dominate the record received. In Figures 10.12 and 10.13 the original correlated Vibroseis record is shown at the top.

The phase velocity of the pseudo-Rayleigh waves varies with frequency (Dobrin, 1951) and, for a given set of velocity and density parameters, a set of phase velocity versus frequency curves can be computed for a series of depths. This set is plotted with a log frequency scale (Figure 10.14). It now remains to determine the phase velocity versus frequency characteristics of the field data, which is done by using a series of bandpass zero-phase filters on the field data. It is well known that the narrower the bandpass, the more spread out in time the energy. A compromise therefore has to be made so that the time width of the filter will allow identification of the filtered data with the more discrete event on the primary record but at the same time allowing a reasonably discrete frequency to be assigned to the phase velocity measured. The filters shown were nominally 3 Hz wide.

Provided that this can be done, there is little difficulty in measuring the time

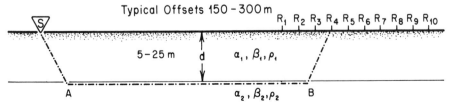

FIGURE 10.11   Typical spread and source configuration for recording pseudo-Rayleigh waves in the weathered layer.

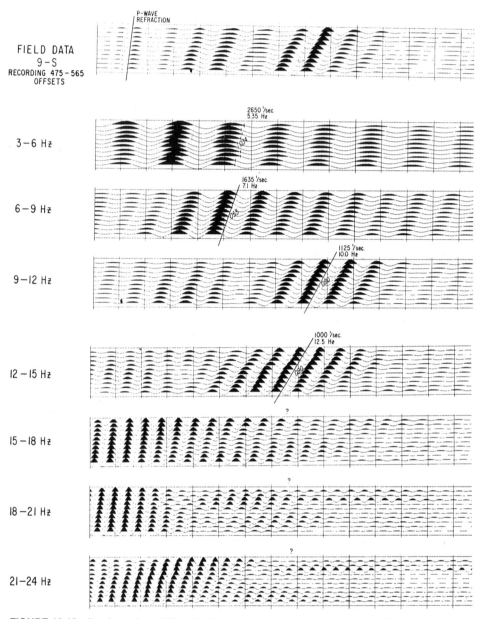

FIGURE 10.12 Bandpass data, 3-Hz wide from a particular set of pseudo-Rayleigh-wave field data.

differential across the record, hence the phase velocity. Examples of easily interpreted arrivals appear in the first four filters in Figure 10.12, but higher frequencies tend to be associated with the P-wave refraction event marked or with other unidentified arrivals. A more consistent set of filtered results is shown in Figure 10.13.

These experiments have demonstrated that it is necessary, even for normal depths of weathering and normal velocities, to be able to generate energies in the

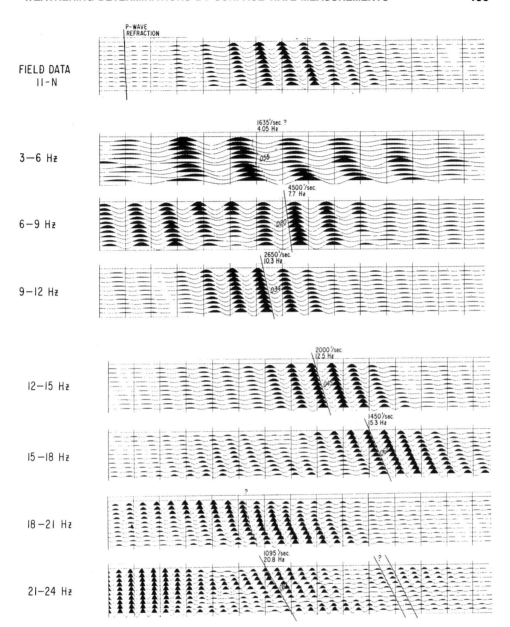

FIGURE 10.13 Bandpass data, 3-Hz wide from a more organized second set of pseudo-Rayleigh-wave field data.

low-frequency range (for exploration seismology) of 3 to 20 Hz. In good areas it is not difficult to identify the corresponding curve of the phase velocity-frequency plot, but the following points must be remembered:

1. Phase velocity curves are chiefly sensitive to the shear wave velocity in weathering and subweathering.

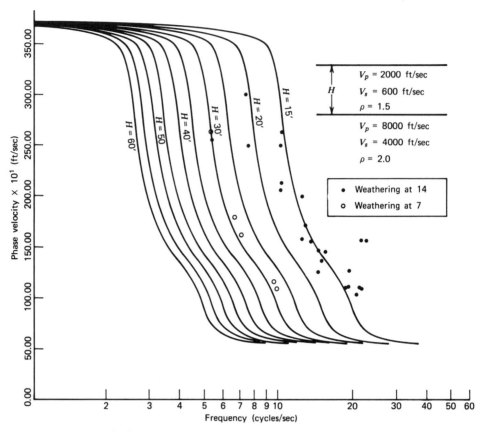

FIGURE 10.14 Phase-velocity-frequency theoretical curves for a model of a single layer over an infinite half-space. Some field data have been superimposed to show the typical scatter.

2. The assumption is made that the weathered layer is a layer of constant thickness with plane boundaries.

Programs have been written that iterate to obtain the set of parameters that allows a best fit to be made to the experimental data. Their success depends on the quality of the field data.

10.6 TREATMENT OF THE NEAR-SURFACE LAYERS AS A GENERALIZED FILTER

There appears to be some evidence that the action of near-surface layers, particularly that of the weathered layer, is not simply a delay mechanism. In the worst areas the best determinations of delay times do not work well, although they are acceptable in most. This has led to speculation that the effect on the phase of the constituent frequencies is not just a linear function of frequency with a zero intercept of 0 or π but a generalized filter. A simple example that shows a condition under which a generalized filter is generated is obtained when the reflection coefficient at the base of a single weathered layer and equal delay times for all frequencies are considered but multiples are included. Figure 10.15 illustrates this concept. If the filter at each end is represented by a transfer function pulse the total filter will be that obtained by convolving the two transfer functions. It is not difficult to see that if R_W is the coefficient of reflection at the base of the weathering the total transfer function may have its highest peak later than the time that corresponds to the one-way delay time of the weathering at each end.

Using the z-transform concept, we obtain the transfer function at the source end (for deep reflections):

$$F_1(z) = 1 - R_W z^N + R_W^2 z^{2N} - \cdots \qquad (10.12)$$

At the receiver end it is

$$F_2(z) = 1 - R_W z^M + R_W^2 z^{2M} - \cdots \qquad (10.13)$$

Both are referenced to the base of the weathered layer. By multiplying them we have the total transfer function:

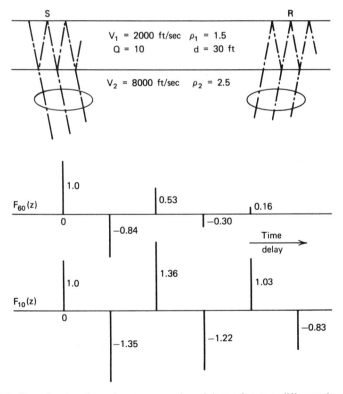

FIGURE 10.15 Transfer functions due to a weathered layer for two different frequencies. The difference is due solely to attenuation of multiple reflections at source and receiver ends. Here $Q = 10$, $d = 30$, and $R = 0.74$.

$$F(z) = 1 - R_W(z^M + z^N) + R_W^2(z^{2N} + z^{N+M} + z^{2M})$$
$$- R_W^3(z^{3N} + z^{2N+M} + z^{N+2M} + z^{3M})$$
$$+ R_W^4(z^{4N} + z^{3N+M} + z^{2N+2M} + z^{N+3M} + z^{4M})$$
$$+ \text{higher terms} \tag{10.14}$$

Let us take the case in which the following parameters hold:

1. The two-way time in the weathered layer *at each end* is equal to $N \Delta t$, where Δt is the sample time.
2. R_W is derived from the velocities and densities $V_1 = 2000 \text{ ft/s}$; $V_2 = 8000 \text{ ft/s}$; $\rho_1 = 1.5$; and $\rho_2 = 2.5$; and is approximately equal to 0.74. This, then, gives rise to the transfer function

$$F(z) = 1 - 1.48z^N + 1.64z^{2N} - 1.62z^{3N} \dots \tag{10.15}$$

This series is slowly convergent because of the high value of the reflection coefficient R. It is much more realistic, however, to include attenuation in the weathered layer. Unfortunately it is frequency-dependent and it can easily be seen that the *effective* reflection coefficient (for a given frequency) should be

$$R'(f) = R \exp(-2\alpha d)$$

where

$$\alpha = \frac{\pi f}{Q V_1}$$

and, say,

$$Q = 10; \quad d = 30 \text{ ft}$$

which gives

$$R'(60) = 0.42 \quad \text{and} \quad R'(10) = 0.673$$

We can then write the corresponding z transforms:

$$F_{60}(z) = 1 - 0.84z^N + 0.53z^{2N} - 0.30z^{3N} + \cdots$$
$$F_{10}(z) = 1 - 1.35z^N + 1.36z^{2N} - 1.22z^{3N} + 1.03z^{4N} - \cdots$$

We may cavil at the propriety of treating the z transforms in this manner, but it provides a good illustration of the difference in the rate of decay of multiples in the combined source and receiver transfer function as a function of frequency. These two transfer functions are illustrated in Figure 10.15, in which it is obvious that the *center of gravity* is not at the first unit pulse, as it is usually assumed to be. The methods related to CDP combined with a straight delay measurement therefore measure incorrect values. The problem is, of course, that the characteristics, velocity, density, and Q of the weathered and subweathered layers are not constant.

The amount of sediment above bedrock in the valleys is usually greater and can be water-saturated, whereas on the hills the aeration of the weathered layer is greater, the layer is generally thinner, and the reflection coefficient R_W is higher. The transfer functions therefore change shape constantly.

We can deal with this situation with any of the previously described CDP methods, but for illustration we choose the one that uses constant-offset traces. Instead of cross correlation to establish a time delay between consecutive traces of constant offset,

$$\tau = \tau_{S_1} - \tau_{S_2} + \tau_{R_1} - \tau_{R_2} + \tau_{\text{dip}} + \tau_N \qquad (10.16)$$

we go to the frequency domain in which each trace $f_i(t)$ is represented by a complex function of frequency $g_i(i\omega)$. (See Gol'din and Mitrofanov, 1975.)

Trace $i + 1$ is then divided (in the frequency domain) by trace i to obtain

$$T_{i, i+1}(i\omega) = \frac{g_{i+1}(i\omega)IR_{i+1}(i\omega)}{g_i(i\omega)IR_i(i\omega)} \qquad (10.17)$$

Now we can assume that over the window chosen the reflections are constant. That is to say, the impulse response of the earth's layering is constant for two near sources and two near receivers. It must be admitted, however, that time delays due to dip are still included in $g_K(i\omega)$ and the differences in depth of the reflections appear in the transfer function calculated. There seems to be some point, now, in choosing a reasonably shallow window with a good reflection energy/noise ratio and with smaller dip (on an expectation basis) than the deeper reflections:

$$T_{i, i+1}(i\omega) = \frac{g_{i+1}(i\omega)}{g_i(i\omega)} = \frac{f_{S_2}(i\omega)f_{R_2}(i\omega)}{f_{S_1}(i\omega)f_{R_1}(i\omega)} \qquad (10.18)$$

where S_2 and R_2 correspond to the trace $i + 1$ and S_1 and R_1 correspond to the trace i.

Now if logarithms are taken (very carefully) we will find

$$\ln T_{i, i+1}(i\omega) = \ln fs_2(i\omega) - \ln f_{S_1}(i\omega) + \ln f_{R_2}(i\omega) - \ln f_{R_1}(i\omega) \qquad (10.19)$$

an equation that corresponds closely to the form of (10.16), except that we have chosen to ignore noise in this development. The noise contributes to each of the terms in (10.19).

The interesting point is that we have one of these equations for every value of ω (every Fourier component), hence must deal with N times as many equations as we would have with the original constant-offset method. We can proceed one stage farther.

If complex numbers like $f_{S_1}(i\omega)$ are expressed in the form $|R_{S_1}|e^{i\theta(\omega)}$ and natural logarithms are taken we can write formally

$$\ln f_{S_1}(i\omega) = \ln |R_{S_1}| + i\theta(\omega) \tag{10.20}$$

and by analogy other terms will transform the same way. In this case

$$\ln T_{i,\,i+1}(i\omega) = \ln |R_{S_2}| - \ln |R_{S_1}| + \ln |R_{R_2}| - \ln |R_{R_1}|$$
$$+ i[\theta_{S_2} - \theta_{S_1} + \theta_{R_2} - \theta_{R_1}] \tag{10.21}$$

It is understood that all terms on the right-hand side are functions of ω. Now, the same geometrical scheme adopted for the constant-offset time-delay system can be used immediately to generate successive differences of $\ln |R_{i+1}/R_i|$ and

$$\theta_{i+1} - \theta_i$$

Thus if a source and a receiver at a particular surface point have the same transfer function the individual surface point transfer functions can be generated and used in combination to correct all the field traces.

It is noted that if a shallow reflection zone is chosen this method should end up by correcting phases and amplitudes of all reflections related to a *constant-time* shallow zone. Therefore it will combine the effects of the weathering with the isochron method mentioned earlier.

10.7 NEAR-SURFACE PROBLEMS ASSOCIATED WITH PERMAFROST

An unusual near-surface problem, which occurs only in the Arctic, Siberia, and Antarctic (?) regions, is the development of the permafrost complex. Thus usually consists of an active zone near the surface which is subject to cyclic freezing and thawing and below which is a permanently frozen zone that extends as deep as 600 m. It also has some *talik* zones of water-filled sediments under pressure and, finally, the unfrozen sedimentary layers beneath. The velocity of the sediments that contain ice is high, varying with salt and air content, salinity, pressure, and temperature. Timur (1968) investigated these relationships and found velocities that ranged from about 3000 to 5000 m/s (9840 to 16,400 ft/s). Normally this would present no difficulty, except that the bottom surface is not level where the permafrost comes in contact with the normal water-laden sediments. The irregularities appear to be related in a complex way to the presence of lakes or lagoons on the land mass or to the presence of the sea at shorelines. Under the sea, where its depth is more than a few tens of feet, the permafrost is limited to isolated thin patches, whereas under lakes, which thaw in summer, the *base* of the permafrost often rises abruptly.

Perched talik zones are kept water-filled, for, like overpressured zones in repidly deposited sediments of the Gulf of Mexico, the water is prevented from escaping to the surface by impermeable zones; it therefore bears some of the weight of the overburden. Figure 10.16 is a highly diagrammatic illustration of the permafrost complex. At the coastline (not illustrated) the permafrost zone can thin rapidly from several hundred to a few tens of feet in the space of a few kilometers. This rapid thinning causes the reflections to appear to deepen and provides false dip toward the coast.

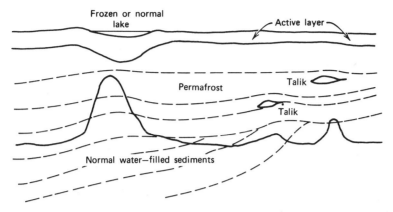

FIGURE 10.16 Diagrammatic representation of the permafrost region. The frozen zones transgress the normal sedimentary boundaries.

Because of the unusual condition of high-velocity surface material and the absence of the usual low-velocity weathered zone, the conventional refraction weathering methods fail. Resort therefore must be made to the *indirect* methods of assessing near-surface corrections based on CDP *redundant* observations. Correction has, of course, to be made for a thick near-surface layer, the overall velocity of which may change rapidly.

10.8   SUMMARY AND CONCLUSIONS

The correction of seismic reflection records for the influence of near-surface or low-velocity layers is one of the most important to be made but also one of the least successful of all data-processing methods.

Early corrections, based on up-hole times, were successful as long as the subweathering velocity remained constant. Geologically, subcrops can change in lithology and elastic parameters. Elevation changes (and consequent stress changes in the subweathering rocks) induce velocity changes that have to be monitored closely by drilling and surveying deep holes or methods must be devised to take these subweathering velocities into account.

The CDP methods of reflection prospecting produce much redundant data that are used in the stacking process to improve the signal/noise ratio of the reflections. Near-surface correction time information can, however, be derived from the reflection data themselves. The methods depend on the assumptions made. Three conceptually simple methods have been given that need a great deal of bookkeeping in the computer programs:

1. CDP NMO-corrected traces are cross correlated to determine time differences between individual traces.

2. Common source or common receiver traces, NMO-corrected, are cross correlated to determine time differences. Surface consistency of the time delay is assumed, no matter whether the surface point is occupied by a source or receiver.

3. Common-offset traces are used to avoid errors due to incorrect NMO removal. Cross correlations give time differences.

In all of these programs averages or other devices are used to separate out the surface-consistent residual statics. One of the most prominent difficulties is that it is impossible to distinguish between long wavelength (along the surface) changes in statics from changes in reflection time due to geological structure.

To obtain greater accuracy in near-surface corrections and velocity, iterative methods have been devised. The convergence of these methods under all conditions of signal/noise ratio has not been established. A method proposed by Wiggins, Larner, and Wisecup (1976) expands the input data in the form of orthonormal functions of distance along the surface, and the accuracy and number of iterations necessary to converge to a given precision can be established a priori. High precision in near-surface corrections is most important for high-frequency reflection work.

The dispersion of surface waves and the use of the *ABC* refraction method are two possibilities for establishing an absolute value for the weathering thickness. These values may be helpful when checking CDP statistical methods. In making surface wave dispersion measurements, however, they must usually be taken at the very lowest bands of the usual exploration seismology frequency spectrum, and some estimates of the P-wave and S-wave velocities and density must be available for the weathered and subweathering layers. The results are most sensitive to S-wave velocities and are only secondarily affected by density and P-wave velocity.

A method of treating near-surface effects as a generalized filter has been given and the procedure outlined.

It appears that good progress has been made in solving the problem of calculation and correction of near-surface effects. More advancement is needed, however, particularly in high-frequency reflection work, and there appears to be a good chance of additional progress. It is emphasized that modeling the weathered layer as a simple, constant-velocity, variable-thickness layer is untenable. Therefore the problem of corrections for the near-surface is one of the most important and complex in exploration seismology.

APPENDIX 10A: INTERPOLATION OF SEISMIC TRACES USING THE SIN x/x METHOD

We first prove that a time function, known to be limited in frequency to f_N (the Nyquist frequency) on the high end, can be represented by a series of sin x/x-type functions whose amplitudes are the sample values of the trace at the known sampling rate; $f(t)$ is given by $\int_{-\omega_N}^{\omega_N} g(\omega)e^{i\omega t}\, dt$. We sample it at a series of points separated by τ, where $\tau = 1/2\omega_N$. The sampled trace is then

$$f_S(t) = \sum_{j=\infty}^{\infty} \int_{-\infty}^{\infty} a_j\delta(t - j\tau)\, dt \qquad (10A.1)$$

an infinite series of delta functions, the amplitudes of which are those of the trace at the sample points.

Note that to filter this trace we can filter each delta function in turn and sum to obtain

$$\int_{\infty}^{+\infty} \delta_F(t - j\tau) \, dt = \frac{1}{2\omega_N} \int_{-\omega_N}^{\omega_N} 1 \cdot e^{i\omega(t - j\tau)} \, d\omega = \frac{1}{2\omega_N} \left(-\frac{ie^{i\omega(t - j\tau)}}{t - j\tau} \right)_{-\omega_N}^{\omega_N}$$

$$= \frac{\sin \omega_N(t - j\tau)}{\omega_N(t - j\tau)} \tag{10A.2}$$

and

$$f_F(t) = \sum a_j \frac{\sin \omega_N(t - j\tau)}{\omega_N(t - j\tau)} \tag{10A.3}$$

Thus the trace can be interpolated by evaluating the expression on the right at the needed sample points. Because $[\sin \omega_N(t - j\tau)]/\omega_N(t - j\tau)$ is an infinite function, it must be truncated for use in the computer process.

REFERENCES

Booker, A. H., Linville, A. F., and Wason, C. B. (1976), "Long Wavelength Static Estimation,"*Geophysics*, Vol. 41, No. 5, pp. 939–959.

Dobrin, M. B. (1942), "An Analytic Method of Making Weathering Corrections," *Geophysics*, Vol. 7, pp. 393–399.

Dobrin, M. B. (1951), "Dispersion in Seismic Surface Waves," *Geophysics*, Vol. 16, No. 1, pp. 63–80.

Dusha, L. (1963), "A Rapid Curved Path Method for Weathering and Drift Corrections," *Geophysics*, Vol. 28, pp. 925–947.

Gol'din, S. V., and Mitrofanov, G. M. (1975), "Spectral-Statistical Method of Calculating Surface Inhomogeneities of Reflected Waves in an Iterative Tracing System," *Geologiya i Geofizika*, No. 6, pp. 102–111.

Handley, E. J. (1954), "Computing Weathering Corrections for Seismograph Shooting," *World Oil*, Vol. 139, pp. 118–128.

Hileman, J. A., Embree, P., and Pflueger, J. C. (1968), "Automated Static Corrections," *Geophysical Prospecting*, Vol. 16, No. 3, pp. 326–358.

Larner, K. L., Gibson, B., Chambers, R., and Wiggins, R. A. (1979), "Simultaneous Estimation of Residual Statics and Crossdip Time Corrections," *Geophysics*, Vol. 44, No. 7, pp. 1175–1192.

Patterson, A. R. (1964), "Datum Corrections in Glacial Drift," *Geophysics*, Vol. 29, pp. 957–967.

Rice, G. W., and Burnett, T. M. (1978), "*Orthogonal Decomposition—Recomposition Methods and Their Application in Minimizing Static Anomalies on Seismic Data*," paper presented at the 48th International S.E.G. Convention, San Francisco.

Saghey, G., and Zelei, A. (1975), "Advanced Method for Self-Adaptive Estimation of Residual Static Corrections," *Geophysical Prospecting*, Vol. 23, No. 2, pp. 259–274.

Timur, A. (1968), "Velocity of Compressional Waves in Porous Media at Permafrost Temperatures," *Geophysics*, Vol. 33, No. 4, pp. 584–595.

Wiggins, R. A., Larner, K. L., and Wisecup, R. D. (1976), "Residual Statics Analysis as a General Linear Inverse Problem," *Geophysics*, Vol. 41, No. 5, pp. 992–938.

Wine, R. L. (1964), *Statistics for Scientists and Engineers*, Prentice-Hall, Englewood Cliffs, New Jersey.

Ziolkowski, A., and Lerwill, W. E. (1979), "A Simple Approach to High Resolution Seismic Profiling for Coal," *Geophysical Prospecting*, Vol. 27, No. 2, pp. 360–393.

ELEVEN

The Interpretation Problem

11.1 INTRODUCTION

At this stage most of the seismic reflection principles, field acquisition systems, and data-processing methods have been outlined. The role they play in reaching the overall goal had formerly been played down, but we can now reach out and use any of these tools, hopefully with a good deal of understanding, in our quest for the location of oil and gas fields.

The location of oil and gas fields is, logically, a purely (geo)physical problem. By this we mean that, given the properties of hydrocarbons—that they reside in the pore spaces of rocks and that gas, oil, and water have densities arranged in that increasing order—the entire process of finding them can be regarded as the application of seismic reflection methods only.

Unfortunately for the ego of geophysicists, this is not the case. Geologists, of varying special talents, are also employed in this task:

1. Because the location of fields must be economical.
2. Because of the limited resolving power (ability to see all the detail necessary).

The initial tasks of geologists lie in the direction of evaluating the available information (from surface outcrop information to detailed lithology and geochemical data available from well logs) to determine the areas in which geophysical work can be done to maximize the chance of finding oil and gas while minimizing the cost of drilling exploratory holes. This geophysical work may include methods other than reflection seismology.

The reflection seismograph was neither the earliest nor the sole method, nor is it capable of finding all the hydrocarbon deposits that still remain. A few million holes have been drilled around the world and the information from these holes, descriptive, geochemical, mineralogical, stratigraphic, and physical, has in the hands of geologists played a most important part in the location of oil fields and in producing *leads* that optimize the economical employment of the reflection seismograph. It is difficult, indeed, to assess the fruitfulness of reflection seismic work because it is so dependent on the state of exploration existing at the time. It has been most fruitful where all the other conditions for hydrocarbon generation and accumulation have been established (or taken for granted) and where the reflection seismograph was responsible for the delineation of the necessary anticlinal structure. Although this is by no means its only utility, it is the best known.

As a backdrop against which we can view the current use of seismic exploration methods it is useful to have a list of the expectations of exploration management. We can then judge to what extent these expectations can be fulfilled and what deficiencies will remain. The latter are then temporary hindrances to the overall exploration program or permanent hiatuses in knowledge that must, if possible, be filled elsewhere.

The objectives for seismology are, at least, the following:

1. To obtain adequate knowledge of the structure versus depth within the prescribed area. The structure of the major discontinuities must be described within this framework.
2. To ascertain the lithological sequence within the boundaries.
3. As a corollary of (1) a knowledge of the locations and types of any faulting that exists.
4. To locate, by direct hydrocarbon indicators, the concentrations of gaseous hydrocarbons.
5. To obtain a knowledge of sedimentation types as a function of depth, to assess, to the limit of the resolution of the seismic information, the possibility and location of changes of facies and isolated porous bodies such as sand bars, and to locate abnormally porous fracture zones and determine their boundaries.
6. To forecast pressure versus depth, in particular the depth to overpressured zones or other hazardous features that might threaten the safety of drilling operations.
7. To recommend optimum well locations for initial field discovery and, later, at the exploitation stage, those locations that have the greatest potential for hydrocarbon production.

These objectives are likely to be realized in several stages. In the beginning seismic work is guided by geologic principles—only in the most general terms. Later, after some wells are drilled, *local* knowledge of lithology, porosity, and velocities are obtained, but seismic results are still the principal bases for extrapolation of geology to areas surrounding the exploratory holes.

Finally, even after a field has been found, there is a need for detailed examination of seismic characteristics (which depend, in part, on the geologic characteristics of sedimentation) to avoid dry holes and to optimize the economics of production.

11.2 STRUCTURAL INTERPRETATION—THE EARLY YEARS

It was only when areas that could be explored by surface geology were nearing exhaustion—or the projections from surface outcrops were shown to be ambiguous (if not entirely erroneous)—that a need was felt for the reflection seismograph. The assignment was, of course, to map as accurately as possible the subsurface structure and tectonics, even though these features occurred in several cycles. These cycles were often unrelated because of erosional or depositional periods that gave rise to unconformities.

In some areas the seismic reflection method of 1940 (say) performed extraordinarily well, and most anticlinal structures found from detailed mapping with a closure of 50 ft had a high probability of being productive. It is no wonder that this led to the adoption of a tool with which, for the expenditure of a few thousand dollars, an oil field worth millions could be almost confirmed before the first hole was drilled. These areas are chiefly characterized by the following:

1. Consistent near-surface conditions for which the primary corrections can be made accurately from up-hole time and elevation information.
2. Persistent reflections at the requisite depths over large areas—the ability to correlate reflection events from point to point over the area under investigation. This now means invariability of layer thicknesses and rock properties.
3. The presence of gentle anticlinal structures and the (fortuitous) presence of source rocks and other rocks with reservoir characteristics, capped by impermeable layers.
4. The lack of steep dips and/or major faulting.

At a time when use of the reflection seismograph was regarded as an art and results where obtained only by experience it was fortunate that areas with these characteristics were widespread enough. Confidence in the method had to be built up before some of the hazards made their presence widely known. It is not necessary to dwell on these halcyon days. It is sufficient to say that simple procedures—correcting the times, transforming to depth by the use of a velocity that matched the well information, and mapping the subsea data for as many different reflections as could be handled—were adequate. The method prospered.

Faulting was inferred by finite isolated reflection correlations across a more-or-less linear zone on a map or, in many cases, by such negative evidence as the deterioration of reflection quality. Structural (anticlinal) traps and those that relied partly on faulting were drilled and added to the general information for the area, even if they did not produce.

11.3 STRUCTURAL INTERPRETATION—PROCESSING IN A MORE SOPHISTICATED ERA

The time of disillusionment soon came. False structural anomalies—owing to time anomalies caused by velocity changes in the near surface or in the subsurface as a whole—were soon found. Deterioration of reflection quality sometimes presented problems and resulted in false structural indications due to miscorrelations. Greater emphasis was then placed on improving record quality—first by the elimination of seismic artifacts or nonreflection events—by patterns of sources and receivers and brute-force averaging and then, later, by the invention of methods that achieved some degree of consistency in the source and receiver characteristics.

In retrospect, the methods of structural interpretation changed only in minor ways—institutional prejudices or preferences largely being responsible—even up through these major seismic revolutions. The development of the Vibroseis method during the early 1950s coupled the use of surface patterns of geophones and source positions with a reproducible source and recording on magnetic tape. By the time the use of the Vibroseis method was routine in the early 1960s a further addition—phase control of the vibrators—achieved a degree of consistency that had long been sought. The use of CDP recording, in spite of its unwelcome increase in unit cost, was demanded because of its utility in giving better data quality—but largely for structural interpretation. It is no surprise to practitioners that there are still some areas that, despite the existence of a layered sedimentary system, still refuse to yield adequate seismic reflections on which to base a reliable structural interpretation.

In the CDP and Vibroseis methods on land and in the use of air or gas guns in marine areas the seeds of an additional revolution in reflection seismic interpretation were laid. The source output could be made consistent for the first time, and some control could be exercised over the presence of multiple reflections. The innovation of redundant data gathering allowed seismic velocities to be calculated accurately *and often*. Basically, the X^2 versus T^2 method of velocity determination had been in existence for many years and had been used sparingly to provide velocity information in the absence of well velocity surveys. In marine work the achievement of consistency was relatively easy because of the constant characteristics of the medium in which the source and receivers were immersed.

Because our task is to teach the principles of interpretation as practiced today, the combination of a consistent source and CDP methods is taken as the basis for further discussion. We summarize the *prime needs* for interpretation:

1. A seismic reflection record section that has established a high signal/noise quality by CDP methods. The degree of redundancy must be determined by the need for such quality.
2. The removal of as many nonreflection (including multiple-reflection) events as possible.
3. The use of consistent source outputs and receiver input and recording characteristics (in other words, changes on the record section, trace by trace, must be due to a change in geology and not to method characteristics).
4. The effects of near-surface abnormalities (at this stage, chiefly time delays) must be removed as accurately as possible.
5. In the first instance, the time section should be produced by carefully following the principal primary data processes. In this context we list the processes that may be needed:
   a. Muting
   b. Energy equalization
   c. Expansion
   d. NMO and static corrections (preliminary)
   e. Stacking
   f. Predictive deconvolution (before and/or after stack)
   g. Filtering
   h. Preliminary velocity determinations

The seismic interpreter must assume responsibility for setting up the processing schedule, which includes the determination of the processing parameters that must be known before the digital processing is done. The need is to obtain one specimen section from each area, which will allow the determination of the residual problems before the data processing is finalized for the area as a whole. This role of the interpreter is often the most difficult and time-consuming one he will have. The prime requisite is to obtain data that in the time domain *is truly representative of the geology and minimally affected by artifacts*.

The disturbing events to be removed (if possible) fall in the following categories:

1. *Residual interference.* This source-generated noise cuts across the reflection events at various angles, depending on the phase velocity along the surface. This should have been removed by patterns of sources and receivers but may, in particular parts of the area under investigation, become very strong because of changes in the weathering or surface. If the events (best seen on the primary or 100% records) can be picked, this residual interference can often be removed by two-dimensional filtering.

2. *Lack of alignment of reflections* (because of inadequate near-surface corrections). The cycle of residual weathering corrections and velocity determination, followed by NMO removal, can be repeated. A final stack may give considerable signal/noise imporovement (see Figure 10.10 for a good example).

3. *Noise.* Random noise should have been removed during recording by using an adequate redundancy in the CDP work and by a proper choice of source and receiver patterns. It is often the result of natural sources that generate waves which arrive at the receivers from random directions. Radial patterns, although more expensive to use in the field, may be preferable to the linear patterns commonly used to reject source-generated interference. Nevertheless, given the fact that the noise has been recorded on each of the primary records, the following procedures may help to reduce its effect.

   a. Study the effect of successive narrow-band filters on the data. The visual coherency of events on these filtered records often shows which frequency bands should be retained. Simple frequency filtering can be used to improve the signal/noise ratio.

   b. Some improvement may be obtained by a weighting process in the stacking procedure; for example, if the primary records are *very* noisy, the individual records can be weighted individually in inverse proportion to the total energy (assumed to be noise energy). This process of diversity stacking is often helpful in noisy areas (e.g., recording near highways or in towns).

   Events such as diffractions and crossing reflections due to highly curved reflecting surfaces have not been included, for they are indicative of geological features and in any case will be minimally disturbing after a later process of digital migration.

4. Multiple reflections, however, are dangerous to interpretation when they are the long-period type. Discussion of the usefulness of short-period or peg-leg reverberations which act to improve the signal/noise level of deeper reflections

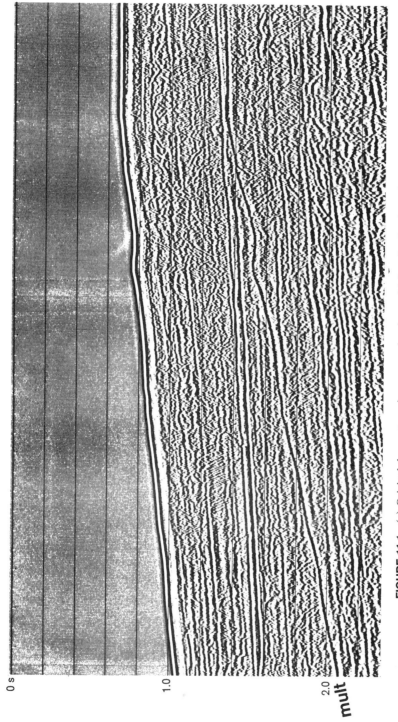

FIGURE 11.1 (*a*) Original deepwater seismograms showing multiple reflection from the sea bottom. (*Source:* Courtesy of Prakla-Seismos, GmbH.)

0 s

1.0

2.0

mult

423

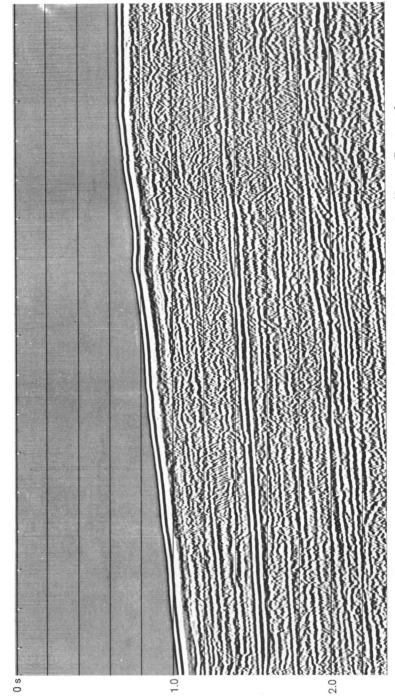

FIGURE 11.1 (*b*) The effective removal of sea-bottom multiple by data processing. (*Source*: Courtesy of Prakla-Seismos, GmbH.)

has already been given. The periods of the long-period multiples can be ascertained by autocorrelation of the seismic traces; the multiples then show up as ghosts—subsidiary pulses at multiples of the multiple time delay. If more than one major long-period multiple is present the ghosts, which consist of differences in individual multiple delay times as well as occurrences of the multiple times themselves, are more complex.

The most promising ways of removing such ghosts are the following:

a. *Gapped deconvolution.* The operator is obtained by assuming a desired output which consists of a unit pulse followed by a number of zeros (out to the desired time) followed by finite amplitudes out to the length of the operator desired.

b. *The Backus (1959) operator.* This consists of convolution with the inverse operator to a simple series of pulses, the amplitudes and signs of which are known. If isolated, very prominent multiples are present, these amplitudes and signs will be known from the trace autocorrelation function.

c. *The use of filtering capability in the p-τ domain.* It is now well known that multiple reflections are exactly periodic in the p-τ domain, whereas they are not in the X-t domain. As a consequence the autocorrelations of traces in p-τ are much better for the indication of multiples and better filters can be derived for the removal of long-period multiples.

Examples of sea-bottom multiple removals are given in Figure 11.1 (courtesy of Prakla-Seismos, GmbH). The final processing desired is thus established. Beyond this, there should always be a relative amplitude section and a migrated section—made with the best velocity information available—provided that the necessary signal/noise ratio has been established. There are areas for which the interpreter will decide that it is a waste of time to make the latter two sections.

11.4 THE MECHANICS OF MAKING A STRUCTURAL INTERPRETATION

There is no way in a summary to deal with all the idiosyncracies of individual areas. All that is attempted here is to lay out some of the methods, the information needed, some of the problems that will be encountered and, hopefully, some solutions.

The first step is to gather all available information on the area—well logs, correlated and with the major horizons marked and named. Among these logs (if indeed *any* logs are available) should be at least one sonic or well velocity log (preferably of the long-interval type) and one density log. It is obvious, from our earlier considerations, that the sonic log(s) should be run over as long an interval of the well as possible, ideally from the base of the weathering to the total depth, although the upper limit usually depends on the surface casing set, hole diameter, and velocity of the sediments near the surface. Synthetic records should be made for all of them by using an input pulse with a frequency spectrum similar to the final playback filter adopted and a phase spectrum determined by the field method and processing. A deconvolved impulsive seismogram and a correlated Vibroseis record are both nominal records that use zero-phase pulses. SAIL synthetic records should be produced.

The field effort—with which the interpreter should have been coordinated—should have included one field line that passes as near as possible to one of these wells.

Even in 1985–1986 the number of VSPs is still comparatively small. If, however, one is available in a nearby well—or is even thought likely to be available in a future well in a known location—the implications for much more informed seismic identification of reflecting horizons make a well tie imperative.

Otherwise, the identification of reflections and the determination of the time lag to use between field and synthetic data is accomplished by a straightforward visual correlation between the synthetic SAIL trace and SAIL field traces taken near the well. This correlation is often easy to determine on a zone-to-zone basis. Slight shifts in time are sometimes necessary between zones because of incorrect well velocity measurement (invaded zones or shale alteration, coupled with an inadequate sonic tool length).

In this manner a basic grid of geological–seismic arrival time pairs is made for a number of geological horizons over the area. Even though only deep horizons may be required it is good practice to produce at least one shallow and one intermediate horizon correlation, correlated from trace to trace on the entire set of seismic cross sections and tied from one to the other at line crossing points. This procedure seems simple but can be beset with difficulties, some of which are now considered.

1. *Faulting.* The continuity of the reflection is interrupted. It should be noted whether this fault is substantiated by the presence of a diffraction. The picked event must be ascertained by correlation across the fault. It is desirable to be able to correlate between two events a few traces from the fault location on either side. This is done most easily when the bandwidth of the record section is as wide as possible, consistent with a high signal/noise ratio. Any faults picked needs to be coordinated from cross section to cross section to make them geologically feasible. Faults are most often clearer on migrated sections because of the proper placement of the diffracted events. In Figure 11.2, an excellent marine seismic section migrated by Western Geophysical, I have had no difficulty in marking the discontinuities due to some of the faults. Horizons *A* and *B* can clearly be correlated across the majority of the faults, but what happens to *A* as we proceed to the right is the subject of dispute. In a complexly faulted area the interpreter must not lose sight of the fact that dips may not only exist in the plane of the cross section but may come about by reflections from the side of the plane. For this reason the lines of profile in complex areas are usually run down the direction of maximum dip (if known). The lines should then be placed in locations in which the geology appears to be least complex.

It is a mistake to assume that faulting, even of geological strata with a high contrast in acoustic impedance with the neighboring rocks, is invariably noticeable. We have already shown that a discontinuity gives rise to a diffraction that changes sign at the break. Then, for two halves of a broken formation, the

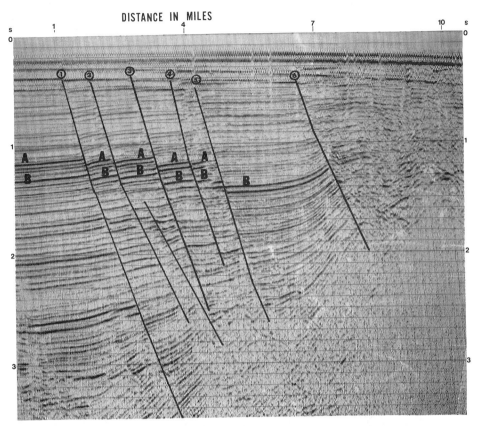

FIGURE 11.2 A portion of a finite difference-migrated section showing faulting. Even in this excellent example, taken by itself, the question of what happens to some horizons (such as *A*) as they cross a fault cannot be resolved. Is *B* marked in the correct position to the right of fault 5? (*Source:* © 1975, Western Geophysical Company of America.)

positive half from one side is superimposed with a time delay on the negative half from the other side. If the delay time (due to the throw of the fault) is small or equal to one period of a particular frequency the diffraction at that frequency will be nearly eliminated and will be small for neighboring frequencies. The proper frequency analysis of diffracted events may give a clue to the throw of the fault, although we have no knowledge of an attempt. In any event, comparison has to be made with the frequency analysis of the reflected event to eliminate the effects of stratum thickness. Often, however, after faulting, one stratum is displaced into a position in which little contrast exists and the diffracted event is eliminated.

A question always arises of the likelihood of reflected events from the fault face. It really should be asked whether diffracted events from individual scatterers are persistent enough along the fault plane to organize themselves into a coherent reflection. Usually this means a set of scattering points along a break where *one* material is placed opposite a different unique material over a distance of several wavelengths of the seismic wave. This distance can be taken to be hundreds of meters in the usual seismic case; hence a fault of considerable throw is involved.

In marine work offshore South Africa a number of indisputable fault plane reflections can be seen clearly. One example of a reflection is given in Figure 11.3, which has been migrated by early finite difference methods that show dispersion. This dispersion has been largely removed in data processing by the frequency domain method (Figure 11.4); 80% of the stacking velocity is used as migration velocity. In these areas throws of many hundreds of meters are not uncommon, bringing Paleozoic sediments against a near homogeneous basement rock. Although the reflection coefficient is probably not constant along the basement rock contact, it must always be strong. For other areas in which the break is not so well defined it is better to talk about a distribution of scatterers on a fault surface than about a reflection from the surface. Even if rock distortion and fracturing have resulted in a rubble zone with different fluid contents and acoustic impedance from the rocks laterally in contact with it, *reflection* is probably too precise a

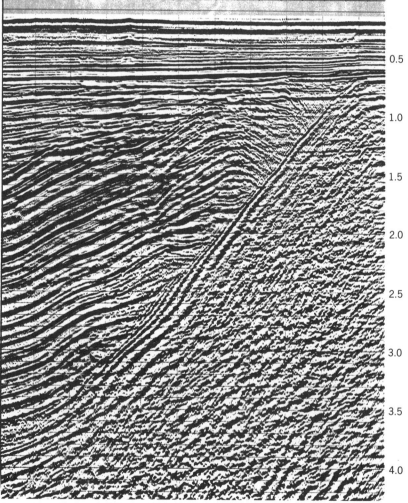

FIGURE 11.3 A large fault shown in marine data offshore South Africa. Note the clearly visible dispersion along the fault face (near 2.7 s) on this finite difference-migrated section. (*Source*: Data from Offshore South Africa, courtesy SOEKOR.)

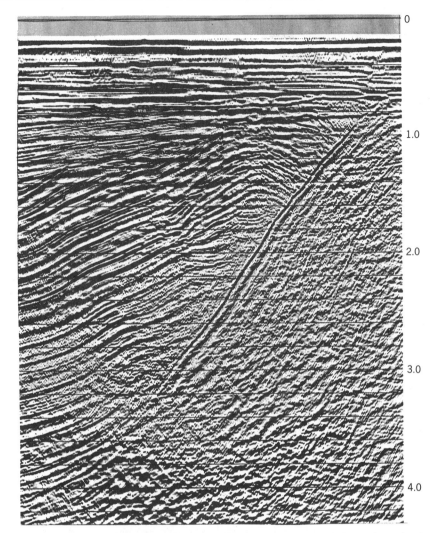

FIGURE 11.4 The same data as the original for Figure 11.3, but this time migrated using ω-k migration. (*Source*: Data from Offshore South Africa, courtesy SOEKOR.)

description to fit the occurrence and the distribution of scatterers is to be preferred. Migration of the data, if done with the correct velocities, causes the energy of the diffractions to pile up at the fault zone as long as the scattered points are evenly distributed about the vertical plane of the seismic cross section. Lack of even distribution, e.g., scattering points predominantly on one side of the plane of section causes the diffracted events to have an improper geometry to be condensed to the point-scattering pattern, and this results in smearing the fault zone.

2. *Truncation of strata by unconformities.* An unconformity is a surface of erosion or nondeposition that separates younger strata from older rocks. It is axiomatic that, as far as the seismic reflection method is concerned, this

surface will not be distinguishable from a bedding plane if, locally, the planes of deposition of the younger and older rocks are parallel. In a local area in which an *angular* unconformity exists the upper rock layers are likely to be continuous and (see Figure 11.5) the older ones are truncated by the unconformity. If the lower rocks were subject to peneplanation (erosion to an almost smooth surface) before the younger rocks were laid down the effects are due largely to the wedges of formations as they are gradually truncated. Rapid subsidence of an immaturely eroded surface, however, may have resulted in an uneven unconformity surface and consequently the probability of diffracted events.

If the angular unconformity has a small angle the truncation results in a gradual loss up-dip of the younger formations. By taking the simple view of a constant-velocity formation, gradually reduced in thickness, it is evident that the seismic record will progress from a full rendition of the layer reflection by changes in the shape of the reflection as the layer thins and finally (for a given frequency band) to a gradual loss of amplitude. The more complex cases of variable velocity zones within a given lithological zone can be dealt with by the methods of synthetic record construction for a sequence of different velocity functions. There are, however, two important points for the interpreter to recognize:

1. The amplitude of a reflection from a truncated formation diminishes to a level below the noise level (or below the side lobe level for neighboring reflections) before the formation reaches zero thickness. Thus the mapped zero edge from seismic work is still down-dip from the actual zero-thickness line. The distance varies, depending on the acoustic impedance contrast of the layer being truncated.
2. If the truncation is gradual there will be no indication of the zero edge by diffraction phenomena.

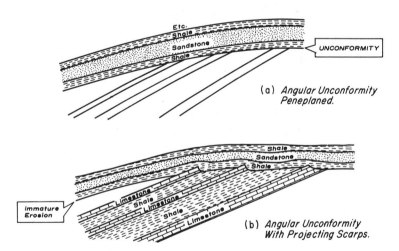

FIGURE 11.5 Two possible types of angular unconformity: (*a*) peneplaned, diffraction effects minimized; (*b*) immature erosion, with diffractions at the scarps. There may be some draping of shallower formations over the scarps.

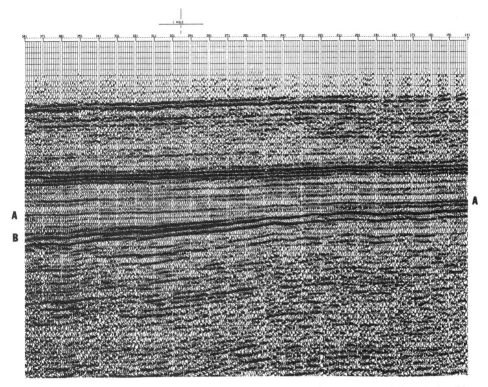

FIGURE 11.6   Truncation of the Hunton formation (B) by Mississippian rocks (A) as depicted in a modern, unmigrated seismic cross section. (*Source*: Courtesy CONOCO, Inc.)

Whenever the erosion of the older formation is immature scarp formation formed by projections of the more resistant rocks may occur. Now we have a case in which the reflections from a layer can terminate abruptly and diffractions are formed. The up-dip limb of the diffracted event must be recognized for what it is; otherwise the reflection will be continued and will show reverse dip. Decisions regarding the nature of a reflection-diffraction event are best made in conjunction with all the evidence from the area; that is, by a decision that an unconformity does or does not exist. Modeling programs in which synthetic records can be made for two-dimensional changes in stratigraphy that include the possibility of diffractions are described later. These programs may be of some help in making an otherwise difficult decision.

In mapping, of course, the reflection terminates at the up-dip limit of the truncated formation. Ties between lines must be made down-dip, where the formation is likely to have full thickness. One truncation—of the Hunton formation in Oklahoma—is shown graphically in Figure 11.6.

11.5   SEISMIC DISPLAYS AND COMPUTER MODELING

The method of display of intermediate and final results has gradually become important. With modern computer terminals the tendency has been to devise displays on a high-resolution color (TV) screen which is flexible enough to show great detail when required or to display the surrounding seismic results in lesser detail. For this purpose, of course, a versatile graphics software package is needed to accept keyboard information that will allow the selection from a massive memory of the correct information for display. Furthermore, the display, boundaries, and scale(s) must be variable.

Data acquired over a finely sampled surface grid, commonly called 3-D, has become much more common, especially in marine areas that have high potential for oil production. This data must be regarded as meriting special data-processing treatment because linear cross sections may eventually be made in any desired direction after the final 3-D migration processes have been performed. The density of information is great enough that for the first time it has been possible and helpful to plot subcrop maps of the reflection data. These maps are usually horizontal slices of the seismic data but there is no need for them to be horizontal. They could, instead, be slices taken just below an unconformity, provided that their time (depth) values are known at each migrated record location.

The effectiveness of these displays is often enhanced by the use of color, which is now almost universally used to provide superior discrimination over the earlier gray scale.

Although the individual interpreter may be effective in the use of a TV screen on which to trace events (and have them permanently recorded) with a hand-operated device (a mouse!) that drives a moving dot on the screen, thereby simulating the picking of events on paper cross sections, the final need for permanent hard copies is inescapable. In many of the terminal systems these permanent copies may, at will, be in black and white or color and in a number of different scales. Examples have been given in Figures 11.7 and 11.8a and b (see color insert in the middle of the book for color representation), which are, respectively, a combination of two vertical sections and a horizontal one (a *cube*) in black and white and the same combination in color. The displays are made by Polaroid photography on 35-mm color slides or, for greater size, on a Versatec color printer. The latter, of course, is a dot matrix device that derives its information from the computer memory and not from the display on the terminal screen.

The interpretation of a seismic cross section in geological terms (e.g., the sequence of lithologies, thicknesses, and fluids that fill the pore spaces) is not possible in an unambiguous sense. It has already been pointed out that for thin layers, compared with the wavelength of seismic waves, it is the product of the layer thickness and the contrast in properties that determines the seismic reflection output. Many possibilities are available, and it is the function of the exploration team to decide the most probable combination of parameters. Some incontrovertible pieces of evidence are the seismic reflection cross sections and the well logs available in the area. Putting the two together and then interpreting the most likely conditions that exist at places in which there are no wells are sometimes aided by the ability to forecast what the seismic output should be from a hypothetical set of geological and seismic parameters. This is the sequence of steps called *modeling*. It is usually a process of examining the seismic data in detail after selecting a time zone or a depth zone known to be of interest in oil

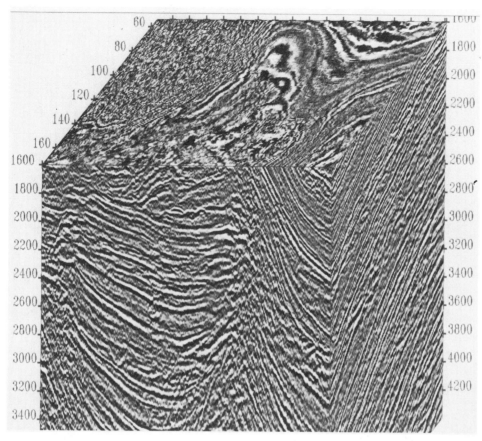

FIGURE 11.7   A black and white illustration of a *cube* of data obtained from a 3-D marine survey. The top face represents the subcrop of reflections (taken at a time of 1.600 s); the sides are vertical cross sections in two orthogonal directions. (*Source*: Courtesy CONOCO, Inc.)

production. It may be a truncated zone in which the thicknesses and possible fluid contents of the formation change, or we may (later) be considering the stratigraphic trap problem in which the lithology, thickness, and fluid contents may change simultaneously.

The basic tool is the synthetic seismogram and the basic equipment is the interactive computer work station, a time-shared computer terminal equipped for inputting and outputting reflection data and logs. The methods of making synthetic seismograms have already been described (Chapter 4). From the point of view of the explorationist user the computer programs should be *friendly*; that is, easy to understand and use, with help available within the program in difficult cases. Figure 11.9 is a schematic of an entire terminal complex. It has an overall function that allows the interpreter to interact with the computer and to obtain the best fit possible of the geological input with the known seismic output of the earth.

The terminal consists of the following visible parts:

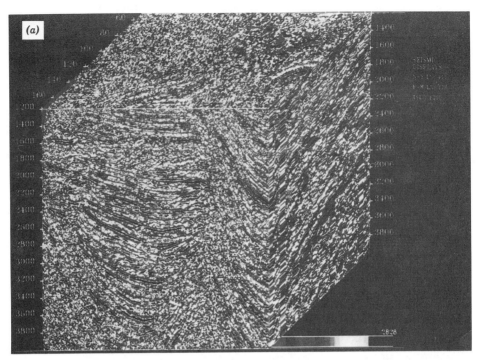

FIGURE 11.8 (*a*) The same data and presentation in Figure 11.7, except that a color scale is used to represent amplitudes. See color section in middle of book for color representation. (*Source*: Courtesy CONOCO, Inc.)

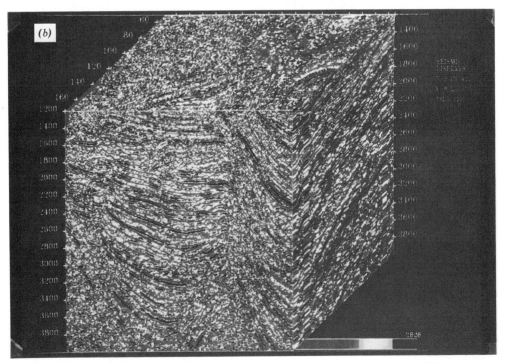

FIGURE 11.8 (*b*) The same data and presentation as Figure 11.7, except that a different color palette has been selected. Note the psychological importance of the subjective choice of the color palette, which can be made to suit the individual preference. See color section in the middle of the book for color representation. (*Source*: Courtesy CONOCO, Inc.)

434

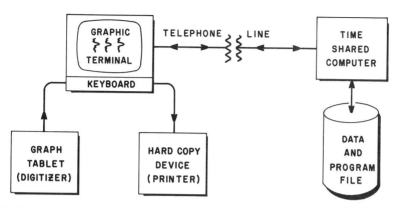

FIGURE 11.9 A schematic of a typical interactive computer terminal used in exploration seismic modeling.

1. A keyboard and a visual monitor [a cathode ray tube to display graphical (and alphanumeric) output and to retain it long enough for the interpreter to decide whether it should be permanently recorded].
2. An input graph tablet that allows the easy and comparatively fast input of logs and model descriptions of all kinds.
3. A means of obtaining a permanent record (hard copy) of any results displayed on the screen.
4. A connection device that allows access to a fast digital computer operating in a time-share mode. By this method the available computer time is allocated to each terminal in small time slices. Return is so rapid that the user has the impression that the computer is under complete control.

The invisible part, the programs that control the computer, is complex. It consists of monitor program(s) that control computer action as a whole and user programs to which the computer has access to complete the special jobs the interpreter requires.

Each institution's modeling system is different, depending on the philosophy of the originator and to some extent on the type of main computer and peripheral equipment available. Usually, however, the system of programs is modular; access to and exit from any particular program is under the control of an executive program. The function of the executive is to gather together initially all the general parameters that are input and communicate them to the individual modules as needed. This facilitates entry into any desired module and, on exit, makes it possible to enter another module to obtain a hard copy of results and/or to exit in a tidy manner from the modeling system. All this is concealed from the user, who normally supplies answers to questions posed by the system and displayed on the graphic terminal screen. An ideal modeling system, once the user has been led through it and the terminal controls have been explained, requires only geophysical knowledge and an understanding of its overall capabilities. Figure 11.10 shows how the executive controls the individual modules and isolates the user from technical knowledge of the computer monitor program.

The one-dimensional modeling system needs little comment in addition to the description of synthetic trace construction given in Chapter 4. The purpose,

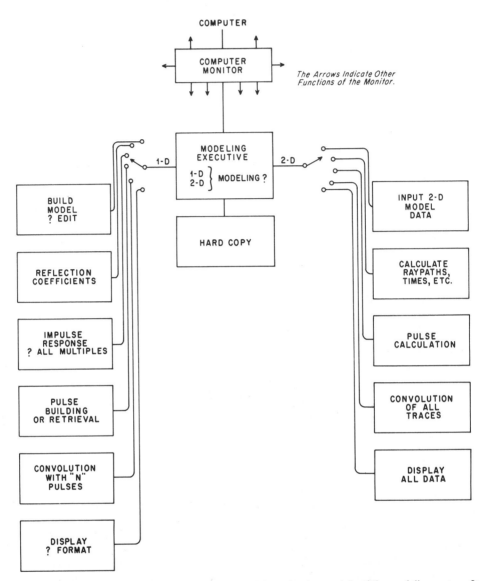

FIGURE 11.10   A schematic showing the function of the executive module of the modeling system. It allows the interpreter to order the execution of several possible functions without exact knowledge of computer requirements.

however, is to help the interpreter make a decision in regard to what the characteristics of the velocity and density logs have to be to give a seismic output that is already on hand. Thus only a small section of these logs is used. Facilities must be available to make modifications quickly (editing) and to display the resulting effect on the seismic section. Multiple reflections *within the section being examined* are important only if the section is likely to contain strong reflection coefficients (e.g., a gas-filled sand in conjunction with a shale and/or a water-filled sand). Multiple reflections (including reverberations or peg-leg multiples)

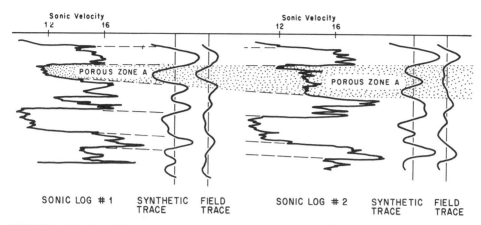

FIGURE 11.11 Two different sonic logs show different thicknesses of the porous zone *A*, with their synthetic traces and field recorded traces. The field traces were integrated and filtered 9 to 30 Hz and an empirical 30° phase correction was applied.

that occur in the remainder of the log, including the near-surface layers, can best be taken into account by changing the input pulse shape. Examples of the output from a one-dimensional modeling system have already been given (Figures 8.5 and 8.6).

These examples, however, were heuristic models designed to teach a specific point concerning the output of thin beds, and the velocity log was not realistic. In Figure 11.11 is a more realistic example. Details concerning the log and the pulses making up the section are given in the legend.

One further remark must be made about modeling. The fact that a modeling program is available does not mean that it is possible to obtain a more detailed description of the lithology that gives rise to a given seismic output. The amount of detail possible is controlled by the bandwidth and frequency range of the seismic output. The proper use of a modeling program, however, ensures that the lithology determination is *one* possible answer to the question of matching that output. If the interpreter plays the interative terminal game thoroughly, even when he restricts himself to the confines of the parameter boundaries, several possible answers emerge. His choice has to be based on other factors that can be brought to bear on the problem. One may be the existence of shear reflection records. The modeling program is, of course, available for use with shear reflections and only the velocity log need be changed. The method of achieving a fit between P-wave and S-wave reflection data to make a better determination of the lithology, porosity, and pore fluid content is in its infancy and is not pursued here.

11.6 TWO-DIMENSIONAL MODELING—THE RAY PATH METHOD

The one-dimensional model, of course, relies on the assumption of horizontal plane-parallel layering. It gives an idea of the seismic output from a geological section if structure is of no consequence; that is, if there are no interfering reflections from dipping formations to the side and no abrupt strata terminations give rise to diffracted events.

To gain a better insight into the conditions that exist in practice the next step is to consider a two-dimensional section, shown in two slightly different forms in Figure 11.12. In these idealizations of truncation it is necessary to compute actual ray paths because some of them are not vertical (even for a common source-receiver point). In addition, it is desirable to be able to consider diffractions from terminating reflectors. Most two-dimensional modeling programs have facilities for calculating offset traces, with limited facilities for including multiples and, by extension, the possibility of creating *n*-fold stacked records. Ray-path plots for models A and B are shown in Figure 11.13*a,b*. The edge diffraction events have been suppressed in the plot to avoid confusion.

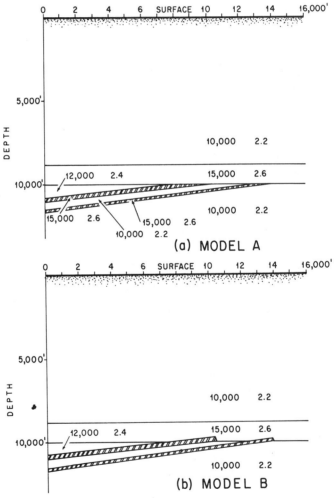

FIGURE 11.12 Two different models, with and without peneplanation of the unconformity. Velocities and densities for models *A* and *B* were identical.

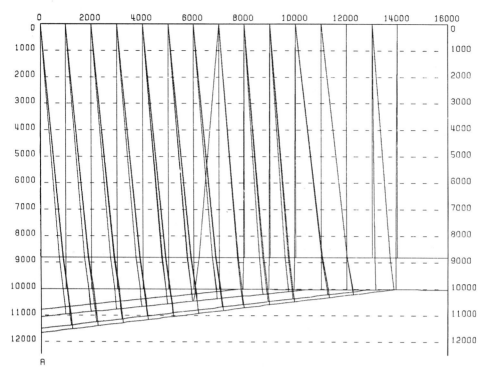

FIGURE 11.13 (*a*) Ray paths for model *A*. Ray paths associated with diffractions are calculated but not included in this display.

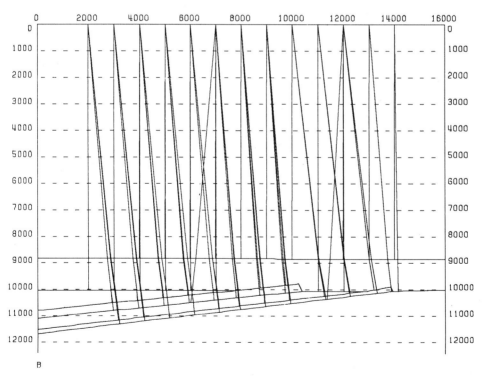

FIGURE 11.13 (*b*) Ray paths for model *B*. Ray paths for diffractions are calculated but not included in this display.

The impulse response for each source-receiver combination is calculated with the two-dimensional program, and it is important that proper consideration be given to the form of the impulse response for different types of event which are controlled by the type of stationarity of the time path involved. We have already seen that for a reflection from a flat surface there is no change in wave characteristic and the response to a delta-function input is also a delta function at the appropriate reflection time. This is not the case, however, for other events. By use of diffraction theory it is possible to calculate the impulse response (Stolt, 1975).

1. *Travel time minimum* (*normal reflection*). This is the type of reflection from flat, convex, or slightly concave surfaces in which the focal point, if any, must lie above the surface observation point. In the model the form of the output is a delta function whose amplitude is controlled by the radius of curvature and the depth of the reflector. Spherical divergence (or convergence) from the reflecting surface is thereby accounted for.

2. *Travel time maximum* (*buried focus*). If the surface is so concave that the wave reaches a focus below the surface a travel-time maximum will result. This gives rise to a characteristic shape of the form $(t - t_0)^{-1}$, which is an amplitude time shape with a 90° phase. The intensity is also proporational to the inverse of the curvature of the travel-time curve; therefore the deeper the focus, the smaller the amplitude of the event. (See Figure 11.14a.)

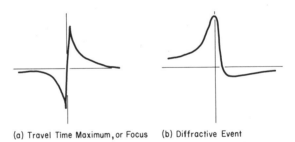

(a) Travel Time Maximum, or Focus   (b) Diffractive Event

(c) Edge Diffraction

FIGURE 11.14 Different types of pulse arise from different types of seismic event. Only reflections from flat or convex surfaces (compared with the incident wave front) give a delta-function reflected pulse for a delta-function incident pulse. (*Source*: Stolt.)

3. *Diffracted event.* The travel-time minimum and maximum correspond to points at which the slope of the travel-time curve, measured along the reflecting surface, is zero. Higher order derivatives of the travel time also produce events. The principal event is caused by a minimum (in absolute value) of the travel-time derivative. Its characteristic shape is shown in Figure 11.14*b*. Its amplitude depends on the first and third derivatives of travel time at the diffraction point. The event may be expanded or contracted in time, depending on the values of the first and third derivatives but again most strongly on the first. Thus at points at which the first derivative of travel time is near zero there is a strong peaked response. The diffractive event can be thought of as a maximum and minimum occurring at the same point. When moving from trace to trace along the seismic section we often see diffractive events split into a maximum and minimum. Although the minima tend to move along the source-receiver surface with the source-receiver midpoint and the maxima tend to move in the opposite direction, the diffraction points tend to remain stationary as long as they exist.

4. *Edge effects.* Whenever a reflecting surface terminates (or abruptly changes slope) it gives rise to an edge-effect diffraction. The amplitude of this diffraction varies inversely with the first derivative of the travel time at the edge, and the speed at which it dies out depends on the first and second derivatives at that point. The general shape shown in Figure 11.14*c* was derived in Chapter 7. At the edge along which several boundary segments join the total response is the sum of the responses from each segment.

These effects, which are incorporated in the modeling program, are of no concern to the interpreter. When needed they are incorporated in the impulse response for each source-receiver combination.

A two-dimensional modeling system normally incorporates NMO removal programs to facilitate stacking the model traces and a convolution program that allows filtering in any manner necessary to try to match the filter characteristics of field traces. This question of the filter characteristics of seismic records is most important in modeling and in interpretation in geological terms. Besides the question of resolution, however, there is the representation of a geological boundary and the ease of conversion of the reflection results into geology.

Although this point has been approached before, it is important enough to be reemphasized in the modeling context. As we know well by the time the seismic record, in a simple no-multiple form, can be envisaged as the sum of pulses, all of the same shape, of different amplitudes and arrival times. If these constituent pulses are long and complex in form the effect of one boundary will be inextricably confused with those of other boundaries. There are two forms of pulse that can be used to the best advantage. These pulses, in practice, are obtained from Vibroseis records after the cross-correlation process. Each symmetrical pulse corresponds to a reflection coefficient or, more exactly, to one impulse of the reflection impulse response. Proper deconvolution of other types of record, however, produces pulses of the same shape. Another pulse we have used is the integrated form of the symmetrical pulse, derived in the SAIL process. The latter has the advantage of making a boundary look like a boundary. In the models constructed for the illustration of the two-dimensional modeling system only these two types have been used. Figure 11.15 shows the seismic output from model A which uses an integrated pulse with a nominal bandwidth of 7 to 90 Hz and a taper of 4 Hz at the low end and 20 Hz at the high. Tapers are centered on the bandwidth frequencies given.

Disturbances and confusion in the vicinity of 1.95 s for traces 6, 7, 8, and 9 could have been anticipated, but the actual form would not have been known without the modeling. At these frequencies it is possible to discern the boundaries of the 200-ft (61 m) and 100-ft (30.5 m) limestone beds truncated along the 10,000-ft (3048 m) depth line. The sense of the velocity change at all boundaries is graphically illustrated. Although these same things can be seen by the interpreter when a normal, nonintegrated pulse is used for convolution (in Figure 11.16), the tie-in with geological boundaries is not quite so pronounced. In both presentations the location of diffracted events is unclear.

The composition and mechanics of the two-dimensional modeling system must not be left without a reminder that it is still an artificial model of the earth. The assumption is made that the model cross section is constant in the direction perpendicular to the plane of cross section. A comparison with field records can be made only if they were obtained on a line directly down-dip. In areas in which dipping events exist, a unique direction (for shallow and deeper zones) may not exist and it is indeed fortuitous if a seismic line runs in the required direction. The

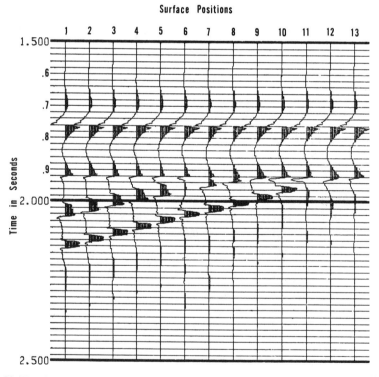

FIGURE 11.15 The output of model A for a 7- to 90-Hz integrated pulse. The disturbance and confusion of traces 6, 7, 8, and 9 near 1.95 reflection time could have been anticipated because of the juxtaposition of similar materials.

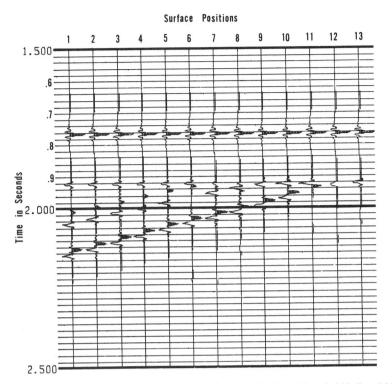

Surface Positions

FIGURE 11.16 The output from model *A* with a nonintegrated pulse of bandwidth 7 to 90 Hz. The tie-in with geological boundaries is not quite so pronounced as in Figure 11.15, although the positions and the pulse senses show the boundary positions.

effect of reflections coming from out of the plane of cross section must therefore be considered.

It is only to be expected that the older acoustic 2-D modeling programs have lately been supplemented by the *elastic* 2-D. With the availability of much faster computers the elastic programs can help to determine the influence of mode conversion on expected seismograms for a given geologic situation. Except for rather simple geometric situations, however, there is still a limitation (for the elastic case) to plane, parallel layering, a limitation that is particularly frustrating when stratigraphic traps are the exploration targets.

11.7 STRATIGRAPHIC TRAP LOCATION—TRANSFER FUNCTION USAGE

Application of the seismic reflection method to the location of stratigraphic oil traps has been a favorite subject for discussion and papers. Unfortunately there appears to be no common understanding of what constitutes a stratigraphic trap. It is usually defined from the point of view of geologists; that is, a stratigraphic trap is one that does not rely on structural or tectonic processes to trap hydrocarbons. As a corollary the process involved must involve changes in the permeability and porosity of the host sediments achieved by sedimentation or metamorphic processes. The examples usually given are those in which porosity in a sandstone body silts up until the pore spaces are not large enough to allow the passage of hydrocarbons or a change from limestone to dolomite by the action of magnesium-bearing salts in groundwater causes porosity because the specific volume of dolomite is lower than that of calcite.

If the stratigraphic trap does not involve structure or faulting from the seismic point of view it is not possible to find a stratigraphic trap simply by measuring changes in the arrival time of the *associated* reflection. Note that this does not preclude indirect evidence of the possibility of a stratigraphic trap by time interval changes in which the interval includes the trapping formation. Although this definition undoubtedly has some holes and is specialized to information the geologist does not always have, it is adopted here because it leads to a clear-cut break between structural and stratigraphic traps. With the latter we simply abandon one piece of evidence the seismograph easily provides. What seismic parameters are left? The easy ones, of course, are time differences and comparative reflection wave shapes. The cross section, which has been flattened by using a strong, consistent shallow marker as a reference time—made constant by shifting all traces—is a seismic analog of the isopach cross section of the geologist and is just as useful. Emphasis must be placed on the criterion of consistency for the reference reflection. If it is known or can be assumed to have been derived from a constant-thickness, constant-lithology-type layer then changes in its character or time of arrival must be due to near-surface layering. In the critical analysis of seismic data for changes due to changes in lithology or bed thickness effects of the surface and near-surface must be removed. This has been done by the use of transfer functions.

These transfer functions are simply filters, usually digital filters, by the use of which the character of a particular zone of one trace A can be made to look *exactly* like the same zone of a second trace B. We simply convolve trace $A(t)$ with the transfer function $T(t)$ to arrive at $B(t)$:

$$A(t) * T(t) = B(t) \qquad (11.1)$$

Leaving, for the moment, the manner in which $T(t)$ is determined, we can see that if the shallow portions of two records are made to look alike and if, in fact, they *are* geologically similar the records would have to have been produced with input pulses that are alike. If the early portion of trace A had to be treated with the transfer function T to make it look like B it is plausible that the later portion of A must also be treated with T to be able to compare A with B in the later zones. This filter takes into account the phase and amplitude spectrum deviations of the source pulse A from that of B and corrects any time and amplitude spectrum errors.

The determination of T is made directly from the sequences $A(t)$ and $B(t)$ early in both records. It is easiest to visualize in the frequency domain; therefore it is assumed that

$$A(\omega)T(\omega) = B(\omega) \qquad (11.2)$$

and

$$T(\omega) = \frac{B(\omega)}{A(\omega)} \qquad (11.3)$$

where A, B, and T = complex functions of ω. There is a corresponding matrix method for the time domain determination of a convoluting pulse.

The operation is particularly necessary when several different vintages and methods of shooting have been used in the area under investigation. It should be noted that the two traces are brought into conformity only within the common bandwidth and, at best, the bandwidth will be left the same with considerable likelihood that it will be reduced. In the determination of $T(\omega)$ any values of $A(\omega)$ that approach zero cause instability and $T(\omega)$ must be (artificially) set to a number determined by other considerations. In calculating $T(\omega)$, it is only the smoothed amplitude and phase spectra that are to be made similar because any rapid variations in amplitude and phase are due to the relative positions of reflected events and are not a function of the primary pulse.

11.8 STRATIGRAPHIC TRAP LOCATION— GENERAL METHODOLOGY

Additional seismic evidence of stratigraphic trap determination is provided by measurements of the seismic interval velocity and by comparison of consecutive seismic traces by the various methods (Section 7.7) of measuring trace similarity (Waters and Rice, 1975). Both parameters are reduced in effectiveness by the fact that they are averaged over a substantial time interval compared with the time thickness of most stratigraphic oil traps. If the record section has adequate bandwidth the role of *local* (interval) velocity indicator is taken over by the SAIL-type trace, for, as previously pointed out, the trace deflection is, within limits imposed by the frequency spectrum, roughly proportional to the product of the velocity, density, and thickness. The answers sought about changes in lithology, porosity, and pore fluids are all in the SAIL section—as far as they can be determined by the seismic method with this spectrum. Thus the similarity coefficients and statistically measured interval velocities can be plotted and their gradients will fulfill the role of attention getters. However, the absolute value of velocity for a portion of the section plays an important role in deciding on the type of section; for example, clastic versus carbonate or evaporite or normal pressured versus overpressured.

Because the stratigraphic traps so far discovered have little in common geologically, it is difficult to describe a set of invariant methods to discover them. The general methodology is nevertheless the same, with variations in interpretation caused by different geological conditions. The general procedures are the following:

1. Obtain a set of seismic cross sections with maximum bandwidth and a maximum signal/noise ratio by the use of highly redundant CDP procedures.
2. Process these records in such a manner that reflection events are symmetrical pulses of maximum bandwidth. It is important that at least a set of relative amplitude-preserved sections be obtained.

3. Integrate the processed records to obtain SAIL-type sections that may also be relative-amplitude.

4. If, on inspection, it is found that there is enough structure to influence section complexity (e.g., synclines that would give a focus below the surface), then the best migrated sections possible should be made from the relative amplitude-processed and SAIL cross sections. These form the first part of the seismic evidence.

5. Prepare a contoured section of the subsurface interval velocity of the proper scale to overlay the seismic cross sections. These are often filtered in two dimensions to remove noise due to low-accuracy velocity determinations.

6. Prepare a contoured similarity coefficient overlay of the correct scale to overlay the seismic cross sections.

The initial data are now at hand. Because stratigraphic trap location is of the same order of difficulty as looking for a needle in a haystack, there must be a preliminary winnowing of this material to determine which zone offers the best chance of finding a stratigraphic trap. If anything, other than the seismic data, is known about the area under investigation it is normally a task for geologists who can examine well logs and samples to grade the different depth zones for source and reservoir potential. If this is not possible the sieving procedure devolves on the seismic interpreter.

The most obvious indication of source rock potential is the change in intensity of a reflection when a reservoir contains gas, with or without water (see Figure 11.17). The change in velocity caused by the replacement of the water in the reservoir sand (Section 7.5) by the gas gives rise to such secondary indicators as delays in the deeper reflections and sudden attenuation of energy on these reflections. Diffractions and changes in reflection character near the boundary of the reservoir provide additional evidence of gas occurrence. Even within this small subset of hydrocarbon location problems there is some danger. Domenico (1975) showed that even a small (5%) amount of gas in the reservoir is sufficient to lower the mixture velocity suddenly. Thus the sudden increase in reflection amplitude, or bright spot, and associated phenomena may come from a small amount of gas in the sand. Such gas may be biogenic, that is, derived from the direct decomposition of plant material, or it may even be carbon dioxide. Economically, the use of bright spots may be fraught with the danger of overestimation, but as a source rock indication there could be little better evidence. Some care is still necessary, however, because of the possibilities of noncommercial gas such as the examples listed. In basins in which direct hydrocarbon indicators (DHI) are not seen on seismic sections the presence of abundant shale can be regarded as the best indirect indicator. Shales are usually indicated by low seismic velocities. The presence of large sections with interval velocities in the range of 2000 to 3500 m/s (6500 to 11,000 ft/s) points to clastic sections that may have both source beds and reservoirs.

For the likelihood of reservoirs within the clastic section, however, there are zones in which the seismic traces have little similarity because of the coming and going of sands with acoustic impedance different from the average. Location of stratigraphic traps in carbonates can be eliminated by this procedure, but they have to be dealt with separately in any case.

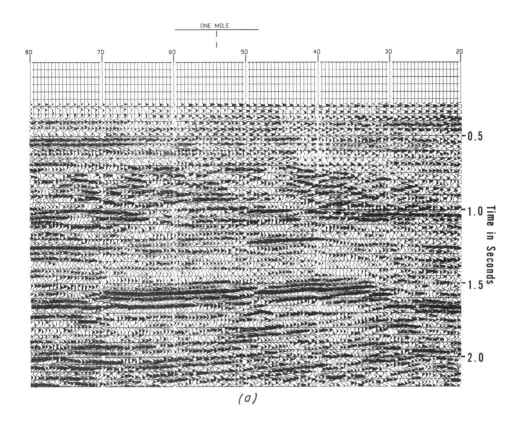

(a)

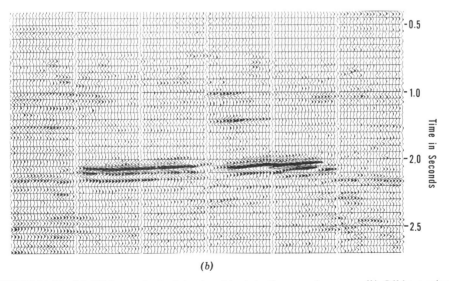

(b)

FIGURE 11.17 (*a*) Offshore seismic data played back in the normal manner. (*b*) Offshore seismic data relative-amplitude-processed. (*Source*: Courtesy of CONOCO, Inc.)

447

As a result of the foregoing qualification work, some traps may have been discovered—they are evidenced directly by the location of bright spots. The possibility of others occurring in small geographic areas and in smaller time zones on the record sections have been indicated, and the more detailed work of obtaining a corresponding geological model can begin.

There appear to be two possibilities: a direct and an indirect way of obtaining this geological model:

1. The direct method uses seismic traces converted to acoustic impedance logs— with the important qualifications that the latter be made by incorporating a knowledge of the statistical velocities obtained and that the acoustic impedance logs be calibrated in terms of actual velocities in one well. The use of statistical velocities involves the following steps:

    a. Make a steplike log of interval velocities. These are determined at times of good reflections (see Figure 11.18).

    b. This velocity log must be Fourier-analyzed over the same time interval as the seismic trace. This gives amplitudes and phases at the same intervals of frequency as those obtained for the seismic trace.

    c. Compare the amplitudes and phases for the velocity log and integrated seismic record over the overlapping frequency band and determine a scaling factor. It is hoped that this will be a real number. Otherwise, a complex scaling factor can be used to adjust the velocity log amplitude and phase and these values for the low-frequency interval must be added in.

    d. The completed Fourier analysis is then synthesized and shifted by a constant velocity to take care of the dc shift, which has not been determined.

    e. If necessary the velocity values can be inverted to interval time values (e.g., microseconds per foot). This has been done by Lindseth (1975).

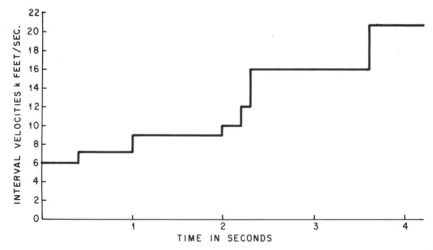

FIGURE 11.18 An interval-velocity plot from the statistical analysis of CDP data. It is the low-frequency components of this curve that are needed to supplement the (higher) frequencies determined directly. In this manner SAIL traces can be made to give pseudovelocity logs.

Note that the band-limited acoustic impedance log involves reflection coefficient contributions by changes in velocity and density. Before incorporating the statistical velocity information it is best to remove this duplication on some physical basis. One basis that has some support is that the density and velocity contributions to the reflection coefficients are equal. If this is true there will be no problem because an arbitrary scaling factor is involved in step c. If, however, the contributions are not linearly related the compensation must be made before step c is undertaken.

Thus each integrated seismic trace is forced into becoming an approximate velocity log which is used for correlation in the same manner as other well logs. The advantages are the following:

a. The changes in lithology and/or porosity can be seen in detail *between* wells.
b. The calibration is in velocity.

It must be borne in mind, however, that even with the broadest band seismic section obtainable the details of the change from one lithology to another are blurred. The existence of long-period multiple reflections is an ever-present danger.

Some pseudodiagraphies or approximate acoustic impedance logs obtained by Grau, Hemon, and Lavergne (1975) are given in Figure 11.19, where their excellent correlation qualities can be seen. These traces were 50 m (164 ft) apart.

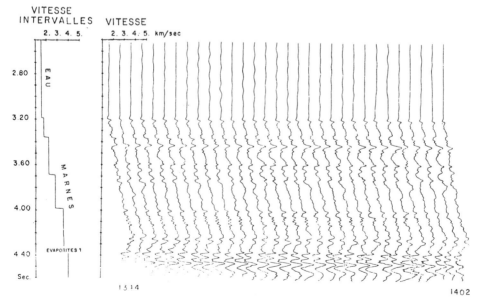

FIGURE 11.19   Pseudovelocity logs spaced 50 m apart. Interval velocities are shown on the left to the same scale. [*Source*: Grau, Hemon, and Lavergne. Courtesy of the World Petroleum Congresses (1975).]

2. The indirect method involves modeling in which an assumption is made in regard to the layering thicknesses, densities, and velocities; the two-dimensional or one-dimensional interactive modeling system is used to predict the seismic record(s) coming from a defined section, and alterations are made until the model fits the field trace(s). Given a starting point at a well or, better still, a tie into a well at each end of a line, the modeling can be reasonably fast, but the direct model from the acoustic impedance logs may even then be a help in providing a first model at each position along the line.

Two-dimensional modeling, because of its greater complexity, is probably used only when steeper dips or structural complications affect the ability of the interpreter to predict a model under complex conditions. The effect of diffractions is difficult to predict and they can sometimes be misinterpreted as lithofacies changes.

In both direct and inverse methods it is important to remember that only one model is obtained that fits the seismic data. Other inputs are necessary to decide whether this is the best model to be obtained.

11.9 TWO-DIMENSIONAL MODEL USAGE IN COMPLEX VELOCITY PROBLEMS

The reader has been alerted several times in this book that false structural anomalies can arise due to lateral changes in velocity, either in the near surface or the deeper subsurface. Some excellent examples of false anomalies have been given by Tucker and Yorston (1973).

These anomalies range all the way from multiples due to strata of uneven thickness near the surface to the simple turnover into a fault plane, which may be due to slower velocities on the downside of the fault. In the past these pseudotraps could be tested only by drilling, but today CDP velocity information can be obtained readily in great detail. Of course, these velocities are subject to experimental error but to avoid gross errors the practice of smoothing the velocities (for a constant depth or to a constant reflection) can be used to minimize experimental error. A smooth curve of known algebraic form can always be fitted to the observations by the method of least squares. Suitable curves may be single low-order polynomials to fit the data over its entire range or a series of cubic splines may be constructed to show the relation between the velocities at different intervals along the profile. Each piece of the curve has a cubic relation of the form

$$f(x) = c_{0i} + c_{1i}(x - x_{i-1}) + c_{2i}(x - x_{i-1})^2 + c_{3i}(x - x_{i-1})^3 \qquad x_{i-1} \leqslant x \leqslant x_i$$

$$(11.4)$$

The values of $f_0, f_1, f_2, \ldots, f_N$ and the slopes f_0' and f_N' are assumed to be known. The second derivations $f''(0)$ and $f''(N)$ are assumed to be zero. At the end points of any interval of length $h_i = (x_i - x_{i-1})$ the coefficients of the cubic polynomial must satisfy the equations

$$c_{0i} + c_{1i}h_i + c_{2i}h_i^2 + c_{3i}h_i^3 = f_i$$

$$c_{0i} = f_{i-1}$$

$$c_{1i} + 2c_{2i}h_i + c_{3i}h_i^2 = f_i'$$ (11.5)

$$c_{1i} = f_{i-1}'$$

To determine the slopes f_i' (except f_0' and f_N') the equations of equality of the second derivatives at the knots are written

$$h_i f_{i+1}' + 2(h_i + h_{i-1})f_i' + h_{i+1}f_{i-1}'$$

$$= \frac{3}{h_i h_{i+1}} [h_i^2(f_{i+1} - f_i) + h_{i+1}^2(f_i - f_{i-1})]$$ (11.6)

$$i = 1, 2, 3, \ldots, N - 1$$

By using all these conditions we find the coefficients c_{0i}, c_{1i}, c_{2i} and c_{3i}. These unique coefficients ensure that the values and function derivatives are continuous at the knots (i.e., in going from one spline to the next).

With a modern two-dimensional modeling process it is possible to specify horizontal gradients of velocity. Thus in cases like interpreting structure under overthrusts it is possible to construct the estimated model and, after a short time, decide whether it gives the required seismic output. In many cases this procedure is complex and time-consuming—chiefly in setting up the model. Nevertheless the problem today is tractable, whereas 10 years ago it would have been unthinkable in a very detailed form.

11.10 THE USE OF STATISTICAL METHODS IN STRATIGRAPHIC TRAP DETERMINATION

In maturely drilled areas more geological knowledge is available. It may consist of sonic logs in producing wells and/or dry holes. These sonic logs can be used for comparison of field records, and some discrimination between the types of record in which sands are present and those in which they are not can be obtained by eye. A more quantitative method of combining geological knowledge with detailed seismic information was reported by Mathieu and Rice (1969). Their procedure makes use of a seismic time zone referenced to a geological marker depicted in the seismic cross section. Each seismic trace provides a suite of variables that is later analyzed according to multivariate statistical procedures. The quantitative variables used in the example were directly measured by taking amplitudes of the seismic trace at specified time intervals referenced to a reflection event. Other quantifiable variables relating to the acoustic properties of the suspected anomaly can also be used in place of or in addition to the direct digital values. Examples include the spectral properties of the zone, measured times of the zero crossings from the reference time, and interval thickness.

It must be remembered that the data are band-limited; therefore there is a sampling interval, determined by the highest frequency present. If the sampling frquency is increased above this Nyquist sampling rate the samples taken will not be independent variables and will cause instability problems in subsequent statistical work.

The method assumes that a set of these data can be measured for all wells in the area or collected from seismic data in the immediate vicinity of the wells.

Linear discriminant analysis is used to establish classification functions for groups of traces defined a priori. In the example given two groups (sand and no sand) were constructed from synthetic seismograms. Linear functions of the variables were defined in the following manner:

$$Z^k = \lambda_1^k x_1 + \lambda_2^k x_2 + \lambda_3^k x_3 + \cdots + \lambda_n^k x_n + \lambda_{n+1}^k \tag{11.7}$$

where $x_1, x_2, x_3, \ldots, x_n$ = measured variables

$\lambda_1, \lambda_2, \lambda_3, \ldots, \lambda_{n+1}$ = estimated coefficients that maximize the following function of the two groups

$k = 1$ or 2, indicating the sand and no-sand conditions

$$D = \frac{(\bar{Z}^1 - \bar{Z}^2)^2}{\Sigma_{i=1}^{ns} (\bar{Z}^1 - Z_i)^2 + \Sigma_{j=1}^{nn} (\bar{Z}^2 - Z_j)^2} \tag{11.8}$$

where ns and nn are the number of sand and no-sand examples.

This statistical process separates the two groups optimally by maximizing differences between group means while minimizing differences among samples of the same group. The mode of action can be most easily visualized by taking the sample case of dependency on only three variables. In this case the individual samples can be plotted in three dimensions, as shown in Figure 11.20. If these samples are looked at in a plane parallel to the X-Y axis their projections will be intermixed. By the proper choice of a plane, however, the projections from the different samples are optimally separated, and at the same time the distance apart of samples within the same group is overall as small as possible. In Figure 11.20, this optimal plane was chosen to be perpendicular to the X-Z plane, making an angle θ with the X axis. The results of using this plane are shown. In general, of course, the plane necessary to effect an optimal separation is achieved only by rotation of the original axes through two different angles. The important point is that *rotation* of axes is involved. In the more general (n-dimensional) case the rotations necessary are achieved by matrix manipulations. Tests can be made, subject to the usual statistical distribution assumptions, that will determine whether the group means are distinguishable from one another at the desired confidence level. Nonlinear discriminants are available but are more difficult to use.

Essentially, the characteristics of this method are the following:

1. The known samples are used to determine a statistical discriminant function whose power to distinguish between two (or more) groups can be ascertained.
2. These groups must be decided before calculation is made. This is relatively simples in the sand–no-sand problem but may be much more difficult in more complex problems.

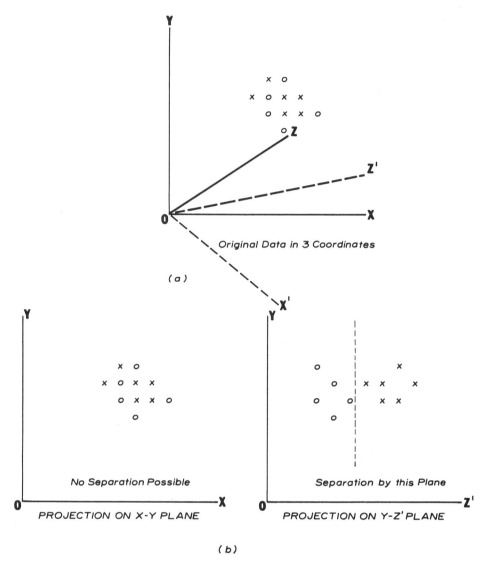

FIGURE 11.20 Discrimination achieved by the rotation of axes.

3. New and unclassified samples—that is, new seismic traces—can now be
   classified into one or another of the established groups.

It can be said that although the method was successful (see Figure 11.21) in
testing preconceived ideas other significant stratigraphic trends may not be
detected.

It now seems to be desirable to formulate a method that removes some of the
restrictions imposed in the early method.

The new technique identifies underlying patterns in a fixed window of seismic
data and, wherever possible, associates a geological meaning with each pattern.
The identification of basic components in a set of quantitative data has been

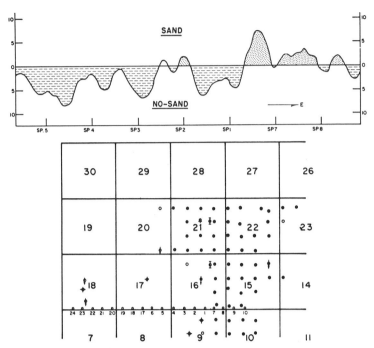

FIGURE 11.21 Classification of field data into one of two groups (sand, no sand) from discriminatory analysis. Classification is corroborated by wells shown on the map at the bottom of the figure. [*Source*: Mathieu and Rice (1969). Reprinted with permission from *Geophysics*.]

discussed by Griffith, Pitcher, and Rice (1969). It is a problem to be tackled by factor analysis—a multivariate statistical technique. These investigators used factor analytic techniques to delineate the number of environmental facies that existed in the area of analysis and associated each sample with the proper (geological) facies. Similar techniques are applicable for quantitative variables derived from a zone of seismic data and can be used effectively to outline the underlying patterns mentioned earlier. Other clustering techniques, such as those described by Anderberg (1973), can be used to augment factor analysis methods.

The original data are digitized at the usual sample interval, and these amplitude sequences which correspond to the set of trace segments, constitute the original data. We first search for underlying constituents that are orthogonal to one another (see Section 10.4) and that explain a large fraction of the data. The technique is one of determining the characteristic components and the weights to be given to each one; consequently a major portion of the energy of the traces is explained. The characteristic components (sometimes called eigenvectors) and their weights (eigenvalues) are determined by standard matrix operations. Each trace can then be constructed (almost) by combining the characteristic function in the proper proportions. The number of characteristics is determined largely by the degree to which the interpreter wishes to approximate the data. The factor analysis, given this degree of approximation, determines how many eigenvectors are needed and what their form will be. It is important that each eigenvector be entirely independent, and this is what is meant by orthogonality. In mathematical terms

$$E_1(x_i) = a_1 x_1 + a_2 x_2 + a_3 x_3 + \cdots + a_n x_n$$
$$E_2(x_i) = b_1 x_1 + b_2 x_2 + \cdots + b_n x_n \qquad (11.9)$$

Then $a_1 b_1 + a_2 b_2 + \cdots + a_n b_n = 0$ for orthogonality. In the sense of n-dimensional vectors the eigenvectors are perpendicular to one another.

By using these eigenvectors as axes each trace can be expressed in terms of them; for example,

$$T_1 = A_{11} E_1 + A_{12} E_2 + \cdots + A_{1m} E_m$$
$$T_i = A_{i1} E_1 + A_{i2} E_2 + \cdots + A_{im} E_m \qquad (11.10)$$

We can imagine that the traces are displayed in n-dimensional space with the eigenvectors as coordinates, just as the samples are shown in three-dimensional space in Figure 11.20.

After the basic patterns have been established in the data it is desirable to be able to associate some physical or geological meaning with them. A more rigorous procedure than visual comparison is needed for cases that require simultaneous consideration of several complex variables. One possible approach is the application of discriminant analysis techniques to groups established by factor analysis techniques.

The resulting functions can be used to classify new seismic data or synthetic seismograms derived from the acoustic impedances found from well logs. Corresponding tests of significance can also be computed to give a quantitative assessment of the uniqueness of the groups that have been defined.

Because of the sensitivity of this method to reflection character, which we know is controlled by the type of input pulse (as seen after filtering by the recording system), it is essential that all data be of the same vintage, collected in the same manner with identical instruments, or that different data types undergo treatment with transfer functions to convert them to one basic type. The choice of standard is optional, except that the transfer functions cannot convert data with one bandwidth to data with a wider bandwidth. The choice is open, but it invariably leads to degradation of all data to the characteristics of the lowest acceptable set. But we always have the choice of throwing some away.

The results of applying this more general method have been given by Waters and Rice (1975). The original data, with the zone of interest (A-B) marked, is given in Figure 11.22. Below the main data is shown the interval concerned, flattened on the zero crossing above the lower reflection. Five different characteristic wave forms were present and the proportion of S_1 (the factor chiefly associated with gas production) is shown in the lower part of the figure that with some shading identifies the chief constituent of each trace segment.

Figure 11.23 shows the association of the characteristic waveforms (end members), the average waveform for a group, and the synthetic waveforms from producing wells and dry holes. The upper characteristic waveform is most indicative of production.

Finally, we must remark that this *method* does not, by itself, constitute a method of increasing resolution—rather it organizes the task of producing the best information that can be obtained from the bandwidth available.

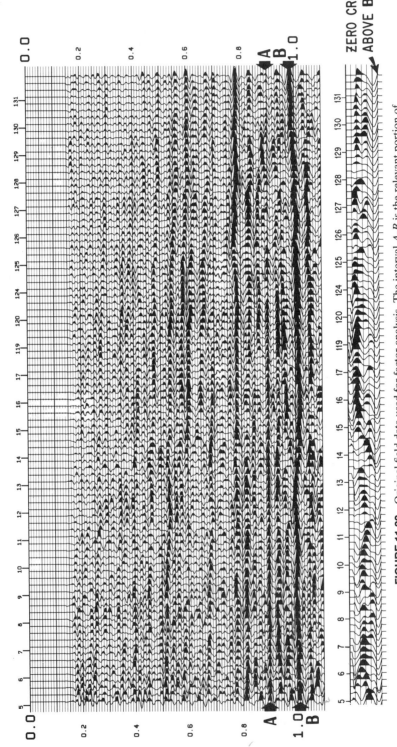

FIGURE 11.22 Original field data used for factor analysis. The interval *A-B* is the relevant portion of the cross section, reproduced below the seismic section by flattening on the zero crossing above *B*. [*Source:* Waters and Rice. With permission from the Ninth World Petroleum Congress (1975).]

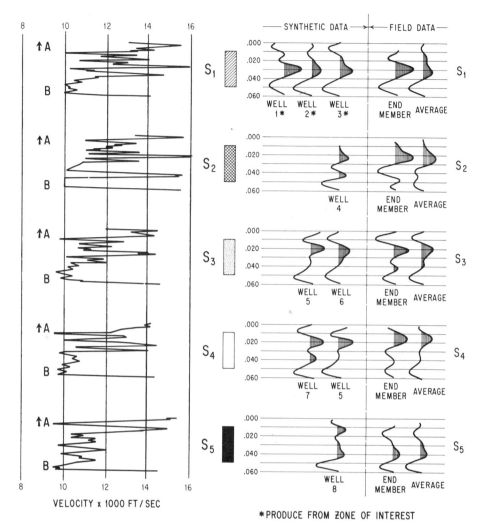

FIGURE 11.23 Average and end-member waveforms for the factor types and associated synthetic seismograms classified by discriminant analysis. Type logs for each group are shown on the left. [*Source*: Waters and Rice. Courtesy of the Ninth World Petroleum Congress (1975).]

The effect of bandwidth is significant in the test example and can be quantified by statistical procedures. Because three vintages of seismic data were available in the area, with different bandwidths, groups derived from dynamite coverage were processed with bandpass filters equivalent to the Vibroseis data. Figure 11.24 illustrates the waveform differences for the five groups.

Some loss of ability to distinguish between the groups is associated with loss of high-frequency content. The loss is not proportional to bandwidth reduction, and in the example the loss of the 45- to 52-Hz band is much more critical than the loss of the 52- to 80-Hz band. When a bandwidth of 15 to 52 Hz is used the average probability of misclassifying a sample is approximately 1.3 times greater than when the 15- to 80-Hz data are used. The 15- to 45-Hz data increase the

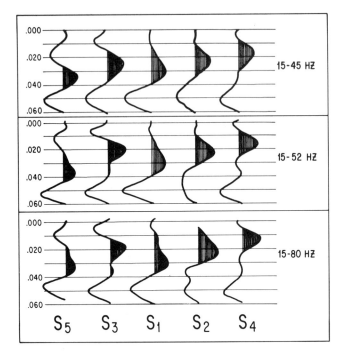

FIGURE 11.24 Waveform differences associated with different filters for the five groups shown previously. [*Source*: Waters and Rice. Courtesy of the Ninth World Petroleum Congress (1975).]

probability of misclassification by a factor of 3 over the 15- to 80-Hz data. Thus a substantial improvement in successful classification can result from a relatively small bandwidth increase.

11.11 INTERPRETATION OF SEISMIC REFLECTIONS FROM COAL SEAMS

Coal seams of economic importance usually occur in the first two or three thousand feet, although deeper production is known. In the United States many mines are located at 200 to 700 ft (60 to 210 m). The interpretation of seismograms taken for coal mining purpose differs, to a considerable degree, from those taken for the *location* of oil and gas deposits. In the first place the presence of coal seams is usually known and only anomalies in their mode of occurrence is of interest.

To secure the maximum information the original field data must contain high-frequency energy. Coal seams are almost invariably thin compared with even the wavelength of the highest viable frequency. As a consequence, interpretation must rely to a large extent on comparative amplitudes and every effort must be made not to obscure them in the field or change them in the data processing.

The challenge may be likened to that of obtaining clearly visible '*bright spots*' in the search for gas fields, in spite of additional obstacles imposed by the following:

1. Shallow targets limit the offset distances for seismic traces.
2. Topography can be very rough compared with total depth. In Pennsylvania or West Virginia, for example, the elevation differences between ridges and valleys may exceed 50% of the coal-seam depth.
3. Near-surface corrections may be difficult because of variable velocities and large elevation changes.
4. Near-surface slumping may exist over old mines with consequent changes in near-surface velocities.
5. The high frequencies (up to 250 Hz) necessary for the best possible resolution consistent with not-too-rapid attenuation require adequate static and dynamic (NMO) corrections.
6. Although the high reflection coefficients associated with the coal/country rock boundaries give a good reflection from the uppermost seam, they often cause the lower ones to be obscured.

Less difficult conditions prevail in flatland areas, and in western United States, England, France, and parts of Germany and this is where the most progress has been made. Ziolkowski and Lerwill (1979) have shown some remarkably good, high-frequency sections (from explosive sources in holes) that include reflections from coal seams at depths of 500 to 700 m. Their area is one in which consistently good reflections can be obtained with small (0.12 kg to 0.25 lb) charges of dynamite, single geophones, and 24-fold stack. A typical geophone spread is between 470 and 564 m long, with only 10 m between geophone stations.

The data processing must compensate for any unavoidable change of stack due to permit problems. A slow change of gain would be admissible but only if it is constant for all changes in the cross section.

The principal effects on which interpretation is based are changes of amplitude, diffracted or scattered energy from sudden changes in reflection quality or displacements in depth of the reflectors, and/or effects attributable to sharp changes in curvature.

Some of these effects can best be discovered by examining stacked sections in which relative amplitudes have been preserved and others by a good migration program. After migration has restored the position of scatterers and reflectors to their proper places a modeling system is essential to check the validity of any interpretation made. Care must be taken in areas of multiple coal seams to compare stacks that use near-offsets with special stacks with long offset traces because on the latter P − S or S − P reflections may produce unwanted noise.

Because all the important information is generated by anomalies of a (seismically) thin bed of low velocity and density sandwiched between rocks of much higher velocity and density, it is of considerable interest to examine the *character* of a reflection from a single coal seam. As an example we take a seam with a velocity of 7000 ft/s (2134 m/s), a density of 1500 kg/m^3 in a *country rock* of 14,000 ft/s (4268 m/s), and a density of 2500 kg/m^3. The thickness is 6.5 ft (2 m).

The reflection impuse response is a single negative *spike*, followed by a series of positive spikes at equal intervals and of rapidly diminishing amplitude. The interval is the two-way time of seismic waves in the coal seam (0.00094 s). Figure 11.25*a* shows the reflected pulse from a delta function input and Figure 11.25*b*

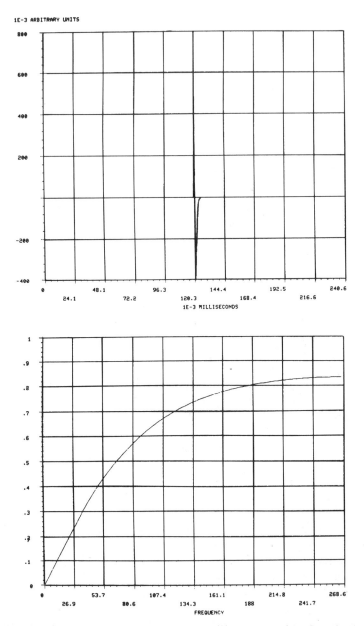

FIGURE 11.25   (*a*) Reflected pulse from a thin coal seam. (*b*) Spectrum of a reflected pulse in Figure 11.25*a*.

shows the amplitude spectrum over the frequency band 0 to 268.6 Hz (this is the high-frequency limit for the time sample 0.00094 s).

As expected, the transmitted pulse (Figure 11.26*a*) has no negative values in the rapidly decaying sequence, and the spectrum (Figure 11.26*b*) complements that of the reflected pulse, being a low-frequency pass system. As a consequence, in a cyclical deposited system a sequence of coal seams occurs, each of which

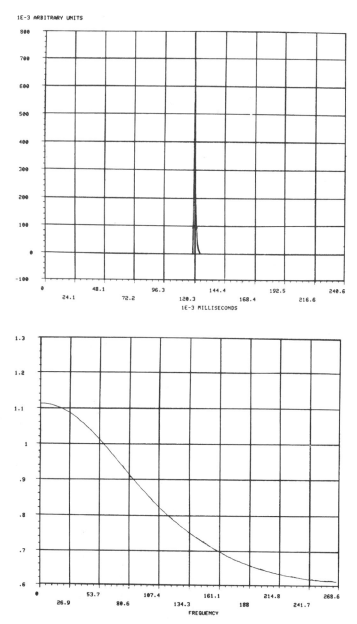

FIGURE 11.26 (*a*) Pulse transmitted through a thin coal seam. (*b*) Spectrum of a pulse transmitted through a thin coal seam.

reflects a predominantly high-frequency signal but cannot do so because the more coal seams passed through, the lower frequency is the energy to be reflected and (in cascade) the coal seams above will preferentially only *pass* low frequencies as the pulse returns to the surface.

With a good knowledge of the velocity modern migration methods for raw data can be used to produce stacked migrated sections that must be carefully compared with the *brute stacked* records to confirm the existence of faults and old distributory channels. When an area is detailed, with lines close together or, preferably, with modern three-dimensional techniques that produce several lines of data *from the same sources*, these anomalies can often be traced from line to line. If it can be proved that they are not associated with a surface or near-surface lineation their interpretation is straightforward and often of great importance. Faults may have throws exaggerated by a change of velocity across the fault.

Finally, once the anomalies are located, the relatively small number of possible causes makes it easy to speculate on a number of geological changes that might be involved; discrimination is assisted by some of the modeling techniques already described.

We know from the studies of Fertig and Müller (1978) that in multiple groups of coal seams it may be quite difficult to separate out the groups after the first.

11.12 SUMMARY AND CONCLUSIONS

In this chapter the methods of interpreting seismic reflection data are examined. Although the task of finding hydrocarbons should be purely geophysical, the resolving power of the tool has still to be proved sufficient to determine exactly enough of the physical data necessary. Geologists of varying special disciplines are needed to make exploration by seismograph economical and to supply a knowledge of likely values of other parameters at present beyond the reach of the geophysicist.

Structural interpretations, although relatively straightforward, are still subject to errors due to unsuspected velocity variations, inadequate migration procedures, and basic limitations in resolution at truncations.

Stratigraphic trap interpretations are limited by resolution, both in thickening and thinning thin beds and by changes in their other physical characteristics. Imaginative construction of one-dimensional models, together with interactive modeling programs, can be used to determine whether the models can be related to the seismic reflection results. At best, however, a single answer is found for what, in view of limited resolution, is probably a multianswer problem.

In cases of structural complexity resort must be made to two-dimensional modeling in which all types of seismic events are included. These programs are slow and expensive to run and, of course, are valid only when the earth structure under the line of section really is two-dimensional. Reflections coming from the side of the line (side swipe) are not dealt with. We would have to rely on two-dimensional data gathering with some discrimination in the processing to remove, or display separately, reflections from outside the line of section.

In all types of interpretation, but particularly in stratigraphic trap modeling, reliance is placed on consistent reflection character when consistent geology is present. To achieve this, if more than one vintage of seismic record is present, resort has to be made to achieving consistency by the use of transfer functions. Unfortunately, although all seismic reflection data are put on an equal footing, the level is that of the set of records with the least bandwidth.

An important clue to traps that contain gas is the relative amplitude-processed section in which the gas zones stand out by virtue of the large change in acoustic impedance between normal and gas-filled sands. The presence of large-amplitude anomalies, evidence of high attenuation, and change in phase at the ends of the high-amplitude reflection are not necessarily indications of *commercial* concentrations of gas.

The direct method of obtaining a *first guess* geological model is to use the SAIL trace modified to include low frequencies by making use of statistical velocity information. Some success has been achieved in contouring the values of inverse velocity (in microseconds per feet) to show up low-velocity zones that indicate porosity in carbonates.

Statistical methods have now been developed for use in maturely drilled areas, by which characteristic wave shapes can be calculated from the seismic data for the zone of interest. These eigenvectors can be used to express the waveforms of all traces, and a discriminant analysis divides them into identifiable groups for a given confidence level. Association can then be made between the groups defined and certain properties of the section—such as oil production. This process works best in areas in which consistent, high-bandwidth records are available and in which numerous wells, some producers, and some dry holes have been drilled. It is really an exploitation rather than an exploration tool. At present the tool is limited to a consideration of zones of constant-time interval.

Coal exploration poses its own problems of interpretation, not the least of which is due to the property of thin beds with high coefficients of reflection at the boundaries, to reflect predominantly high frequencies and to transmit preferentially low frequencies. Thus the deeper seams are obscured by the effects of the shallower ones. It is essential to employ migration and modeling programs and to leave relative amplitudes alone. The identification of trends between lines, or *through* three-dimensional data, is probably as important as the identification of an anomaly on a single line.

For P waves the reflection coefficient differential between a coal seam and a flooded mine working is not great. For optimum *old mine boundary* mapping SH-wave reflection lines are recommended in addition to P-wave work.

APPENDIX 11A: LOCATION OF DEPTH OF OVERPRESSURED ZONES

In all geological sections the internal pressure of the fluids increases with depth. This increase is brought about by the fact that part of the weight of the sediments above a particular depth is borne by the fluids and part by the rock matrix. If the fluid in a porous rock is connected even tenuously with the surface by the permeability of the rocks or by faults and fissures this pressure can be somewhat relieved and reduced to a pressure consistent with a column of water equal to the depth being considered.

In some areas, however, thick sections of clastics may have been laid down very quickly and may have retained water within an impermeable matrix. Under these conditions the water in the shales helps to bear the overburden load and may have considerably greater pressure than normal. It is of considerable importance before drilling a hole to be able to predict the depth to the top of any overpressured zones so that a warning can be given to the drilling crew that additional care is necessary when these depths are approached. Producing sands do occur within overpressured zones and, when found, are often prolific. They, and the possible sands below overpressured shales, are the economic incentive for continuing to drill.

Although reflection seismology cannot provide exact depths to the top of overpressured shales, the accurate prediction of average or interval velocities as a function of depth can often provide a depth within a few hundreds of feet. Figure 11A.1 shows diagrammatically the form of the average velocity versus depth curve for a location in which overpressured shales are present. The shales themselves have low P-wave velocities and this causes a rapid decrease, even of average velocity, with depth. The *knee* of the velocity depth curve is gradual, however, and can at best be determined within 100 m (300 ft).

Some care is necessary because many statistically determined velocity depth curves show zones in which the average velocity appears to be constant or decreases slightly. These zones may be due to causes other than overpressure, such as multiple reflections and diffracted events.

There is reason to believe that some overpressured zones, which give rise as they do to gradual changes in acoustic impedance (transition zones), should be observable from reflection data. Transition zones are characterized (Berryman, Goupillaud, and Waters, 1958) by having a higher reflection coefficient at low frequencies. In addition, any slow-velocity medium delays reflections that come from layers deeper in the section.

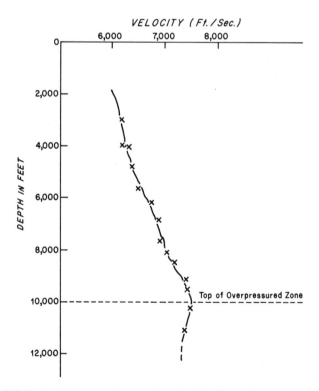

FIGURE 11A.1 A velocity-depth curve, showing the effect of an overpressured zone.

Overpressured zones have been found at markedly different depths on opposite sides of some faults, and a contoured, partly smoothed, interval velocity section superimposed on the normal seismic section can provide excellent clues to their pressure and depth.

REFERENCES

Anderberg, M. R. (1973), *Cluster Analysis for Applications*, Academic Press, New York.

Backus, M. M. (1959), "Water Reverberations, Their Nature and Elimination," *Geophysics*, Vol. 24, No. 2, pp. 223–261.

Berryman, L. H., Goupillaud, P. L., and Waters K. H. (1958), "Reflections from Multiple Transition Layers—Theoretical Results," *Geophysics*, Vol. 23, pp. 223–243.

Domenico, S. N. (1975), "Effect of Brine-Gas Mixture on Velocity in an Unconsolidated Sand Reservoir," *Proceedings of the 9th World Petroleum Congress, Tokyo, Japan.*

Fertig, J., and Müller, G. (1978), "Computations of Synthetic Seismograms for Coal Seams with the Reflectivity Method," *Geophysical Prospecting*, Vol. 26, No. 4, pp. 868–883.

Grau, G., Hemon, C., and Lavergne, M. (1975). "Possibilities Nouvelles pour la Sismique Stratigraphique," *Proceedings of the 9th World Petroleum Congress, Tokyo, Japan.*

Griffith, L. S., Pitcher, M. G., and Rice, G. W. (1969), "Quantitative Environmental Analysis of a Lower Cretaceous Reef Complex," SEPM Special Publication 14, Tulsa, Oklahoma.

Lindseth, R. O. (1975), "Interpretation of Seismic Data from Derived Velocity Logs," paper presented at the 45th Annual Meeting of the Society of Exploration Geophysicists, Denver, Colorado.

Mathieu, P. G., and Rice G. W. (1969), "Multivariate Analysis Used in the Detection of Stratigraphic Anomalies from Seismic Data," *Geophysics*, Vol. 34, No. 4, pp. 507–515.

Stolt, R. L. (1975), personal communication.

Tucker, P. M., and Yorston, H. J. (1973), "Pitfalls in Seismic Interpretations," Society of Exploration Geophysicists, Monograph No. 2.

Waters, K. H., and Rice, G. W. (1975), "Some Statistical and Probabilistic Techniques to Optimize the Search for Stratigraphic Traps on Seismic Data," *Proceedings of the 9th World Petroleum Congress, Tokyo, Japan.*

Ziolkowski, A., and Lerwill, W. E. (1979), "A Simple Approach to High Resolution Seismic Profiling for Coal," *Geophysical Prospecting*, Vol. 27, No. 2, pp. 360–393.

BIBLIOGRAPHY

Anderson, T. W. (1957), *An Introduction of Multivariate Statistical Analysis*, Wiley, New York.

Collins, R. E. (1968), *Mathematical Methods for Physicists and Engineers*, Reinhold, New York.

Dixon, W. J. (1968), *BMD Biomedical Computer Programs*, University of California Press, Berkeley and Los Angeles, Calif.

Dixon, W. J. (1969), *BMD Biomedical Computer Programs—X Series Supplement*, University of California Press, Berkeley and Los Angeles, Calif.

Harman, H. H. (1960), *Modern Factor Analysis*, University of Chicago Press, Chicago.

Marriott, F. H. C. (1974), *The Interpretation of Multiple Observations*, Academic Press, New York.

Press, S. J. (1972), *Applied Multivariate Analysis*, Holt Rinehart and Winston, New York.

TWELVE

New Tools in the Making

12.1 INTRODUCTION

Reflection seismology has never been a static science and it is no surprise, given a fixed time for the termination of any discussion, that new concepts and new methods are in an active stage of development. These methods do not form coherent material for an additional chapter, yet some coherence attends the ever-present need to increase the amount of detailed information derived about subsurface rocks. The reader who has arrived at this point, however, should be able to place these anomalous subjects in the proper order to allow each facet of reflection seismology to have its own developing phase that will fit in with the overall goal. This is not a comprehensive catalog of research and development being carried on in industry, government, and the universities and published every three years in *Geophysics*. Instead, we present the main advancing phases of which we have some knowledge and place them in context so that their future probable importance can be assessed. If that means a discussion of the problems still to be solved, then that is done.

12.2 THE USE OF SHEAR WAVES TO PROVIDE SUPPLEMENTARY DATA

In Chapter 2 it was shown that two types of elastic wave can travel through the body of an elastic medium; compressional (P waves) and shear (S waves). The physical characteristics of these two waves have already been described.

Because shear waves were known to exist, there has always been some effort by the exploration geophysics industry to determine how they could be produced and fitted into the overall exploration picture—to supplement knowledge of structure and stratigraphy gained from the well established methods that use compressional waves. Progress has been rather slow, and not the least reason for this has been the overwhelming success of the *standard* compressional wave method, yet there have been other barriers to the implementation of shear waves.

Early research by Jolly (1956), White and Sengbush (1963), and others gave little encouragement to the reflection use of shear waves. Their research showed that many earth materials behaved in an anisotropic manner; that is, the velocity of waves in one direction is not the same as in a direction at right angles. In fact, this phenomenon, already known to exist for compressional waves, was much enhanced for shear waves. The atmosphere was definitely discouraging.

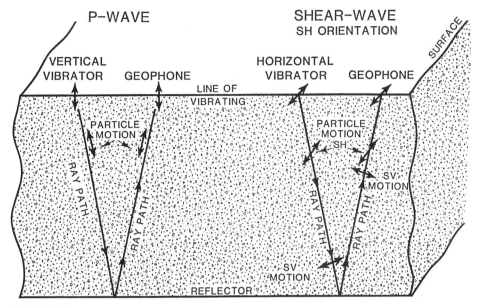

FIGURE 12.1 The relationship between source and geophone motions and waves designated P, SV and SH.

Nomenclature in Simple Cases

To follow some of the later discussion it must be understood that the standard (earthquake) designations for polarized shear waves are strictly limited in application. It is assumed that exploration geophysics is dominated by applications, for which the model is one in which there are plane, parallel, *horizontal* boundaries between the rock layers. Under these conditions only the reflection path (whether P or S) must lie in a vertical plane that contains the source and receiver, as shown in Figure 12.1. In that illustration the P-wave reflection is obtained from a wave whose particle motion is along the ray path (perpendicular to the wave front if the material is isotropic). However, there are shear waves that can have particle motions at right angles to the ray path. It is the normal convention to split up arbitrarily oriented shear waves into SV and SH components which have motion *in* the vertical plane of propagation and normal to that plane, respectively. Thus the SV component will have a component of particle motion in a vertical direction (perpendicular to the horizontal reflectors), whereas the SH waves will have particle motion in a horizontal plane.

Figure 12.1 shows how SH waves are generated by a horizontal vibrator whose motion is perpendicular to the plane of survey. These waves are reflected and refracted from a plane interface with no change of type, provided that the dip of the reflector is aligned with the survey line. Under all other conditions the incident waves are no longer purely SH and SH; SV and P reflections and refractions are obtained.

The reader should understand that when the designations SV and SH are given they are subject to the conditions specified. In general, these designations form a useful, but imprecise, way of communicating the intended characteristics of the surface sources, receivers, and line of survey.

Practically, nearly all the results we shall be discussing have to do with SH waves. The reasons are twofold:

1. It is known the P waves and/or SV waves will be partly converted into one another on striking any reflecting interface at an angle other than normal incidence. For normal incidence there are, by symmetry, no SV waves. The Zoeppritz equations (Chapter 2) govern the distribution of energy in an elastic medium after an incident P or SV wave strikes a boundary. Snell's laws, of course, apply, namely,

$$\frac{\sin \theta_1}{\alpha_1} = \frac{\sin \phi_1}{\beta_1} = \frac{\sin \theta_2}{\alpha_2} = \frac{\sin \phi_2}{\beta_2}$$

where ϕ_1, ϕ_2 = reflected and refracted shear wave angles
θ_1, θ_2 = reflected and refracted P wave angles

This means that, in general, after a few boundaries have been encountered, the situation becomes hopelessly confused and the sorting out of possible ray paths becomes a practical impossibility.
2. It is easier to generate SH waves alone by using a horizontal force at right angles to the direction of the surface traverse, thereby minimizing the conversions of type at near horizontal boundaries.

A number of terms are used in this discussion that may for convenience be summarized here. These terms are strictly applicable only when the earth materials are isotropic. Anisotropy is discussed, at length, in a later section.

1. We deal with perfectly elastic bodies through which only P waves and S waves are possible.
2. The earth materials are characterized by moduli that describe the strain associated with compressional or shear stresses. Hooke's law states that

$$\text{stress} = \text{modulus} \times \text{strain}$$

K represents the bulk modulus, μ, the shear modulus.
3. Together with the density, these two moduli determine the speeds of propagation of P and S waves:

$$V_p = \alpha = \sqrt{\frac{K + \frac{4}{3}\mu}{\rho}} \qquad V_s = \beta = \sqrt{\frac{\mu}{\rho}}$$

4. In addition there is a (dependent) elastic constant called Poisson's ratio. In an inverse sense it may be regarded as describing the malleability of a material; for example, a soft, easily manipulated metal, like lead, has a Poisson's ratio of 0.4, whereas a brittle rock, like quartzite, has a Poisson's ratio of 0.25. It can be shown that this ratio (σ) can be calculated if both V_p and V_s are known:

$$\sigma = \frac{1}{2} \frac{\left[\left(\frac{V_p}{V_s}\right)^2 - 2\right]}{\left[\left(\frac{V_p}{V_s}\right)^2 - 1\right]}$$

It is actually defined as the negative of

$$\frac{\text{transverse strain}}{\text{longitudinal strain}}$$

Rock Velocities, Velocity Ratios, and Poisson's Ratio

Much of the discussion of shear waves that follows makes use of arguments based on the fact that the velocities of the two different wave types have a (unique) ratio

TABLE 12.1
Poisson's Ratio

Lithology		R	V_s/V_p	Remarks
Evaporate	Anhydrite	0.50	0.33	Benzing et al. (1978)
Carbonates	(General)	0.41 to 0.50	0.40 to 0.333	Benzing et al. (1978)
	Chalk "A"	0.48	0.35	V Shallow
	Chalk "B"	0.47	0.36	Little difference fissured or unfissured
Sandstones	(General)	0.47 to 0.60	0.35 to 0.22	Benzing et al.
	Bunter SS "A"	0.36	0.43	Newman & Worthington (1982)
	Bunter SS "Mail"	0.40	0.40	Newman & Worthington (1982)
	Bunter SS "B"	0.41	0.40	Newman & Worthington (1982)
Mudrocks	Shales (general)[b]	0.30 to 0.76[a]	0.45 to −0.17[a]	
	Greenhorn shale	0.487	0.34	Jones and Wang (1981)
	California and	0.44	0.38	Benzing et al. (1978)
	argillaceous clay	0.30	0.452	One sample only
	Wet clay (USSR)	0.22 to 0.07	0.47 to 0.497	
	Limits (theory)	0.00 to 0.700	0.50 to 0.0	

[a]This experimentally determined value corresponds to a rock that expands laterally as it extends longitudinally, due to longitudinal stress. A few esoteric substances do this and have a negative Poisson's ratio, but no rocks are among them.
[b]It should be noted that shales can cover the entire range for other consolidated rocks and the actual value depends on the degree of consolidation and actual composition (e.g., limey shale and siltstone).

TABLE 12.2

V_s/V_p versus Lithology

	Lithology	V_s/V_p
CONOCO Lab measurements	Carbonates	0.42 to 0.50
	Sandstones	0.51 to 0.58
	Shales, claystones	0.58 to 0.65
Schlumberger	Limestone, dolomite, and anhydrite in various proportions	0.53 to 0.55
	Sandstones and conglomerates with minor carbonates	0.60 to 0.63
Welex (Kithas, 1976)	Limestone	0.53
	Dolomite	0.56
	Sandstone	0.59 to 0.63
	Shale	0.56 to 0.59
	Dolomite, anhydrite	0.52 to 0.55

that changes with rock type and probably with other parameters, such as porosity, pressure, and fracture density.

The velocities given in Table 12.1 are average (sometimes single-value) estimates and considerable care must be exercised in using them. Although P-wave velocities can be measured routinely by the careful application of sonic logs, the measurement of S-wave velocities can be made only with the most modern sonic logging equipment (as discussed in Chapter 7), and even then it is not always easy to define which S-wave velocity (in anisotropic materials) is being measured. Table 12.2 lists some V_s/V_p ratios derived by various sources and cataloged by rock type. It is evident that considerable disagreement exists in regard to the values to be adopted for discrimination between the different gross rock types. The problem is probably not one of errors of measurement or even of failure to establish in situ conditions during laboratory measurements but simply an inadequate listing of all of the physical factors that can effect V_p and V_s.

Reasons for Desiring Shear-wave Information

Compressional wave seismic work, which has been practiced for many years, has been successful in most areas in furthering geologic structural interpretation. The reasons for seeking additional information must therefore have to do with providing—by shear wave seismology—new and independent information. Alternatively, new information may come from the nature of shear waves to make better structural or stratigraphic information available. The principle reasons are as follows:

1. The compressional and shear waves travel at different speeds. As a consequence the two wave types respond to changes in elastic parameters and density that are *entirely independent*. Reflection coefficients at the boundaries between layers are different for P waves than for S waves.

2. The transit times of waves in a given layer of material are different. By taking the ratio of the transit times (t_p/t_s) a quantity V_s/V_p, related to the elastic constants of the rock, could be at least partly diagnostic of the type of rock, including the mineralogic composition, the pore fluids, and the amount of porosity or fracturing. To do this *corresponding P and SH* reflections must be available and *correlatable*. We must note, however, that *there may not be corresponding reflections* and/or *they may not have the same polarity*.

3. Because shear waves travel slower than P waves, for a given frequency the shear wavelengths are shorter than the corresponding P wavelengths. Hence, because rock boundaries are separated by a *distance*, the shear waves of a given frequency will give greater resolution than compressional waves. This is fundamentally true—for perfect materials—but shear waves are attenuated faster than P waves of the same frequency and, in practice, this phenomenon works against greater S wave resolution.

4. Reflections are always recorded against a background of noise and interference. Shear-wave reflections from a given boundary must arrive later than the corresponding P-wave reflection and might therefore arrive under lesser conditions of interference.

   It is known that the near-surface guided waves (Rayleigh waves for P and Love waves for S) *do* travel at about the same speeds because both are primarily controlled by the shear-wave velocities in the near-surface. Furthermore, a Rayleigh wave is always present even if the surface material is faster than the material below it—but the same is not true for Love waves—only a surface S wave is present (Cherry, 1962).

   Therefore there is a chance that some areas, particularly those with high-velocity surface materials, may yield better S-wave records than P-wave records.

5. Changes in reflectance *of a layer* with porosity susceptible to secondary processes (e.g., dolomitization or silting up) might easily show up better on S-wave records. This is a rather complex process because it depends on the resolution afforded by P and S waves and, in turn, on the realizable spectrum of energy that can be injected into the earth.

6. Gas-filled zones may differ from other high-contrast zones (e.g., basalt layers or coal beds) in their response to S and P waves. A fluid layer does change its velocity for P waves, depending on whether the fluid is a gas or a liquid. There are no *gas-caused* bright spots for S waves. On the other hand, a coal bed or a basalt intrusion would give a bright spot for both S and P.

7. If anisotropy and attenuation could be measured for P and S waves they might provide additional controls for the determination of lithology or stratigraphy.

Shear-Wave Recording and Display

At the present time, for economy of recording channels, shear-wave ground motion is usually measured and recorded by groups of single, horizontal geophones, all pointing in the same direction. Three component geophones have been used for experimental purposes and have usually been an orthogonal group (three geophone units perpendicular to one another) of two horizontal geophones

and a single vertical one. It is perhaps redundant to point out that all these
geophones should have the same spectral response. This, however, is difficult to
accomplish when the vertical geophone has to counter the weight of the coil with
the spring assembly, whereas this affect is minimal in the horizontal geophones. It
is, moreover, a practical fact that the horizontal geophones are sensitive to
leveling. This particular arrangement of two horizontal geophones and one
vertical one is not the only possible means of measuring the components of
motion. For the future it appears that a large number of recording channels may
be available routinely. Then it may be commonplace to record the three motions
simultaneously. It is to be hoped that this will lead to the use of three component
units *in which all geophone elements are similar.* They are arranged within the
case, again orthogonally, but in the manner of the legs of a tripod, each geophone
inclined to the vertical by an angle of 54.7°. Combinations of the outputs from
these three geophones, after gain adjustment, give the vertical and horizontal
motions desired. Basically, shear (SH) waves are recorded with the same P-wave
recording field arrangements; that is, CDP methods are used in which the sources
and receivers may be groups. Beyond this, however, some differences are due to
the polarization of horizontal sources and receivers.

There is first the question of internal recording consistency; CDP arrangements
allow for the source to lead, to follow, or to be in between the recording
spread(s). Usually the SH vibrator trucks are *pointed* in the direction of movement of the traverse, and it must be arranged that a POLARIZATION MARK
on the geophone case is consistently to one side (say the left) of the line as it
progresses. This is true *whether the line progresses to the east, west, north, or south
or in between.*

The next problem is that of consistent polarity between P and SH records; both
must give a reflection in the same direction from an interface that has a positive
change of acoustic impedance downward—for both wave types.

We note that the reflection coefficients are consistent:

$$R_{SH} = \frac{\beta_1 \rho_1 - \beta_2 \rho_2}{\beta_1 \rho_1 + \beta_2 \rho_2} \qquad \beta = \text{shear velocity}$$

$$\rho = \text{density}$$

$$R_p = \frac{\alpha_1 \rho_1 - \alpha_2 \rho_2}{\alpha_1 \rho_1 + \alpha_2 \rho_2} \qquad \alpha = \text{P-wave velocity}$$

In the P wave a step function downward (called a positive displacement) is
reflected as a step function of displacement upward (negative). On reaching the
surface, it is reflected with the same sign and the net motion of the geophone
upward is doubled. This, in turn, conventionally (SEG), gives rise to a downward
motion of the seismic trace on a horizontal record or to the left on a seismic
section where the traces are downward. However, we would like the indication of
a positive reflector to be to the *right*, where it will be emphasized by the
conventional shading. It is necessary, then, to change the phase by 180°.

In the shear wave case the SH motion is perpendicular to the plane of the
diagram. A positive initial displacement (however that is defined) is converted
into a negative displacement by the reflector and, in turn, is reflected as a
negative displacement of twice the amplitude by the free surface. We *make the*

convention that this will give a trace deflection to the left on a cross section and now a 180° phase shift is necessary to give a deflection to the right for a reflection from a positive reflector. *These are important matters and should not be left to chance.* Eventually we shall be comparing SH and P records and want to make sure that the conventions are the same for sections from both types of energy. In VIBROSEIS, of course, we are talking about trace deflections after the correlation procedure. There is really no excuse for having to produce both positive and negative polarity cross sections because all of this polarity consistency should have been made at an earlier stage.

A more artificial change in the recording and display techniques occurred after the initial experience with shear-wave recording (Cherry and Waters, 1968). It was then determined that the apparent average ratio of shear-wave velocity to compressional-wave velocity in the sedimentary section is nearly 0.5. For this reason, because we wish to have the *same wavelengths* in the earth for both types of wave, *it is customary to make the sweep signal for the SH vibrators range through frequencies that are one-half of those used for the P-wave vibrators.* To a first approximation the layers, given the same velocity contrast, should behave in the same way for both types of wave.

Using the same experimental information, displays are altered as follows:

1. The same horizontal scale is used for the cross sections.
2. A time scale for SH time sections is used which is 2/1 compressed compared with the P-wave section (as an example: 2 s SH = Z(depth) = 1 s P).

Impulse Shear-Wave Sources

Shear waves can be generated by explosive methods, although perhaps not so purely as with some other methods. The explosive method now sometimes used (SYSLAP®) follows essentially the methods of Puzyrov et al. (1967) and Brodov et al. (1968) in the USSR who are credited with first making use of a multiple hole method with explosions to generate S waves while reducing to a minimum the effects of the P waves that result from the same explosions. Four holes were originally used but this has been reduced to three for SYSLAP®. (Trademark CGG.) In Figure 12.2 the center hole (B) is *sprung* by an explosive charge that weakens the right-hand side of hole (A) and the left-hand side of (C). Then, shooting in (A) will generate a force +F to the left, giving rise to SH records recorded on horizontal geophones along the spread, together with residual P waves generated by the expansion of the earth near (A). Similarly, shooting in hole (C) will generate a horizontal force −F to the right that will give rise to negative polarity S waves recorded on the spread and a nearly equal P-wave set. Subtraction of these two records will (with or without normalization) nearly cancel the P-wave set and augment the S-wave recordings:

$$\text{symbolically—}(SH)^+ + P - (SH)^- - P = 2(SH)^+$$

Primacord has also been used in trenches as a variant of the SYSLAP® method.

MARTHOR® (trademark I.F.P.), a mechanical method, has been developed and tested by the Institute Française de Petrole (Figure 12.3). It relies, as have

SYSLAP ®

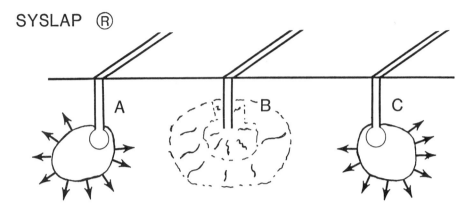

FIGURE 12.2 Diagrammatic explanation of the SYSLAP® process for producing SH waves by explosive charges. (® Registered trademark, courtesy C.G.G.)

many other experimental tests [e.g., see Beeston and McEvilly (1977)] on the use of a horizontal impulsive force, generated by a large pivoted hammer that strikes a plate fixed to the ground. Here, again, the plate may be struck from either side to yield S waves mixed with P waves. After normalization the two opposite polarity records can be subtracted to give almost compression-free S-wave records. The disposition of the recording spread with respect to the source will decide whether the waves recorded can be regarded as SH or SV.

Impulsive methods have the general weakness that the frequency content of the impulses generated cannot be easily controlled and that these source signa-

FIGURE 12.3 The impulsive shear-wave surface source MARTHOR® ("MARTHOR" is a registered trademark © I.F.P., courtesy C.G.G.)

tures may vary according to the placement of the source and other factors. The amount of variation cannot be mitigated by the use of phase control and feedback.

VIBROSEIS Methods

The vibrator principles are well illustrated by Figure 12.4 which shows the reaction mass in the form of a 2500-lb steel mass, machined out to form the cylinder for a servohydraulic system. The development parallels many vintages of vertical vibrator. The double-acting piston is formed from a steel bar, 5 in. in diameter, attached rigidly to two supports fastened, in turn, to a baseplate. Contact to the ground is provided by large steel inverted pyramids bolted (or welded) to the baseplate. As usual in servohydraulic vibrators the lift system provides a means of bringing the weight of the truck to bear down, through vibration-isolating air bags, on the baseplate, thereby forcing the pyramids into the surface soil or rock. If there is any tendency to *wallow out* of the pyramid holes (by vibration) it is counteracted by the weight of the truck driving the pyramids further into the ground. As usual, the motion of the baseplate is monitored and fed back into the phase compensation system, thus causing the baseplate to follow the action dictated by the control trace. Of course, the field traces produced by these vibrators must be correlated with the control signal to produce the seismograms.

In the last 10 years the early vibrators have been superseded by more powerful shear-wave vibrators that have raised the signal/noise ratio for SH records considerably. The specifications of the CONOCO Model 13 vibrators are listed in Table 3.1.

The main advantage of vibrators over impulsive sources is the possibility of injecting energy over a relatively long period of time so that very large forces are not converted into nonlinear strains. The linear region of the stress-strain curve for surface initiation of SH waves is small. It is easy to stress the surface beyond

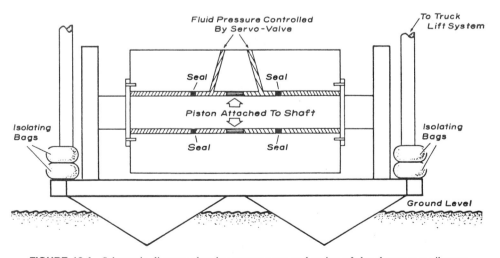

FIGURE 12.4 Schematic diagram showing components and action of the shear-wave vibrator.

the yield point. When this happens the pulverized surface absorbs energy that is needed for reflections. Later, correlation effectively compresses the available energy into a short pulse, but without the hazards of nonlinearity. Even so, coupling to the earth does represent a substantial problem. The large pyramids that are forced into the earth create large holes that limit the areas in which the vibrators may work.

SH-Waves from Vertical Vibrators

To avoid this problem of coupling Edelmann (1981) suggested the use of two *vertical* vibrators, out of phase and situated side by side across the line of traverse. It is well known that the output of the vertical vibrator is symmetrical about a vertical axis and consists of P waves, S waves, and Rayleigh waves. The use of two vertical vibrators out of phase immediately reduces the P-wave and Rayleigh-wave amplitudes (for long distances of travel and, preferentially, for low frequencies) and, on a vertical axis midway between the two vibrators, produces a substantial S wave whose polarization is perpendicular to the line of traverse. Field tests of this method (called SHOVER by Prakla-Seismos, GMBH) have shown that a shear-wave reflection can be made this way but at present the signal/noise ratio attainable on these sections is much poorer than that provided by the standard SH vibrators. The reflections on the SHOVER section are earlier than those on the horizontal vibrator section by approximately 0.2 s. The reason for this time difference is not known.

Torsional Vibrators, Sources, and Receivers

Considerable speculation has been made about the utility of pure SH sources (with a vertical axis of oscillation) and the corresponding receivers, usually called torsional sources and receivers. An early U.S. Patent No. 3280935 that covered most of the basic vibrator methods was issued to Graydon L. Brown of CON-OCO, Inc. in October 1966. Since that time proposals (both of the impulse type) have been made by Applegate (1974) and Won and Clough (1981). Theoretical properties of such a source situated on the surface of a plane-parallel laminated medium have been treated exhaustively by Dorn (1980, 1984).

Despite its apparent simplicity, the torsional source has some drawbacks when compared with the now conventional *linear* horizontal vibrator. It can be seen, from symmetry, that no linear motion is produced down the vertical axis, and at points near the vertical axis the motions resulting from source points almost cancel; there are always corresponding regions of the source diagonally opposite one another and moving in antiparallel directions. The distances from these corresponding regions of the source to a slightly off-axis point differ only by second-order quantities.

Physical intuition suggests that although the motion of a single particle on, or near, the Z axis is near zero the *rotation* is not. The angle of rotation is the linear motion divided by the distance from the center of rotation; hence the rotation can remain finite as the axial distance decreases to zero. Following this line of reasoning, detectors that respond to rotary motion could also be used at short offsets to detect reflections of this rotary motion. As far as we know, this has not been tried experimentally or even tested mathematically.

There are, of course, no good generators of oscillatory rotary motion in nature and the noise level in the field could be very small.

Shear-Wave Exploration in Water-Covered Areas

Tatham and Stoffa (1976) proposed that water-covered areas are susceptible to shear-wave exploration methods in spite of the difficulty caused by the zero shear strength of the surface. They also suggested that use could be made of mode conversion at the water-bottom/consolidated rock boundary. The energy would start as P waves and be converted to SV waves after refraction into the solid earth. Conversion efficiency has been found (in models) to be good at angles slightly beyond the critical angle. These conversions have been found and analyzed by Tatham and others (1981) who noted that, for sea-bottom P-wave velocities between 2000 and 2743 m/s (6500 to 9000 ft/s), the mode conversion is efficient both upward and downward (Figure 12.5). These hard-water bottoms, however, have sudden velocity and density changes and more experiments are needed to determine whether conversion is just as efficient when the change is transitional, as in the major portion of the northwest Gulf of Mexico. Experience with one-dimensional acoustic synthetic records from multiple transition layers (Berryman, Goupillaud and Waters (1958))—and intuition—show that this is unlikely. Of course, a double conversion is needed for these PSSP reflection recordings.

By the use of τ-p transform methods filters can be devized that make it easier to differentiate between the required PSSP waves and normal P reflections. Tatham et al. (1982) claim some success in the actual reception of these waves in an offshore Florida area. The sea-bottom P-wave velocity was 2286 m/s. The

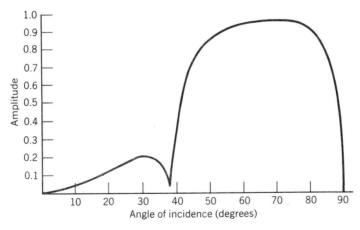

FIGURE 12.5 Amplitude for a *double* mode converted wave (PSSP) as a fraction of P-wave amplitude at all angles of incidence on a hard sea bottom. Water, density 1000 kg/m³; P-wave velocity 1650 m/s. Limestone, density 2,500 kg/m³; P-wave velocity 2700 m/s. Critical angle 37.67°.

resulting cross sections were, by normal P-wave standards, poor but might be the first evidence that this rather exotic process works.

More recently, additional evidence has been produced that this is not an isolated instance (Sheline, 1985). Filter design is aided by computer 2-D modeling (both acoustic and elastic) from which the shear arrivals on the τ-p sections can be discriminated from P events. The model parameters in the first instance are obtained from P-wave cross sections that originate in the same field data.

Figure 12.6 is a synthetic CMP gather from the elastic model that shows both

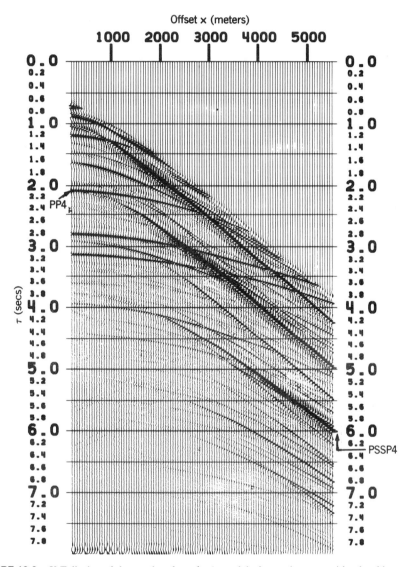

FIGURE 12.6 *X-T* display of the results of an *elastic* model of a marine area with a hard bottom and several subsurface layers. The model was constructed from P-wave data. The PSSP event from boundary 4 is partly obscured in this display. The PP4 event is indicated on the left. (*Source*: Courtesy CONOCO, Inc.)

compressional (PP) and marine shear wave (PSSP) reflections. The reflection time is in seconds and the offset, in meters. The position of the PP and PSSP reflections caused by layer 4 are indicated.

Figure 12.7 shows the same data as Figure 12.6, except that it is transformed into τ-p coordinates to separate PP from PSSP reflections. For this horizontally layered model the angle of incidence at the water bottom is given by arc sin (p. 1500). Positions of the PP and PSSP reflections caused by layer 4 are indicated.

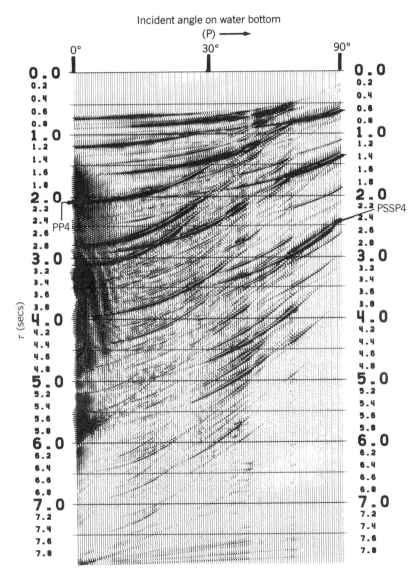

FIGURE 12.7 p-τ display of the results from the same elastic model in Figure 12.6. The PP4 and PSSP4 events are labeled and the latter is much clearer. Note how with this display it is possible to draw an approximate boundary to separate the two types of event. (*Source*: Courtesy CONOCO, Inc.)

Because of the need for high angles of incidence to give the necessary high conversion efficiency, the marine work for this purpose has to use longer-than-usual offsets. Geometrical considerations alone show that to reach a depth of 2000 m, below a 200-m water layer, with shear wave (SV) reflections the spread length would have to be about 4100 m.

The Effects of Anisotropy

In two recent publications (Crampin et al. 1984a; Crampin, 1984b) it has been pointed out that for both isotropic and transversely isotropic materials the wave motion in the plane that contains the P, SV, and Rayleigh wave motion (the sagittal plane) decouples from the SH wave motion. A horizontally layered medium with plane-parallel layers, each of which is transversely isotropic (approximating the conditions usually encountered in exploration geophysics), allows this decoupling, and by suitable experimental design the shear-wave component of any raypath can be designated as SH or SV. The appropriate velocity can then be determined and applied.

It has been customary in compressional wave exploration to ignore the fact that in some rocks (or rock groups) the velocity of seismic waves is not constant but dependent on direction of travel. Most analyses of velocity (including programs for computing NMO or RMS velocities from timed-offset reflections) have assumed that the rocks were isotropic. In these circumstances it is relatively easy to assume geometrical paths for the waves from source to reflecting point to receiver. Then by Dix's equation

$$T_X^2 = T_0^2 + \frac{X^2}{V^2}$$

where X is the half-offset and V is the appropriate RMS velocity.

When the layering is cyclic at a high rate, compared with the wavelengths of the waves involved, the rock assumes a pseudoanisotropic characteristic (transversely isotropic) in which the horizontal velocity is different from the vertical velocity and Dix's equation may be modified to give

$$T_X^2 = T_0^2 + \frac{x^2}{A^2 V^2}$$

where A is the anisotropic coefficient.

The wave front in a transversely isotropic medium is ellipsoidal and the ray path is no longer orthogonal to the wave front.

Even if the rock layer consists of a single rock type it is possible that, due to preferred orientation of the crystalline components, anisotropy is still present. For P waves in shales the anisotropic effect is not large but may cause an 8% increase of horizontal velocity over vertical velocity.

The standard experiment that involves change of horizontal offset will therefore not give a measure of the vertical velocity through the sediments but *the horizontal velocity*. As Postma originally observed, the anisotropic ratio can never be obtained from surface seismic work alone. If VSPs for both P and SH waves are available, however, the anisotropic ratio can be obtained.

J. E. White and coworkers (1983) measured the anisotropy of the Pierre shale with a rectangular arrangement of four geophones (at a time) in two nearby holes. The assumption was made that the Pierre shale is transversely isotropic and theory states that five elastic constants are required. In the standard notation of Love, Stoneley (1949), and White (1965) the elements of stress are related to the elements of strain by

$$
\left.\begin{aligned}
P_{xx} &= A e_{xx} + (A - 2N) e_{yy} + F e_{zz} \\
P_{yy} &= (A - 2N) e_{xx} + A e_{yy} + F e_{zz} \\
P_{zz} &= F e_{xx} + F e_{yy} + C e_{zz} \\
P_{xy} &= N e_{xy} \\
P_{yz} &= L e_{yz} \\
P_{xz} &= L e_{xz}
\end{aligned}\right\}
\tag{12.1}
$$

where *five* constants (Moduli), A, C, F, N, L are required and the phase velocities of P, SV, and SH waves are given as

$$
\left.\begin{aligned}
V_{\mathrm{P}} &= \left[\frac{(p + q)}{2\rho} \right]^{1/2} \\
V_{\mathrm{SV}} &= \left[\frac{(p - q)}{2\rho} \right]^{1/2} \\
V_{\mathrm{SH}} &= \left[\frac{(N \sin^2 \gamma + L \cos^2 \gamma)}{\rho} \right]^{1/2}
\end{aligned}\right\}
\tag{12.2}
$$

where $p = A \sin^2 \gamma + C \cos^2 \gamma + L$

$q = \{[(A - L) \sin^2 \gamma - (C - L) \cos^2 \gamma]^2 + 4(F + L)^2 \sin^2 \gamma \cos^2 \gamma\}^{1/2}$

ρ = density

γ = angle between the normal to the plane of constant phase and the vertical axis

Relations (12.2) allow the interpretation of $(c/\rho)^{1/2}$ as the velocity of vertically traveling P waves, $(L/\rho)^{1/2}$ as the velocity of vertically traveling S waves, and $(A/\rho)^{1/2}$ and $(N/\rho)^{1/2}$, the velocity of *horizontally* traveling P waves and *horizontally* traveling SH waves, respectively.

The particle motion of P waves is not exactly along the phase-velocity direction nor is the SV motion perpendicular to it. The angle between the particle motion and the vertical (in White's experiment) is given by

$$
\theta_p = \tan^{-1}\{[(F + L) \sin \gamma \cos \gamma / (\rho V_p^2 - A \sin^2 \gamma - L \cos^2 \gamma)]\}
\tag{12.3}
$$

$$
\theta_s = \theta_p + 90°
$$

It has been shown by Berryman (1979), Levin (1979, 1980) and White (1982) that a difference in magnitude and direction exists between the phase velocity and group velocity.

In a homogeneous transversely isotropic medium *energy* moves with group velocity V_g in a direction ϕ. Locally, planes of constant phase travel with another velocity V_{ph} in the direction γ. For P or SV waves the direction of motion θ_p or θ_{SV} may differ from the directions parallel (or orthogonal) to the wave-front normal (γ) or ray (ϕ). *The motion of SH is normal to the vertical plane of propagation, parallel to the wave front.* Thus even in transversely isotropic materials the SH waves behave in the simplest manner.

White et al. have used arrival times from offset sources to four geophones in a rectangular pattern to establish phase velocities (V_{ph}) and directions (γ) at the center of each four-point pattern:

$$V_{ph} = \left[\left(\frac{1}{V_H} \right)^2 + \left(\frac{1}{V_V} \right)^2 \right]^{-1/2}$$

$$\gamma = \tan^{-1} \left(\frac{V_V}{V_H} \right)$$

(12.4)

Eight source points separated from 200 $-$(200) to 1400 ft were used. Geophones were at 200 $-$(200) to 1000 ft in depth in each of two holes, 100 ft apart.

The determination of the five elastic constants is complex and reference to this detail should be made in the original paper.

This set of experiments, beautifully designed and apparently well executed, was performed in an area in which there is a massive rock formation as homogeneous as exists for experimentation. Although the results apply only to this particular location, the methodology should be well known and may be performed anywhere that appears to be appropriate. The need for two shallow holes close together should not prove to be much of a barrier, although there is a need for deep information (say in the Gulf Coast sedimentary section). In addition the expense may prove to be a difficult hurdle to overcome.

The overall average vertical P-wave velocity from 450 to 950 ft was 2191.5 m/s (sonic log gives 2343 m/s). Measured vertical shear-wave velocity was 938 m/s and V_s/V_p was 0.428. The ratio of the horizontal-to-vertical V_p was 1.1023 and the ratio of the horizontal V_{SH} to vertical V_{SH} was 1.1417.

Therefore in this case the SH-wave anisotropy was larger than the P-wave anisotropy but not alarmingly so. There have been numerous reports of shear-wave anisotropy of 40% or more and there appears to be some need to make in situ experiments to determine anisotropic ratios under conditions of high temperature and pressure.

Crampin, in a long series of publications from 1981 to 1985, has suggested that shear-wave trains contain much more information than a corresponding P-wave train but are much more complex and difficult to interpret. Because cracking or fracturing is now known to lead to a more general anisotropy than the transverse isotropy we have discussed, there is a strong incentive to use three-component, shear-wave seismology to obtain the requisite information. The same incentive should also motivate theoretical research into the most fruitful method for examining the data and extracting the required information in its clearest form.

Near-Surface Waves

Most of the near-surface waves (guided waves controlled by the free surface, base of the weathered layer, or other media boundaries) that plague P-wave prospecting have twin events that create problems in S-wave exploration. The catalog is not quite symmetrical, however. The characteristics of true Rayleigh waves, pseudo-Rayleigh waves, Love waves and *head* waves were examined in Chapter 2.

Although the Love waves are generated in a layer by multiple reflections, beyond the critical angle of SH waves and therefore parallel the (more complex) generation of pseudo-Rayleigh waves, there is no shear-wave analogue of the true Rayleigh wave unless we regard the surface-traveling SH wave as such an analogue. The retrograde elliptical motion does not exist.

All wave forms, P, SV, and SH, have their own head waves that travel with appropriate velocities in the lower, higher velocity medium. It is easy to generate and study all of them, side by side, with appropriate applications of the vertical and horizontal vibrators. Although the use of the vertical vibrator generates waves that are symmetrical about a vertical axis, the horizontal vibrator generates waves that are SH (Love waves, SH head waves, SH reflections) when examined along the direction at right angles to the vibrator motion, whereas the waves are SV (and P) along the direction of the vibrator motion. Thus in this direction pseudo-Rayleigh waves, SV head waves, and SV (and P) reflections are obtained. The recorded waves along the azimuths in between are a complex mixture of all.

The presence of a groundwater level within the near-surface, low-velocity layer has a profound effect on the characteristics of the surface waves set up. In the porous rock of the near-surface the presence of connate water changes the velocity and density of the rock appreciably. Although there is a P-wave velocity change at the top of the water table, there may be little variation in S-wave velocity because the velocity for shear waves is scarcely affected by the fluid in the pore space except insofar as it changes the density.

Furthermore, the same effects exist in a more complex form when changes of porosity occur within the near-surface layer, due to local changes of clay, sand, or gravel content and consequent variations in capillary rises in the water level. Similar effects occur in leaching of water-soluble salts that bind the rock particles together. Then the effect is a change of impedance for a wave traveling through the medium. Not only are variations in vertical weathering corrections initiated but scattering of surface waves also occurs, thus giving rise to interference.

Because these surface waves turn out to be important to the problem of obtaining SH (or SV and P) reflections and may even alter the *relative quality* of the SH and P reflection sections, the two cases (for high-velocity and low-velocity surface materials) are now examined in more detail:

A. Surface-Layer Lower Acoustic Impedance

1. **Vertical Source**
    a. Pseudo-Rayleigh waves are generated by interference of multiply-reflected SV and P waves past the critical angle in the layer. These waves travel in a dispersive manner and the phase velocity and group velocity are related as shown in Figure 12.8a. Multiple modes exist.
    b. There is no P wave along the surface.
    c. Four sets of refracted waves travel along the lower boundary, continually radiating to the surface. Phase velocity is that of the lower medium mode. (See Figure 12.8b.) Which are actually generated depend on generalized Snell's laws.

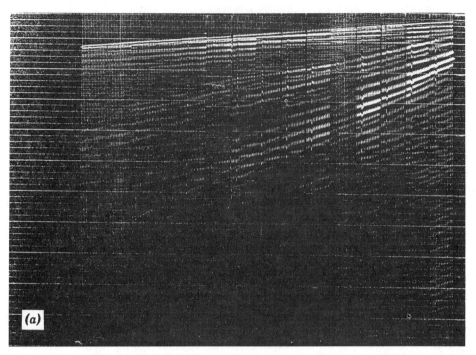

FIGURE 12.8 (*a*) Display of interference (over a 7000-ft spread) from a vertical vibrator with vertical geophones. Note the P-wave refraction and early P-wave refledtions as well as dispersive Rayleigh waves. This is taken from an area in southwestern Texas (*Source*: Courtesy CONOCO, Inc.)

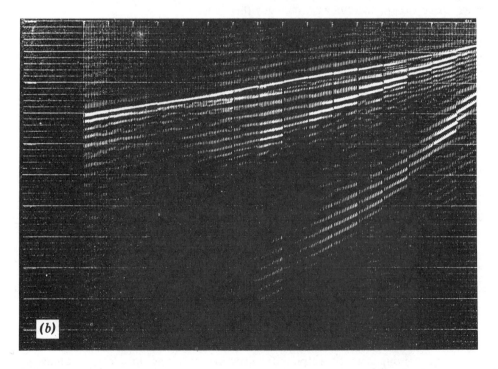

FIGURE 12.8 (*b*) Display of interference (over a 7000-foot spread) from an SH vibrator in the same area. Note the much slower SH refraction. There are no visible indications of SH reflections, but Love waves are strong. (*Source*: Courtesy CONOCO, Inc.)

2. **Horizontal Source**

a. SH waves are transmitted along the surface in a direction perpendicular to source polarization.

b. SH waves are totally reflected beyond the critical angle and combine, by interference, to give rise to Love waves dispersive with group and phase velocities related as shown in Figure 12.8b. Multiple modes exist.

c. One refraction is formed SH, SH, SH whose phase velocity is the S velocity in the lower medium.

d. Along the SV direction the same set of refractions is formed as for vertical source but the amplitudes are changed.

e. Pseudo-Rayleigh waves are set up when the original SV wave strikes the lower boundary beyond the critical angle. Multiple reflections generate Rayleigh waves.

B. **Surface Layer Higher Acoustic Impedance**

1. **Vertical Source**

a. The conditions for true Rayleigh waves are met for high frequency.

b. The conditions for pseudo-Rayleigh waves are *not met*. All energy can eventually seep into the main body of the material and be available for reflections; but, *be careful*, the reflection coefficient at the base of the layer may be high but it is never equal to unity.

c. The conditions for refractions could be met when

$$\frac{\sin \theta_2}{\sin \phi_1} > 1$$

that is, when $\alpha_2 > \beta_1$ P-wave refractions with SV vertical paths are possible.

2. **Horizontal Source**

a. SH wave along the surface.

b. SH waves under conditions $\beta_1 > \beta_2$ cannot be *totally* reflected; therefore Love waves are diminished and confined to one or two bounces. They are, however, dispersive.

c. Under the same conditions SH refractions are not possible.

d. For SV conditions for pseudo-Rayleigh waves are not met.

e. Conditions for true Rayleigh waves and SV along the surface are met, particularly for high frequencies.

f. As for the vertical vibrator, conditions for refractions could be met when $\alpha_2 > \beta_1$.

In practice the contrast between the weathered layer velocity and the subweathering velocity is not necessarily the same and is usually much greater for shear waves than for P waves; for example, the *usual* contrast (V_w/V_1) for P waves falls between 0.333 and 0.222, whereas for S waves it can be 0.11 when the water content is high and the shear strength of the weathered layer is low. For rigid rock surface layers these contrasts may be nearly unity. Therefore the reflection coefficient at the base of the weathered layer may be high for shear waves or very low. For a low-velocity weathered layer Love waves are usually well organized and strong, whereas for high-velocity surface layers the surface waves may have low amplitude.

Measurements of Shear-Wave and P-Wave Velocities

The measurement of wave velocity in rocks has been made in at least four different ways:

1. Ultrasonic measurements have been made on specimens cut from cores. King (1965, 1966) and Benzing (1978) used these methods. A problem is encountered in that measurements are not made under the same conditions of matrix and pore-fluid pressure, temperature, and pore-fluid constitution as experienced by the undisturbed rock unless special precautions are taken.
   To the extent that (a) the process of change is reversible, (b) pressure, temperature, and salinity are known, (c) ultrasonic velocities are the same as velocities for the long wavelengths used in seismic exploration, (d) the specimen is undisturbed by the preparation (i.e., not fractured or dried out), this method may establish approximate values for the velocities sought but it is not certain. Certain soft rocks are not amenable to this treatment.

2. By the use of modern velocity measuring tools (see Chapter 7) the refracted wave can be expected to travel well into the formation that surrounds a drilled hole and thus measure the velocity associated with unaltered rock, but even the drilling of the hole changes the environment of the material:
   a. The formation may be altered by the drilling fluid.
   b. The pressure and temperature are certainly disturbed.

3. The standard *velocity spectrum* method of obtaining velocities from seismic profiles [Taner et al. (1969)] is difficult to use for interval velocities associated with individual rock types unless thick homogeneous formations are present. Root-mean-square velocities to a set of predetermined depths are usually accurate enough for stacking and other purposes, but the interval velocity accuracy falls off rapidly as the depth intervals decrease. The almost invariable presence of anisotropy renders the evaluations as *vertical velocities* questionable, except, perhaps, for the SH measurements.

4. Down-hole measurements in vertical seismic profiles have added a new method in which reflection times (after deconvolution) for two different wave types and measured depths may be correlated. Some uncertainties, however, are still due to the dispersion associated with attenuation. Nevertheless this method offers the best chance of obtaining velocities that can be used to calculate V_s/V_p values and correlate them with lithology.

Association of Velocities and Velocity Ratio with Rock Properties

With these methods of measurements and their frailties in mind we can turn to one desired goal of shear-wave exploration, the ability to associate velocities, or velocity ratios, with rock lithology, porosity, fracturing, or connate fluids. Many attempts have been made to do just that. In view of the uncertainties and the number of factors affecting them, it is not surprising that these correlations have been contradictory at times and ambiguous in interpretation at best. In the following few paragraphs an attempt has been made to summarize the results:

A. **Unconsolidated Rock Velocities Versus Density or Velocity Ratio**

1. Gardner's equation relates specific gravity to compressional wave velocity by $\rho = 229.5 V_p^{1/4}$, where V_p is in m/s:

    for example $V_p = 6000$ ft/s $\rho = 2020$ kg/m^3
    $V_p = 10,000$ ft/s $\rho = 2295$ kg/m^3
    $V_p = 18,600$ ft/s $\rho = 2680$ kg/m^3

2. Ludwig et al. (1970) found a rather complex relationship between V_s, V_p, V_p/V_s, and density. It is clear that in the low density range of 1600 to 2000 kg/m^3 the individual V_p and V_s, as well as the V_p/V_s ratio, are extremely sensitive to density (or consolidation).

3. M. Yoshimura et al. (1982) have suggested a relation

$$\frac{V_s}{V_p} = 0.576 - \frac{0.139}{V_s}$$

    where V_s, V_p are in km/s. This holds for velocities from $V_s = 0.4$ to 2.0 km/s.

4. Clays and soils display large values of Poisson's ratio (more than 0.4) and V_p/V_s can take on values as high as 6.0. For Gulf Coast sediments Anderson et al. (1972) and Tixier et al. (1973) have derived a direct proportionality empirically between what they call a *Shaliness Factor* (q) and Poisson's ratio (σ):

$$q = \frac{\phi_z - \phi_e}{\phi_z}$$

    where ϕ_z is the total pore space between grains that support the overburden (porosity from the density log) and ϕ_e is the porosity available to water and hydrocarbons (porosity from the sonic log). $\sigma = 0.125q + 0.27$. For a shaliness index of zero (liquids can fill all the pore space) $\sigma = 0.27$ ($V_s/V_p = 0.5613$). For $q = 0.4$, $\sigma = 0.32$ ($V_s/V_p = 0.5145$).

B. **Consolidated Rock Velocities Obtained from Well Logs, Versus Lithology**

A number of suggestions have been made by Pickett (1963), National (1974), and Kithas (1976) that a relationship exists between V_s/V_p and rock type. Pickett has shown (see Figure 12.9) that when velocity parameters are plotted on a graph whose axes are shear-wave and P-wave slownesses a separation is obtained between carbonates and sandstones. The carbonate points all have V_s/V_p ratios lower than 0.56, whereas the sandstone points have V_s/V_p values greater than 0.56. Furthermore, the limestones have a lower average value (0.526) than the dolomites (0.555). This hypothesis for discrimination has been supported by more recent data from Leslie and Mons (1982) from wells in the North Sea (Figure 12.10).

$$\frac{V_p}{V_s} = \frac{\Delta t_s}{\Delta t_p} = \frac{1}{R}$$

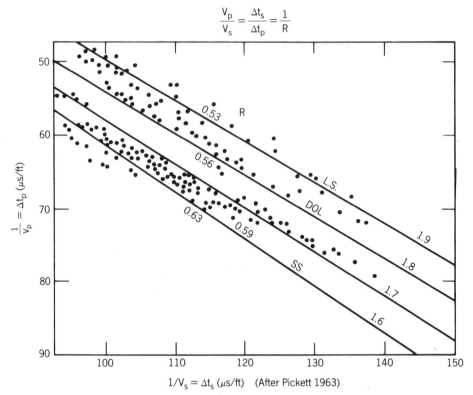

FIGURE 12.9 Display of numerous measurements of P-wave slowness versus S-wave slowness. [*Source*: Pickett (1963).]

C. Consolidated Rock Velocities Obtained from Laboratory Experiments with Simulated Rock and Fluid Pressures

Benzing's measurements (Figure 12.11) were made with a rock pressure of 10,000 psi, a fluid pressure of 4350 psi, and 100% water saturation. The nearly closed boundary shown encloses all the sandstone points. At low velocities the $V_s/V_p = 0.50 \pm 0.08$, whereas at higher velocities in the quartzite range V_s/V_p are more nearly 0.55 ± 0.05. The carbonates measured has V_s/V_p values that average 0.46 ± 0.03. The separation between carbonates and sandstones does not appear to be so clear as it is on Pickett's chart.

D. Determination of Velocity Ratios from Field Reflection Experiments and from VSPs

If corresponding reflections can be obtained from S- and P-wave reflections obtained in the field a direct determination of V_s/V_p $(=t_p/t_s)$ can be make. Well logs would then supply the lithology. The problem is that it is *not always easy—perhaps not possible—to find corresponding reflections*. This problem may be alleviated if a VSP is available somewhere in the area, provided that the reflection continuity is good enough to follow the desired

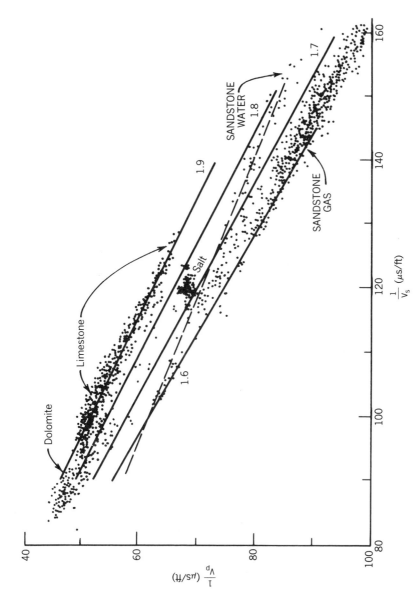

FIGURE 12.10 Display of numerous measurements of P-wave slowness versus S-wave slowness for wells in the North Sea. The dashed line transgressing the other trend lines of V_p/V_s is based on Pickett's work.

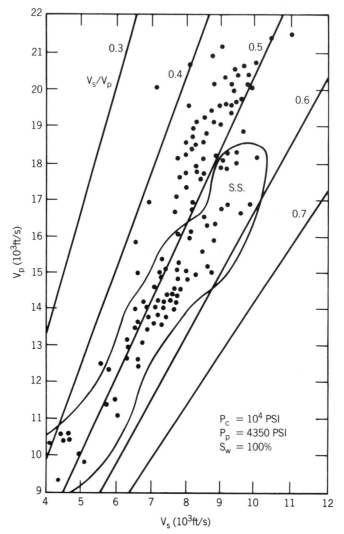

FIGURE 12.11 Measurements of V_p plotted against V_s for numerous specimens measured in laboratory apparatus with simulated temperatures and pressures. After Benzing (1978). The separation between sandstones and carbonates given by the almost closed boundary is based on physical description but would be hard to duplicate on the basis of a constant V_s/V_p zone.

events from one location to another. Certainly the VSP should show (after deconvolution and other suitable processing) the exact point on the seismic record that should correspond to the lithology boundary. Lack of seismic resolution, however, may vitiate this determination.

As this is one of the major hopes for shear-wave seismology, the determination of V_s/V_p in different areas and its relation to lithology or other rock parameters are pointed out on all suitable examples. For some rock types, such as shales, salt and other *soft* rocks, the use of modern sonic logs and VSPs is really the only satisfactory way to obtain accurate V_s/V_p values. It is a pity that there have been few opportunities, so far to get them.

Data Processing—Shear Waves

Although the same principles are used in data processing for shear waves as for P waves, in practice there are some minor differences.

The main process of correlation is exactly the same, although the shear-wave records are longer for a given penetration depth. The sampling rate usually remains the same as for P waves [although theoretically it could be halved (e.g., 0.008 instead of 0.004 s) without danger]. Thus the correlation process could take twice as long.

Second harmonic distortion due to large amplitude ground movement at the source does not appear to be quite so troublesome as for P waves (vertical vibrator). A second reason for second harmonic generation, however, is due to the inability to generate horizontal motion on the surface without the generation of some rocking motion; the latter causes pressure on the ground at twice the sweep frequency.

John Hopkins (personal communication) has proved the existence of second harmonic P-wave generation by recording with *vertical* geophones the energy from an SH source operating at half-normal frequencies and using a normal frequency control trace for correlation purposes. Figure 12.12a shows the record section which may be compared with a normal frequency P-wave section (vertical vibrators, vertical geophones) in Figure 12.12b. The close correlation between the two examples is conclusive of the existence of rocking motion, but the poor S/N ratio for section (a) shows that the rocking motion is produced at a low level. The importance of these second harmonics is reduced still further by SH geophones.

In the muting process used in compressional wave processing to reduce or eliminate the influence of P-wave refractions these particular refractions are not present at high amplitude for S-wave work. The high amplitude Love waves that are present for SH-work are just as troublesome as the Rayleigh waves in P-wave reflection work and must be reduced as much as possible in the field. A low-velocity mute, combined with a well designed velocity filter, is sometimes helpful.

One troublesome aspect of S-wave reflection work is the presence of much larger weathering corrections than in P-wave work. It is well known also that the static and dynamic corrections are interactive; error in one causes large errors in the other. For the same offset distance NMO corrections for S-wave reflections are also greater than for P-wave reflections because of the slower velocities for S waves. Figure 12.13 illustrates the comparison of static corrections over a short reflection line.

Investigation of the *statics* problem for S-waves in particular is still underway. It is still necessary to know whether, in general, the *delay* is frequency-sensitive (i.e., not just a time delay) or whether deconvolution (to remove ringing in the weathered layer) plus a static time shift is adequate. Methods such as MISER® which make use of matrix methods to determine *residual static* and *residual dynamic* corrections simultaneously need a much better initial estimate for S waves than for P waves. The program that does the calculation must also allow more delay tolerance in searching for matching trace sections.

At the present (1986) stage the S-wave records are almost always lower in S/N ratio than the corresponding P-wave records. This means that more effort is still

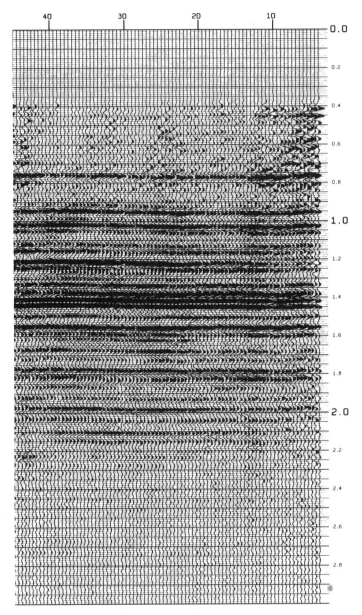

FIGURE 12.12 (*a*) Record obtained in a particular area using half-normal frequencies, an SH vibrator and vertical geophones. Correlation is with normal frequency control trace, which picks up only second harmonics. The close correlation between 12.12*a* and 12.12*b* is conclusive evidence of rocking motion of the SH vibrator. (*Source*: Courtesy CONOCO, Inc.)

required in the field as well as in data processing; this includes methods of stacking to produce CMP traces. Consideration must therefore be given to methods of weighting the individual components of the stacked traces. Although traces may still be weighted to balance lack of source numbers or power, it is desirable to give more weight to those traces that have a high S/N ratio. If low energy and S/N ratio for a trace is due to the fact that some of the vibrators have

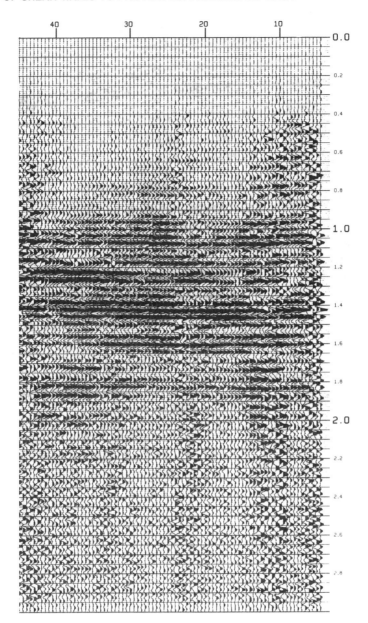

FIGURE 12.12 (*b*) Record obtained in the same area as 12.12*a* with normal frequencies, a P-wave vibrator, and vertical geophones. Correlation is with normal frequency control trace. The increase in S/N ratio over that for 12.12*a* shows that the SH vibrator rocking motion produces only low level P radiation. (*Source*: Courtesy CONOCO, Inc.)

been placed on an area of loose surface sand or low-strength clay these components must have low weight compared with the other high S/N ratio components. An extra stage in data processing is called for and extra recording (without field compositing) may be necessary.

Shear-wave records are generally more *ringy* [i.e., they have a narrower spectrum (in octaves)] than P-wave records. This makes the process of spiking

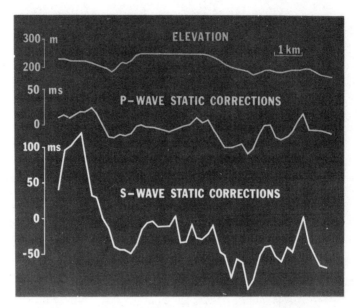

FIGURE 12.13 A comparison between P-wave and SH-wave static corrections over a short line. (*Source*: Courtesy C.G.G.)

deconvolution more critical because it is used to widen the bandwidth subject to noise restrictions. It also performs the function of adjustment of phase to make the individual wavelets as near zero phase as possible.

Provided that the initial control traces for S and P, respectively, are related in frequency by a ratio usually taken to be 0.5 but otherwise have the same bandwidth, the low-frequency S waves are reduced during the generation process due to mechanical characteristics of the vibrators compared with earth characteristics, whereas the high frequencies of both S and P waves are reduced by attenuation. If the rule

$$\frac{Q_s}{Q_p} = \frac{4}{3}\left(\frac{\beta^2}{\alpha^2}\right)$$

and the empirical observation $\beta/\alpha \simeq \frac{1}{2}$ are true (Chapter 7) then

$$Q_s = \tfrac{1}{3} Q_p$$

and it can be shown that for plane waves of *comparable wavelengths* in the earth (meaning $f_p = 2f_s$) the ratio

$$\frac{A_{s,x}}{A_{p,x}} = \frac{A_{s,0}}{A_{p,0}} \epsilon^{-2\pi f x/3 Q_s \beta} \qquad f = \text{shear-wave frequency}$$

decreases with distance and frequency. Thus the P waves hold up in amplitude better than the shear waves in the same (general sedimentary) medium. As an example for $f = 20$, $Q_s = 17$, $Q_p = 51$, $\alpha = 2133.5$ m/s, $\beta = 1066.8$ m/s,

$$\text{for } X = 500 \text{ m}, f_{(20)} \text{ ratio} = 0.32, f_{(40)} \text{ ratio} = 0.10$$

$$\text{for } X = 1000 \text{ m}, f_{(20)} \text{ ratio} = 0.10, f_{(40)} \text{ ratio} = 0.0099$$

This shows that the bandwidth of shear waves decreases more rapidly than the bandwidth of P waves, and deconvolution is limited by the smaller bandwidth and decreased S/N at high frequencies.

This may indicate that there should be some preemphasis of high frequency amplitudes at the source, although its effectiveness is usually limited.

Practical Accomplishments

The use of shear waves in seismic exploration has not yet reached total acceptance, although from the expenditures in 1984 ($80 million—S.E.G. report) it must be well on its way. Examples that are provided in this chapter, however, are some of those that have helped the oil exploration industry decide on its usefulness in the primary goal: the provision of supplementary information. Numerous oil companies, and companies in related activities, have participated in these evaluations. Some of them have done their own interpretations of the *group-shoot* data of 1977–1978 in areas of particular interest to themselves. Therefore this short account cannot be taken as anything other than a sampling of the *experimental* activity. It must be inferred that a great deal more proprietary work has not been released or revealed.

Example 1

Figure 12.14 shows corresponding S and P cross sections taken in southwestern Wyoming. The Tertiary Wasatch, Upper Cretaceous Mesaverde, and the Jurassic Morrison are all reasonably correlatable. Insufficient source power (1975) was available and the shear-wave section fails to show the Weber and Madison reflections, quite plainly visible on the P-wave section. Because this is a gas-producing area it is conceivable that the P-wave reflection is partly a bright spot but the depth (18,000 ft) must have caused severe attenuation of the high frequencies that appear to prevail in the Madison reflection.

The Fort Union reflection is rather broken up on the S-wave record but not so much on the P records. As a consequence the proper correlation there is unclear. The Morrison pick on the P-wave record could be 2.050 s (as shown in the figure), but the peak of the energy band is at 2.100 s and that corresponds nicely with the S-wave Morrison pick shown, to give an overall V_s/V_p ratio of 0.55. The section (Wasatch to Morrison) is predominantly clastic.

Example 2

Figure 12.15 shows a P and SH comparison for a well in southwestern Texas (1976). This is an excellent example of shear-wave quality and relative ease of correlation. Although we could quibble about a few of the correlative picks, those

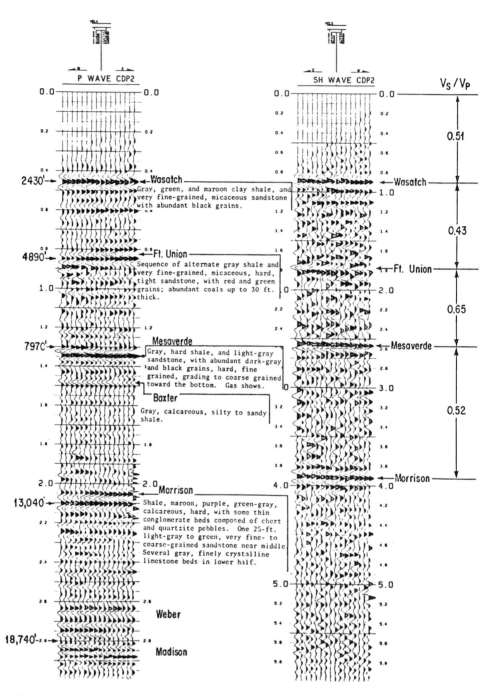

FIGURE 12.14 Comparative SH- and P-wave cross sections are shown for an area in southwestern Wyoming (1975). Major formation names as well as some sample descriptions are given. Calculated V_s/V_p values are shown on the right. (*Source*: Courtesy CONOCO, Inc.)

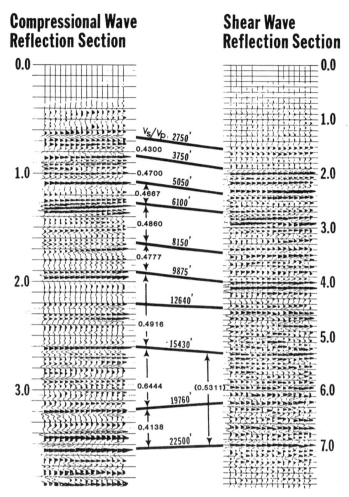

FIGURE 12.15 Comparative SH- and P-wave records are shown for a well in southwestern Texas (1976). Probable correlations are shown and the resulting V_s/V_p rations are shown (*Source*: Courtesy CONOCO, Inc.)

down to 10,000 ft are known by check shots to be well established. There is a remarkable smooth change in the V_s/V_p ratio from 0.4300 at 3250 ft to 0.4916 from the 9875 to 15,430-ft interval, but this is followed by 0.5311 in the overall 15,430- to 22,500-ft interval. The lower part of this interval (19,760 to 22,500) consists of carbonates $V_s/V_p = 0.4138$, whereas the upper part consists of mud-rocks and has a V_s/V_p ratio of 0.6444. The correlation quality at 19,760 is poor, however, and the diagnostic reliability must surely be questioned.

Example 3

In this West Texas well, besides a set of check shots derived from a dynamite source, a series of check shots made with an SH vibrator source recorded with a nonclamped, three-unit, orthogonal geophone gave some high-quality arrivals on

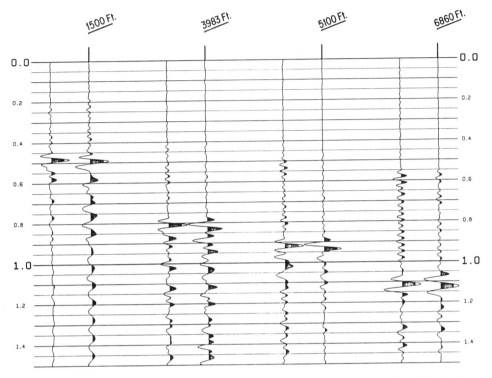

FIGURE 12.16   Records taken on two orthogonal horizontal geophones at the indicated depths in a west Texas well. A horizontal vibrator was used near the top of the well (1975). (*Source*: Courtesy CONOCO, Inc.)

the horizontal pair of geophones. A few of these arrivals have been shown in Figure 12.16. Note that the SH arrivals must be in-phase or out of phase, because they represent components along their axes of the real S-wave motion.

The comparison of SH and P records (Figure 12.17) is unremarkable. Depths that appear beside the records are calculated from stacking velocity values, hence are subject to considerable error. The only correlation that can probably be made unaided by other information is the strong reflection near 12,000 ft (SH) or 12,520 ft (P). Calculations of V_s/V_p ratios from well velocity data show values that average 0.516, and it is remarkable that with such good quality data no better correlation is achieved with the 2:1 reduction in time scale for the SH-wave record. The entire section consists of carbonates, evaporites, and shales.

Example 4

In Figure 12.18 comparable compressional and shear-wave records have been placed spliced together at the midway point (SP 355) to achieve a correlation as close as possible of reflection below the Tonkawa. This involves a shift of 0.1 s (shear wave time) upward for the SH record. The standard 2:1 compression of the SH time scale has been used. This comparison is remarkable in that, even though a good correlation at $V_s/V_p = 0.5$ has been obtained for the pre-Tonkawa section, from this point to the surface correlations are difficult to make with

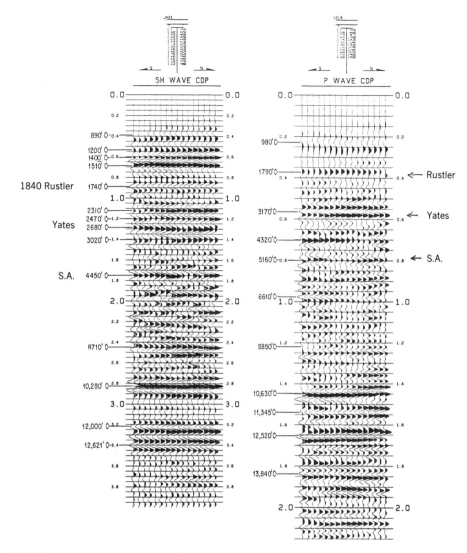

FIGURE 12.17 A comparison of SH- and P-wave records near the same west Texas well (Figure 12.16). Although the individual records are good, detailed correlation is difficult. Depths shown were calculated from well-velocity survey data. (*Source*: Courtesy CONOCO, Inc.)

assurance. Part of the difficulty lies in the presence of *unpaired shear- and P-wave reflections*. Some of these unpaired reflections have been marked.

Example 5

This project, which compares SH and P reflection data over a reef oil field in West Texas, unfortunately ran into high Love-wave interference. The final comparison nevertheless is instructive.

Figure 12.19 shows the P-wave data along the line. The formation names penetrated by some of the wells, marked by vertical lines on the section, are

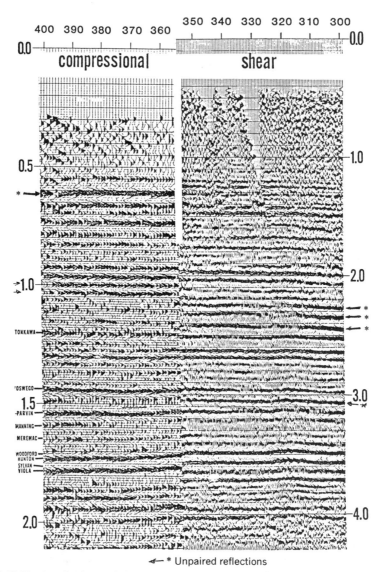

FIGURE 12.18   A spliced record section, one-half P wave and one-half shear wave, is shown. This comes from an area in the Anadarko Basin, Oklahoma, an example given to illustrate *unpaired* reflections (marked with an *). (*Source*: Courtesy CONOCO, Inc.)

shown on the right. On this section not only is the reef self-evident (from SP 95 to 150 s approximately) but other structural features such as faulting in the Mississippian–Devonian is clearly shown.

Figure 12.20 shows the corresponding SH-wave data along the same line. Apart from a fairly clear reflection at the Atoka-Mississippian contact at the east end of the line, graduating to poorly defined at the west end, there is little of interest on the section. Careful scrutiny, however, will reveal the reflection at the top of the reef, albeit very ringy and revealing no details of the reef contrast.

It is clear from the lack of reflection energy on the SH section and from the

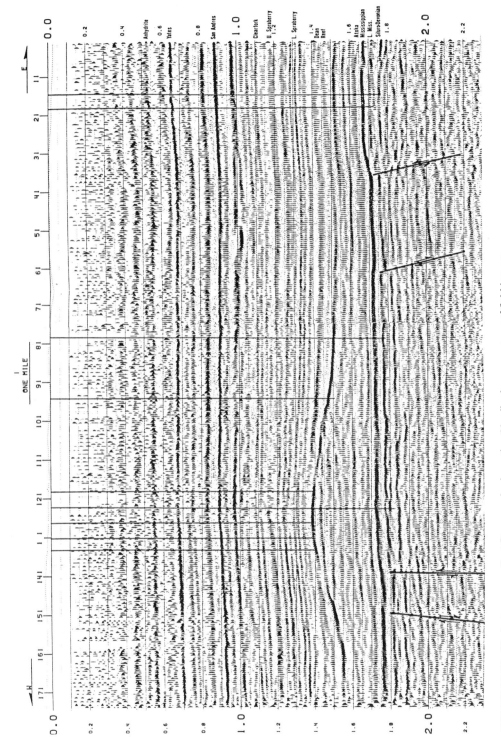

FIGURE 12.19 Illustrates an excellent P-wave seismic section 11 miles long across a reef field in west Texas. Formation names are shown on the right at approximately the correct record times. (*Source:* Courtesy CONOCO, Inc.)

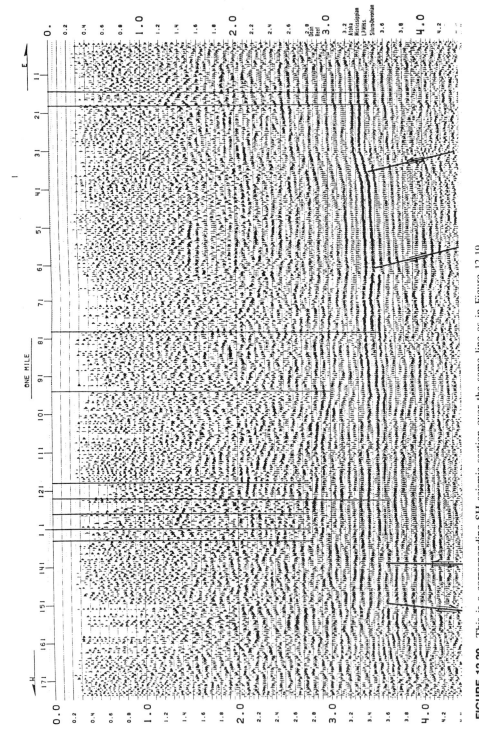

FIGURE 12.20 This is a corresponding SH seismic section over the same line as in Figure 12.19. Long arrays were used (660 ft for sources and receivers to try to cancel strong surface waves). (*Source:* Courtesy CONOCO, Inc.)

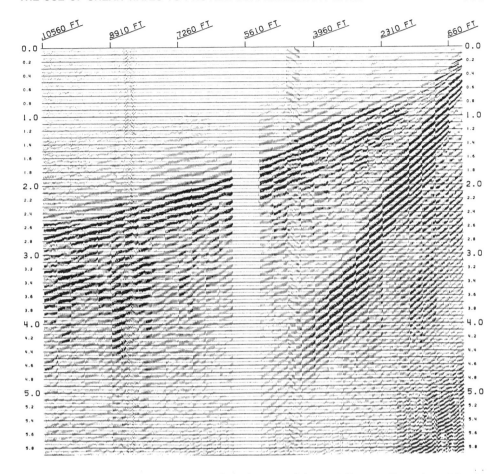

FIGURE 12.21 A walk-away noise analysis in the area of the preceding two illustrations. Like the SH-reflection record of Figure 12.20, this noise analysis is band-limited. It shows both near-surface refracted events and Love waves. Bad missties at the ends of the individual component records are evidently due to large changes in weathering below the successive source locations. (*Source:* Courtesy CONOCO, Inc.)

noise analysis shown in Figure 12.21 that the failure in this area is due largely to a high-contrast, near-surface layer combined on the west end with difficult surface source conditions, such as thick, loose sand, that affect the SH effort much more than the P work. Evidently, it is also not possible to predict SH wave quality in an area from a knowledge of P-wave reflection quality.

Example 6

Only a few tens of miles from Example 5 another P to SH comparison has been taken, this time to try to illustrate the truncation of a 1000-ft clastic wedge by the Midland Basin shelf carbonates. An interference test suggested that interference conditions would not dominate the SH-wave recording as much as they did in Area 5. Accordingly, the source and receiver arrays were shortened from 660 to 440 ft.

Figure 12.22 is a montage that places the relevant time zones of the P and SH sections, one above the other. It is immediately obvious that the SH record

COMPRESSIONAL WAVE CDP PROFILE

SHEAR WAVE CDP PROFILE

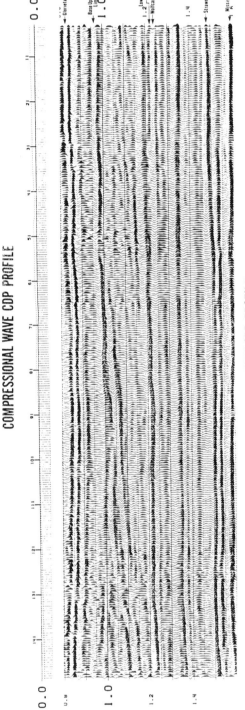

FIGURE 12.22 A montage places the relevant time zones of P- and SH-record sections above one another. The area is Midland Basin in west Texas. Shorter arrays (440 ft) than in Figures 12.19 and 12.20 were used. (*Source: Courtesy CONOCO, Inc.*)

504

SHEAR

COMPRESSIONAL

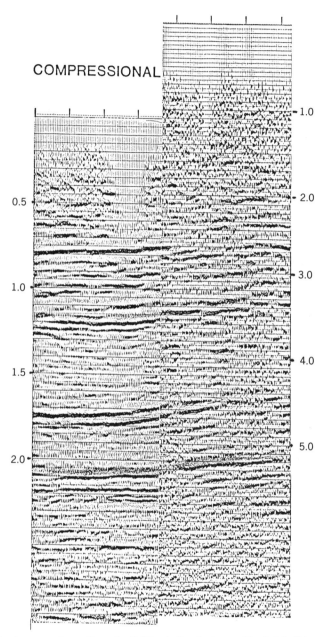

FIGURE 12.23 A composite section of a California line is shown. The left half is compressional, the right, SH, and the line is joined at the common shot point. There is the usual 2 : 1 compression of the SH records, and, to give a reasonable correlation for the lower half of the records, it was necessary to shift the shear-wave record up by 1.1 s. (*Source*: Courtesy CONOCO, Inc.)

section has a much narrower (octave) bandwidth than the P record section and therefore the P section is superior in definition of events. Even so, the truncation is obvious on both sections, being at the basal Upper Clearfork Lime at approximately SP 51-61. Some lack of clarity on the left end of both sections is due to decreased redundancy caused by the absence of permits and/or suitable source locations.

Example 7

To illustrate further that the use of shear-wave reflections is still in the learning stage Figure 12.23 shows a composite section—left-half compressional, right-half shear—joined at a common shot point. The data from California show that with the usual 2:1 compression of the S-wave records it is necessary to shift the shear-wave record upward by the large amount of 1.1 s to achieve a reasonable correlation. This fact points, without question if this correlation is believed, to a deep, low shear-wave velocity section near the surface. The reflection at 0.75 s, P-wave time, corresponds to one at 2.6 s on the S-wave record. The V_s/V_p ratio is 0.2885 (a clay?). It is evident that small changes in water content of these materials will have a profound effect on reflection times and, if the near-surface moisture content changes slowly, the apparent structure shown by the subsurface reflections becomes suspect. This emphasizes that the interpreter must be aware of these vagaries of SH velocities, particularly in recent, unconsolidated sedimentary sections. It is necessary to be uninhibited when it comes to correlation but careful about apparent structural indications.

If these deep weathering (?) corrections change rapidly by the large amounts indicated in some of the examples discussed (including this one) another danger exists to the use of large arrays to cancel surface waves.

12.3 PRESENT STATUS OF SHEAR-WAVE (SH) USAGE

With the advent of large shear-wave sources, impulsive as well as vibrators, the opportunity exists to achieve SH reflection recording in many areas with the same (if not better) quality as P recording. There are, however, a few reservations about the acquisition of SH data:

1. There may be some areas in which surface conditions make it difficult to generate deep SH reflections.
2. The presence of deep and rapidly varying weathering may cause changeable corrections and this means that large arrays are inadvisable.
3. Because the more rapid decay of SH waves with distance of travel than the *corresponding* P-wave frequencies, the SH records become more narrowbanded at depth, thereby losing definition.
4. Another way that peaked spectra develop (therefore ringiness) in SH records seems to be associated with areas having a near-surface layer of high shear-wave contrast. A clue is afforded in preliminary experimentation by the presence of high-amplitude Love waves.
5. Present sources can cause surface damage that limits the acquisition of permits.

On the other hand, to offset those possible problems a number of positive features have emerged:

1. In a limited number of areas it has been found possible to produce SH records with *better* continuity than the P-wave records. Although this phenomenon is not well understood, it appears to exist in those areas in which a high-velocity rock layer is present on the surface. It is well known, then, that pseudo-Rayleigh waves cannot exist, although, depending on the layer thickness, true Rayleigh waves can. Love waves should not exist because of leakage. Research on this valuable asset for S waves is mandatory.

2. Enough work has been done to establish the SH-wave tool as valuable for the removal of ambiguities regarding the origins of *bright spots*.

3. Once SH energy has penetrated the weathering shear-wave reflections can be expected, although there may not be correspondence (pairing) of reflections in P and SH records. This can be an asset.

4. Once correspondence has been rigorously established (e.g., by the use of VSPs for P and shear waves) V_s/V_p ratios can be calculated and, given good continuity in the corresponding reflections, V_s/V_p changes can be followed throughout the surveyed area.

Lithology typing within sections bounded by corresponding P- and S-wave reflections is not in a satisfactory state chiefly because of insufficient data on in-situ P and S velocities for different lithologies, fracture densities, porosities, and pore fluids. The acknowledged presence of anisotropy adds complications. Furthermore, the change in the V_s/V_p indicator from such basic lithologies as carbonates to clastics (sandstones) is not well known and the value of V_s/V_p for mudrocks varies over the entire range.

It now seems (1986) *unlikely* that a simple two-parameter (say V_s/V_p and V_p) discriminant between rock types can be found. It is more likely that statistical discriminant techniques will be based on multiple variables that may include

V_s/V_p (or Poisson's ratio)
V_p or V_s separately
Time of occurrence on records (or, better still, depth burial)
Anisotropic ratio
Attenuation constant (Q)

Nevertheless detection of *lithology change* within the same stratigraphic unit is now often possible.

12.4 THE USE OF CONVERTED WAVES

The SH method of shear-wave usage, as just demonstrated, relies on the possibility of finding a source that will shake a small area of the earth's surface horizontally while generating a minimum of compressional-wave seismic energy. It means using some form of attachment to the low-strength, near-surface soils and rocks that will allow rather large horizontal forces to be applied without shearing off or otherwise invoking nonlinear strains. To a certain extent that has proved to be possible, particularly with vibrators that will allow substitution of generation of lower stresses over long periods of time for the more direct generation of high impulsive stresses. If the near-surface is prestressed by forcing it to hold up a heavy piece of equipment higher oscillatory stresses can be applied without scientifically disastrous consequences. This does not, however, mean that the results are not environmentally disastrous. The use of the large steel inverted pyramids that are forced into the ground to permit large horizontal stresses to be applied are therefore discouraged by most landowners over open land and, of course, are prohibited on the seismologist's normal path—the unpaved or paved highway.

In an effort to overcome this environmental problem other methods of generating shear waves have been invented, some of which were discussed in Section 12.2. In the last year or two there has been a renewed interest in the composite reflection. This simple concept, investigated in the early years of applied seismology by Ricker and Lynn (1950), involves a ray path for compressional waves in one direction and shear (SV) waves in the other direction. Either the downward or the upward path of the composite reflection may be P.

It is known from theoretical studies by Miller and Pursey (1954) and others (see Chapter 3) that a vertical force on the surface produces not only P waves but SV waves as well. The radiation patterns for the two types of wave were given in Section 3.5. After reflection the seismic waves (P and SV) can be picked up by vertical and horizontal geophones, the latter oriented along the direction to the source (unless substantial cross components of reflector dip exist), in which case three-component geophones may be mandatory.

As a consequence, assuming that enough recording channels are available, the output can consist of two separate sets of raw records. If it is assumed that the shear-wave velocity is, simplistically, one-half the P-wave velocity in the sedimentary section, then (only approximately) the P–SV composite reflection from a given reflector will arrive at 1.5 times the P-wave reflection time.

Garotta (1982) has pointed out that the location of the CDP (we can no longer talk about CMP) depends on the ratio of P and S velocities. Moreover, the CDP does not remain on the same vertical line as the depth changes.

Furthermore, the normal move-out must take into account the ever-changing P-wave and S-wave velocity functions.

In the field the swept frequency signal for the vertical vibrators must be wide enough to supply comparable frequencies for P and S recordings and the same wavelengths are generated in the earth. Experience has shown that a frequency ratio of $2:1$ is needed. Thus a 5- to 40-sweep would supply 5 to 20 Hertz for S-wave reflections and 10 to 40 for P-wave reflections, each with a bandwidth of two octaves. In the frequency domain the separation of the two sets of data may be done by correlation with separate sweeps (over the appropriate frequencies and times): for example, this sweep may have been activated at $2.5\,\mathrm{Hz/s}$; therefore the individual control traces should be

5 to 20 Hz over 6 s
10 to 40 Hz over 12 s

Separation of the individual records (P–P, P–SV, and SV–SV) can then be accomplished by the proper frequency band, the proper geophone component, and the proper NMO removal before stacking. An additional control is available with the angle of incidence as a filter. This is best accomplished in the τ-p domain.

Because the velocity of SV waves may not be the same as SH waves, it is conceivable that some knowledge of anisotropy of individual zones of sediments can come from extended experimentation with converted waves and SH-wave reflections. At the present time it is too early in the history of these exotic forms of exploration to predict the degree of realization of these dreams. It is a rare occurrence in research, however, that we can get "something for *almost* nothing"; therefore it is to be expected that two-component and possibly three-component recording will become more routine.

12.5   INCREASED RESOLUTION BY THE USE OF HIGHER FREQUENCIES

For several years in marine prospecting impulses with a higher frequency content than normal have been generated by special equipment to obtain much higher resolution between the near-water bottom layers than could be obtained with conventional marine seismic equipment. A survey of this equipment was made by Saucier (1970). The sound energy in the water is produced by a variety of means—piezoelectric or magnetostrictive transducers (relying on a change in volume of a material when an electric or magnetic field is applied), electromechanical types, sparkers, arcers, gas guns, and air guns. The type of source and the power used must be based on the discrimination and the depth of penetration required because it is rare that a single type of transducer is efficient over a wide range of frequencies. In these discussions we consider only devices developed for obtaining geological information below the sea floor and we therefore disregard an enormous amount of work done on the initiation of pulses in the water for military purposes.

The generation of surface waves is negligible enough (because of the zero velocity of shear waves in a liquid) that it is rare that an array of sources or receivers is used specifically for the cancellation of such waves. Arrays are used, however, for increasing the input energy by using multiple sources and decreasing the ambient noise level. With single sources and receivers results such as those shown in Figure 12.24 can be obtained. In this example an input of several hundred joules of electric power can give resolution of near-bottom sediments down to about a 1-ft thickness, which is equivalent to the reflection of frequencies up to 5000 Hz. The penetration, however, is limited to about 100 ft. Although these are reflection means of obtaining information, they do not use most of the data enhancement principles and methods developed for reflection seismology. The principle of using an extended, unique signal and a correlation receiver was employed by Clay and Liang (1962, 1964) and, even though not always an integral part of the equipment, the principle can always be used if the outgoing pulse shape is known (e.g., recorded on magnetic tape) and the reflected signals are recorded.

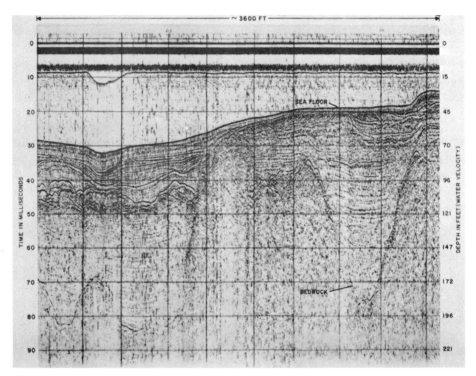

FIGURE 12.24 An example of higher resolution sparker results in thin sediments. (*Source*: By permission of E.G. and G. International).

An increase in power and a lowering of the useful frequency spectrum but still staying with the single sparker source and receiver system lowers the resolution but increases the depth of penetration, as shown in Figure 12.25. Frequencies of 700 to 800 Hz can be identified down to penetration depths of 1000 ft, and the clarity of faulting and erosional unconformities in near-surface sediments is quite evident. Clearly these devices are extremely useful for gaining information about the sediments near the sea floor and examples are known that show gas-filled sediments angling upward to the sea bottom and even seeping into the water.

For depths of penetration useful in exploring for oil and gas deposits offshore, however, a further decrease in the upper limit of the frequency spectrum and a further increase in high-frequency power are required. In some areas the evidence of a gain in resolution compared to the conventional 10- to 45-Hz records can be spectacular. Figure 12.26 shows a record section that was made with a 62 KJ sparker array and recorded on several channels so that a 24-fold stacked section could be obtained. The clarity of the faulting is striking and clear evidence can be obtained for coherent 100-Hz events at a record time of 1.5 s, which corresponds roughly to 5200 ft.

As with all marine records, some care has to be taken with multiple reflections from the ocean bottom. Some are clearly evident here, with double the dip and double the reflection time of the water bottom. It is clear that much progress has been made with the use of higher frequencies in marine oil and gas prospecting.

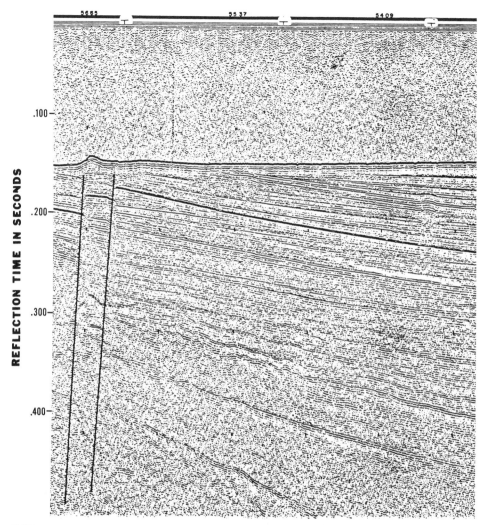

FIGURE 12.25   High-frequency marine data from marine sparker. Frequencies of 700 to 800 Hz can be identified at reflector depths of 1000 ft. (*Source*: Courtesy of the Teledyne Exploration Company).

The general problem of resolution versus penetration has been indicated and put on a firmer footing later. We must first consider the progress that has been made on land.

Seismic records with usable P-wave reflection energy with frequencies up to 130 Hz have been made by Vibroseis methods. In addition to the one shown in Figure 8.22 over a coal mine, high-frequency P-wave and S-wave reflection cross sections have been made for other purposes (Dunster and Miller, 1968). Two appear in Figures 12.27a and b. The structure shown has not been corrected for surface weathering thickness or for elevation changes. These were taken in north-central Oklahoma and the decrease in average frequency with depth is quite pronounced, even though the waves travel through Paleozoic, Permian, and Pennsylvanian formations.

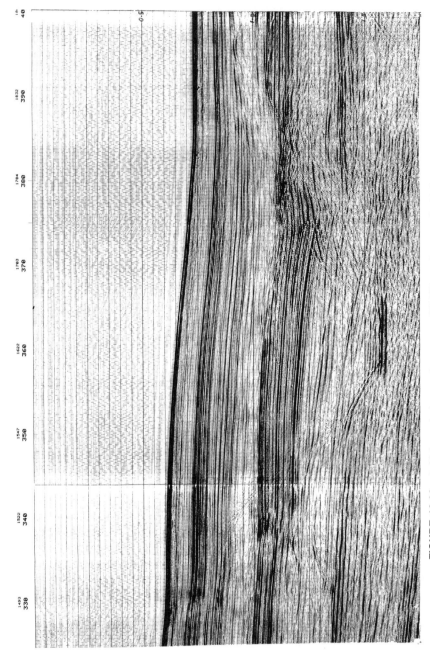

FIGURE 12.26 A high-resolution marine system (62 KJ sparker array), 24-fold stacked. Note the clarity of the faulting. (*Source:* Courtesy of Digicon, Inc.)

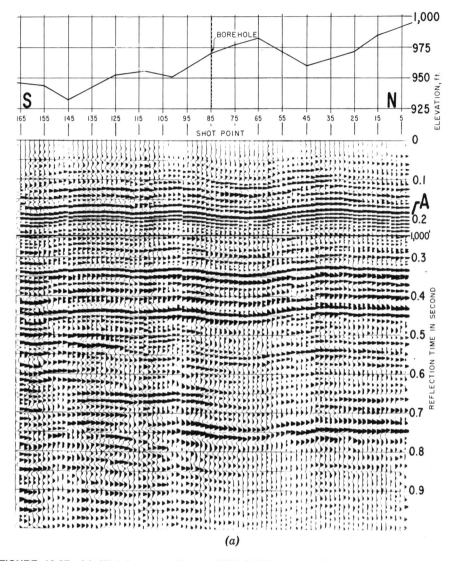

(a)

FIGURE 12.27 (a) High-frequency P-wave VIBROSEIS section. The frequency range is 32 to 130 Hz, and the data are fivefold stacked. No corrections have been made for near-surface variations. (*Source*: Courtesy of the U.S. Corps of Engineers.)

Substantial progress has been reported by Chapman et al. (1981). The effort, a concerted one, involved a completely new design for a vibrator in addition to a search for all possible sources of high-frequency loss in the field effort. Once found, they were eliminated. Figure 12.28a is an example of a line over a stratigraphic sand channel with a 58- to 12-Hz sweep. Figure 12.28b shows the same line but uses a 90- to 25-Hz sweep. The channel at a 4700-ft depth is shown and it is evident that 90 Hz energy has penetrated to this depth.

In areas in which the water table is close to the surface some success with high frequencies has recently been reported (Figure 12.29). The system uses explosives

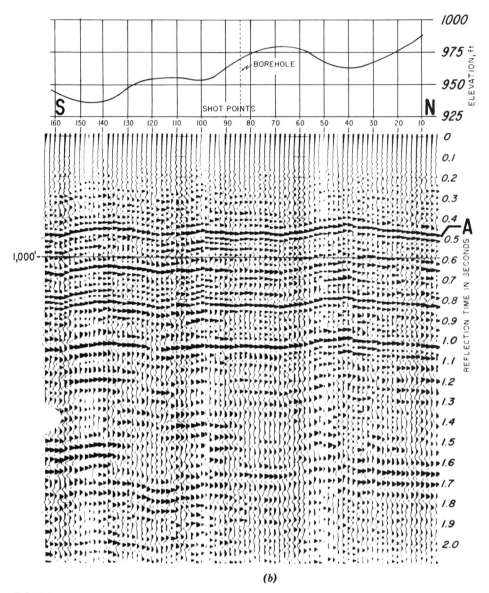

(b)

FIGURE 12.27 (*b*) High-frequency SH-wave VIBROSEIS section over the line in Figure 12.27*a*. Frequency range is 15 to 50 Hz. (*Source*: Courtesy of the U.S. Corps of Engineers).

and special hydrophones in shallow (4 ft deep) holes filled with water. Explosives follow the hydrophones in hole usage, and CDP methods can be used for data enhancement. The fault shown has been confirmed by drilling and the throw varies from 12 ft at a depth of 200 ft to 36 ft at a depth of 800 ft. It is evident that the success of this method, in these areas, stems from the fact that the weathered layer is saturated and evinces neither the variability nor the attenuation for seismic waves that is a feature of a normal weathered layer.

Apart from the weathered layer which, for higher frequencies, can destroy

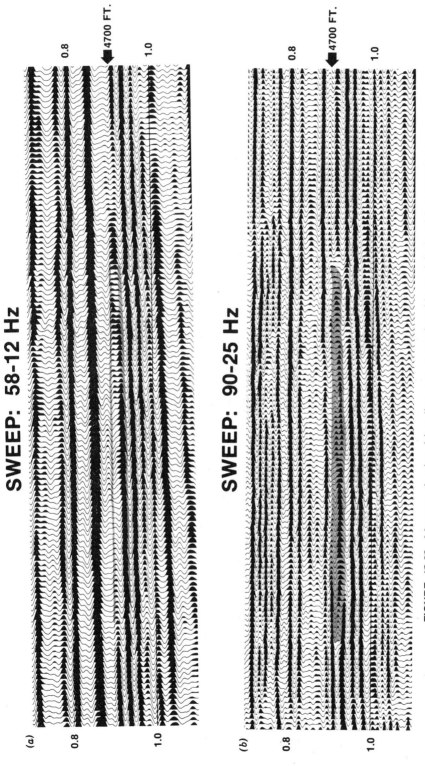

FIGURE 12.28 Montage showing (*a*) a line over a stratigraphic sand body with conventional frequencies (58 to 12) and (*b*) the seismic section over the same line with new high-frequency (90 to 25) vibrator and field techniques. The sand body is shaded. (*Source:* Courtesy CONOCO, Inc.)

SWEEP: 58-12 Hz

SWEEP: 90-25 Hz

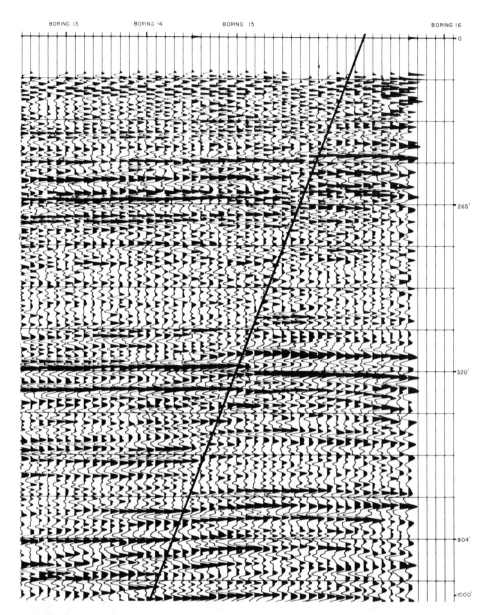

FIGURE 12.29 A high-frequency (65 to 500 Hz) land record section showing a subsidence fault whose existence is supported by drilling (*Source*: © 1975, Western Geophysical Company of America.)

coherence and scatter and give rapid, unknown delays, which are all difficult to counter, the problem of pursuing the goal of higher resolution to greater depths is, of course, attenuation—as explained in Section 7.4. The attenuation due to *primary* reflection scattering from thin beds is not frequency-sensitive. To understand this it is simply necessary to remember that the amplitude loss in passing through a series of boundaries (in both directions) is

$$\frac{A_F}{A_0} = \prod_{j=1}^{N} (1 - R_j^2)$$

where A_F = final amplitude
$\quad\quad A_0$ = initial amplitude

This is a function of the reflection coefficients only, which are not frequency-dependent—and the high frequencies pass through the same boundaries as the low frequencies.

If the individual reflection coefficients are high, as in the boundaries of a coal

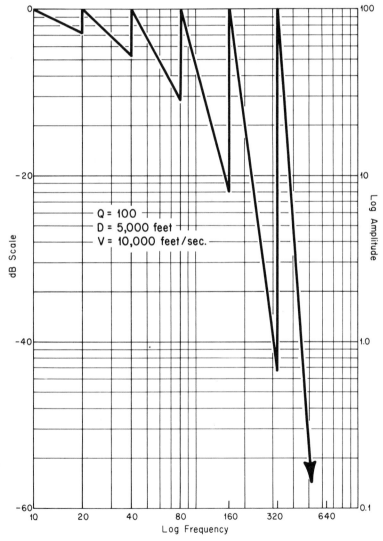

FIGURE 12.30 A variation in amplitude of a reflection across successive-frequency octave bands: $Q = 100$, depth $= 5000$ ft, $V = 10,000$ ft/s.;

seam or a gas reservoir, the effect of the multiple reflections must be taken into account and, as shown in Section 11.11, dispersion occurs. Transmission both ways through a strongly reflecting thin bed attenuates higher frequencies. Generally, the effects of peg-leg multiples have caused a loss of high frequencies with depth. It is convenient, but not correct, to lump these multiple reflection effects with material attenuation, in which the loss of energy to heat is an exponential function of depth. The exponent is proportional to both distance traveled and (approximately) to the first power of the frequency.

To try to record a wide-frequency band during a single experiment involves a dynamic range problem as well as the problem of the signal/noise ratio for the higher frequencies. To solve the dynamic range problem it is possible to record the required frequency bandwidth in several sequential bands. For convenience we use octaves so that for each octave the resolution increases by a factor of 2. During each octave recording the signal input can be increased and the amplitude stays at the same level for some reference point (say, the starting frequency of each band). Figure 12.30 shows the situation graphically. An end point for this process is provided by the exponential increase necessary in input power or by the rapidly increasing rate of decay in amplitude returned across each octave band. The decibel (log amplitude) range across the octave bands is proportional to the frequency bandwidth in hertz. There is a point at which the higher frequencies of an octave eventually dip below the noise level.

In practice this method of recording different bands and adding the results after inverse filtering to flatten the contributions of each spectrum necessitates an overlap from one octave to the next to judge the weights with which the spectra are to be added.

It is to be expected that the barriers of obtaining wide-band, high-frequency records will gradually be forced back, but as can be seen from the preceding analysis, the problems rise exponentially with frequency and an economic limit is soon reached.

A final note should be added. Figure 12.30 was computed by using a Q of 100, a depth of 5000 ft (two-way distance, 10,000 ft), and a velocity of 10,000 ft/s. The decibel loss across an octave is proportional to the linear frequency range and depth. It is obvious that a greater extension of the frequency range can be expected to be feasible for shallow reflection targets than for deep ones. In a general sense therefore frequency increase is likely to be more useful in coal applications than in seismic prospecting for oil and gas.

12.6 THREE-DIMENSIONAL SEISMIC DATA

In spite of the great strides made in the quality of linear seismic cross sections, isolated profiles are at times limited in their usefulness for two main reasons.

First, events are recorded from out of the vertical plane of the cross section, thereby yielding noise as far as geological interpretability is concerned. To sort out and eliminate these anomalous events some form of three-dimensional control is needed.

Second, unless the type of sedimentation responsible for the accumulation of hydrocarbons is known before the accumulation itself is found, an unlikely supposition, there is no way in which the trend of the accumulation can be known. This makes the disposition of the available seismic effort and the assignment of line positions difficult. A sandbar can cross over the seismic line at almost any angle (many oil fields are controlled by fluvial sands in meandering ancient stream channels), and it is difficult to make an assessment of the value of an isolated anomaly. Thus the detail of seismic reflection work along the line should be matched by equal detail across the line. Several methods exist, such as the seismic swath method, which give good CDP coverage for this purpose. However, to preserve the high redundancy needed for good data, it must be possible to have a recording system capable of recording a large number of surface positions simultaneously—a greater number than is common at present. In addition, the arrangement on the surface of the earth is usually awkward for coaxial cabling or other cabling methods. Moreover, if radio communication is considered the number of separated channels is prohibitive and, in many land areas in which seismic work is done the line-of-sign (recorder to receivers) conditions demanded by ultrahigh-frequency radio communication are just not present.

The technology gap that existed five years ago has now been filled by the development of recorders that are self-contained units with large but inexpensive memories (magnetic cassette, solid state, or bubble memories) and are controlled by low-frequency, pulse-coded radio initiation signals. Accurate internal clocks control the sampling of signals coming from a plurality of geophone groups. These recorders can be made to *dump* their accumulated recordings onto tape whenever it is convenient. Several other types of recording system are available for spatially scattered geophone groups.

A real gap exists in the condensation of information, data processing methods, and the presentation of data in a form that allows easy assimilation. Anyone who has worked with three-dimensional data knows that the last of these should not be taken lightly. Significant advances can be made within a few years by computer calculations of changes in patterns recorded on two-dimensional cross sections as these sections are moved perpendicular to their planes. It is essential to remove from consideration geological indications that remain the same or change very slowly.

12.7 NEW OR DEVELOPING DISPLAY METHODS AND METHODS FOR INTERPRETATION

Over the last 15 years there has been a gradual emergence of the use of color displays for seismic data (Anstey, 1973). Levin et al. (1976) noted: "Color presentations were tried nearly 20 years ago, but advances in technology have eliminated difficulties that delayed the use of color as a practical aid to the interpreter."

With the introduction of individual work stations equipped with color monitors color graphics became possible and now appear to be available almost everywhere. A few examples have been included in this book. The latest (see color insert in the middle of the book for a color representation of Figure 12.31), is a display of a box with sides opened up and with the top at a time of 1120 ms. The side sections have a time depth of 200 ms.

The reader may decide for himself whether this use of color makes the data

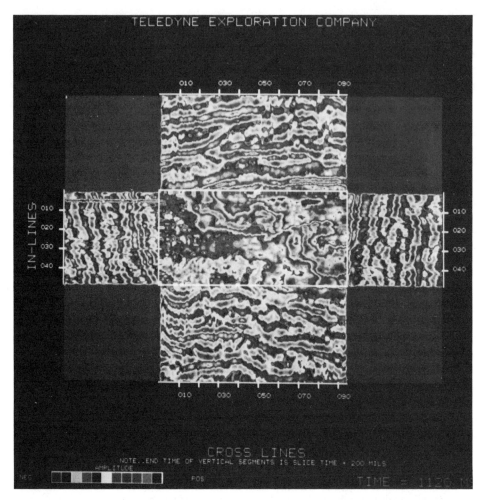

FIGURE 12.31   A portrayal in the form of an opened-up box. The box top is a horizontal time slice at 1120 ms and the four hinged, vertical-section sides show 200 ms of added section. See color insert in the middle of the book for color representation. (*Source*: Courtesy Teledyne Exploration Company.)

clearer for interpretation. The chief appeal (for the noncolor-blind) seems to be a combination of an immediate standout of the color difference and the fact that the eye can distinguish many more changes of color than it can discriminate different shades in the gray scale.

Although the original color presentations were for single attributes of the seismic section (e.g., amplitude, instantaneous frequency, and interval velocity), there has been a growing tendency to use the color *triangle* as a means of presenting several different attributes simultaneously. Each characteristic can contribute an arbitrary amount of its assigned primary color (plus white for changing intensity). An alternative for integrating the presentation of different attributes has been the effective use of pale (pastel shades) as a background for the normal seismic display. Interpreters accustomed to looking at seismic sections can immediately discern structure, but they are made aware of another parameters also; for example, V_s/V_p (hopefully lithology related) can be a color overprint for a seismic cross section.

The use of these aids is subjective and there are psychological overtones in the selection of the color palette. An economic justification probably has to be made on an individual prospect basis because the addition of color to structure hunting contributes little, whereas it may be valuable in exploration for stratigraphic traps.

Although the deduction of geology is the standard practical use of reflection seismic data, it is rare that analyses between data in different basins are attempted. Vail, Mitchum, and Todd (1975) have shown that some geological features, notably unconformities caused by worldwide sea-level changes, can be inferred from seismic record data. This represents a startling advance in seismic reflection interpretation and is undoubtedly the beginning of a considerable expansion in seismic data use by geologists.

It is always tempting to try to derive still more detailed information from seismograms. One of these brave attempts is the interpretation of depositional facies from the seismic data. Sangree and Widmier (1979) and Roksandic (1978) have given detailed descriptions of the classification and identification of *seismic facies* elements on actual reflection seismograms. It seems preferable to proceed in two steps from the original seismic data to the seismic facies thence to the geological interpretation of the patterns, as the word *facies* implies much more in geology than we can hope to infer from seismic information alone. Thus we postulate an intermediate seismic facies that can possibly be inferred from the seismic data.

Before discussing the basic types of reflection configurations it must be always borne in mind that the following are the key seismic characteristics of the survey:

1. Reflection line down dip.
2. Consistency of input pulse.
3. Consistency of reflection output transfer function.
4. Optimum resolution characteristics achieved through data processing.

The patterns to be sought in the seismic data are micropatterns that can very easily be disturbed. In most cases the reflection coefficients are small and close together in reflection time. The distinctive patterns that geologists can see from drill logs or from canonical outcrops are, with the reflection seismograph, seen "through a glass, darkly"—much of the detail obscured and distorted.

Thus the basic elements of seismic facies analysis consist of first giving a definition of all possible reflection configurations, breaking down the simple layered, complex layered, and chaotic (noise) classification still further to (Roksandic, 1978) the following list:

1. Parallel
2. Convergently
3. Cross
4. Oblique
5. Sigmoid
6. Chaotic
7. Chaotic with diffractions (lumpy!)

In locations in which seismic data are available from two orthogonal directions (or, preferably, three-dimensional data) it may be possible to predict directions of paleotransport or to rotate cross sections until they present cross bedding in its most favorable aspect. In clastic sediments the reflection coefficients are likely to be small unless the analyses are vitiated by the pressure of gas. In this case large reflection coefficients will exist, but the sedimentary patterns are not likely to be preserved.

We must not forget the search for carbonate reservoirs, however. In these cases the reflection coefficients are likely to be large, although rough surfaces and random porosities of reefs may make the reflection energy bands somewhat unpredictable. One predictable case in which the field seismogram analysis was confirmed by the synthetic model traces obtained from a postulated model is shown in Figure 12.32. Here the carbonate bank buildup and limestone clinoform are indicated by the field seismograms.

The utility of the seismograph in exploration for stratigraphic trap oil fields will depend almost entirely on the ability of the geophysicists to produce clean, consistent, wideband seismograms and the ability of geologists to analyze and classify the information provided.

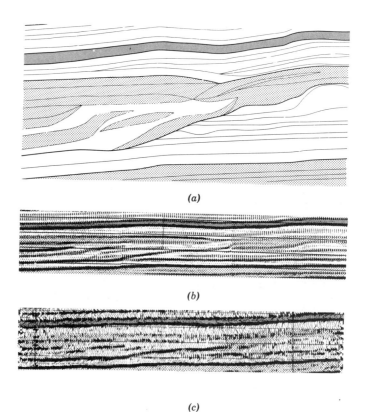

(a)

(b)

(c)

Figure 12.32 (*a*) A geologic model of a clinoform and carbonate bank buildup, (*b*) predicted seismic reflections, and (*c*) actual seismic reflections. (*Source*: Courtesy CONOCO, Inc.)

12.8 NEW REFLECTION SEISMOLOGY APPLICATIONS

Although the use of reflection seismology increased in the area of coal exploration during the decades of the 1970s and 1980s, particularly in Europe, petroleum exploration continues to be the chief consumer by a large margin. The 1984 slump in oil prices produced a significant drop in seismic-crew utilization and has continued to the present day.

Although they do not fill the gap created by the decline in petroleum usage, three other users have appeared on the scene:

1. Hydrothermal energy.
2. Prospecting for uranium sources.
3. Use in South Africa for detailing structure in the Precambrian Witwatersrand Basin as an aid to gold mining.

All appear to be rather specialized. In the first case the temperature anomalies beneath the earth's surface can be sought by geophysical methods much more directly related to the high temperature required; for example, electrical resistivity anomalies related directly to the high conductivities of hot brines are more economical to run. Small self-potential anomalies have been reported (Corwin and Hoover, 1979) and may be attributable to a voltage gradient produced by the maintenance of a temperature gradient (thermoelectric coupling) or to the flow of fluid through a porous medium (electrokinetic coupling).

Nevertheless, although the seismic work has been devoted to measuring the noise produced by the reservoir or characterizing the elastic rock parameter variations in the neighborhood of the geothermal field (Majer and McEvilly, 1979), some direct reflection surveying has been reported by Goupillaud and Cherry (1977) which has increased our understanding of the problems of P-wave and S-wave reflection work in these environments. Their test ground, the East Mesa Geothermal Area is in the Imperial Valley trough, filled with recent sediments resting on a thin basaltic basement. Variation in the materials and rates of sedimentation over the last five million years has produced layering capable of giving both P-wave and S-wave reflections. Although the quality of the P-wave results is not exceptional, they can be used after proper processing to show the location of the fractured zone (by lack of reflections). The shear-wave data are much poorer but can be interpreted in the same general manner. This work must be regarded as an interesting piece of pioneering in a field that has barely been touched.

Another pioneering attempt to use the reflection seismography in another energy source investigation has been undertaken by Hajnal and Reilkoff (1980) in the northern part of the Athabaska Basin. The energy source sought is uranium in one of its secondary mineral forms. Although the primary mineralization is in the Precambrian basement rocks, the occurrence is in fractures in the Athabaska Sandstone, in which the leached uranium minerals are redeposited. Seismic work has been done in the 60- to 130-Hz spectral range with Mini-Sosie® equipment. In some respects—the shallow depth (500 to 700 m) and high velocities (4000 to 5000 m/s) of the geologic section—the problems are similar to those of coal investigations. The velocity and static problems are unusually severe and have required some new, unconventional methods to solve them. Improvements in data quality have resulted in the ability to map the basement and the fracture patterns in Athabaska sandstone. Provided that these breakthroughs are further substantiated, it may be possible to trace the uranium mineralization from present workings deeper into the basin.

The latest use of conventional reflection seismology has been in an unlikely location, namely, within the confines of the Witwatersrand ("Wits") Basin, a basin that contains sediments originally deposited 2500 million years ago. The basin (see D. A. Pretorius, 1984), approximately 350 by 200 km, is active on the northwest side; the southeast side is subject to a gentle downwarping. The distribution of goldfields is intimately related to a pattern of folding that produced structural depressions and domes of basement granite. The folds are accentuated by major faults parallel to the folding.

The goldfields are located in the downwarps between the domes and all take the form of fluvial fans, much reworked as the regressive history progresses; thus the heavy minerals are concentrated in a number of different horizons in the stratigraphic column.

It is important to realize that the reflection seismograph is not being used to *find* the gold concentrations; there is no basis for that. Rather it is a tool that allows more direct structural mapping at much less expense than the traditional diamond drilling. The question just a few years ago was whether reflection seismology would work in a typical sequence of lavas, quartzites, dolerites, and dolomites. Sonic and density logs were taken during 1982 and 1983 and from the calculated reflection coefficients it appeared that reflections could exist if the rock boundaries were sufficiently smooth over sufficient distances.

Campbell and Peace (1984) have described some of the problems and solutions that have made this application extremely interesting.

Velocity and density values for rocks sampled lie in the high region, compared with most rocks encountered in oil prospecting. Typical values in a stratigraphic column are (Campbell and Peace, 1984)

Dwyka tillite	4800 m/s	2650 kg/m^3
Diabase sill	6400 m/s	2950 kg/m^3
Malmani dolomite	6800 m/s	2850 kg/m^3
Malmani chert	6000 m/s	2.65 kg/m^3
Ventersdorp acid lava	6000 m/s	2.70 kg/m^3
Ventersdorp basic lava	6400 m/s	2.84 kg/m^3
Ventersdorp shales	5900 m/s	2.90 kg/m
Ventersdorp shales and sandstone	5700 m/s	2.70 kg/m^3
Witwatersrand		
Upper quartzites	5700 m/s	2.67 kg/m^3
Lower silts	5900 m/s	2.76 kg/m^3
Lower quartzites	5800 m/s	2.68 kg/m^3

Faulting is common in most areas; throws of several hundred meters are not uncommon. The exact location of these faults is quite critical, often marking the economic mining boundary.

A key to successful VIBROSEIS operation in many areas, particularly where shallow and deep reflection data are both desired, is the fact that Rayleigh waves are not strong and can be made tolerable *without using long arrays* by using a swept frequency signal with a lower cutoff just above 20 Hz.

The almost constant high velocities of the rock layers, although causing relatively simple wave paths, make CMP methods poor protection against multiple reflections. Some strong ones are sometimes seen in processed seismic cross sections.

It is now felt that, given a thorough appreciation of the limitations imposed by the unusual rock parameters and intense folding and fracturing, the seismic reflection method is capable of being put to use to select locations for the expensive diamond drill holes that will locate and allow examination of the gold-bearing conglomerate *reefs*.

REFERENCES

Anstey, N. A. (1973), "The Significance of Color Displays in the Direct Location of Hydrocarbons," paper presented at the 43rd Annual International Meeting of the Society of Exploration Geophysicists, Mexico City.

Applegate, J. K. (1974), "*A Torsional Seismic Source*," Ph.D. thesis, Colorado School of Mines, Golden, Colorado.

Berryman, J. G. (1979), "Long-wave Elastic Anistropy in Transversely Isotropic Media," *Geophysics*, Vol. 44, No. 5, pp. 0896–0917.

Berryman, L. H., Goupillaud, P. L., and Waters, K. H. (1958), "Reflections from Multiple Transition Layers," Parts I and II, *Geophysics*, Vol. 23, No. 2, pp. 223–252.

Benzing, W. (1978), personal communication.

Beeston, H. E., and McEvilly, T. U. (1977), "Shear Wave Velocities from Down-Hole Measurements," in *Earthquake Engineering and Structural Dynamics*, Vol. 5, Wiley, Chichester, England, pp. 181–190.

Brown, G. L. (1966), U.S. Patent 3,280,935.

Bycroft, G. N. (1956), "Forced Vibrations of a Rigid Circular Plate on a Semi-infinite Elastic Space and an Elastic Stratum," *Philosphical Transactions of the Royal Society* (*A*), Vol. 248, pp. 327–368.

Campbell, G., and Peace, D. G. (1984), "Seismic Reflection Experiments for Gold Exploration," paper given at the 1984 Annual Meeting of the European Association of Geophysicists, London.

Chapman, W. L., Brown, G. L. and Fair, D. W. (1981), "The VIBROSEIS system, A High Frequency Tool," *Geophysics*, Vol. 46, No. 12, pp. 1657–1666.

Cherry, J. T. (1962), "The Azimuthal and Polar Radiation Patterns Obtained from a Horizontal Stress Applied to the Surface of an Elastic Half-space," *Bulletin Seismological Society of America*, Vol. 52, No. 1, pp. 27–36.

Cherry, J. T., and Waters, K. H. (1968), "Shear-wave Recording Using Continuous Signal Methods," *Geophysics*, Vol. 33, No. 2, pp. 229–239.

Clay, C. S., and Laing, W. L. (1962), "Continuous Seismic Profiling with Matched Filter Detector," *Geophysics*, Vol. 27, pp. 786–795.

Clay, C. S., and Laing, W. L. (1964), "Seismic Profiling with a Hydroacoustic Transducer and Correlation Receiver," *Journal of Geological Research*, Vol. 69, No. 16, pp. 3419–3428.

Corwin, R. F., and Hoover, D. B. (1979). "The Self-Potential Method in Geothermal Exploration," *Geophysics*, Vol. 44, No. 2, pp. 226–245.

Crampin, S. (1981), "A Review of Wave Motion in Anisotropic and Cracked Elastic Media," *Wave Motion*, Vol. 3, pp. 343–391.

Crampin, S. (1984a), "Effective Anisotropic Elastic Constants for Wave Propagation Through a Cracked Solid," *in* Proceedings of the First International Workshop on Seismic Anisotropy, *Geophysical Journal of the Royal Astronomical Society*, Vol. 76, pp. 135–145.

Crampin, S. (1984b), "Anisotropy in Exploration Seismics," *First Break*, Vol. 2, pp. 19–21.

Crampin, S. (1984c), "Anisotropy and Transverse Isotropy," *Geophysics*, submitted.

Crampin, S. (1984d), "The Behavior of Shear waves at the Free Surface," *Geophysics*, submitted.

Crampin, S. (1985), "Evaluation of Anisotropy by Shear-wave Splitting," *Geophysics*, Vol. 50, No. 1, pp. 142–152.

Domenico, S. N. (1977), "Elastic Properties of Unconsolidated Porous Sand Reservoirs," *Geophysics*, Vol. 42, No. 7, pp. 1339–1369.

Domenico, S. N. (1984), "Rock Lithology and Porosity Determination from Shear and Compressional Waves," *Geophysics*, Vol. 49, No. 8, pp. 1188–1195.

Dorn, G. A. (1980), *Radiation Impedance and Radiation Pattern of Torsionally Vibrating Seismic Source*, Ph.D. thesis, University of California, Berkeley.

Dorn, G. A. (1984), "Radiation Patterns of Torsionally Vibrating Seismic Sources," *Geophysics*, Vol. 49, No. 8, pp. 1213–1222.

Dunster, D. E., and Miller, D. E. (1968), "A Research Study to Determine the Occurrence and Structural Competency of Deep Rock Formations by the Use of Surface Vibrators," U.S. Army Engineers Waterways Experiment Station, Vicksburg, Mississippi, Contract No. DACA-39-67-C-0053.

Edelmann, H. A. K. (1981), "Shover—Shear-wave Generation by Vibration Orthogonal to the Polarization," *Geophysical Prospecting*, Vol. 29, No. 4, pp. 0541–0549.

Erickson, E. L., Miller, D. E., and Waters, K. H. (1968), "Shear-wave Recording Using Continuous Signal methods—Part II. Later Experimentation," *Geophysics*, Vol. 33, No. 2, pp. 240–254.

Garotta, R. (1982), "*Comparisons between P, SH, SV and Converted Waves*," C.G.G. Technical Series 521.82.08.

Garotta, R. (1985), "*Routine Two-Component Acquisition*," C.G.G. Technical Series 534.85.04.

Geyer, R. L., and Martner, S. T. (1969), "SH-waves from Explosive Source," *Geophysics*, Vol. 34, No. 6, pp. 893–905.

Gregory, A. R. (1976), "Fluid Saturation Effects on Dynamic Elastic Properties of Sedimentary Rocks," *Geophysics*, Vol. 41, No. 5, pp. 895–921.

Goupillaud, P. L., and Cherry, J. T. (1977), "Utilization of Seismic Exploration Technology for High Resolution Mapping of a Geothermal Reservoir," Final Report SAN (1249-1) on Contract EY-76-C-1249 for ERDA by Systems, Science and Software, La Jolla, California 92038.

Hajnal, Z., and Reilkoff, B. (1980), personal communications, M.S. Thesis, University of Saskatchewan, B. Reilkoff.

Jolly, R. N. (1956), "Investigation of Shear Waves," *Geophysics*, Vol. 21, No. 4, pp. 905–938.

Jones, L. E. A., and Wang, H. F. (1981), "Ultrasonic Velocities in Cretaceous Shales from the Williston Basin," *Geophysics*, Vol. 46, No. 3, pp. 288–297.

King, M. S. (1966), "Wave Velocities in Rocks as a Function of Changes in Overburden Pressure and Pore Fluid Saturants," *Geophysics*, Vol. 31, No. 1, pp. 50–73.

Kisslinger, C., Mateker, E. J., and McEvilly, T. V. (1961), "SH Motions in Soil," *Journal of Geophysical Research*, Vol. 66, pp. 3487–3496.

Kithas, W. (1976), "Lithology, Gas Detection and Rock Properties," paper given at SPWLA Annual Meeting.

Levin, F. K. (1979), "Seismic Velocities in Transversely Isotropic Materials (Part I)," *Geophysics*, Vol. 44, pp. 918–936.

Levin, F. K. (1980), "Seismic Velocities in Transversely Isotropic Materials (Part II)," *Geophysics*, Vol. 45, pp. 3–17.

Levin, F. K., Bayhi, J. F., Dunkin, J. W., Lea, J. D., Moore, D. B., Warren R. K., and Webster, G. H. (1976), "Developments in Exploration Geophysics 1969–1974," *Geophysics*, Vol. 41, No. 2, pp. 209–218.

Leslie, H. D., and Mons, F. (1982), "Sonic Waveform Analysis Applications," paper given at SPWLA 23rd Annual Logging Symposium.

Majer, E. L., and McEvilly, T. V. (1979), "Seismological Investigations at the Geysers Geothermal Field," *Geophysics*, Vol. 44, No. 2, pp. 246–269.

Pretorius, D. A. (1984?), "The Nature of the Witwatersrand Gold-Uranium Deposits," Wits University, Johannesburg, South Africa, Referred to by Campbell and Peace (1984).

Ricker, N., and Lynn, R. D. (1950), "Composite Reflections," *Geophysics*, Vol. 15, No. 1, pp. 30–49.

Roksandic, M. M. (1978), "Seismic Facies Analysis Concepts," *Geophysical Prospecting*, Vol. 26, No. 2, pp. 383–398.

Sangree J. B., and Widmier, J. M. (1979), "Interpretation of Depositional Facies from Seismic Data," *Geophysics*, Vol. 44, No. 2, pp. 131–160.

Saucier, R. T. (1970), "Acoustic Sub-bottom Profiling Systems—A State-of-the-Art Survey," U.S. Army Engineers Waterways Experiment Station, Technical Report S-70-1.

Sheline, H. (1985) Personal Communication.

Stoneley, R. (1949), "Seismological Implications of Aeleotropy in Continental Structure," *Monthly Notices of the Royal Astronomical Society*, Vol. 5, pp. 343–353.

Tatham, R. H. (1982), "V_p/V_s and Lithology," *Geophysics*, Vol. 47, No. 3, p. 336.

Tatham, R. H., Keeney, J. W., and Noponen, I. (1982), "Application of the Tau-P Transform (Slant Stack) in Processing Seismic Reflection Data," paper presented at the 52nd International S.E.G. Meeting, Dallas, Texas.

Tatham, R. H., and Stoffa, P. L. (1976), "V_p/V_s—A Potential Hydrocarbon Indicator," *Geophysics*, Vol. 41, pp. 837–849.

Vail, P. R., Mitchum, R. M., and Todd, R. G. (1975), "Global Unconformities from Seismic Sequence Interpretation," paper presented at the 45th Annual International Meeting of the S.E.G. Denver, Colorado.

White J. E. (1965) "*Seismic Waves: Radiation, Transmission and Attenuation*, McGraw-Hill, New York.

White, J. E. (1982), "Computed Waveforms in transversely Isotropic Media," *Geophysics*, Vol. 47, pp. 771–783.

White, J. E., Heaps, S. N., and Lawrence, P. L. (1956), "Seismic Waves from a Horizontal Force," *Geophysics*, Vol. 21, No. 3, pp. 715–723.

White, J. E., Martineau-Nicoletes, L., and Monash, C. (1983), "The Measurement of Anisotropy in Pierre Shale," *Geophysical Prospecting*, Vol. 31, pp. 709–725.

White, J. E., and Sengbush, R. L. (1963), "Shear Waves from Explosive Sources," *Geophysics*, Vol. 28, pp. 1001–1019.

Won, I. J., and Clough, J. W. (1981), "A New Torsional Shear-Wave Generator," *Geophysics*, Vol. 46, No. 11, pp. 1607–1608.

Yoshimura, M., Fujii, S., Tanaha, K., and Morita, K. (1982), "On the Relationship Between P and S Wave Velocities in Soft Rock," paper given at S.E.G. Annual International Meeting, Dallas, Texas.

Index

529